THERMODYNAMICS OF POLYMER BLENDS

HOW TO ORDER THIS BOOK

BY PHONE: 800-233-9936 or 717-291-5609, 8AM–5PM Eastern Time

BY FAX: 717-295-4538

BY MAIL: Order Department
Technomic Publishing Company, Inc.
851 New Holland Avenue, Box 3535
Lancaster, PA 17604, U.S.A.

BY CREDIT CARD: American Express, VISA, MasterCard

BY WWW SITE: http://www.techpub.com

POLYMER THERMODYNAMICS LIBRARY, VOL. 1

Thermodynamics of Polymer Blends

YURI S. LIPATOV
ANATOLY E. NESTEROV

Academy of Sciences of Ukraine

TECHNOMIC PUBLISHING CO., INC.
LANCASTER · BASEL

Polymer Thermodynamics Library–Volume 1
a **TECHNOMIC**® publication

Published in the Western Hemisphere by
Technomic Publishing Company, Inc.
851 New Holland Avenue, Box 3535
Lancaster, Pennsylvania 17604 U.S.A.

Distributed in the Rest of the World by
Technomic Publishing AG
Missionsstrasse 44
CH-4055 Basel, Switzerland

Printed in the United States of America
10 9 8 7 6 5 4 3 2 1

Main entry under title:
Polymer Thermodynamics Library–Volume 1: Thermodynamics of Polymer Blends

A Technomic Publishing Company book
Bibliography: p.
Includes index p. 443

Library of Congress Catalog Card No. 97-62285
ISBN No. 1-56676-624-9

This Technomic Publishing Company, Inc. edition is produced under license between Technomic Publishing Company, Inc. and ChemTec Publishing.

Table of Contents

Preface

Polymer alloys and blends belong to one of the most dynamic sectors of the polymer industry. This explains the great interest in the studies of their physical and mechanical properties, their structure, and the processes of their formation and manufacture. Thermodynamics of polymer blends is based on the classical theories of polymer solutions developed by Flory, Huggins, Prigogine, Patterson, Sanchez, and others as well as on the theories of the phase equilibrium and phase separation in solids developed by Cahn, Prigogine, and other authors. There are many books and reviews on the subject of the thermodynamics of polymer solutions. After the fundamental works by Flory,[1] Huggins,[2] and Tompa,[3] the following publications have been dedicated mainly to the problems discussing excluded volume. Theoretical investigations regarding this subject, performed before 1970, were systematically summarized as the so-called two-parameter theory in the book by Yamakawa[4] and compared with an extensive experimental data. This first stage of research was immediately followed by a new theoretical approach, namely, the renormalization group theory. These developments, achieved until recently, can be found in books by Freed[5] and des Cloizeaux and Jannink.[6] However, the recent experimental data point out to the deficiencies in the two-parameter theory. These data were summarized and discussed by Fujita.[7] There have been published also some reviews.[9,10] The new developments and the fundamental contribution to the theory of polymer solution come from de Gennes[11] discussed in his famous book *Scaling Concepts in Polymer Physics*.

However, if presently there exists a voluminous scientific literature dedicated to various aspects of the physics and chemistry of polymer blends, there are no many books dedicated to the thermodynamics of mixing and thermodynamic behavior of polymer mixtures. This subject was presented in such fundamental works as *Polymer Blends* edited by Paul,[12] books by Olabisi, Robeson, and Shaw,[13] Manson and Sperling,[14] Utracki,[15] and some other authors.[16,17]

Thermodynamic behavior of polymer blends determines the compatibility of the components, their morphological features, rheological behavior, microphase structure, and in such a way the most important physical and mechanical characteristics of blends.

This book is dedicated to the detailed analysis of the thermodynamics of polymer- polymer systems. This book is not intended as an introduction in the field but the knowledge of the main principles of thermodynamics is assumed. We attempted to consider the modern state of thermodynamics of mixing and phase separation and to discuss both theoretical and experimental aspects of the problem. Chapters 1 and 2 discuss thermodynamics of polymer solutions and the most important theories describing the processes of dissolution and phase separation. These chapters give the reader a fundamental understanding of the present state of knowledge on the thermodynamics of polymer blends. Analyzing the thermodynamic behavior of polymer blends, we pay a special attention to the processes of phase separation and to the formation of an interphase between coexisting phases. The thermodynamic properties of an interphase play a very important role in the mechanical behavior of the blends.[18] One may say that formation of the interphase distinguishes the properties of blends from properties of solution. After all, the rheological and mechanical properties of polymer blends are determined by the compatibility of the components and by the degree of microphase separation that implies the formation of an interphase.

Some words about terminology. Although the concept of "compatibility" is widely used in the literature, we prefer to follow the definition given by Olabisi, Robeson, and Shaw.[13] We cite: "compatibility has been used by many other investigators involving various studies of polymer-polymer blend behavior to describe good adhesion between the constituents, average of mechanical properties, etc. The term "miscibility" has been chosen to describe polymer-polymer blends with behavior similar to that expected of a single-phase system. The term miscibility . . . does not imply ideal molecular mixing but suggests that the level of molecular mixing is adequate to yield macroscopic properties expected of a single-phase material."

Another term that should be defined is: polymer alloys and blends. Both alloys and blends are produced by mixing two polymers in various conditions. Utracki[15] considers all the mixtures as blends that may be either miscible or immiscible. Alloys are formed, after Utracki, only after some modification of immiscible blends which allows to improve the compatibility, i.e., alloy is the result of some compatibilization. This compatibilization does not imply that the system becomes miscible and is only connected with some improvement of interactions at the interface between the two phases of immiscible mixture (a typical example is compatibilization obtained by the introduction in the homopolymer blend of diblock-copolymers composed of blocks corresponding to two homopolymers).

We propose another terminology[18,19] based on the thermodynamic considerations and including the analysis of the phase diagrams of binary mixtures as the basis for differentiation.

It is known that the most common and convenient method of mixing two polymers is by melt blending, which is typical for metallic alloys. Generally, alloys are macroscopically uniform substances obtained by fusion of two or more metals, nonmetals, and organic compounds. In general, alloys are not obtained by simple mechanical mixing of the components. By fusing, the components may form mixtures of various phases. The phase state of an alloy in equilibrium can be determined from the phase diagram. It is known also that, for linear polymers, the phase diagrams are of two types, with upper and lower critical solution temperatures (UCST and LCST). Taking the phase diagrams as a basis, we give the following definitions:

- Alloys of linear polymers are binary or multicomponent systems that, when mixed in the molten state, they are situated in the region of the phase diagram corresponding to the mutual miscibility of the components and to the formation of one-phase solution. This means that the system is thermodynamically stable (at equilibrium) in the molten state. By cooling the melt, a structure develops which depends on the thermodynamic state at a given temperature. If by cooling, the system with UCST enters the region of unstable states in the phase diagram (immiscibility or thermodynamic incompatibility arises), then the structure becomes the two-phase structure and it is determined by the conditions of phase separation.

  The structure depends on the degree of phase separation and on the mechanism of phase separation (nucleation or spinodal decomposition). The ratio and composition of two phases is determined by the kinetics and mechanism of phase separation. For binary systems with UCST, the mixing of the alloy components should be conducted in the region above spinodal and for the systems with LCST – below spinodal, i.e., in the range of one-phase solution. For systems with LCST, one-phase structure is preserved by cooling, i.e., a miscible system is maintained.
- Blends of linear polymers are such binary systems that, by mixing in the molten state, they are not miscible and do not form one-phase system (they are not thermodynamically miscible). The components forming a blend may also have UCST or LCST. For systems with UCST, the formation of blend proceeds at temperatures below the binodal and for systems with LCST above the binodal. The transition from two-phase to one-phase state for systems with LCST is practically impossible by lowering temperature because of the high viscosity of the melt and the slowness of the mutual dissolution process. Thus, the structures of blends are deter-

mined by the degree of dispersion of one or both components achieved during mixing in the melt.

In such a way, depending on the temperature range of mixing, relative to the position of the binodal or spinodal, the same polymer pair may form both alloys and blends. The realization of both possibilities depends on the relationship between the temperature of phase separation for a given composition of a mixture and the glass transition temperatures or melting points of both components. Because the formation of blends or alloys by fusing the components (mixing in the molten state) is always followed by temperature reduction, the definitions given above allow to distinguish between the structural features of polymer alloys and blends. The analysis made above allows to conclude that two-phase or multi-phase blends are typical of systems with UCST, whereas one-phase alloys for systems with LCST.

The principle feature of polymer alloy consists of the incomplete phase separation in the system. By cooling a melt of two polymers the thermodynamic incompatibility of two components arises which causes the incomplete phase separation of the system. This incompleteness of the phase separation causes the development of both the microphase separation regions of various composition and transition or interphase zone between coexisting microregions. The system with incomplete microphase separation is not in the state of thermodynamic equilibrium. A segregated structure develops in the bulk because of these processes with a complex of specific properties: appearance of the regions with different density, composition and mechanical properties, appearance of the internal interphase boundaries, etc.

The same definition may be applied to the reactive mixtures capable of the formation of interpenetrating polymer networks. The transition from the initially one-phase mixture of components to cured two-phase system is connected with the phase separation and the phase-separated system may be attributed to alloy.

However, in the present book, where we consider the thermodynamic behavior of the polymer-polymer systems and do not deal with the peculiarities of their structure in a solid state, we will use the words "blends" and "mixtures" as a general, equivalent, and habitual terms. In reality, when we mix or blend two polymer components we do not know beforehand their thermodynamic behavior.

It is worth noting that there are no either fully miscible or fully immiscible mixtures. All depends on the phase diagram of the system, i.e., on the composition and temperature. At the same time, we fully agree with the statement[13] that "the two-phase system must be defined and contrasted with miscible system to delineate the two subjects and to establish the criteria." One of the aims of this book is the analysis of conditions of miscibility and immiscibility.

These authors understand that the field of thermodynamics of polymer blends is too extensive to be covered in one book. There are many questions that have not been considered or were considered only briefly. However, we hope that this volume will further stimulate interest in this branch of physical chemistry of polymers.

Authors had worked with full comprehension. Each of us thanks the other for the mutual support and encouragement. We wish also to express our sincere gratitude to Dr. G. Wypych who initiated this work and who waited patiently for its result.

Institute of Macromolecular Chemistry,
National Academy of Sciences of Ukraine,
Kiev, June 1997

REFERENCES

1. P. J. Flory, *Principles of Polymer Chemistry*, Cornell Univ. Press, Ithaca, N.Y., 1953.
2. M. L. Huggins, *Physical Chemistry of Polymers*, Interscience, N.Y., 1958.
3. H. Tompa, *Polymer Solutions*, Butterworth, London, 1956.
4. H. Yamakava, *Modern Theory of Polymer Solutions*, Harper and Row, N.Y., 1971.
5. K. F. Freed, *Renormalization Group Theory of Macromolecules*, Wiley-Interscience, N.Y., 1987.
6. J. des Cloizeaux and G. Jannink, *Polymers in Solutions*, Oxford Univ. Press, Oxford, 1990.
7. H. Fujita, *Polymer Solutions*, Elsevier, Amsterdam, 1990.
8. T. Hashimoto, *Current Theories in Polymer Science*, R. M. Ottenbrite, L. A. Utracki, and S. Inoue, Eds., Hanser, N.Y., v. 2, p. 197 (1987).
9. J. D. Gunton, M. San Miguel, and P. S. Sahni, in *Phase Transitions and Critical Phenomena*, D. Domb and J. L. Lebowitz, Eds., Academic Press, London, v. 8, p. 267 (1983).
10. T. M. Aminabhai, R. H. Balundgi, and S. S. Joshi. *J. Sci. Ind. Res.*, **47**, 375 (1988).
11. P. de Gennes, *Scaling Concept in Polymer Physics*, Cornell Univ. Press, Ithaca-London, 1979.
12. *Polymer Blends*, D. R. Paul and J. W. Marlow, Eds., Academic Press, N.Y., 1978.
13. O. Olabisi, L. Robeson, and M. Shaw, *Polymer-Polymer Miscibility*, Academic Press, N.Y.-London, 1979.
14. A. Manson and L. Sperling, *Polymer Blends and Composites*, Plenum Press, N.Y., 1976.
15. L. A. Utracki, *Polymer Alloys and Blends*, Hanser Publishers, Munich-Vienna-New York, 1989.
16. *Polymer Alloys-Properties and Application* (in Japanese), Society of Polymer Science Japan, Tokyo, 1981.
17. S. Akiyama, T. Inoue, and T. Nishi, *Polymer Blends* (in Japanese), *CMC*, Tokyo, 1981.
18. Y. S. Lipatov, *Polymer Reinforcement, ChemTec Publ.*,Toronto, 1995.
19. Y. S. Lipatov, *Int. J. Polymer Mater.*, **17**, 91 (1992).

Chapter 1

BASIC PRINCIPLES OF THERMODYNAMICS OF POLYMER SOLUTIONS

1.1 THE MAIN THERMODYNAMIC CHARACTERISTICS OF SOLUTION

The main thermodynamic functions of any solution are the changes in enthalpy of its formation, ΔH, entropy changes, ΔS, and Gibbs free energy, $\Delta G = \Delta H - T\Delta S$. At constant temperature and pressure, the dissolution is a spontaneous process proceeding in the direction of diminishing Gibbs free energy of mixing. The solution has lower free energy as compared with the sum of free energies of solution components (polymer and solvent):

$$G_{solution} < \sum_i n_i G_i^0 \qquad [1.1]$$

where G^0 is the free energy of i-th component.

The excess free energy determined as

$$\Delta G = G_{solution} - \sum_i n_i G_i^0 \qquad [1.2]$$

by spontaneous dissolution is negative. Correspondingly, the chemical potential of a component in solution, μ_i, should be lower than its value before dissolution, $\mu_i < \mu_i^0$ i.e., $\Delta\mu_i < 0$. Values $\Delta G < 0$, $\Delta\mu_i < 0$ are the criteria of the affinity between components. Their values are determined by the absolute magnitude of the differences ΔG or $\Delta\mu_i$.

The solution in which all molecules of components equally interact with each other are considered as ideal solutions. Their properties are the additive sum of the properties of components. Such solutions obey the Henry and Raul laws:

$$p_2 = Kx \qquad [1.3]$$

$$p_1 = p_1^0(1 - x) \qquad [1.4]$$

where p_1 and p_2 are partial pressure of components, x is the molar fraction of component in solution. Because

$$\Delta\mu_i = RT \ln p_i/p_i^0 \qquad [1.5]$$

we find that

$$\Delta\mu_i = RT \ln x_i \qquad [1.6]$$

It means that the chemical potential of any component is only determined by its molar fraction in solution. In the majority of cases, solutions do not obey Eqs 1.3 to 1.6. They are thus considered non-ideal. The deviations from the ideality ($p_i/p_i^0 = x_i$) may be positive ($p_i/p_i^0 > x_i$) or negative ($p_i/p_i^0 < x_i$). To estimate the degree of deviation from the ideality, the concept of excess quantities is introduced as an excess of properties in comparison with the ideal magnitude. These are:

$$\Delta\mu_i^{exc} = \Delta\mu_i - \Delta\mu_i^{id} = RT \ln \gamma_i \qquad [1.7]$$

$$\Delta\overline{S}_i^{exc} = \overline{S}_i - S_i^{id} = -RT \ln \gamma_i - RT\left(\frac{\delta \ln \gamma_i}{\gamma T}\right)_p \qquad [1.8]$$

$$\Delta\overline{H}_i^{exc} = -RT^2\left(\frac{\delta \ln \gamma_i}{\delta T}\right)_p \qquad [1.9]$$

$$\Delta\overline{G}_i^{exc} = R \ln \gamma_i \qquad [1.10]$$

where γ_i is the activity coefficient of component i in a real solution ($\gamma_i = a_i/x_i$). If the activity is expressed as $RT \ln a_i = \mu_i - \mu_i^0$, then we have:

$$\Delta G_i^{exc} = \sum x_i \Delta\mu_i^{exc} = RT \sum x_i \ln \gamma_i \qquad [1.11]$$

$$\Delta H^{exc} = \sum x_i \Delta\overline{H}_i^{exc} = -RT^2 \sum x_i\left(\frac{\delta \ln \gamma_i}{\delta T}\right)_p \qquad [1.12]$$

$$\Delta S^{exc} = \sum x_i \Delta\overline{S}_i^{exc} = -R\sum x_i \ln \gamma_i - RT\sum x_i\left(\frac{\delta \ln \gamma_i}{\delta T}\right)_p \qquad [1.13]$$

1.2 REGULAR SOLUTIONS. HILDEBRAND-SCATCHARD EQUATION

According to Hildebrand, a solution is called regular if by its formation $\Delta H \neq 0$. It is accepted that the distribution of mixed molecules is fully random, i.e., the entropy of mixing equals to the ideal value, whereas mixing proceeds without changing volume. The energy of mixing of pure liquid components, ΔE_{mix}, equals to the difference between intermolecular energy of one mole of solution, E_{sol}, and the sum of intermolecular energies of pure liquid components, E_i^0:

$$\Delta E_{mix} = 2\pi N_0 V\varphi_1\varphi_2\left(\frac{2}{V_1V_2}\int_0^\infty \varepsilon_{12}\,\rho_{12}\,r^2dr - \frac{1}{V_1^2}\int_0^\infty \varepsilon_{11}\,\rho_{11}\,r^2dr - \frac{1}{V_2^2}\int_0^\infty \varepsilon_{22}\,\rho_{22}\,r^2dr\right) \quad [1.14]$$

where V is the volume of one mole of solution, ε_{11}, ε_{22}, and ε_{12} are the interaction energies of two molecules in a pair, ρ_{11}, ρ_{22}, and ρ_{12} are the probabilities that any arbitrary chosen molecule in the volume unit is the molecule of the 1, 2, etc. kind, r is the distance from the arbitrary chosen central molecule, φ_1 and φ_2 are volume fractions of components that are constant in a solution: $\varphi_1 = V_1(1-x)/[V_1(1-x) + V_2x]$; $\varphi_2 = V_2x/[V_1(1-x) + V_2\,x]$. Using expression for ε_{ij} as $-k_{ij}/r^6$, where k_{ij} is the interaction constant for molecules of the type i and j, and the rule of the geometric average $k_{12} = (k_{11}\,k_{22})^{1/2}$, one can obtain Eq 1.14 in the form:

$$\Delta E_{mix} = V\varphi_1\varphi_2[(E_1^0/V_1)^{1/2} - (E_2^0/V_2)^{1/2}]^2 \quad [1.15]$$

This is the Hildebrand-Scatchard equation for the regular solutions. The ratio E_i/V_i is called the density of cohesion energy. The value $\delta = (E_i/V_i)^{1/2}$ was called by Hildebrand the coefficient or solubility parameter of the component. The difference of magnitudes in the square brackets (Eq 1.15) determines a deviation of the solution from the ideal behavior. The shortcoming of the theory of regular solution consists of the assumption that the behavior of molecules in solution does not depend on the heat effect of mixing, or that the entropy of mixing is ideal, $\Delta H \neq 0$. This theory is only applicable to solutions containing weakly polar or non-polar components, having the molecular field of spherical symmetry and mixing with a low change in volume.

The non-uniform distribution of molecules is the basis of the theory of strictly regular solutions. The theory uses the lattice model, based on the following assumptions:

- the solution structure is considered quasi-crystalline (short range order) and the particles of the solute and solvent have one and the same coordination number, z
- only the interaction between neighboring molecules is considered

- the potential energy of the system is divided into configurational energy (the energy of the disposition of all the molecules into the lattice site) and acoustic energy (the vibration energy near the equilibrium center). The vibration energy of the molecule does not depend whether the molecule is in solution or in a pure liquid.
- each molecule occupies one lattice cell (therefore molecules should have the same shape and volume).

The potential energy of the molecular exchange of various liquids is taken as:

$$E_{11} + E_{22} - 2E_{12} = -2\Delta w_{12} \qquad [1.16]$$

where w_{ij} is the interaction energy of two molecules. Due to exchange, the number of bonds, z, of the type 1-1 with energy E_{11} and z bonds of the type 2-2 with energy E_{22} are destroyed and 2z bonds of the type 1-2 are formed with energy E_{12}. If the exchange energy is zero, the system is athermic, whereas, for the non-zero energy, the system is called non-athermic. The zero approximation of the theory of regular solution gives the relationship:

$$\Delta H = N_0(1 - x)x\left(\Delta w_{12} - T\frac{\delta \Delta w_{12}}{\delta T}\right) \qquad [1.17]$$

and

$$\Delta G = N_0 x(1 - x)\Delta w_{12} \qquad [1.18]$$

1.3 PHASE EQUILIBRIUM IN SOLUTIONS. BINODALS AND SPINODALS

A system of unperturbed equilibrium consists of only one phase. Any perturbation leads to the appearance of some amount of a new phase. The intensive properties of this new phase (partial molar volume, composition, etc.) may differ from the properties of the equilibrium system either by infinitely small or finite value. Various situations may arise:[1]

- the initial phase is stable in the relation to all other phases, independently on the difference in properties is infinitely small, or it has a finite value
- the initial phase is stable in relation to all phases that differ infinitely small from it, but there is at least one phase in the system in relation to which the initial phase is not stable
- the initial phase is not stable in relation to the phases with infinitely small difference.

In the first case, the phase is called stable, in the second metastable, and in the third non-stable. If the phase is non-stable in relation to other phases with infinitely small difference, this phase disappears and one or more new phases appear. The new phase has properties very similar to the initial phase. This process continues until the formation of the phase that is stable in relation to all adjacent phases.

When the metastable phases are present, the system may stay in the equilibrium state for an indefinitely long time without the appearance of a new phase. If the nuclei of a new, more stable, phase are introduced into the system, the transition to the stable equilibrium state occurs. For solutions, the stability of the system depends on the diffusion leading to the appearance of the non-uniformity of composition in the initially uniform binary system. For all the systems, stable in the relation to diffusion, $\delta^2 g/\delta x_i^2 > 0$ (g is the molar Gibbs free energy). This condition is equivalent to the condition $\delta\mu_1/\delta x_1 > 0$. Because

$$\delta\mu_1/\delta x_2 = -x_2\delta^2 g/\delta x_2^2 \quad [1.19]$$

we have

$$\delta\mu_1/\delta x_2 < 0 \quad [1.20]$$

$$\delta\mu_2/\delta x_1 < 0 \quad [1.21]$$

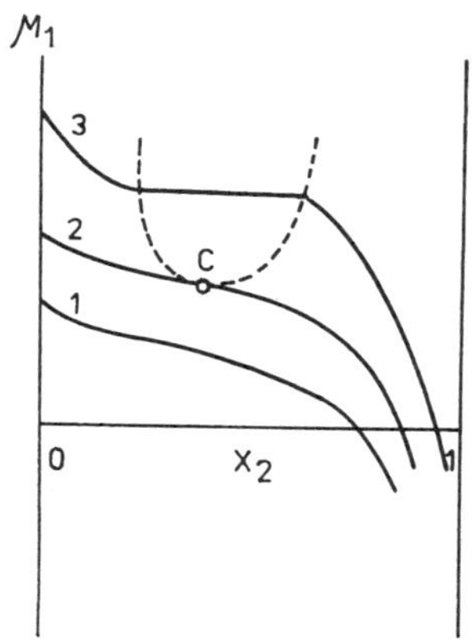

Fig. 1.1. Dependence of chemical potential on the composition of the system n-hexane-nitrobenzene with P and T constant. Adapted, by permission, from I. Prigogine and R. Defay, *Chemical Thermodynamics*, Longmans Green and Co., London, New York, Toronto, 1954.

These conditions of equilibria were illustrated[1] in relation to diffusion for the hexane-nitrobenzene system (Fig. 1.1). Above 292 K (curve 1), there is only one phase and conditions of Eqs 1.20 and 1.21 are always fulfilled. Below 292 K, the curve (curve 3) consist of three parts: the first corresponds to the layer enriched with nitrobenzene, the second to the layer enriched with hexane, and the third is the horizontal line tying these parts and characterizing the presence of two phases simultaneously. The curve at 292 K forms the boundary between these two types of curves. The horizontal part of this curve is degenerated into one inflection point C. This point is characterized by the conditions:

$$\delta\mu_1/\delta x_2 = 0; \quad \delta^2\mu_1/\delta x_2^2 = 0 \qquad [1.22]$$

The critical state is stable because $(\delta^3\mu_2/\delta x_2^3) < 0$. If to expand the chemical potential μ_1 into series in the region adjoining the critical point, then, neglecting the terms of higher orders, we obtain:

$$\mu_1 - \mu_{1,crit} = \frac{1}{3!}\frac{\delta^3\mu_1}{\delta x_2^3}(x_2 - x_{2,crit})^3 \qquad [1.23]$$

It follows from Fig. 1.1 that the sign of $(\mu_1 - \mu_{1,crit})$ is opposite to the sign of $(x_2 - x_{2,crit})$ and thus

$$\frac{\delta^3\mu_2}{\delta x_2^3} < 0 \qquad [1.24]$$

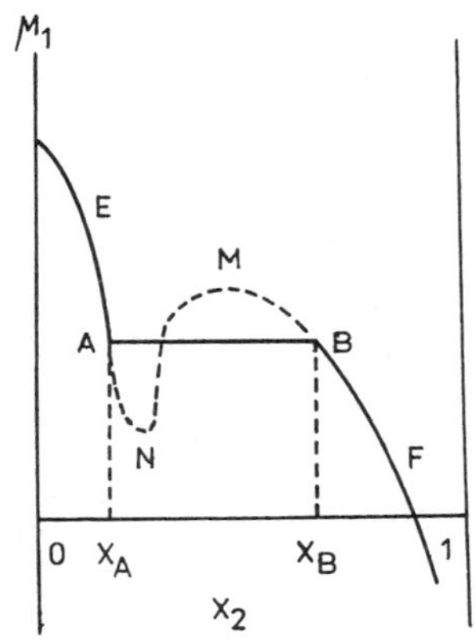

Fig. 1.2. Dependence of chemical potential on the composition of the system n-hexane-nitrobenzene with T = 298K and P = constant. Adapted, by permission, from I. Prigogine and R. Defay, *Chemical Thermodynamics*, Longmans Green and Co., London, New York, Toronto, 1954.

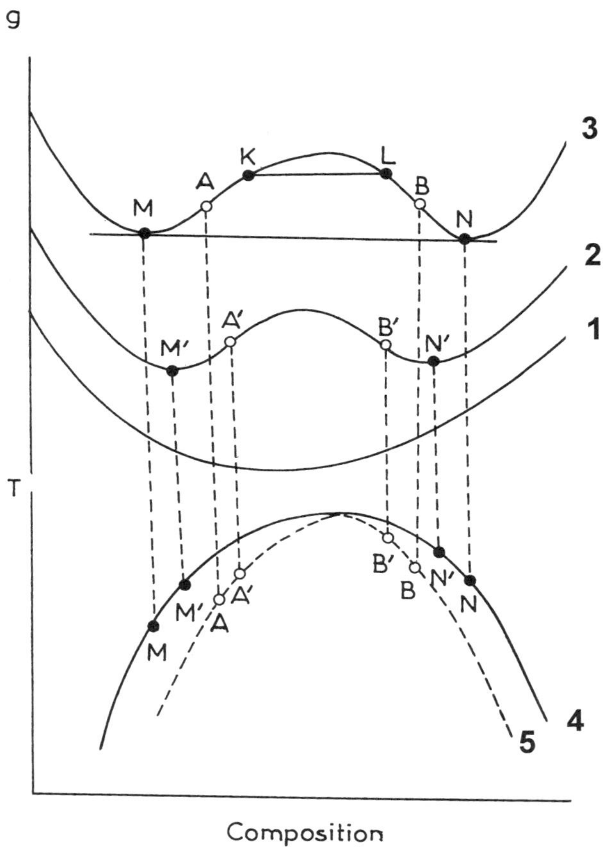

Fig. 1.3. Dependence of the Gibbs free energy on the composition with P and T constant (1-3), curve 4-binodal, curve 5-spinodal.

The transition from one-phase to two-phase state is presented in Figs. 1.2 and 1.3. Increasing temperature above 292 K, one can transit from the layer enriched with hexane to the layer enriched with nitrobenzene. No stage of this transition is accompanied by the appearance of a new phase and therefore both parts of the curve 3 (Fig. 1.1) may be considered as the segments of the continuous curve FBMNAE, Fig. 1.2. The states between M and N are unstable because $\delta\mu_1/\delta x_2 > 0$, whereas BM and AN correspond to the metastable states. The boundary between the metastability and instability is determined by the point at which $\delta\mu_1/\delta x_2 = 0$. The conditions of stability (Eqs 1.20 and 1.21) for $(\delta\mu_1/\delta x_2)_{P,T} = -x_2(\delta^2 g/\delta x_2^2)_{P,T}$ can be given as

$$\frac{\delta^2 g}{\delta x_2^2} > 0 \qquad [1.25]$$

This inequality has very simple geometric meaning. The function g vs. x_2 at the constant T and P should be turned down by the convexity for stable systems (Fig. 1.3, curve 1). If the curve has the shape 2 or 3 and between some values of x_2 there is a part with the convexity turned up, in this region (AB) the system cannot be in the state of stable equilibrium and decomposes into two phases. The molar fractions x'_2 and x''_2 in these two equilibrium phases can be calculated in the following way. According to Gibbs-Duhem, $g = \mu_1 x_1 + \mu_2 x_2$. Differentiation of g by x_2 gives $(\delta g/\delta x_2)_{P,T} = \mu_2 - \mu_1$ or $\mu_2 = g - x_1\,(\delta g/\delta x_1)_{P,T}$. The condition of a true equilibrium in relation to the distribution of the component 2 between two phases is $\mu'_2 = \mu''_2$, or

$$g' - x'_1\frac{\delta g'}{\delta x'_1} = g'' - x''_1\frac{\delta g''}{\delta x''_1} \qquad [1.26]$$

$$g' - x'_2\frac{\delta g'}{\delta x'_2} = g'' - x''_2\frac{\delta g''}{\delta x''_2} \qquad [1.27]$$

Inserting in Eq 1.26 value, $x_1 = 1 - x_2$ and $\delta g/\delta x_1 = -\delta g/\delta x_2$, we have

$$g' + \frac{\delta g'}{\delta x'_2} - x'_2\frac{\delta g'}{\delta x'_2} = g'' + \frac{\delta g''}{\delta x''_2} - x''_2\frac{\delta g''}{\delta x''_2} \qquad [1.28]$$

Subtracting Eq 1.27 from Eq 1.28, we find the condition of the true equilibrium:

$$\left(\frac{\delta g'}{\delta x'_2}\right)_{P,T} = \left(\frac{\delta g''}{\delta x''_2}\right)_{P,T} \qquad [1.29]$$

$$g' - g'' = \left(\frac{\delta g'}{\delta x'_2}\right)_{P,T}(x'_2 - x''_2) \qquad [1.30]$$

The Eqs 1.29 and 1.30 have a simple geometric meaning. Values x_2 corresponding to two phases in the state of equilibrium, i.e., x'_2 and x''_2, have such magnitudes that functions g′ and g″ have the common tangent MN (Fig. 1.3). The segments MA and NB meet the state of the metastable equilibrium and are inclined to the transformation into two-phase system. In the interval between two inflection points the condition $\delta g/\delta x_2^2 < 0$ is valid. Therefore any solution, having had the fluctuation of concentration in these composition intervals, may spontaneously decompose at the current point K and L. Such decomposition will be enhanced until the initial solution is separated into two solutions with compositions corresponding to the point M and N. The curve 4, called binodal, is the geometric place of points M and M′, N and N′, etc. for the curves $g = f(x_2)$ corresponding to various temperatures. Binodal separates the regions of the stable and metastable states of the system. Each point of binodal meets the condition $\delta^2 g/\delta x_2^2 > 0$ and $\mu'_2 = \mu''_2$. Curve 5, called spinodal, represents the geometric place of points A and A′, B, and B′, etc. Spinodal separates the regions of metastable and unstable states. At each point of spinodal $\delta^2 g/\delta x_2^2 = 0$. In accordance with condition 1.19, Eq 1.22, valid at the critical point, may be presented as

$$\frac{\delta^2 g}{\delta x_2^2} = 0; \quad \frac{\delta^3 g}{\delta x_2^3} = 0; \quad \frac{\delta^4 g}{\delta x_2^4} > 0 \qquad [1.31]$$

1.4 THE FLORY-HUGGINS THEORY

The polymer molecules in solution, which are the chains constructed of the same chemical units, are characterized, in many cases, by the same interactions as the molecules of low molecular weight solvent. Therefore, if the chemical nature of the polymer segments and the solvent molecules are the same, the interaction energy between polymer molecule and solvent in such a solution is close to that for the solution of low molecular weight solute. The heat of mixing here should be close to zero. Such solutions are athermic and their thermodynamic properties are determined only by the changes of entropy of mixing, as distinct from the regular solutions with ΔS^{exc} equal to zero.

For the athermic solutions formed by the components with molecules strongly differing by their dimension (namely, the solutions of polymers in low molecular weight solvents), the entropy of mixing is not equal to its ideal value. In many cases, in such solutions, ΔS^{exc} is very high, showing a great deviations from the ideality. Because the athermic solutions are a simple example of polymer solutions, the theories of polymer solutions were developed firstly for the case of athermal mixing of polymer with low molecular weight liquids. The theories were proposed independently by Flory and Huggins[2,3,4] who developed the principles of the statistical thermodynamics of polymer solutions. They have derived the expression for the entropy of mixing known as the combinatorial entropy by Flory-Huggins. This equation plays the same role for the polymer solution as the van der Waals equation for the real gases.

1.4.1 ENTROPY OF MIXING

To calculate the entropy of mixing, Flory has used the lattice model of liquid (solution). This model was modified for polymer solution in such a way that the polymer molecule occupied the row of adjacent cells in the lattice, whereas the solvent molecule occupies only one cell. The segment of the polymer molecule was placed in such a cell that its dimension corresponds to the dimension of the solvent molecule. After this, the number of conformations available to each chain was calculated by a successive introduction of segments of the chain into the lattice. If the lattice has already N polymer chains or the equivalent number of the solvent molecules, then for the last elements there are free $(n_0 - rN)$ cells. Therefore, the first segment $(N + 1)$ of the chain may be placed by the $(n_0 - rN)$ modes. With the fixed first segment, the second segment is placed by the $[z(n_0 - rN)/n_0]$ modes, the third by the $[(z - 1)(n_0 - rN)/ n_0]$ modes, etc., where z is the coordination number of the lattice. The total number of the possible conformations for the chain consisting of r segments, is defined as

$$\nu_{N+1} = \frac{1}{2}\frac{(n_0 - rN)z(n_0 - rN)(z-1)^{r-2}(n_0 - rN)^{r-2}}{n_0 n_0^{r-2}} \qquad [1.32]$$

or, after some transformation,

$$\nu_{N+1} = \frac{1}{2}(z-1)^{r-1}(n_0 - rN)^r n_0^{1-r} \qquad [1.33]$$

As the next step, the thermodynamic probability of the solution state, W, is calculated as the product of the number of modes of placing of all N chains and all solvent molecules:

$$W = \frac{1}{N!}\prod_{N=1}^{N} \nu_N \qquad [1.34]$$

Here N! is the number of permutations between the polymer chains or between the solvent molecules. These permutations are excluded because they do not give any new ways of the molecule placing. Inserting Eq 1.34 into Eq 1.33, using the Stirling formula and after some transformation, we obtain:

$$W = \left(\frac{(z-1)}{e}\right)^{(r-1)N}\left(\frac{1}{2}\right)^N \frac{(n + rN)^{n+N}}{n^n (rN)^N} \qquad [1.35]$$

Finding the logarithm of Eq 1.35 gives the value of the entropy of solution, S_{sol}. The same calculation is made to obtain the number of conformation in the initial state, i.e., for pure amorphous polymer. Generally, the entropy of the amorphous polymer is calculated using the same Eq 1.35 if to suppose that either only polymer chains are placed on the lattice ($n = 0$), or only solvent molecules ($N = 0$). In this case

$$\Delta S = k(r - 1)N[\ln(z - 1) - 1] - k\ln 2 \qquad [1.36]$$

The entropy of mixing is expressed as

$$\Delta S^M = S_{sol} - \Delta S = -k[(n\ln(n/(n+rN)) +$$

$$+ (N\ln(Nr/(n+rN))] \qquad [1.37]$$

where $Nr/(n + Nr)$ and $n/(n + rN)$ are equated with the volume fractions of polymer and solvent, respectively, i.e., φ_1 and φ_2. For one mole of the sites in the lattice

$$\Delta S^M = -R(\varphi_1 \ln\varphi_1 + \frac{\varphi_2}{r}\ln\varphi_2) \qquad [1.38]$$

Here, it is supposed that the molecules are placed on the lattice at random, the mixing is not accompanied by the volume changes and the value r is determined as the ratio of the molar volumes of polymer and solvent, $r = V_2/V_1$.

1.4.2 THE HEAT AND FREE ENERGY OF MIXING

In the lattice model the heat of mixing for athermal mixtures is calculated as the exchange contribution that is added to the combinatorial entropy of mixing. The existence of the heat of mixing is determined by the difference in the total energy of intermolecular interactions in pure components and in solution. Because the interactions in liquids drop very fast with the distance between interacting molecules, one can take into account only the closest neighbors by considering a pair of interactions in usual liquids. Let w_{11}, w_{22}, and w_{12} to be the interaction energy of the corresponding pairs of molecules. By the formation of one contact of the type 1-2 (polymer-solvent), the change in energy is the following:

$$\Delta w_{12} = \frac{1}{2}(w_{11} + w_{22}) - w_{12} \qquad [1.39]$$

The heat of mixing is expressed as

$$\Delta H^M = \Delta w_{12} p_{12} \qquad [1.40]$$

where p_{12} is the number of contacts 1-2. Since each molecule of the solvent has $z\varphi_2$ of the neighboring polymer molecules, the number of solvent molecules is N_1, we have $p_{12} = zN_1\varphi_2$. Flory has introduced a dimensionless parameter

$$\chi_1 = \frac{z\Delta w_{12}}{kT} \tag{1.41}$$

that characterizes the excess energy of interaction in solution per one molecule of the solvent. Value $kT\chi_1$ represents the difference of the solvent molecule energies immersed in polymer and pure solvent. The heat of mixing of N_1 of the solvent molecules with N_2r segments of the polymer can be found from the equation:

$$\Delta H^M = z\Delta w_{12}N_1\varphi_2 = kT\chi_1 N_1\varphi_2 \tag{1.42}$$

For one mole of the lattice site, the change in free energy by mixing is:

$$\frac{\Delta G^M}{RT} = \varphi_1 \ln \varphi_1 + \varphi_2 r^{-1} \ln \varphi_2 + \chi_1\varphi_1\varphi_2 \tag{1.43}$$

Differentiation of Eq 1.43 gives the following expression for the chemical potential of the solvent:

$$\mu_1 = \mu_1^0 + RT[\ln(1 - \varphi_2) + (1 - r^{-1})\varphi_2 + \chi_1\varphi_2^2] \tag{1.44}$$

Because the activity of the solvent is $\exp[(\mu_1 - \mu_1^0)/RT]$, by the expansion the logarithm in Eq 1.44 in series by φ_2 we can obtain

$$\ln a_1 = \ln X_1 + \frac{\mu^E}{RT} = -\frac{\varphi_2}{r} - \left(\frac{1}{2} - X_1\right)\frac{\varphi_2}{2} - \sum_{j=3}^{\infty}\frac{\varphi_2^j}{j} \tag{1.45}$$

It is easy to show that value φ_2/r approaches asymptotically zero by infinite dilution ($X_1 \rightarrow 1$). Thus, by the infinite dilution, the solution obeys the Raul Law whereas the excess chemical potential μ_1E depends only on the second and higher powers of concentration. From this point of view, it is worth to determine another excess potential, $\mu_1^E = RT(\ln a_1 + \varphi_2/r)$ without consideration of the particular form of $\ln a_1$. It will differ from μ_1^E by the terms φ_2^2/r that are obtained by the expansion $\ln X_1$. Because the osmotic pressure $\pi = (\mu_1 - \mu_1^0)/V_1$, one may find the expression for π making the changes in Eq 1.36; $\varphi_2 = c_2\bar{v}$, $\varphi_2/rV_1 = c\bar{v}/rV_1 = c_2/M$ (where $\bar{v}$ is the partial specific volume of a polymer and M is its molecular mass):

$$\frac{\pi}{c_2} = RT\left[\frac{1}{M} + \frac{\bar{v}_2^2}{V_1}((1/2) - \chi_1)c_2 + \frac{\bar{v}_2^3}{3V_1}c_2^2 + \dots\right] \quad [1.46]$$

The non-ideality of the solution is reflected in this equation by the presence of terms following 1/M.

1.5 DILUTE SOLUTIONS OF POLYMERS

The combinatorial calculation giving Eq 1.38 is acceptable only for concentrated polymer solutions because only in this case the distribution of segments in solution approaches uniform distribution on microscopic level. Such approach is valid when chains in solution are overlapped. For a very dilute solutions, the lattice model should be inadequate because in such solution the swollen macromolecular coils are not touching each other and there are regions sharply differing in segment concentration: some regions of the solution contain only pure solvent and some swollen coils. This inadequacy of the lattice model led Flory to the theory of diluted solutions that accounts for non-uniform structure of the solution. In both cases, the macromolecular coil is characterized by the spherically symmetric distribution of segments in relation to the gravity center. In the model considered above, the uniform density of segments in the coil is supposed to be the case,[5] whereas the second model accepts a more realistic Gaussian distribution of segments around the mass centre.[6] The coils are supposed to be interpenetrable whereas the energy of interaction between two coils is calculated as a function of the distance between their centers of mass. For this purpose the lattice models are used for small regions where the segment distribution is taken as uniform because of the coil overlapping. In each microregion, the energy of interaction between segments is proportional to the square of the segment density. The calculations have given the distribution function that, as in the statistical theories of the real gases, leads to the appearance of the second virial coefficient in the equation of state, or equation of the osmotic pressure. Consider how this equation may be derived for the dilute solutions.

Let the molecule be a sphere. Two spheres cannot approach each other at the distance $r < d$ (d is the sphere diameter). For the sphere there is an excluded volume $u = 4\pi d^3/3$ where $d = 2r$. The number of modes of the placing in the solution is: for the first molecule $\approx V$, for the second $\approx V - u$, for the third $\approx V - 2u$, for the i-th molecule $\approx V - iu$. The number of the modes of placing all molecules is

$$Q = \text{const} + \prod_{i=0}^{N_2-1} (V - iu) \quad [1.47]$$

If to account only for pair of contacts between molecules (N_2! permutation of N_2 molecules is excluded; in a given case it is included in the constant), the energy of mixing can be calculated from the equation:

$$\Delta G = -kT \ln Q = -kT \sum_{st} = -kT[N_2 \ln V + \sum_{i=0}^{N_2-1} \ln (1 - (iu/V))] + \text{const} \qquad [1.48]$$

(there is no ΔH in the equation as it is accounted for by the excluded volume). We consider here only the dilute solution where the excluded volume is small. Therefore we can expand ln (1 – iu/V) into series:

$$\Delta G \approx -kT\left[N_2 \ln V - \frac{u}{V}\frac{N_2^2}{2}\right] + \text{const} \qquad [1.49]$$

Taking into account that $N_2/V = cN_A/M$, we have:

$$\pi = RT\left[\frac{c_2}{M} + \frac{c_2^2}{M^2}\frac{N_A u}{2} + \dots\right] = RT[A_1 c_2 + A_2 c_2^2 + \dots] \qquad [1.50]$$

where $A_1 = 1/M$, $A_2 = N_A u/2M^2$. Such expansion of the osmotic pressure into series is called virial expansion and coefficients are virial coefficients. This general relation is valid for low molecular weight substances as well. If the interaction between molecules is absent (ideal solution) there is no excluded volume, i.e., $u = 0$. Generally, $A_2 \neq 0$ and it is a measure of intermolecular interaction.

The same analysis, as for the combinatorial entropy of mixing, is used to derive the expression for free energy of interaction of all segments of two macromolecules in a small element of the solution volume, δv, which are at the distances r_k and r_l from the inertia center of macromolecules. The only distinction is that only restricted parts of macromolecules are present in the volume element, not the whole macromolecules. Thus, to place segments into the lattice cell, it is necessary to take into account that for any segment one of the cells of the first coordination sphere is already occupied by the preceding segment. If it is necessary to place δx segments in δn_0 cells of the volume δv, the mean probability to find free cell for (j + 1)th segment should be $(\delta n_0 - j)/\delta n_0$ and the mode number of placing of all segments is

$$w_{\delta v} = \prod_{j=0}^{j=\delta x} (z-1)\frac{\delta n_0 - j}{\delta n_0} \qquad [1.51]$$

or

$$w_{\delta v} = \frac{\delta n_0!}{(\delta n_0 - \delta x)!}\left[\frac{z-1}{\delta n_0}\right]^{\delta x} \qquad [1.52]$$

The sequence of operations that was used to transit from Eq 1.34 to Eq 1.38 leads to the relation:

$$\delta(\Delta S^M) = -k\delta n_1 \ln \varphi_1 \qquad [1.53]$$

where δn_1 and φ_1 are the number of molecules and volume fraction of the solvent in the volume element δv.

The free energy of mixing the polymer segments with the solvent in the volume, δv, can be obtained by adding to $-T\delta(\Delta S^M)$ the term $\chi_1\delta n_1\varphi_2$. This gives

$$\delta(\Delta G^M) = kT[\delta n_1 \ln(1-\varphi_2) + \chi_1\delta n_1\varphi_2] \qquad [1.54]$$

where volume fraction of polymer, φ_2, is only related to the volume δv. Differentiation of Eq 1.54 gives the change in chemical potential of solvent:

$$\Delta\mu_1 = RT[\ln(1-\varphi_2) + \varphi_2 + \chi_1\varphi_2^2] \qquad [1.55]$$

At small φ_2, expanding $\ln(1-\varphi_2)$ into series by φ_2 we have:

$$\Delta\mu_1 = -RT\left[\left(\frac{1}{2} - \chi_1\right)\varphi_2^2 + \frac{\varphi_2^3}{3} + \ldots\right] \qquad [1.56]$$

Preserving only quadratic term of Eq 1.56 and accepting that $(\chi_1 - 1/2) = k_1 - \Psi_1$ we find that

$$\Delta\mu_1 = RT(k_1 - \Psi_1)\varphi_2^2 \qquad [1.57]$$

where k_1 and Ψ_1 are the enthalpy and entropy parameters, respectively, $\Delta H_1 = RTk_1\varphi_2^2$, $\Delta S = R\varphi_2^2\Psi_1$. Further the concept of the "ideal temperature" is introduced,

$$\Theta = \frac{k_1T}{\Psi_1} = \frac{\Delta H_1}{\Delta S_1} \qquad [1.58]$$

Then we have

$$\Delta\mu_1 = -RT\Psi_1\left(1 - \frac{\Theta}{T}\right)\varphi_2^2 \qquad [1.59]$$

At $T = \Theta$, we have $k_1 = \Psi_1$, $(-1/2 + \chi_1) = 0$, i.e., $\Delta\mu = 0$. Therefore, at the Θ-temperature, the behavior of the solution becomes ideal. At the Θ-temperature, the macromolecules approaching each other put some restrictions on the placing new molecule which is entropically unfavorable but energetically favorable (the energy of polymer-polymer contact is higher than the energy of polymer-solvent contact). At the Θ-temperature, the entropy factor is compensated by the energy factor, i.e., the compensation of the forces of repulsion and attraction occurs. The polymer coil acquires the unperturbed dimensions.

Solvent, in which the entropy factor is compensated by the energetic factor (i.e., when the compensation of the attraction and repulsion forces occurs), is called "poor solvent" (or ideal solvent obeying the Henry-Raul Law). In the poor solvent, the second virial coefficient (Eq 1.50), $A_1 = 0$. On the contrary, the solvent is called "good", if in a given solvent the entropy factor is prevailing (repulsion forces prevail over attraction forces). In a good solvent $A_1 > 0$. The names "poor" and "good" are conventional, depending on conditions (above or below Θ-temperature). A good solvent may become a poor solvent (by approaching the Θ-point) and *vice versa* a poor solvent may become good solvent by moving off the Θ-temperature, i.e., at $T > \Theta$.

Flory derived the equation for the entropy of mixing by the successive placing of non-correlated but bound monomeric units on the lattice, whereas Huggins used statistical scheme for calculations that accounts for the short-range correlations connected with the individual bonds. The approximation by Huggins differs from Flory's approximation by additional contribution to entropy that depends on the coordination number, z. The lack of knowledge of the corresponding value z for real polymer systems and a greater simplicity of the one-parameter theory by Flory resulted in its wide application in the form of Eq 1.43. The theory is called after Flory-Huggins in accordance with the independent contribution by Huggins.

1.6 PHASE SEPARATION IN POLYMER SOLUTIONS

By the quenching the solution of non-crystallizable polymer, one can reach temperature at which the solution separates into two phases: one enriched with polymer and the second is dilute solution. Further quenching makes enriched phase more concentrated and dilute phase more diluted. That implies that the system has the regions of insolubility that are narrowed with increasing temperature up to the upper critical solution temperature, UCST. At this temperature, the heat and entropy of mixing is positive and excess chemical potential, μ_1^E, decreases with increasing temperature. In the lattice theory, the exchange contribution, $RT\chi_1\varphi_2^2$, may exceed the negative combinatorial entropy of mixing. However, at the increasing temperature, the entropy of mixing becomes prevailing because χ_1 diminishes with temperature. One has to bear in mind that the critical concentration for polymer solutions is low and decreases with the increase in the molecular mass of the polymer, whereas the critical temperature grows (Fig. 1.4). The calculated binodals have been obtained from the experimental data on the critical points, using the Flory-Huggins theory. From the condition for critical point, i.e., $(\delta\mu_1/\delta\varphi_2)_{P,T} = 0$ and $(\delta^2\mu_1/\delta\varphi_2^2)_{P,T} = 0$ by differentiation of Eq 1.44, we find:

$$1 - (1 - r^{-1} + 2\chi_1\varphi_2)(1 - \varphi_2) = 0 \quad [1.60]$$

$$1 - 2\chi_1(1 - \varphi_2)^2 = 0 \quad [1.61]$$

Solving Eqs 1.60 and 1.61, we obtain the critical composition:

$$\varphi_{2,crit} = (1 + r^{1/2})^{-1} \quad [1.62]$$

The equation connecting the critical temperature with Θ-temperature, parameter of interaction entropy, Ψ_1, and the chain length (number of segments, r) was obtained by Flory and Fox and has the form:

$$\frac{1}{T_{crit}} = \frac{1}{\Theta}\left[1 + \frac{1}{\Psi_1}\left(\frac{1}{r^{1/2}} + \frac{1}{2r}\right)\right] \quad [1.63]$$

The plot of T_{crit}^{-1} vs. $(1/r^{1/2} + 1/2r)$ should be linear. From the slope and the length of segment of this dependence one obtains Θ and Ψ_1. Inserting these values to Eq 1.44, the binodals can be calculated, i.e., the temperatures and compositions at which different phases are in the state of equilibrium. The procedure is simple but tiresome because the equilibrium compositions cannot be clearly expressed as a function of temperature.

Fig. 1.4 shows the shift of the critical points to the region of low concentrations by increasing molecular mass in accordance with Eq 1.62, whereas the dependence of T_{crit} on molecular mass meets Eq 1.63 in a qualitative way. The calculated binodals are in agreement with experimental data, although the region of solubility is much narrower and $\varphi_{2,crit}$ is lower than experimental values.[7] Eq 1.63 predicts that Θ-temperature is the limiting value of T_{crit} for the infinitely long chain. The experimental data confirm the fact that at this temperature the second virial coefficient A_2 is equal to zero.[3,8,9] It was shown[10] that the limiting behavior of Eq 1.63 can be obtained only if to suppose that the virial expansion near the critical point is valid.

In spite of good agreement between Θ and limiting critical solution temperature, value Ψ_1 obtained from the critical points is

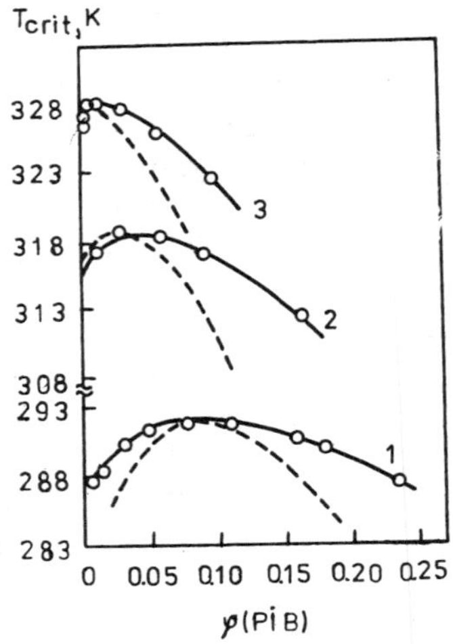

Fig. 1.4. Precipitation temperatures, T_p, for polyisobutylene fractions in diisobutyl ketone plotted against concentration expressed as volume fractions of polymer. The broken lines are the corresponding theoretical curves: 1 - M = 2.27x10^4, 2 - M = 2.85x10^5, 3 - 6x10^6. Adapted, by permission, from A. R. Shultz and P. J. Flory, *J. Amer. Chem. Soc.*, **74**, 4760 (1952).

much larger than one found from the temperature dependence of A_2. The reason is not clear, presumably because the third virial coefficient is not taken into account. Stockmayer has shown[11] that the slope of the dependence $1/T_{crit}$ vs. $1/r^{1/2}$ includes the ternary molecular cluster contributions that do not vanish at $A_2 = 0$.

The shaded lines in Fig. 1.4 show spinodals, i.e., the limits of the metastability in one-phase region. They determine the positions of the points for which $(\delta\mu_1/\delta\varphi_2)_{P,T} = 0$ (or $\delta^2\Delta G/\delta\varphi_2^2 = 0$). Values μ_1 or ΔG can be calculated from Eqs such as 1.44 or 1.43. Fig. 1.4 shows maxima of the calculated spinodals, situated at lower concentrations as compared with the experimental results.

The following attempts to improve the theoretical calculations of the phase diagram on the basis of the lattice model were directed to the correction of the parameter χ_1. For this purpose the parameter χ is usually replaced by the general, concentration-dependent parameter of the free energy $g(T, \varphi_2)$ that includes both enthalpy and entropy contributions:

$$\frac{\Delta G}{RT} = \varphi_1 \ln\varphi_1 + \varphi_2 r^{-1} \ln\varphi_2 + g(T,\varphi_2)\varphi_1\varphi_2 \qquad [1.64]$$

Parameter χ_1 is connected with g by the relation:

$$\chi_1(T,\varphi_2) = g - (1-\varphi_2)(\delta g/\delta\varphi_2)_T \quad [1.65]$$

The concentration dependence of the parameter g is approximated by the polynomial over φ_2 (or w_2). Koningsveld[12] has used the following expression for g:

$$g = \sum_{j=0}^{n} g_i w_i^j \quad (j = 0,1,2,...,n) \quad [1.66]$$

where each coefficient g_i may be the function of temperature:

$$g_j = g_{j1} + g_{j2}/T + g_{j3}T + g_{j4}\ln T \quad [1.67]$$

For the systems where there are no polar or hydrogen bonds, in Eq 1.66, n = 0 and g does not depend on w_2. If the temperature dependence is weak, the calculated values of the free energy of mixing are in a good accordance with experiment.[12] For the systems with intersegmental polar interactions, parameter g should depend both on T and w_2.[13,14] For the systems with polar interactions Eq 1.66 may be used with n = 1:

$$g = g_0 + g_1 w_2 \quad [1.68]$$

where g_0 and g_1 depend on temperature. The temperature dependence of parameters can be written as

$$g_j = g_{j1} + g_{j3}T \quad [1.69]$$

where g_{j1} can have both negative and positive values. Using Eqs 1.68 and 1.69, one can calculate the values of ΔG at various temperatures from Eq 1.64. Double differentiation of the curves $\Delta G = f(\varphi_2)$ allows to determine binodals of spinodals (Ch. 6). Six types of binodals have been obtained:[15]

- If $g_0 \neq f(T)$ and $g_1 = 0$, the cloud point curve (CPC) should show the different maximum at $w_1 < 0.1$ and have narrow region of amorphous phase separation (Fig. 1.5a). The typical example is the system PE-diphenyl ether.
- If $g_0 \neq f(T)$ and $g_1 \neq f(T)$, CPC shows very broad region of amorphous phase separation with poorly distinguished maximum (Fig. 1.5b). An example - cellulose diacetate-benzyl alcohol.
- If $g_0 \neq f(T)$ and $g_1 = f(T)$, CPC should show a very broad region of the phase separation with maximum at $w_1 < 0.3$ (Fig.

1.5c), for example: poly(ethylene-*co*-vinyl acetate) (Elvax-40)-caprolactam.

- If $g_0 = f(T)$ and $g_1 = 0$, CPC shows broad maximum at w_1 between 0.05 and 0.20 (Fig. 1.5d) (PS-toluene).
- IF $g_0 = f(T)$ and $g_1 \neq f(T)$, there should be a broad region of amorphous phase separation with maximum at $w_1 > 0.7$ (Fig. 1.5e).
- If $g_0 = f(T)$ and $g_1 = f(T)$, CPC shows broad region of amorphous phase separation with maximum at w_1 between 0.3 and 0. 7 (Fig. 1.5 f) (PPO-toluene).

Curves I and IV are the main types. Because there can be many combinations of the values g_0 and g_1, maxima at the CPC can be moved to higher or lower values of w_1, whereas the heights of maxima may shift to higher or lower temperatures.

In other works,[8,16] Koningsveld and coworkers have used the equation for g with four parameters (so-called second approximation of the theory):

$$g = \alpha + (\beta_0 + \beta_1/T)(1 - \gamma\varphi_2)^{-1} \qquad [1.70]$$

This equation has proved to be suitable to describe the data on the critical temperature and composition for the PS-cyclohexane system.[8,16] The constants in Eq 1.70 are tying parameters to describe the experimental data. The theoretical curves give a better agreement with experiment as compared with calculated on the basis of the Flory-Huggins theory, using Θ and Ψ_1. For the specimens with a broad molecular mass distribution, the critical points are shifted to the right with increasing M_Z.[8]

The shortcoming of the second approximation is that it does not take into account the effect of the end-groups. The expansion of the theory of the second approximation to the branched polymers has allowed to estimate the effect of the end-groups on the position of calculated spinodal. The dependence of g on M_n was taken as:[17]

$$g(T,\varphi_2,r_n) = \alpha + \frac{\beta_a(T) + \beta_b(T)r_n^{-1}}{1 - \gamma\varphi_2} \qquad [1.71]$$

The introduction of additional terms including g improves the position of the calculated spinodal as compared with experimental values, however has no effect on the shift and slope for the polymers with a broad molecular mass distribution

Parameter g may be used in a broad range of compositions, i.e., from dilute to concentrated solutions, as shown by the theory of the third approximation.[18] It was supposed that

$$g = Pg_{dilute} + (1 - P)g_{conc} = g_{conc} + P(g_{dilute} - g_{conc}) \qquad [1.72]$$

where g_{conc} and g_{dilute} are values of g for the limiting cases of dilute and concentrated solutions and P is the probability that the volume v, equal to the volume of segments of dissolved polymer chain in the solution of the macroscopic volume, V, is not occupied by the segments of N chains:

$$P + (1 - v/V)^N \approx \exp(Nv/V) = \exp(-\lambda\varphi_2) \qquad [1.73]$$

Here λ is the volume fraction of the polymer at which the sum of the volumes of all segments should be equal to the volume, V, if the coils are not overlapped. The volume, v, is chosen as a sphere with a radius equal to the mean-square gyration radius $(R^2_\Theta)^{1/2}$. When the value $g_{dilute} - g_{conc}$ diminishes exponentially with concentration, it may be approximated by the limiting value g* at zero solution concentration. Therefore,

$$g = g_{conc} + g^*\exp(-\lambda\varphi_2) \qquad [1.74]$$

By changing χ_1 for g_{conc}, according to Eq 1.65, the second virial coefficient may be expressed as

$$\frac{A_2V_1}{v_2^2} = \frac{1}{2} - \chi_1 - g^*(1 + \lambda) \qquad [1.75]$$

The theory of excluded volume gives the value of A_2 according to Eq 1.76:

$$A_2 = \frac{N_AuF(z)}{2M^2} \qquad [1.76]$$

In this case

$$g^*(1 + \lambda) = \Psi_1(1 - \Theta/T)[1 - F(z)] \qquad [1.77]$$

The final expression for g is written as

$$g = g_{conc} + \Psi_1(1 - \Theta/T)[1 - F(z)](1 + \lambda)^{-1}\exp(-\lambda\varphi_2) \qquad [1.78]$$

where u is excluded volume and F(z) may be expressed by any theoretical equation

Flory-Krigbaum-Orofino

$$F(z) = \frac{1}{5.73z}\ln(1 + 5.73z) \qquad [1.79]$$

Stockmayer

$$F(z) = 1/(1 + 2.865z) \quad [1.80]$$

Kurata-Yamakawa

$$F(z) = 1 - (1 + 3.903z)^{-0.4683}/1.828z \quad [1.81]$$

Casassa-Markovitz

$$F(z) = [1 - \exp(-5.73z)]/5.73z \quad [1.82]$$

To construct the spinodal curve for the PS-cyclohexane system, Koningsveld *et al.*[18] have used the value g_{conc} from equation with parameters of the critical point taken from the work.[19] From various F(z) functions, the most suitable happened to be the one by Flory-Krigbaum-Orofino, if the constant changes from 2.73 to 5.73. Value λ was taken at about $0.5x_2^{1/2}$. Using the value g, (Eq 1. 78) the spinodal was obtained by the double differentiation of Eq 1.64. The spinodals calculated better coincide with the experimental data regardless of the shape of the function F(z) used.

Koningsveld *et al.*[20,21] using the phase-volume ratio method have determined the critical points for the solutions of PS and PE in a wide range of n-alkanes, n-alcohols, and 1-chloroalkanes as the solvent. The PS-n-alkane and PS-n-alcohol systems showed an optimum of miscibility against the number of carbon atoms in solvent molecule, whereas PE solutions did not (Fig. 1.5).

To describe the effect of chain-length, differences in miscibility of a simple version of the rigid lattice model has been chosen. In particular, if an interchange energy is assigned to the middle and end-groups in the solvent, the expressions for the free energy of mixing of polymer and solvent contain six different interaction parameters for various contacts if the end-groups in the polymer are neglected. The free energy of mixing has been taken as the equation of the Flory-Huggins-Staverman theory:[2,3,22]

$$\frac{\Delta G}{RT} = \frac{\varphi_1}{m_1} \ln \varphi_1 + \frac{\varphi_2}{m_2} \ln \varphi_2 + \varphi_1 \varphi_2 g \quad [1.83]$$

where

$$g = b/Q \quad [1.84]$$

with the following quantities:

$$b = \frac{z_2}{z_1}[((z-1)/m_1)(g_{e_1p} + g_{e_2p}) + (z-2)(1 - 2/m_1)g_{mp} -$$

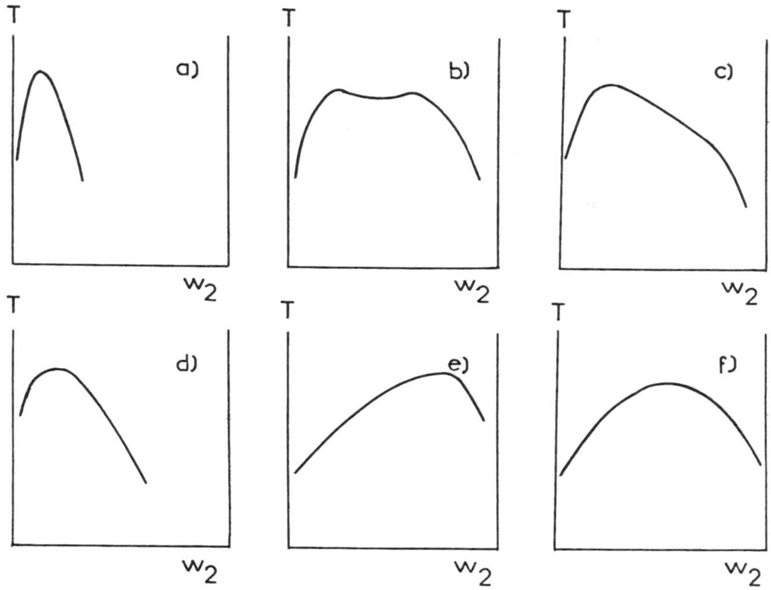

Fig. 1.5. Different types of cloud point curves calculated with $g = g_0 + g_1w_1$ and $g_K = g_{K,1} + g_{K,3}T$. w_1 is the polymer concentration expressed in weight fractions. For further explanation see pages 19-20. Adapted, by permission, from P. T. van Emmerik and C. A. Smolders, *Eur. Polymer J.*, **9**,157 (1973).

$$- (z-1)^2/m_1^2 z_1]g_{e_1e_2} - [(z-1)(z-2)(1 - 2/m_1)/zm_1](g_{e_1m} + g_{e_2m}) \quad [1.85]$$

$$z_k = z - 2 + 2/m_k \quad [1.86]$$

$$Q = 1 - \varepsilon c\varphi_2 \quad [1.87]$$

$$\varepsilon = \frac{1}{m_1} - \frac{1}{m_2} \quad [1.88]$$

$$c = \frac{1}{z_1} \quad [1.89]$$

The interaction parameter for h-k contacts is denoted by $g_{hk} = (g_{hk,S} + g_{hk,H}/T)$. Distinction is made between end-groups (e_1 and e_2), middle groups (m) in the solvent molecule and polymer segments (p). Here z is the coordination number of the lattice and m_i is the number of segments of the species i that is equal in the given case $m_i = M_i/M(C_2H_4)$

where M_i is the molecular mass of the polymer. If two end-groups are identical, as in the case of n-alkanes, $e_1 = e_2 = e$ and $ge_1e_2 = 0$, Eq 1.85 comes to

$$b = \frac{z_2}{z_1}\left[\frac{2(z-1)}{m_1}g_{ep} + (z-2)(1-2/m_1)g_{mp} - \right.$$

$$\left. - \frac{2(z-1)(z-2)(1-2/m_1)}{z_1 m_1}g_{em}\right] \qquad [1.90]$$

The number of parameters has diminished to three (or six, if the temperature dependence is taken into account). The spinodal and critical point are obtained from

$$\frac{1}{m_1\varphi_1} + \frac{1}{m_2\varphi_2} = \frac{2b(1-\varepsilon c)}{Q^3} \qquad [1.91]$$

$$\frac{1}{m_1\varphi_1^2} - \frac{m_z}{m_w^2\varphi_2^2} = \frac{6b(1-\varepsilon c)\varepsilon c}{Q^4} \qquad [1.92]$$

To get quantitative agreement of calculated and experimental spinodal points an additional empirical entropy parameter a is introduced in the expression for g:

$$g = a + b(T)/Q \qquad [1.93]$$

Since there are two equations per critical point, a value of the parameter a for n-alkane system could be determined.

The calculations show the negative value of $g_{hk,H}$, which indicates that PE-n-alkane systems should be expected to exhibit LCST mixing in accordance with experiment.[21,25,26]

Similar calculations have been done also for the solutions of PS in n-alcohols and 1-chloroalkanes, using critical data of PS solutions in n-alkanes. A good agreement with experiment has been found, i.e., the existence of the minimum of the consolute temperature as a function of the number of carbon atoms in the solvent molecule (Fig. 1.6). It was supposed[21] that such behavior of PE and PS solutions in n-alkanes and n-alcohols is determined by breaking up association (n-alcohols) and/or orientational order (n-alkanes) during mixing. Such effects influence both the entropy and the enthalpy of mixing. Whereas the entropy contribution is positive, the enthalpy term is negative. Their magnitude therefore decides whether miscibility increases with the chain length of the solvent or not.

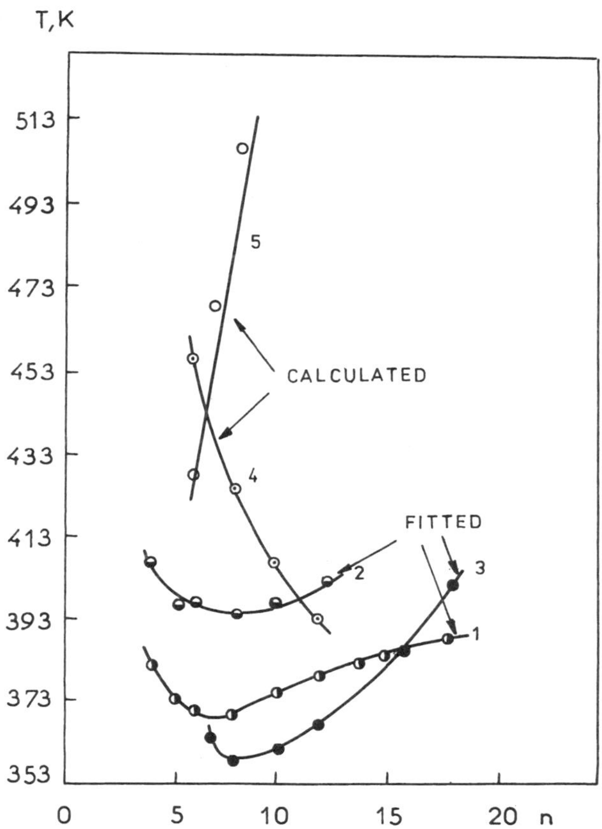

Fig. 1.6. Critical temperatures vs. solvent chain length for various polymer systems: PS (M = 4 × 10^3)—n-alcohol (1); PS (M = 9 × 10^3)—n-alcohol; PS (M = 4 × 10^3)—n-alkane (3); PE (M = 9.2 × 10^4)—n-alcohol (4); PE (M = 17.7 × 10^4)—n-alkane (5). The latter system shows LCST behavior; the other systems show UCST. Adapted, by permission, from L. van Opstal, R. Koningsveld, and L. A. Kleintjens, *Macromolecules*, **24**, 161 (1991).

In the framework of the Flory-Huggins lattice model,[27] it was shown that parameter χ depends on the volume fraction φ_2 of polymer as

$$\chi = (0.5 - \bar{v}_1 q) + \lambda\varphi_2 \qquad [1.94]$$

where $\bar{v}_1$ is the molar volume of the solvent and q characterizes the deviation of the solution behavior from the ideal, based on measure-

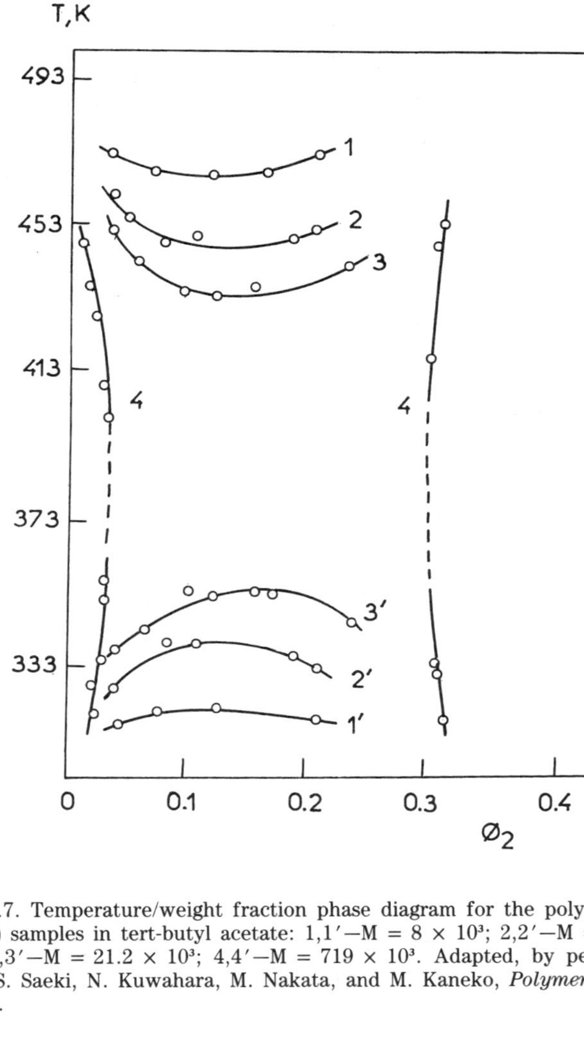

Fig. 1.7. Temperature/weight fraction phase diagram for the poly(ethylene glycol) samples in tert-butyl acetate: 1,1′—M = 8 × 10^3; 2,2′—M = 14.4 × 10^3; 3,3′—M = 21.2 × 10^3; 4,4′—M = 719 × 10^3. Adapted, by permission, from S. Saeki, N. Kuwahara, M. Nakata, and M. Kaneko, *Polymer*, **17,** 685 (1976).

ments by osmometry or other methods. It was found that in the region φ_2 from 0 to 0.25, value λ changes from 0.3333 to 0.4116 that allows to reproduce the observed slight curvature of the experimental dependence of χ on φ_2. The value $(0.5 - v_1 q)$ depends on the polymer-solvent system but does not depend on φ_2. Consequently, value of χ at $\varphi_2 \to 0$ is the number characterizing the given system to be tabulated.

The new theory of diluted solutions, based on the mean-field approximation,[28] allows to show that phase diagrams have a lower critical temperatures and broader two-phase than diagrams corresponding to a simpler Flory theory based also on the same approximation. The critical value of the volume fraction of polymer depends on the polymerization degree, N, because $\varphi_{crit} = N^{-0.40}$ which agrees with experiment. It is also established that in the region of high polymer concentrations, near the phase coexistence line, the molecules have a shape of extended ellipsoid whereas near the curve branch, corresponding to a low polymer concentrations, the molecules have lower dimensions and become more spherical with diminishing concentration.

The lattice models were used also by other authors.[29,30] In its time, the Flory-Huggins theory had played very important role in understanding the processes of dissolution of polymers and contributed to the general theory of solutions. At the same time, this theory suffers from some deficiencies. One of the most important shortcomings of the theory is the neglectance of the volume changes on mixing. It was accepted that any change in the system volume does not influence the entropy and enthalpy of mixing. The theory predicts only positive enthalpy of mixing whereas in many cases the exothermic effects are observed. The most important deficiency of the theory is that it does not predict the existence of the lower critical solution temperature, LCST, that was discovered for a great number of systems,[31-54] for some systems both lower and upper CST were found.[35-41] It was established that the LCST shifts to lower temperatures with increasing molecular mass whereas UCST to higher temperatures (Fig. 1.7). The appearance of the coexistence of CST curves is typical for polymers although much more complicated phase diagrams were observed[33,35,36,48] (Fig. 1.8). In some cases the existence of LCST was explained by the destruction of hydrogen bonds on heating.[38,48,50] However, very often the LCST is not connected with the specific interactions which are situated at high temperatures near the critical temperatures of liquid-vapor for the solvent.[44]

In the systems with UCST, the enthalpy and entropy of mixing are positive. For LCST these parameters should be negative. For the systems with both UCST and LCST, these values should change the sign between UCST and LCST.[51] If the system has two CST, the second virial coefficient A_2 should be equal to zero and by approaching LCST (usually marked as Θ_L like Θ_U), i.e., there is expected some symmetry in properties between UCST and LCST (Figs. 1.9 and 1.10).[52] Near each CST the following relation should be valid:

$$\frac{A'_2V_1}{v_2^2} = \frac{1}{T}\int_{\Theta}^{T}\Psi_1 dT = \Psi_1(1 - \Theta/T); \quad A_2 = A'_2F(z) \qquad [1.95]$$

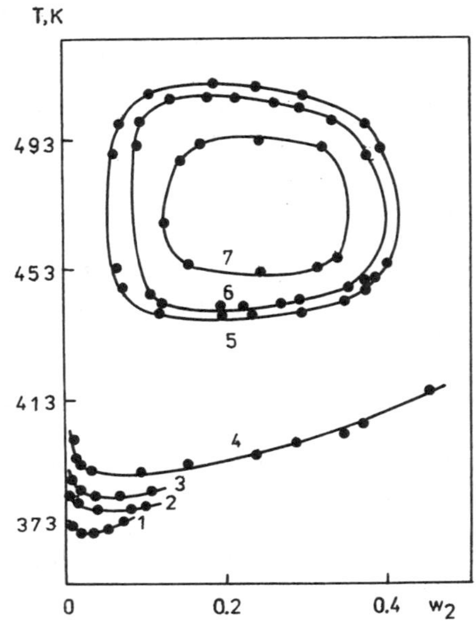

Fig. 1.8 Temperature-weight fraction phase diagram for the poly(ethylene glycol) samples in water: 1—M = 1020 × 10³; 2—M = 21.2 × 10³; 3—M = 14.4 × 10³; 4—M = 8 × 10³; 5—M = 2.29 × 10³; 6—M = 2.27 × 10³; 7—M = 2.18 × 10³. Adapted, by permission, from S. Saeki, N. Kuwahara, M. Nakata, and M. Kaneko, *Macromolecules*, **17,** 685 (1976).

However, between two CSTs this equation is not always correct. Therefore, the variable z cannot be given by a simple equation such as

$$z = (3/2\pi)^{3/2}\sqrt{N}u/A^3 \qquad [1.96]$$

To obtain the correct function z(T), Eichinger[51] has introduced the temperature-dependent parameter, Ψ_1. According to the definition, $R\Psi_1\varphi_2^2$ is the partial, molar entropy of dilution. The integration gives the value of the excess chemical potential:

$$-R\varphi_2^2 \int_{\Theta_U}^{T} \Psi_1(T)dT = -RT\,(V_1/v_2^2)\,A'_2\varphi_2^2 \qquad [1.97]$$

where A'_2 is given by Eq 1.95 at Ψ_1 = const. Eichinger introduced also another parameter $\zeta = T(\delta\Psi_1/\delta T)_P$, taking as the first approximation

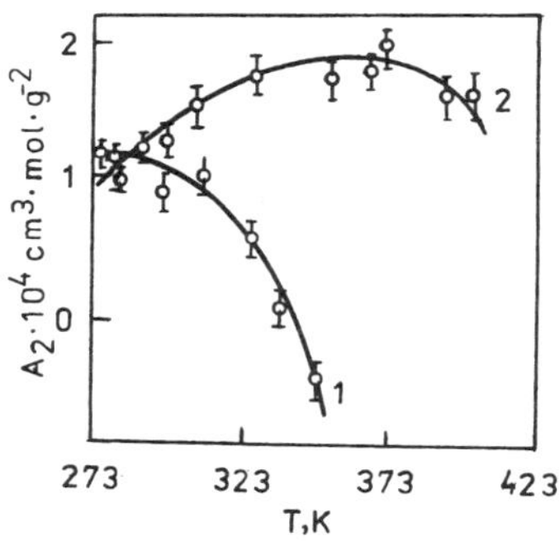

Fig. 1.9. Virial coefficient as determined by light scattering versus temperature: 1-PIB-pentane, 2-PIB-butyl ether. Adapted, by permission, from G. Delmas and D. Patterson, *Polymer*, **7**, 513 (1966).

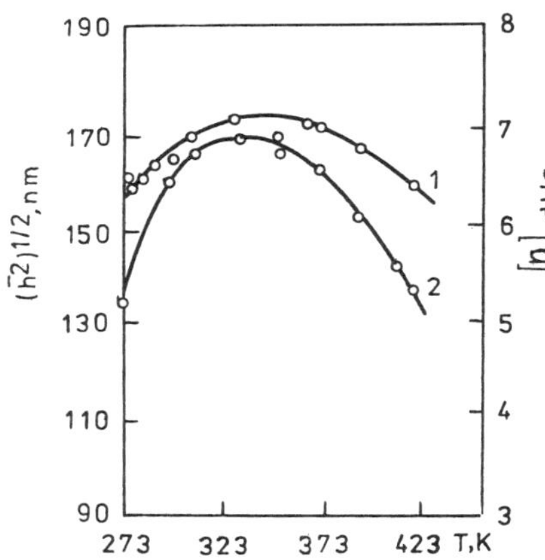

Fig. 1.10. PIB-dibutyl ether: radius of gyration (1) and intrinsic viscosity (2) versus temperature. Adapted, by permission, from G. Delmas and D. Patterson, *Polymer*, **7**, 513 (1966).

the mean value of ζ in the interval between Θ_U and Θ_L. For the temperature-dependent Ψ_1, we have:

$$\Psi_1(T) = \Psi_{1\Theta_U} + \int_{\Theta_U}^{T} (\zeta/T)dT = \Psi_{1\Theta_U} - \zeta \ln(\Theta/T) \qquad [1.98]$$

where $\Psi_1\Theta_U$ is the value Ψ_1 near UCST. Then, inserting $\Psi_1(T)$ in Eq 1.97 after integration in the limits from Θ_U to Θ_L, we can find:

$$\Psi_{1\Theta_B}/\zeta = 1 + (1 - \Theta_U/\Theta_L)^{-1} \ln (\Theta_U/\Theta_L) \qquad [1.99]$$

If Θ_U, Θ_L, and Ψ_1 are known, one can find ζ. Inserting this value from Eq 1.98 to Eq 1.97, gives the following relation:[51]

$$\frac{A'_2V_1}{v_2^2} = \Psi_{1\Theta_U}\left(1 - \frac{\Theta_U}{T}\right) - \zeta\left[\left(1 - \frac{\Theta_U}{T}\right) + \ln\frac{\Theta_U}{T}\right] \qquad [1.100]$$

Parameter z is expressed as

$$z = (3/2\pi\overline{h}^2)^{3/2}(2A'_2M^2/N_A) \qquad [1.101]$$

Then substituting A'_2 from Eq 1.100 in Eq 1.101, we can find the correct temperature dependence of the parameter z. Using Eqs 1.98, 1.99, and 1.100 and experimental values of Θ_U, Θ_L, and Ψ_1 for any system, the theoretical dependence $z = f(1/T)$ can be calculated for any other system. For this purpose, it is necessary to assume that Eqs 1.76 and 1.101 provide the same functional dependence of $A_2M_2/(\overline{h}^2)^{3/2}$ on z for all polymer-solvent systems.

New statistical theories of polymer solutions give more detailed and full description of the thermodynamic properties of polymer solutions having LCST. The basis for the consideration in these theories is the principle of corresponding states.[53]

REFERENCES

1. I. Prigogine and R. Defay in *Chemical Thermodynamics*, Longmans Green and Co., London-New York-Toronto, 1954.
2. P. J. Flory, *J. Chem. Phys.*, **9**, 660 (1941); **10**, 51 (1942).
3. P. J. Flory in *Principles of Polymer Chemistry*, Cornell Univ. Press, New York, 1953.
4. M. L. Huggins in *Physical Chemistry of Polymers*, Interscience, New York, 1958.
5. P. J. Flory, *J. Chem. Phys.*, **13**, 453 (1945).
6. P. J. Flory, *J. Chem. Phys.*, **17**, 1347 (1949).
7. A. E. Nesterov, *Thesis Ph.D.*, Leningrad, 1966.
8. R. Koningsveld and L. A. Kleintjens, *Macromolecules*, **4**, 637 (1971).
9. T. A. Orofino and J. W. Mickey, *J. Chem. Phys.*, **38**, 2512 (1963).
10. A. R. Shultz and P. J. Flory, *J. Amer. Chem. Soc.*, **74**, 4760 (1952).
11. W. H. Stockmayer, *Makromol. Chem.*, **35**, 54 (1960).
12. R. Koningsveld and A. J. Staverman, *J. Polymer Sci.*, **C16**, 1775 (1967).
13. Th. G. Scholte, *Eur. Polymer J.*, **6**, 1963 (1970).
14. G. Renage, D. Moller, and O. Ernst, *Makromol. Chem.*, **88**, 232 (1965).
15. P. T. van Emmerik and C. A. Smolders, *Eur. Polymer J.*, **9**, 157 (1973).
16. M. Gordon, H. A. G. Chermin, and R. Koningsveld, *Macromolecules*, **2**, 207 (1969).

17. J. W. Kennedy, M. G. Gordon, and R. Koningsveld, *J. Polym. Sci.*, **C39**, 43 (1972).
18. R. Koningsveld, W. H. Stockmayer, J. W. Kennedy, and L. A. Kleintjens, *Macromolecules*, **7**, 73 (1974).
19. R. Koningsveld, L. A. Kleintjens, and A. R. Shultz, *J. Polymer Sci. Phys. Ed.*, **8**, 1261 (1970).
20. R. Koningsveld and L. A. Kleintjens, *J. Polymer Sci.*, **C61**, 221 (1977).
21. L. van Opstal, R. Koningsveld, and L. A. Kleintjens, *Macromolecules*, **24**, 161 (1991).
22. M. L. Huggins, *J. Chem. Soc.*, **9**, 440 (1941).
23. A. J. Staverman and J. H. van Santen, *Recl. Trav. Chem. Pay-Bas*, **60**, 76 (1941).
24. J. J. Staverman, *Recl. Trav. Chem. Pay-Bas*, **60**, 640 (1941).
25. F. Hamada, K. Fujisawa, and A. Nakajima, *Polymer J.(Japan)*, **4**, 316 (1973).
26. Y. Kodama and F. L. Awinton, *Brit. Polymer J.*, **10**, 191 (1978).
27. D. M. French, *J. Polymer Sci.*, **C26**, 469 (1986).
28. I. Szleifer, *J. Chem. Phys.*, **92**, 6940 (1990).
29. J. W. Kennedy, *J. Polymer Sci.*, **C39**, 71 (1972).
30. Y. Baba, K. Fujimoto, A. Kagemoto, and R. Fujischiro, *Makromol. Chem.*, **178**, 1439 (1977).
31. I. Freeman and J. S. Rowlinson, *Polymer*, **1**, 20 (1960).
32. S. Saeki, N. Kuwahara, S. Konno, and M. Kaneko, *Macromolecules*, **6**, 246 (1973).
33. G. Allen and C. H. Baker, *Polymer*, **6**, 181 (1965).
34. A. H. Liddell and F. L. Swinton, *Disc. Faraday Soc.*, **49**, 115 (1970).
35. J. M. G. Cowie, A. Maconnachie, and R. J. Ranson, *Macromolecules*, **4**, 57 (1971).
36. K. S. Siow, G. Delmas, and D. Patterson, *Macromolecules*, **5**, 29 (1972).
37. C. H. Baker, W. B. Brown, G. Gee *et al.*, *Polymer*, **3**, 215 (1962).
38. S. Saeki, N. Kuwahara, S. Konno, and M. Kaneko, *Macromolecules*, **6**,1 (1974).
41. S. Konno, S. Saeki, N. Kuwahara *et al.*, *Macromolecules*, **8**, 799 (1975).
42. J. M. G. Cowie and I .J. McEwen, *Macromolecules*, **7**, 291 (1974).
43. D. Patterson, G. Delmas, and T. Somsinsky, *Polymer*, **8**, 503 (1967).
44. G. Delmas, D. Patterson, and T. Somsinsky, *J. Polymer Sci.*, **C57**, 79 (1862).
45. G. Delmas and P. de Sant-Romain, *Eur. Polymer J.*, **10**, 1133 (1974).
46. V. M. Andreeva, A. A. Anikeeva, S. A. Vshivkov, and A. A. Tager, *Vysokomolek. soed.*, **B12**, 789 (1970).
47. S. A. Vshivkov, A. A. Tager, and N. B. Gajfullina, *Vysokomolek. soed.*, **18**, 25 (1976).
48. S. Saeki, N. Kuwahara, M. Nakata, and M. Kaneko, *Polymer*, **17**, 685 (1976).
49. F. E. Bailey and R. W. Callard, *J. Appl. Polymer Sci.*, **1**, 56 (1959).
50. T. E. Lipatova, A. E. Nesterov, V. K. Ivaschenko, and Y. S. Lipatov, *Vysokomolek. soed.*, **A12**,1039 (1970).
51. B. E. Eichinger, *J. Chem. Phys.*, **53**, 561 (1970).
52. G. Delmas and D. Patterson, *Polymer*, **7**, 513 (1966).
53. I. Prigogine, N. Trappeniers, and V. Mathot, *Disc. Faraday Soc.*, **15**, 93 (1953).

Chapter 2

STATISTICAL THEORIES OF POLYMER SOLUTIONS

The expression for the free energy of mixing $\Delta G = \Delta H - T\Delta S$ assumes that only term ΔH depends on the nature of molecules or the type of the intermolecular interactions that are characterized by the empirical parameter χ_1. However, the values of χ_1, experimentally determined from ΔG or ΔH, are not the same. Thus, the value Δw_{12} that determines the interaction parameter (see Eq 1.41) should be reinterpreted in regard to the exchange energy by including both enthalpy and entropy contributions:

$$\Delta w_{12} = \Delta w_{12}(H) - T\Delta w_{12}(S) \qquad [2.1]$$

$$\chi_1 = \chi_H + \chi_S = \frac{z\Delta w_{12}(H)}{kT} - \frac{z\Delta w_{12}(S)}{k} \qquad [2.2]$$

The entropy increases because of mixing and formation of the contacts 1-2 instead of the contacts 1-1 and 2-2. The value $\Delta w_{12}(S)$ is positive and, thus, χ_S is a small negative correction of the parameter χ_1. In the statistical theories of polymer solutions, this correction follows directly from consideration of the volume changes due to mixing.

Earlier attention concentrated on the differences in the cohesion energies between polymer segments and solvent molecules. To distinguish between the molecules of the polymer and the solvent, a new factor was introduced, namely, the difference in their free volumes and temperature dependence of the free volume. The difference in the free volumes is determined by various dimensions of the polymer and solvent molecules. The polymer solution should have similar values of w_{11}, w_{22}, and w_{12} in the chemically identical solvent but different free volumes which determine a new thermodynamic consequences. The concept of the role of free volume in the polymer solution theory was introduced by Prigogine.

A new concept of mixing was proposed that differs from the described by the Flory-Huggins theory. The theory by Prigogine was later developed by Patterson[2-4] and became the theoretical basis for the description of liquid mixtures.[5,6]

2.1 THE THEORY BY PRIGOGINE-PATTERSON

The fundamental concepts regarding the role of free volume have been formulated by Patterson[7] who has given the intuitive introduction to the Prigogine's theory. It is known that the theory of regular solutions models considers them as dispersions of spherical molecules. The polymer solution differs because a solute is formed of segments and chains. The number of ways of placing polymer molecules in the solution lattice is not the same as for spherical molecules and therefore the mixing entropy cannot be ideal. The heat of mixing is taken essentially unchanged from strictly regular solutions so that it is again proportional to the interchange energy, this time, for creation of a contact between the solvent molecule (or segment thereof) and the polymer segment.

It is clear that the polymer-solvent interaction is qualitatively changed because the segments are the parts of the polymer chain. The traditional theories do not explain these changes. Similar, the effects of the excluded volume cannot explain the positive sign of χ_S that corresponds to the negative Δw_S nor an inexplicable large increase of the ordering by formation of polymer-solvent contact.

The polymer-solvent interaction is characterized by the excluded volume, v_0, connected with the energy of the segment interaction, U(r). However, because U(r) cannot be exactly estimated, v_0, according to the old theories, is identified[8] with χ_1 through

$$v_0 = \frac{2V_1(1/2 - \chi_1)}{N_A} \qquad [2.3]$$

where V_1 is the molar volume of the solvent, N_A is the Avogadro's number.

The great difficulties in using traditional theories arise when one attempts to explain the phase separation in solution caused by increasing temperature. The temperature dependence of χ_1 is not monotonously falling function, similar to $\chi_1 = a + b/T$ (in the Flory-Huggins theory $(1/2 - \chi_1) = \Psi_1(1 - \Theta/T)$). New critical solution temperature was called lower CST. The thermodynamic conditions of the appearance of LCST are $\Delta H < 0$ and $\Delta S < 0$. These conditions show that, due to mixing, the system becomes more ordered. Thus, the combinatorial entropy, ΔS_{comb}, may be exceeded by a high negative term $\Delta w_{12}(S)$ (or positive χ_S). This condition is realized in many mixtures of the molecules of large dimensions, for example, in water solutions of amines where the ordering of water molecules around molecules of solute leads to the negative value of $\Delta w_{12}(S)$. At the same time, this effect is typical of many polymer solutions. The reason for the appearance of LCST or positive values of χ_S consists of the

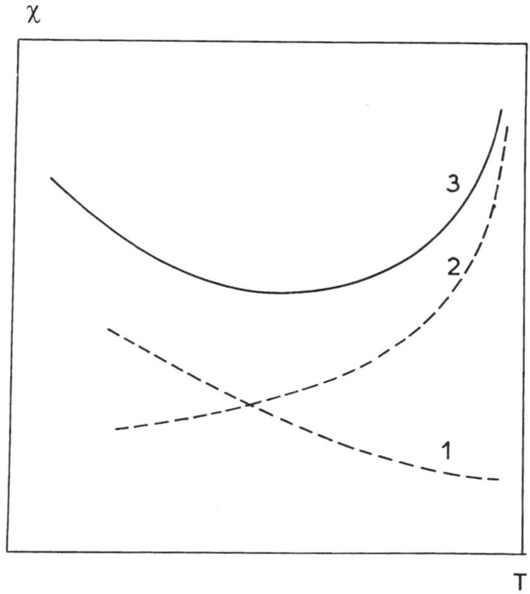

Fig. 2.1. The temperature dependence of the parameter χ: curve 3, total; curve 2, contribution to χ due to free volume dissimilarity between polymer and solvent; curve 1, contribution due to contact energy dissimilarity between polymer and solvent. Adapted, by permission, from D. Patterson, *Macromolecules*, **2**, 673 (1969).

difference of the free volumes of components and, in particular, more dense polymer liquid and more loose solvent. The solvent in the dense polymer medium cannot expand as freely as in pure state. Thus, the free volume of a solution is lower than additive value, i.e., the total volume change by mixing is negative. This fact explains also the exothermic effect of mixing and negative contribution to ΔS.

Each function of mixing represents the sum of two contributions: one connected with the difference in the contact energy (i.e., w_{11} and w_{22}), whereas the second is determined by free volumes of components. The sign of χ_1 or ΔG depends on the relative contributions of χ_H and χ_S (Fig. 2.1). The first contribution (i.e., connected with the difference in the contact energies) can be estimated from the Flory-Huggins theory. It has positive value and diminishes with temperature. However, with increasing temperature, the difference between free volumes of polymer and solvent increases and the new contribution to χ_1 increases, as well. As a result, χ_1 increases and, at the critical value, $\chi_{1,crit}$, the LCST is observed.

Consider the characteristics of LCST. According to the Flory-Huggins, the critical value $\chi_{1,crit}$ is determined by the expression:

$$\chi_{1,crit} = \frac{1}{2}(1 + r^{-1/2})^2 \qquad [2.4]$$

where r is the ratio of molar volumes of the polymer and solvent. $\chi_{1,crit}$ increases with diminishing molecular mass. It means that there is a possibility to fractionate polymers by increasing temperature to the LCST.[9] Because χ_1 increases on approaching LCST, the dimensions of macromolecules and the second virial coefficients decrease.[10]

Fig. 1.10 shows data for the solution of polyisobutylene, PIB, in n-pentane by changing the temperature from 273 to 349K. The change of χ_1 is determined mainly by the difference in the free volumes of components. Since the molecular mass of PIB is very high ($M \approx 9\times10^6$), the LCST should correspond to the value $\chi_1 = 0.5$ and Θ-temperature. If to take butyl ether as a solvent for PIB and consider some difference in the contact energy of interaction between PIB and solvent, then we can observe their contribution to χ_1. The dependence of χ_1 on T is presented by the curve 3 in Fig. 2.1. The dimensions of macromolecules and the second virial coefficient pass through the maximum (Fig. 1.10).

The theory by Prigogine[1] considers the contribution of the differences in the free volumes of components on value of χ_1. The expansion of liquid containing chain molecules, reflects the thermal energy of only some degrees of freedom of macromolecule. These are the external degrees of freedom of rather low frequency and, therefore, of high amplitude. These external degrees of freedom are connected with the forward and rotational motion of the molecules as a whole and their internal rotation around the bonds. The total 3c external degrees have thermal energy 3ckT which is opposed by the cohesive energy. The latter is proportional to the depth of the potential of interaction between neighboring non-bonded segments, ε*, and to the number of the external intermolecular contacts formed by the chain molecule. It may be proportional also to the number of segments, r, or to the number of atoms in the main chain (the segment is chosen in such a way that its molecular volume is proportional to r). All external contacts between segments are characterized by the potential of one form, for example, that of Lennard-Jones:

$$\varphi(R) = 4\varepsilon^*[(\sigma^*/R)^{12} - (\sigma^*/R)^6] \qquad [2.5]$$

where ε* is the contact energy at 0 K, σ* is the distance of this contact, R is the distance between molecules, φ(R) is the energy of intermolecular interaction. The cohesion energy of monomeric liquid per one molecule is proportional to $z\varepsilon^*$ and the cohesive energy of the polymeric liquid is $zq\varepsilon^*$, where z is the coordination number of the lattice and $q\varepsilon^*$ is the energy of the external contacts of segments. For N chain molecules the cohesion energy of liquid is equal to $Nqz\varepsilon^*$, their volume is $Nr\sigma^*$ and entropy Nck. Values $\varphi^*=q\varepsilon^*$, $v^* = r\sigma^*$ and $S^* = ck$ are called molecular units. Usually, it is taken that $q \approx r$. The free volume of the liquid depends on the ratio of two energies mentioned, this ratio is called reduced temperature:

$$\tilde{T} = \frac{U_{thermal}}{U_{cohesion}} = \frac{c}{r}\frac{kT}{\varepsilon^*} = \frac{T}{T^*} \qquad [2.6]$$

In this expression, the ratio c/r represents one third of the number of external freedom degrees per segment. For the series of chemically identical homologues, the c/r diminishes with the increase in r or n. For example, the monomer molecule has three degrees of freedom and c/r = 1. Only two external degrees of freedom are added for dimer as one of the degrees of freedom is connected with a high frequency oscillations of segments and is not taken into account. Therefore, for dimer c/r = 5/6, for a long molecule parameter, c/r depends on the length of the chain part H for which the torsion oscillations do not depend on other parts, i.e., on the chain flexibility (Kuhn segment). The ratio c/r changes linearly with 1/r, i.e., c/r = A + b/r , where A is Kuhn segment. For the spherical molecules (monomers), the reduced temperature depends only on the parameter of cohesion energy, ε^*, because c/r = 1. For the chain liquid, the reduced temperature and free volume depend additionally on the length and flexibility of the chain via parameter c/r (structural parameter).

Schematically the dependence of the reduced temperature on n for two series of homologues is shown in Fig. 2.2. In the first case, the rotation around the bonds is relatively restricted (in alkanes), in the second case the chain is very flexible (dimethyl siloxanes). The difference in the free volumes, at any temperature, is higher for polyethylene and pentane than for poly(dimethyl siloxane) and its homologue with n = 5 (hexamethyl disiloxane). This difference has an effect on the LCST. For the PE-pentane system, the LCST is 363 K and equals to 0.77 of the critical temperature of pentane.[11] For poly(dimethyl siloxane)-hexamethyl disiloxane LCST is 484 K, which is 0.95 of the critical temperature of hexamethyldisiloxane.[11]

According to the Prigogine theory, temperature enters in the partition function only in its reduced form. At the constant reduced pressure (usually equal to zero at the pressure of the saturated vapors), the molar or specific thermodynamic quantities are connected

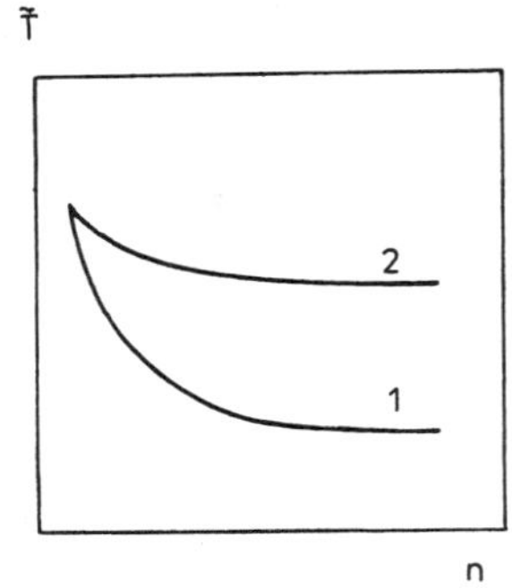

Fig. 2.2. Dependence of reduced temperature, $\tilde{T}$, on n number of atoms in principal chain for two homologous series: (1) with small flexibility, e.g., alkanes; (2) with large chain flexibility, e.g., dimethylsiloxanes. Adapted, by permission, from D. Patterson, *Macromolecules*, **2**, 673 (1969).

with reduced values by the reduction parameters depending only on the chain length:

$$V(n,T) = V_0^*(n)\tilde{V}(\tilde{T}) = rv*\tilde{V}(\tilde{T}) \quad [2.7]$$

$$U(n,T) = U_0^*(n)\tilde{U}(\tilde{T}) = r\varepsilon*\tilde{U}(\tilde{T}) \quad [2.8]$$

$$G(n,T) = \tilde{G}(\tilde{T})U_0^*(n) = r\varepsilon*\tilde{G}(\tilde{T}) \quad [2.9]$$

Value U is the configurational or external intermolecular energy of the liquid and has a negative sign. The reduction parameters can be considered as the intrinsic molecular values at 0 K. They may be represented in the following way:

Energy	$\tilde{U} = U/MU_0^* = U/Nq\varepsilon*$	[2.10]
Volume	$\tilde{V} = V/Nv_0^* = V/Nr\sigma^{*3}$	[2.11]
Entropy	$\tilde{S} = S/NS_0^* = S/Nck$	[2.12]
Pressure	$\tilde{P} = -(\delta\tilde{U}/\delta\tilde{V})_{\tilde{S}} = PV_0^*/U_0^*$	[2.13]
Temperature	$\tilde{T} = (\delta\tilde{U}/\delta\tilde{S})_{\tilde{V}} = TS_0^*/U_0^*$	[2.14]
Heat capacity	$\tilde{C}_p(\tilde{T}) = C_p(T,n)/NS_0^*$	[2.15]

Simha *et al.*[13-15] have established that for many polymer homologues, the data on V(n,T) may be reduced to a simple curve $\tilde{V}(\tilde{T})$ (Fig. 2.3).

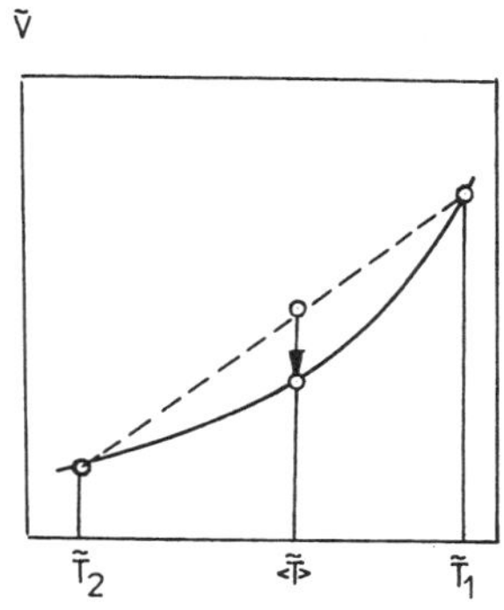

Fig. 2.3. Schematic representation of reduced volume, Ṽ, against reduced temperature, T̃, showing the decrease in the average of Ṽ of the components which takes place on mixing. Adapted, by permission, from D. Patterson, *Macromolecules*, **2**, 673 (1969).

According to Patterson, this curve is universal for all liquids where dispersion forces of interaction are operative.

The theory by Prigogine assumes that the mixture behaves like a pure liquid with the mean reduction parameter for any thermodynamic quantity and mean reduced temperature. In particular:

$$v^* = x_1 v_1^* + x_2 v_2^* \qquad [2.16]$$

$$U^* = x_1 U_1^* + x_2 U_2^* \qquad [2.17]$$

where x_1 and x_2 are the molar fractions of components. In a similar way:

$$q = x_1 q_1 + x_2 q_2 \qquad [2.18]$$

$$c = x_1 c_1 + x_2 c_2 \qquad [2.19]$$

For this case, according to Eq 2.6, the reduced temperature of the mixture can be determined from the expression:

$$\tilde{T} = \frac{T}{T*} = \frac{c}{q}\frac{kT}{\varepsilon*} = \frac{(x_1c_1 + x_2c_2)kT}{(x_1q_1 + x_2q_2)\varepsilon*} = \frac{x_1c_1}{x_1q_1 + x_2q_2}\frac{kT}{\varepsilon*} +$$

$$+ \frac{x_2c_2}{x_1q_1 + x_2q_2}\frac{kT}{\varepsilon*} \qquad [2.20]$$

If to multiply and divide each item by q_1 and q_2, we find:

$$\tilde{T} = X_1\tilde{T}_1 + X_2\tilde{T}_2 \qquad [2.21]$$

where

$$X_1 = \frac{x_1q_1}{x_1q_1 + x_2q_2} \quad X_2 = \frac{x_2q_2}{x_1q_1 + x_2q_2} \qquad [2.22]$$

Values X_1 and X_2 represent the "surface fractions", i.e., the ratio of the number of the external contacts of molecules of components 1 and 2 to the number of all external contacts in solution.

Consider two components:[4] solvent 1 and polymer 2 that are characterized by various parameters but obey the principle of the corresponding states. Three parameters describe the difference between solvent and polymer:

$$\delta = \frac{\varepsilon_2}{\varepsilon_1} - 1 \qquad [2.23]$$

$$\rho = \frac{\sigma_2}{\sigma_1} - 1 \qquad [2.24]$$

$$-\lambda = (c_2q_2/c_1q_1) - 1 \qquad [2.25]$$

The first two parameters do not depend on the molecular mass, whereas the third should be positive and increase asymptotically with increasing difference in the length of polymer chain and solvent. For the polymer with infinitely high molecular mass, this parameter is larger than δ and ρ. If to define the surface fraction by Eq 2.22 for the mixture, we have:

$$S^* = x_1S_1^* + x_2S_2^* \qquad [2.26]$$

$$U^* = x_1U_1^* + x_2U_2^* - \varepsilon_1X_1X_2\nu^2(x_1q_1 + x_2q_2) \qquad [2.27]$$

The non-linear term in Eq 2.27 leads to the diminishing the cohesion energy in the mixture that is determined by the exchange energy of the 1-2 contact formation (δ-effect) and due to the mixing of segments of various length (ρ-effect). Coefficient ν^2 can be presented as the sum of these two effects. Assuming that for ε_{12} the rule of the geometric mean is valid, we obtain

$$\nu^2 \approx \frac{\delta^2}{4} + 9\rho^2 \qquad [2.28]$$

The precise magnitude of the coefficient at ρ^2 is not known. Then

$$1 - \frac{\tilde{T}}{\tilde{T}_1} = (\lambda + \delta)X_2 \approx \tau X_2 \qquad [2.29]$$

where

$$\tau = 1 - \frac{T_1^*}{T_2^*} = \frac{\lambda + \delta}{1 + \delta} \approx \lambda + \delta \qquad [2.30]$$

If $\delta = \rho = 0$, Eq 2.29 is precise. Then, the noncombinatorial part of the Gibbs free energy may be presented as

$$\Delta G = U^*\tilde{G}(\tilde{T},\tilde{P}) - x_1 U_1^*\tilde{G}(\tilde{T}_1,\tilde{P}_1) - x_2 U_2^*\tilde{G}(\tilde{T}_2,\tilde{P}_2) \qquad [2.31]$$

Assuming from Eq 2.31 that P = 0, one can get:

$$\Delta\overline{G} = \chi_1 R T\varphi_2^2 = U_1^*[-\tilde{U}(\tilde{T})]\nu^2 x_2^2 + U_1^*[\tilde{F}(\tilde{T}) - \tilde{F}(\tilde{T}) +$$

$$+ (\tilde{T}_1 - \tilde{T})\delta\tilde{F}/\delta\tilde{T}] \qquad [2.32]$$

$$\Delta\overline{H}_1 = k_1 R\varphi_2^2 = U_1^*[-\tilde{U}(\tilde{T}) + \tilde{T}\tilde{C}_p]\nu^2 x_2 + U_1^*[\tilde{U}(\tilde{T}) - \tilde{U}(\tilde{T}_1) +$$

$$+ (\tilde{T}_1 - \tilde{T})\delta\tilde{U}/\delta\tilde{T}] \qquad [2.33]$$

here it is assumed that $x_2 \rightarrow \varphi_2$ because $q \rightarrow r$.

Both values given by Eqs 2.32 and 2.33 consist of two parts contributing to χ_1 and k_1. The ΔG_1 depends on the parameters of energy and dimensions δ and ρ. There is an analogy of $z\Delta w_{12}$ or χ_1 and k_1 in the Flory-Huggins theory.

The contribution to χ_1 changes with value $\tilde{U}(\tilde{T})$ which may be defined as the heat of the evaporation of the averaged segment of the solution. With increasing solution concentration, the reduced temperature diminishes and the $\tilde{U}$ value increases due to the densening of the solution. Thus, parameter χ_1 (and k_1) increases with concentration of solution.

The second contribution in Eqs 2.32 and 2.33 depends mainly on the structural parameter λ in τ and characterizes the mixture of homologues of the different chain length, where $\delta = \rho = 0$, i.e., $\tau = \lambda$.

Expanding Eqs 2.32 and 2.33 in series by $\tilde{T}$ and using Eq 2.30, we can obtain the expressions for χ_i and k_i:

$$\chi_1 = -\frac{U\nu^2}{RT} + \frac{C_p\tau^2}{2R} \qquad [2.34]$$

$$k_1 = \frac{-U + TC_P}{RT}\nu^2 - \frac{TdC_P/dT}{2R}\tau^2 \qquad [2.35]$$

$$\Psi_1 - \frac{1}{2} = k_1 - \chi_1 = \frac{C_P}{R}\nu^2 - \frac{C_P + T\delta C_P/\delta T}{2R}\tau^2 \qquad [2.36]$$

$$\chi_2 = \frac{C_P}{R}\nu^2\tau - \frac{T^2 dC_P/dT}{3R}\tau^3 \qquad [2.37]$$

$$k_2 = -\frac{TdC_P/dT}{R}\nu^2\tau + \frac{Td^2C_P/dT^2}{3R}\tau^3 \qquad [2.38]$$

where ν^2 is given by Eq 2.28 and

$$\tau^2 = (\delta + \lambda)^2 \qquad [2.39]$$

Each parameter contains two contributions: the first is determined by the difference in the cohesion energies and dimensions of segments of the polymer and solvent and the second is determined only by the structural factor, λ. In the reduced form, the expression for χ_1 has the form:

$$\frac{\chi_1}{c_1} = \frac{\tilde{U}}{\tilde{T}_1}\nu^2 + \frac{\tilde{C}_{P1}}{2}\tau^2 \qquad [2.40]$$

For simplicity, the value ν^2 may be presented as

$$\nu^2 = [1 - (P_2^*/P_1^*)^{1/2}]^2 \qquad [2.41]$$

For the dependence of χ_1 on the pressure, the following expression has been found:[3]

$$\frac{\chi_1}{c_1} = \frac{\tilde{U}(\tilde{P})}{\tilde{T}}\nu^2 + \frac{\tilde{C}_{P1}}{2}\left[\tau + \frac{\beta_{T_1}\tilde{P}}{\alpha_P\tilde{T}}\pi\right]^2 \qquad [2.42]$$

where $\pi = P_2^*/P_1^* - 1$.

2.2 THE ROLE OF VARIOUS CONTRIBUTIONS OF THE MIXING PARAMETERS

Patterson has considered the role of various contributions to the interaction parameters.[4] Values of U, $(-U + TC_P)$ and C_P are positive for any solvent and proportional to the chain length for the series of homologues. Therefore, the contributions of the cohesion energy and dimensions on χ_1 and k_1 are positive, as it should follow from the model of the rigid lattice accepted in the Flory-Huggins theory. If $\delta = \rho = 0$, parameter χ_1 depends only on the structural factor (parameter λ). Simha[13-15] has shown that the data on V(T) for various polymers and oligomers may be presented in the reduced form. Simha obtained the values T_1^*/T_2^* for various polymer-solvent systems. The parameter λ can be found from Eq 2.30.

From Eq 2.34, it follows that a better solvent for the polymer is not oligomer, but the component with $\rho = 0$ and negative value of δ, i.e., with a higher cohesion energy in comparison with polymer. Parameter χ_1 passes minimum when

$$-\frac{U}{T}\frac{\delta}{2} + C_P(\lambda + \delta) = 0 \qquad [2.43]$$

The minimum value of χ_1 is expressed as

$$\chi_1 = \frac{\lambda^2}{4R}\frac{-UC_P/T}{C_P + (-U/2T)} \qquad [2.44]$$

By approaching the liquid-vapor critical point $C_P \to \infty$ and $dC_P/dT \to \infty$. Thus, the structural contribution to χ_1 increases limitlessly and by increasing temperature (for the case of LCST), the phase separation should occur. The structural contribution to k_1 (Eq 2.35) is negative and its value increases with increasing temperature. Thus, the heat of mixing becomes negative at rather high temperatures close to LCST. However, in many cases of volatile solvents, together with small values δ and ρ, temperature 298 K is very close to the liquid-vapor critical temperature and because of it the value k_1 is not always negative.

Another consequence of the LCST is that the total entropy, Ψ_1, i.e., the sum of combinatorial (1/2) and non-combinatorial ($\Psi_1 - 1/2$) entropies of mixing is negative (see Eq 2.36). The negative entropy of mixing was obtained even for the high-molecular weight PIB in pentane at 298 K. For the majority of systems at room temperatures values Ψ_1 are positive, whereas ($\Psi_1 - 1/2$) are negative. The first term of Eq 2.36 is positive and typical of the mixtures of monomeric liquids where the noncombinatorial entropy of mixing is positive. For polymer solutions it contains the negative structural contribution leading to ($\Psi_1 - 1/2) < 0$.

The change in C_P at usual temperatures remote from the liquid-vapor critical temperatures is different for the simple and complicated liquids (i.e., for higher homologues). For simple liquids, the value C_p increases monotonously to infinity at the liquid-vapor critical point, so there is no qualitative distinction between high and low temperatures. For the complicated liquids, d^2C_P/dT^2 is positive at all temperatures. However, firstly C_P diminishes at low temperatures and then begins to increase to infinity. Minimal values of C_P are observed at temperatures above 298 K. Such an uncertainty in the change of C_P

does not allow to predict the change of parameters at low-temperature region even in a qualitative way.[16]

Thus, according to Eq 2.35, value k_1 (i.e., ΔH) is positive at low temperatures and negative at high temperatures, where $dC_P/dT > 0$. The possibility of the positive structure contribution to χ_1 seems to be unexpected. If C_P increases with diminishing temperature and leads to increasing χ_1, there exists the possibility of the phase separation of the system at low temperature even at $\delta = 0$ and $\rho = 0$. This temperature should be UCST or Θ-point. Probably such behavior explains the properties of systems: low molecular PIB-pentane,[17] PS-ethyl benzene,[18] and mixtures of polymer homologues of styrene.[19,20]

The effect of χ_2 on phase separation of solution manifests itself in that the curve of the phase coexistence near Θ-point corresponds to a higher concentration as compared with predicted by the Flory's theory. The term of cohesion energy and macromolecules dimensions in Eq 2.38 is positive and therefore the value χ_2 should be higher for the poor and Θ-solvents. The structural contribution is probably also positive, in particular, at low temperatures, where dC_P/dT may be negative. To predict the sign of k_2 is much more difficult because at $d^2C_P/dT^2 > 0$, the first derivative may be either negative or positive. However, for the mixtures of n-alkanes, $k_2 > 0$ and in these systems the heat of mixing may change the sign from a negative to positive with increasing concentration of alkane with longer chain.[16]

2.3 EFFECT OF SOLVENT CHAIN LENGTH ON THE INTERACTION PARAMETERS

Patterson *et al.*[23] used the thermodynamic functions[21] for the solutions of monomers A and r_B-mers, having $3c = (r_B + 3)$ degrees of freedom. It was assumed that the dimensions of segments A and B are the same, i.e.,

$$r_{AA} = r_{BB} = r_{AB} \quad [2.45]$$

$$\rho = \frac{r_{BB}}{r_{AA}} - 1 = 0 \quad [2.46]$$

For the general case of the mixture of r_A-mers and r_B-mers, the number of the external degrees of freedom is correspondingly $3c_A$ and $3c_B$. The simple averaged potential was used instead of the harmonic potential of the cell model. It was found that

$$\Delta H = \Delta H^{I} + \Delta H^{II} \quad [2.47]$$

where

$$\Delta H^{I} = z\Delta w_{12}(q_A N_A + q_B N_B)X_A X_B; \quad q_A z = r_A(z - r) + 2 \qquad [2.48]$$

Here, X_A is the surface fraction given by the relation

$$X_A = \frac{q_A N_A}{(q_A N_A + q_B N_B)} \qquad [2.49]$$

$$\Delta w_{12} = \frac{1}{2}(\varepsilon_{11} + \varepsilon_{22}) - \varepsilon_{12} \qquad [2.50]$$

The value Δw_{12} is similar to the exchange energy but is estimated at the minimum of potential energy, i.e., at 0 K. If to use the rule of the geometrical mean, i.e.,

$$\varepsilon^*_{AB} = (\varepsilon^*_{AA}\varepsilon^*_{BB})^{1/2} \qquad [2.51]$$

we find that Δw_{12} is the positive value. Using $\delta = \varepsilon^*_{BB}/\varepsilon^*_{AA} - 1$, we obtain:

$$\Delta w_{12} = \varepsilon^*_{AA}\delta^2/8 \qquad [2.52]$$

Value H^{II} is introduced to take into account the expansion of the solution and pure component, i.e.,

$$\varepsilon^*_{AB} = (\varepsilon^*_{AA}\varepsilon^*_{BB})^{1/2} \qquad [2.51]$$

we find that Δw_{12} is the positive value. Using $\delta = \varepsilon^*_{BB}/\varepsilon^*_{AA} - 1$, we obtain:

$$\Delta w_{12} = \varepsilon^*_{AA}\delta^2/8 \qquad [2.52]$$

Value H^{II} is introduced to take into account the expansion of the solution and pure components at a given temperature :

$$\frac{\Delta H^{II}}{2kT(q_A N_A + q_B N_B)} = \left(X_A\frac{c_A}{q_A} + X_B\frac{c_B}{q_B}\right)\tilde{T} - X_A\frac{c_A}{q_A}\tilde{T}_A - X_B\frac{c_B}{q_B}\tilde{T}_B \qquad [2.53]$$

Here $\tilde{T}$ and $\tilde{T}_i$ are the reduced temperatures for the mixture and pure liquids:

$$\tilde{T} = \frac{kT}{2z}\,\frac{X_A\frac{c_A}{q_A} + X_B\frac{c_B}{q_B}}{X_A\varepsilon^*_{AA} + X_B\varepsilon^*_{BB} - 2\Delta w_{12}X_A X_B} \qquad [2.54]$$

$$\tilde{T}_A = \frac{kT}{2z\varepsilon^*_{AA}}\,\frac{c_A}{q_A} \qquad [2.55]$$

$$\tilde{T}_B = \frac{kT}{2z\varepsilon^*_{BB}} \frac{c_B}{q_B} \qquad [2.56]$$

Let be $\varepsilon^*_{AA} = \varepsilon^*_{BB} = \varepsilon^*$ and $\tau = 0$. In this case, Eq 2.53 takes simpler form:

$$\frac{\Delta H^{II}}{2kT(q_A N_A + q_B N_B)} = -\frac{kT}{2z\varepsilon^*}\left(\frac{c_A}{q_A} - \frac{c_B}{q_B}\right)^2 X_A X_B \qquad [2.57]$$

Consider the case when $r_B >> r_A$ and $r_A \neq 1$ and accept that the components have the same flexibility expressed as lim c/r (r→∞) = 1/3. Then we have

$$c_A q_A^{-1} - c_B q_B^{-1} = [1 - 2[3(z-2)]^{-1}]q_A^{-1} \qquad [2.58]$$

If z→∞ we find:

$$(c_A q_A^{-1} - c_B q_B^{-1})^2 = f r_A^{-2} \qquad [2.59]$$

where f is a small numerical constant. If A is a monomer and B is a polymer, from the condition r = 1, c =1 and c/r = 1, it follows that

$$(c_A q_A^{-1} - c_B q_B^{-1})^2 = 4/9 \qquad [2.60]$$

$$\frac{\Delta H}{(r_A N_A + r_B N_B)(\varphi_A \varphi_B / N)} = \frac{\Delta H}{r_B N_B \varphi_A / N} \qquad [2.61]$$

$$\frac{\Delta H}{\bar{x}\varphi_A} = A - B(T/r_A)^2 \qquad [2.62]$$

where A and B are constants given by

$$A = zwN = z\varepsilon^*_{AA}\delta^2 N/8 \qquad [2.63]$$

$$B = 10.5(k^2/z\varepsilon^*)N \qquad [2.64]$$

In this case, the enthalpy Flory parameter is equal to

$$Rk_1 = Ar_A/T - BT/r_A \qquad [2.65]$$

From the other side, when the monomer serves as a solvent ($r_A = 1$), the right hand side of Eq 2.61 can be replaced by ($A - 4/9\ BT^2$). From these equations, it is seen that ΔH depends on temperature and becomes negative at high temperatures.

The noncombinatorial entropy of mixing may be presented[21] as $T\Delta S_{noncomb} = -2\Delta H^{II}$. Therefore from Eq 2.57 at $r_B > r_A$ it follows that

$$\frac{\Delta G_{noncomb}}{(r_A N_A + r_B N_B)(\varphi_A \varphi_B / N)} = A + B(T/r_A)^2 \qquad [2.66]$$

The parameter χ_1 is expressed as

$$RT\chi_1 = \frac{1}{\varphi_B^2}\left[\frac{\delta}{\delta N_B}\Delta G_{noncomb}\right]_{P,T} \qquad [2.67]$$

$$R\chi_1 = A(r_A/T) + B(T/r_A) \qquad [2.68]$$

Again, if $r_A = 1$, the term B is multiplied by 4/9. In Eq 2.68, the dependence of χ_1 on T and the length of the solvent chain differs from expressed by equations:

$$\chi_1 = z\Delta w_{12} r_A / kT \qquad [2.69]$$

$$\chi_1 = \left(\frac{1}{2} - \Psi_1\right) + \frac{\Psi_1 \theta}{T} \qquad [2.70]$$

Value A may be sufficiently large to produce similarity. If $A \neq 0$, Eq 2.68 shows a minimum for χ_1 with changing r_A for constant at $r_0^2 = (B/A)T^2$ or with change of T with r_A constant at $T_0^2 = (A/B)r_A^2$. Above and below T_0, the quality of solvent becomes worse and at the critical temperatures, χ_1 should approach some critical value. Taking the infinitely large r_B, the value 1/2 as the critical value of χ_1, one can find that

$$R\chi_{crit} = A(r_A/T_{crit}) + B(T_{crit}/r_A) = R/2 \qquad [2.71]$$

Solving this equation shows that T_{crit}/r_A does not depend on r_A at $r_A \neq 1$

$$T_{crit}/r_A = R/2 \pm [(R/2)^2 - 4AB/2B]^{1/2} \qquad [2.72]$$

If $A = 0$, i.e., the components differ only by the chain length, the UCST is equal to zero, whereas LCST is determined as

$$\frac{T_{crit}}{r_A} = \frac{R}{2B} \qquad [2.73]$$

If $4AB < (1/2R)^2$ both UCST and LCST exist and if $4AB > (1/2R)^2$ the polymer is not soluble in a given solvent. It is worth noting that the critical state is very sensitive to z. Because of it, the expression for χ_{crit} may be written generally as

$$R\chi_{crit} = \frac{Ar_A}{T_{crit}} + B\frac{T_{crit}}{r_A}\left(\frac{r_A}{q_A}\right)^2\left(\frac{z}{z-2}\right)^2\left(1-\frac{8}{3z}\right)^2 - \frac{1}{2}\left(\frac{z}{z-2}\right) \quad [2.74]$$

Usually r_A is taken equal to (n + 1)/2, where n is the number of carbon atoms in the chain. Values A and B may be found from the heats of mixing, from χ_1, from the data on swelling, LCST, etc.

2.4 NEW FLORY THEORY

Flory and his coworkers[5,6] have used the same cell distribution function as Prigogine, but for describing the configurational energy they have applied the Hirschfelder-Eyring approximation to van der Waals liquid, i.e., $E = -1/\tilde{V}$.

This approximation describes better the experimental dependencies of the configurational energy on the volume as compared with approximation based on the Lennard-Jones potential. Flory has considered the one-dimensional system of particles with length l^* distributed in the configurational space of length L. It was accepted that the particles do not overlap. Then for the total configurational space occupied by the system, the Tonks distribution function was used:

$$\Omega = \frac{(1 - Nl^*)^N}{N!} \approx [(1 - l^*)e]^N \quad [2.75]$$

where l^* is the length of the rigid core of the particle, $l = L/N$ is the space occupied by the particle. The general entropy is introduced through factor e^N. To extend these results on the three-dimensional space, v, of the parameter $(\gamma v)^{1/3} = l$ and $(\gamma v^*)^{1/3} = l^*$ are introduced ($v = V/N$, where V is the total volume of the system, γ is the numerical factor, v^* is the core volume of the particle). For the generalization of the result, N molecules are divided to r segments. Then v^* is identified as the characteristic of core segment volume and $v = V/r$, where V is the molecular volume.

Following Prigogine, Flory introduced parameter c in such a way that 3c represents the number of the degrees of freedom per segment. The distribution function has the following form:

$$\Omega = [\gamma e^3 v^*(\tilde{v}^{1/3} - 1)^3]^{Nrc} \quad [2.76]$$

where $\tilde{v} = v/v^*$. The mean intermolecular energy forming a part of the configurational integral of such system is

$$Z = \Omega \exp(-E_0/kT) \quad [2.77]$$

$$E_0 = -Nrs\eta/2v \quad [2.78]$$

where s is the number of contact sites per segment (proportional to the segment area), N is the number of molecules, $-\eta/v$ is the intermolecular energy per one contact. Introducing the reduced temperature as

$$\tilde{T} = \frac{T}{T^*} = 2v^*ckT/\eta s \qquad [2.79]$$

the distribution function may be found:

$$Z = \text{const}\,(\tilde{v}^{1/3} - 1)^{3Ncr}\exp\,(Ncr/\tilde{v}\tilde{T}) \qquad [2.80]$$

From this function the equation of the state follows:

$$\frac{\tilde{P}\tilde{V}}{\tilde{T}} = \frac{\tilde{v}^{1/3}}{(\tilde{v}^{1/3} - 1)} - \frac{1}{\tilde{v}\tilde{T}} \qquad [2.81]$$

where

$$\tilde{P} = P/P^* = 2Pv^{*2}/s\eta \qquad [2.82]$$

The reduction parameters (or characteristic parameters) P^*, v^*, and T^* are interconnected

$$P^*v^* = ckT^* \qquad [2.83]$$

At $\tilde{P} = 0$, the equation of the state comes to simplified form:

$$\tilde{T} = \frac{(\tilde{v}^{1/3} - 1)}{\tilde{v}^{4/3}} \qquad [2.84]$$

This equation may also be expressed as

$$\tilde{v}^{1/3} - 1 = \frac{\alpha T/3}{(1 + \alpha T)} \qquad [2.85]$$

where

$$\alpha = \left(\frac{\delta \ln v}{\delta T}\right)_{P=0} \qquad [2.86]$$

From these equations, the values $\tilde{v}$ and $\tilde{T}$ may be found if α is known. The reduction temperature T^* is found from Eq 2.79 and v^* from $\tilde{v}$ and v. After differentiation of Eq 2.81 by substituting the corresponding terms of Eqs 2.82 and 2.83, one can find:

$$P^* = \gamma T\tilde{v}^2 \qquad [2.87]$$

where $\gamma = (\delta P/\delta T)_V$. Value P^* estimated from the isothermal compressibility coefficient (Eq 2.87) fully characterizes the pure liquid. The rest of parameters (ε and $s\eta$) may be found from Eqs 2.79, 2.82, and 2.83. For the binary mixture, the distribution function can be written in the form:

$$Z = Z_{comb}\Omega\exp(-E_0/kT) \qquad [2.88]$$

where Z_{comb} is the combinatorial factor for the mixture of two components, Ω is expressed by Eq 2.75. Neglecting the difference between Gibbs and Helmholz free energies for the condensed phase, the Gibbs free energy may be presented as

$$\Delta G = \Delta G_{comb} + G^R \qquad [2.89]$$

where $\Delta G_{comb} = -T\Delta S_{comb}$, G^R is so-called residual free energy following from the distribution function $\Omega\exp(-E_0/kT)$.

The segments of both molecules of the mixture are usually chosen in such a way that they are equal to the core volumes. This choice is not limited because $\Omega\exp(-E_0/kT)$ does not depend on the shape and disposition of molecules. For the mixture, the distribution function has the form:

$$Z = Z_{comb}(\tilde{v}^{1/3} - 1)^{3\bar{r}N\bar{c}}\exp(-E_0/kT) \qquad [2.90]$$

where $N = N_1 + N_2$ and

$$\bar{r} = (r_1N_1 + r_2N_2)/N \qquad [2.91]$$

$$\bar{c} = (r_1N_1c_1 + r_2N_2c_2)/\bar{r}N \qquad [2.92]$$

These relations allow one to interpret values φ_1 and φ_2 as segment fractions (not volume fractions). Then

$$\varphi_2 = 1 - \varphi_1 = r_2N_2/\bar{r}N \qquad [2.93]$$

and Eq 2.92 may be rewritten in the form

$$\bar{c} = c_1\varphi_1 + c_2\varphi_2 \qquad [2.94]$$

The definitions given above allow one to calculate φ_1 in another way. Molar characteristic volumes of components i are equal to $V_i^* = r_iv_i^*$ and in such a way $r_2/r_1 = V_2^*/V_1^*$. Here V_i^* is determined as $\tilde{v}_i = V_i/V_i^*$ and $V_i = r_iv$ is molecular or molar volume. The segment fractions may be expressed in the following way:

$$\varphi_2 = N_2V_2^*/(N_1V_1^* + N_2V_2^*) \qquad [2.95]$$

$$\varphi_2 = w_2v_{sp2}^*/(w_2v_{sp2}^* + w_1v_{sp1}^*) \qquad [2.96]$$

where w_i is the mass of i-th component, $v^*_{sp,i}$ is its characteristic specific volume. The energy of the mixture is taken in the same way as for pure component:

$$-E_0 = (A_{11}\eta_{11} + A_{22}\eta_{22} + A_{12}\eta_{12})/v \qquad [2.97]$$

where A_{ij} is the number of contacts each having the energy $-\eta_{ij}/v$. It follows from the definition (Eq 2.97) that

$$2A_{11} + A_{12} = s_1r_1N_1 \qquad [2.98]$$

$$2A_{22} + A_{12} = s_2r_2N_2 \qquad [2.99]$$

and for the homogeneous mixture

$$A_{12} = s_1r_1N_1\Theta_2 + s_2r_2N_2\Theta_1 \qquad [2.100]$$

The intermolecular energy is described by the expression

$$-E_0/rN = [\Theta_1\eta_{11} + \Theta_2\eta_{22} - \Theta_1\Theta_2\Delta\eta](s/2v) \qquad [2.101]$$

where

$$\Delta\eta = \eta_{11} + \eta_{22} - 2\eta_{12} \qquad [2.102]$$

$$s = (s_1r_1N_1 + s_2r_2N_2)/\bar{r}N = \varphi_1s_1 + \varphi_2s_2 \qquad [2.103]$$

The fractions of sites Θ_1 and Θ_2 are determined by the equation:

$$\Theta_2 = 1 - \Theta_1 = s_2r_2N_2/(s_1r_1N_1 + s_2r_2N_2) = (s_2/s_1)\varphi_2/[\varphi_1 +$$

$$+ (s_1/s_2)\varphi_2] \qquad [2.104]$$

By the analogy to the energy of pure component, for the mixture, its value is

$$-E_0/\bar{r}N = P^*v^*/\tilde{v} = ckT^*/\tilde{v} \qquad [2.105]$$

The comparison of Eq 2.105 with Eqs 2.101 and 2.82 gives

$$P^* = \varphi_1P_1^* + \varphi_2P_2^* - \varphi_1\Theta_2X_{12} \qquad [2.106]$$

$$X_{12} = s_1\Delta\eta/rv^{*2} \qquad [2.107]$$

Eq 2.83 is suitable for the mixtures and therefore one can obtain

$$\frac{1}{T^*} = \left(\frac{\varphi_1 P_1^*}{T_1^*} + \frac{\varphi_2 P_2^*}{T_1^*}\right) / P^* \qquad [2.108]$$

The distribution function for the mixture has the same form as for the pure liquid. Therefore, the equation of state is identical to Eq 2.81. However, the reduced variables $\tilde{P} = P/P^*$ and $\tilde{T} = T/T^*$ for the mixture depend on composition according to Eqs 2.106 and 2.108. For small pressures, the enthalpy of mixing can be substituted by energy:

$$\Delta H = \bar{r}Nv^*[\varphi_1 P_1^*/\tilde{v}_1 + \varphi_2 P_2^* \tilde{v}_2 - P^*/\tilde{v}] \qquad [2.109]$$

or, from the definition P^* by Eq 2.106:

$$\Delta H = \bar{r}Nv^*[\varphi_1 P_1^*(\tilde{v}_1^{-1} - \tilde{v}^{-1}) + \varphi_2 P_2^*(\tilde{v}_2^{-1} - \tilde{v}^{-1}) + \varphi_1 \Theta_2 X_{12} \tilde{v}^{-1}] \qquad [2.110]$$

However, the following expression for the integral heat of mixing:

$$\Delta H_M = BN_1\varphi_2 \qquad [2.111]$$

where B is the constant not depending on concentration, does not correspond to Eqs 2.109 and 2.110. The constant B is usually found from the heat of mixing of the polymer with large excess of the solvent. The value B used to calculate enthalpy according to Eq 2.111 does not come true. Value B determined from Eq 2.111, as follows from Eqs 2.109 and 2.110, should dependent on concentration. The expansion of Eq 2.110 in series by φ_2 for the infinite dilution gives the result:

$$B = \lim_{\varphi_2 \to 0}\left(\frac{\Delta H_M}{N_1\varphi_2}\right) = \lim_{N_2 \to 0}\left(\frac{\Delta H_M}{N_2}\right)(V_1^*/V_2^*) =$$

$$= \left(V^*(V^*/\tilde{v}_1)\frac{1}{\tilde{v}_1}\right)[P^*\left[\frac{\tilde{v}_1}{\tilde{v}_2} - 1 - \alpha_1 T\left(1 - \frac{\tilde{T}_2}{\tilde{T}_1}\right)\right] +$$

$$+ \ (1 + \alpha_1 T)(s_2/s_1)X_{12}] \qquad [2.112]$$

This equation makes calculation of X_{12} easier from the enthalpy of mixing of polymer with excess of solvent. Value B found in such a way cannot be used in Eq 2.111 to calculate H_M at high concentrations. At $\alpha_1 > \alpha_2$, the contribution of the equation-of-state term (in the square brackets, Eq 2.112) is negative if X_{12} is not high enough to exceed the term in the square brackets. The expression for the excess volume of mixing is obtained from the simple relations. The additive volume

$$\tilde{v}_0 = \varphi_1\tilde{v}_1 + \varphi_2\tilde{v}_2 \qquad [2.113]$$

differs from the volume of the mixture by the additional term

$$\tilde{v}^E = \tilde{v} - \tilde{v}_0; \quad \tilde{v}^E/\tilde{v} = \Delta V/V_0 \qquad [2.114]$$

where ΔV is the volume change by mixing and V_0 is the volume before mixing. The additional term is connected with the composition through value T^* and Eq 2.113. The theory under consideration predicts also important contribution to the entropy of mixing due to the difference in parameters of the equation-of-state of components. From the distribution function and Eq 2.109, it follows:

$$S^R = -3(N_1P_1^*V_1^*/T_1^*) \ln [(\tilde{v}_1^{1/3} - 1)/(\tilde{v}^{1/3} - 1)] -$$

$$- 3(N_2P_2^*V_2^*/T_2^*) \ln [(\tilde{v}^{1/3} - 1)/(\tilde{v}^{1/3} - 1)] \qquad [2.115]$$

This residual entropy is added to the combinatorial terms expressed by Eq 2.116 and 2.117:

$$\Delta S_{comb} = -k[N_1 \ln \varphi_1 + N_2 \ln \varphi_2] \qquad [2.116]$$

$$\Delta S_{comb} = -k_1[N_1 \ln \varphi_1 + N_2 \ln \varphi_2] + k(N_1z/2) \ln [1 -$$

$$- 2\varphi_2(1 - 1/r)/z] + krN_2[z/2 - (1 - 1/r)] \ln [1 +$$

$$+ \varphi_1(1 - 1/r)/[z/2 - (1 - 1/r)]] \qquad [2.117]$$

The partial molar quantities may be obtained from Eqs 2.110 and 2.115:

$$\overline{H}_1^R = \overline{H}_1 - \overline{H}_1^0 = \left(\frac{\delta \Delta H_M}{\delta N_1}\right)_{T,P,N_2} = P_1^*V_1^*[(\tilde{v}_1^{-1} - \tilde{v}^{-1}) +$$

$$+ (\alpha T/\tilde{v})(\tilde{T}_1 - \tilde{T})/\tilde{T}] + (V_1^*X_{12}/\tilde{v})(1 + \alpha T)\Theta_2^2 \qquad [2.118]$$

$$T\overline{S}_1^R = -P_1^*V_1^*[3\tilde{T}_1 \ln [(\tilde{v}_1^{1/3} - 1)/(\tilde{v}^{1/3} - 1)] -$$

$$- (\alpha T/\tilde{v})(\tilde{T}_1 - \tilde{T})/\tilde{T}]] + \alpha T(V_1^*X_{12}/\tilde{v})\Theta_2^2 \qquad [2.119]$$

Here α is the thermal expansion coefficient connected with $\tilde{v}$ through Eq 2.85. The combination of Eqs 2.118 and 2.119 leads to the following:

$$(\mu_1 - \mu_1^0)^R = P_1^*V_1^*[3\tilde{T}_1 \ln [(\tilde{v}_1^{1/3} - 1)/(\tilde{v}^{1/3} -1)] +$$

$$+ (\tilde{v}_1^{-1} - \tilde{v}^{-1})] + (V_1^*X_{12}/\tilde{v})\Theta_2^2 \qquad [2.120]$$

The expansion in series gives

$$(\mu_1 - \mu_1^0)^R = (P_1^* V_1^* / \tilde{v}_1)[A^2 \alpha_1 T/2 + Y_{12}]\varphi_2^2 + O(\varphi_2^3) \quad [2.121]$$

where:

$$A = (1 - T_1^*/T_2^*)(P_2^*/P_1^*) - (s_2/s_1)X_{12}/P_1^* \quad [2.122]$$

$$Y_{12} = (s_2/s_1)^2 X_{12}/P_1^* \quad [2.123]$$

In the similar way one can obtain:

$$\overline{H}_1^R = (P_1^* V_1^* / \tilde{v}_1)[Y_{12}(1 + \alpha_1 T) - (2/3)(A\alpha_1 T)^2(1 + $$

$$+ \alpha_1 T)]\varphi_2^2 + O(\varphi_2^3) \quad [2.124a]$$

$$T\overline{S}_1^R = (P_1^* V_1^* / \tilde{v}_1)[Y_{12}\alpha_1 T - (A^2 \alpha_1 T/2)[1 + (4/3)\alpha_1 T(1 + $$

$$+ \alpha_1 T)]\varphi_2^2 + O(\varphi_2^3)] \quad [2.124b]$$

The expression for χ_1 at the zero pressure is given as:[5,6]

$$\chi_1 = \frac{P_1^* V_1^*}{\tilde{v} RT}\left[(s_2/s_1)^2 \frac{X_{12}}{P_1^*} + \frac{\alpha_P T}{2}\left[\frac{P_2^*}{P_1^*}\tau - \frac{s_2}{s_1}\frac{X_{12}}{P_1^*}\right]^2\right] \quad [2.125]$$

To apply the proposed model, it is necessary to assure that the combinatorial entropy is a result of the mixing segments, not molecules. The preceding expressions obtained using lattice models or other methods, may be used to change the volume fractions to the segment fractions. After such replacement, the chemical potential can be presented as the sum of combinatorial and residual parts, i.e.,

$$(\mu_1 - \mu_1^0)/RT = \ln a_1 = \ln(1 - \varphi_2) + (1 - 1/r)\varphi_2 + $$

$$+ [(\mu_1 - \mu_1^0)^R / RT\varphi_2^2]\varphi_2^2 \quad [2.126]$$

where a_1 is the activity of the solvent, φ_2 is the segment fraction. Parameter χ_1 will then be identified as reduced residual chemical potential:

$$\chi_1 = (\mu_1 - \mu_1^0)^R / RT\varphi_2^2 \quad [2.127]$$

If to present the parameter as a series, we find that

$$\chi_1 = \chi_{10} + \chi_2 \varphi_2^2 + \ldots \quad [2.128]$$

The value corresponding to the reduced partial molar residual enthalpy is written as

$$k_1 = \overline{H}_1^E / RT\varphi_2^2 = \overline{H}_1^R / RT\varphi_2^2 \equiv k_{10} + k_2\varphi_2 + \ldots \qquad [2.129]$$

The reduced partial molar entropy is

$$\Delta\overline{S}_1 / R\varphi_2^2 = -\ln[(1 - \varphi_2) + \varphi_2]/\varphi_2^2 + \overline{S}_1^R / R\varphi_2^2 \qquad [2.130]$$

At high dilutions, this expression comes to the following:

$$\Psi_1 = \frac{1}{2} + \lim_{\varphi \to \infty} (\overline{S}_1^R / R\varphi_2^2) \qquad [2.131]$$

where Ψ_1 is the entropy parameter of interaction for dilute solutions. Thus, value $\Psi_1 \to 1/2$ represents the reduced partial molar residual entropy at $\Psi_2 \to 0$.

2.5 MODIFICATION OF THE NEW FLORY THEORY

The application of the new theory by Flory to interpret the experimental data on polymer solution was done by Flory[6] and other authors. The comparison of the predictions of the theory with experiment confirmed the main concepts of the theory, in particular, the existence of LCST and concentration dependence of the parameter χ_1. Simultaneously, in many cases the important discrepancies were observed between the theory and experiment. The reasons for these discrepancies are due to the fact that the Flory's theory was developed under assumption of random mixing, i.e., that each segment (or the solvent molecule) chooses the nearest neighbor in a fortuitous way without any preference. The simplified analysis of the non-random mixing was done for the first time by Renuncio and Prausnitz[24] who proposed to change Eq 2.97 by introducing correction for the local composition to calculate the excess functions of mixing. As a result, two new parameters were introduced in these equations connected with the local fractions of the component sites. This modification gives good agreement between the theory and experiment.

More correct description of the role of the non-random mixing was done by Brandani.[25,26] It was shown that accounting for the non-random mixing changes the form of equation of state proposed by Flory. The reduced equation of state is presented as

$$\frac{\tilde{P}\tilde{V}}{\tilde{T}} = \frac{\tilde{v}^{1/3}}{\tilde{v}^{1/3} - 1} - \frac{1}{\tilde{T}\tilde{v}} - \frac{A}{\tilde{T}^2\tilde{v}^2} \qquad [2.132]$$

This equation differs from the Flory equation by the third term, where

$$A = c\left[\varphi_1\Theta_{11}\Theta_{21}\left(\frac{\nu_{21}}{P^*}\right)^2 + \varphi_2\Theta_{22}\Theta_{12}\left(\frac{\nu_{12}}{P^*}\right)^2\right] \quad [2.133]$$

$$P^* = \varphi_1 P_1^* + \varphi_2 P_2^* + \varphi_1\Theta_{21}\nu_{21} + \varphi_2\Theta_{12}\nu_{12} \quad [2.134]$$

$$c = \frac{P^*V^*}{RT^*} = \varphi_1 c_1 + \varphi_2 c_2 \quad [2.135]$$

where Θ_{ii} and Θ_{ij} are the mean fractions of local surfaces

$$\Theta_{ii} = \frac{\Theta_i}{\Theta_i + \Theta_j\tau_{ij}}; \quad \Theta_{ij} = \frac{\tau_{ji}\Theta_j}{\Theta_i + \Theta_j\tau_{ji}} \quad [2.136]$$

where $\Theta_{ii} + \Theta_{ji} = 1$ (i, j = 1, 2). Eq 2.136 relates the fraction of mean local surfaces to the total surface of fractions (Eq 2.137) and parameters given by Eq 2.138:

$$\Theta_i = \frac{s_i r_i x_i}{\sum_{k=1}^{2} s_k r_k x_k} = \frac{s_i\varphi_i}{\sum s_k\varphi_k} \quad (i = 1, 2) \quad [2.137]$$

$$\tau_{ij} = \exp\left[(1/2)s_i(\eta_{ji}^0 - \eta_{ii}^0)/RTv = \exp(v^*\nu_{ij}\tilde{v}RT) \quad [2.138]$$

Here, φ_i are the segment fractions whereas ν_{ji} is determined by the expression:

$$\nu_{ji} = \frac{1}{2}s_i\frac{\eta_{ji}^0 - \eta_{ii}^0}{v^{*2}} \quad (i, j = 1, 2) \quad [2.139]$$

where η_{ij}/v is the potential energy per one contact of sites i and j. The characteristic pressures for pure components are given as

$$P_1^* = \frac{1}{2}s_1\eta_{11}^0/v^{*2} \quad [2.140]$$

$$P_2^* = \frac{1}{2}s_2\eta_{22}^0/v^{*2} \quad [2.141]$$

From Eqs 2.134 and 2.135, it follows that

$$\frac{1}{T^*} = (\varphi_1 P_1^*/T_1^* + \varphi_2 P_2^*/T_2^*)/P^* \quad [2.142]$$

where

$$T_1^* = \frac{1}{2} s_1 \eta_{11}^0 / Rc_1 v^* \quad T_2^* = \frac{1}{2} s_2 \eta_{22}^0 / Rc_2 v^* \qquad [2.143]$$

It was supposed[27] that to describe satisfactory the properties of solutions in the framework of the Flory theory, the number of the external degrees of freedom should be taken as

$$c = \varphi_1 c_1 + \varphi_2 c_2 - \varphi_1 \Theta_2 (v^*/k)\delta \qquad [2.144]$$

where k is the Boltzman constant and δ is an empirical parameter introduced to account for change of the number of the external degrees of freedom by the solution formation. In the framework of modified Flory theory, there were obtained the values of $\Delta\mu^R$, ΔH_M^∞, and V^E which have been used to compare the theoretical and experimental data. Using the adjusting parameters X_{12},δ and ratio s_2/s_1, a good agreement has been reached.

One more modification of the Flory theory has been done[28] using another rules of the combinations of the mixture parameters. It was assumed that the number of the external degrees of freedom may be changed. To determine the core volume of the mixture, there were taken expressions such as the van der Waals rule of combination:

$$v^* = \varphi_1^{*2} v_1^* + 2\varphi_1^* \varphi_2^* v_{12}^* + \varphi_2^{*2} v_2^* \qquad [2.145]$$

where

$$v_{12}^* = \zeta[(v_1^{*1/3} + v_2^{*1/3})/2]^3 \qquad [2.146]$$

where $\zeta = 1$ for the rigid spheres. By analogy with Eqs 2.82 and 2.145, the characteristic pressures were determined as

$$P_{12}^* = s\eta_{12}/2v_{12}^* \qquad [2.147]$$

$$P^* = \varphi^* \Theta_1 P_1^* (v_1^*/v^*)^2 + \varphi_2^* \Theta_2 P_2^* (v_2^*/v^*)^2 + 2\varphi_1^* \Theta_2 P_{12}^* (v_{12}^*/v^*)^2 \qquad [2.148]$$

The number of the external degrees of freedom was taken as

$$c = \varphi_1^* c_1 + \varphi_2^* c_2 - \varphi_1^* \Theta_2 c_{12} \qquad [2.149]$$

where c_{12} is parameter characterizing the deviation of the number of the external degrees of freedom from the additivity. From Eqs 2.105 and 2.149, the expression for T^* may be found:

$$\frac{1}{T^*} = [\varphi_1^* P_1^* (v_1^*/v^*)/T_1^* + \varphi_2^* P_2^* (v_2^*/v^*)/T_2^* -$$

$$- \varphi_1^* \Theta_{12} k c_{12}/v^*]/P^* \qquad [2.150]$$

The ratio of the excess volume, V^E, to the sum of volumes of pure components, V^0, is expressed as

$$\frac{V^E}{V^0} = \frac{v^* \tilde{v}}{\varphi_1^* v_1^* \tilde{v}_1 + \varphi_2^* v_2^* \tilde{v}_2} - 1 \quad [2.151]$$

It was shown using the solutions of poly(dimethyl siloxane)[28] that two parameters P_{12}^* and c_{12} give a good agreement between theoretical and experimental concentration dependencies of χ_1, ΔH^∞, and V^E/V^0. On the basis of the Flory theory, the equation of state was derived for the parameter of interaction polymer-solvent in infinitely diluted solution,[29] χ_0:

$$\chi_0 = (M_1 \tilde{v}_1/RT)[(3A^2/2P_1^*)[(3A^2/2P_1^*)[(\tilde{v}_1^{1/3} - 1)^{-1} - 3]] +$$

$$+ [(s_2/s_1)\tilde{v}_2]^2(\chi_1 - T\tilde{v}_1 - Q_{12})] \quad [2.152]$$

where

$$A = v_1[P_2^*(\tilde{T}_2 - \tilde{T}_1) + T_1 X_{12}(s_2/s_1)v_2(1 - 4\tilde{T}_1\tilde{v}_1)] \quad [2.153]$$

It was shown that this equation describes experimental data obtained by various authors on the dependence of χ_0 and T, and, in particular, for solutions of PS in toluene and cyclohexane and PIB in benzene. The values X_{12} and Q_{12} are different for the same system in various temperature intervals.

The effect of the polar groups on thermodynamic functions in the new statistical theories has also been considered.[30] There were made attempts to describe the thermodynamics of mixing on the basis of free volume concept but without lattice model.[31] It was assumed that the mixing process is equivalent to the irreversible expansion of the mass center of molecules from the free volume typical of pure components to the free volume of solution. The comparison of this theory with developed by Flory has shown that for the athermic mixture there is a qualitative agreement between the theory and experiment, both theories giving very close values of parameter χ_S. Mixtures of polymer homologues have been analyzed as well.[32]

To describe the properties of solution, it is very important to know the characteristic values of the exchange interaction parameter for some chemical groups. There was made an attempt[33] to derive a simple expression for the contact interaction parameter X_{12} in terms of characteristic exchange parameters X_{ij} between chemically different segments (i, j, ...) composing molecules of components 1 and 2. Each segment or chemical group should have the definite fraction

(α_i, α_j,...) of sites (or surfaces) in relation to the total molecule surface. The expression for X_{12} is given in the form

$$X_{12} \approx \sum_{i>j} (\alpha_{i,1} - \alpha_{i,2})(\alpha_{j,1} - \alpha_{j,2})X_{ij} \qquad [2.154]$$

To estimate the fraction of sites, the Bondi approximation is used.[34] For example, for the groups -CH_2-, -CH_3-, -C_6H_4-, -O- the surface areas are equal to 3.51, 2.24, 3.65, 8.85, and 1.25 nm^2, respectively. To simplify the considerations, the molecule may be presented as consisting of only two types of the surfaces. Aliphatic polyester may have "aliphatic" (al) surface, including methyl- and methylene groups and "ester" (es) surface. For example, 1,2-ethane diolacetate has $\alpha_{es} = 0.39$ and $\alpha_{al} = 0.61$, toluene has $\alpha_{arom} = 0.72$ and $\alpha_{al} = 0.28$, 1,4-dioxane has $\alpha_{al} = 0.79$ and $\alpha_{eth} = 0.22$, etc. Using the calorimetric data, values of X_{12} were estimated[33] for many mixtures of aliphatic esters in toluene and many mixtures of aliphatic ethers in benzene. It was shown that the dependence of X_{12} on α_{es} obtained experimentally may be approximated by the equation:

$$X_{12} = X_{al,es}\alpha_{es}^2 - 0.72(X_{al,arom} + X_{al,es} - X_{arom}\alpha_{es}) + + 0.518X_{al,arom} \qquad [2.155]$$

If values of constituent X are known, such approach can be adopted to polymer solutions. For example, from the calorimetric data for the PIB solution in benzene, it was found that $X_{12} = 42$, whereas the calculated $X_{12} = 40$ J/cm^3 for PS in toluene and ethyl benzene X_{12} are equal to 0.6 and 4.0, whereas calculated values are 0 and 1 J/cm^3 that can be considered as a satisfactory agreement.

2.6 NEW HUGGINS THEORY

Huggins has made an attempt to use the concepts of the thermodynamics of polymer solution to all solutions independent of their nature.[35-39] His aim was to find parameters that are necessary to describe the concentration dependence of the thermodynamic properties of polymer solutions from the experimental data for solutions of low molecular weight compounds. For this purpose the model has been used where each molecule has the surface that is in the mutual contact with the surfaces of the surrounding molecules. If the molecules consist of chemically different segments, each segment can be characterized by the definite surface and contact energy. For example, the molecule of n-alkane can be considered as consisting of two methyl segments and (n – 2) methylene segments. The segments may have contacts with similar and dissimilar segments. This implies that for each type of the segment-segment contact, the mean contact energy per unit area is constant at any given temperature in spite of changes in type and areas of other contacts. The relative areas of contacts should provide a minimal free energy. It is assumed that the relative contact surfaces are described by one or more equilibrium constants that depend on the interaction energy per unit of the contact surface. Using such approximations, Huggins derived the equations connecting the cohesion energy in the system with the number of different segments, average surface of the segment-segment contact and with one or more equilibrium constants.

Let the segment types be designated as α, β, γ, etc. and the types of the contacts as $\alpha\alpha$, $\beta\beta$, $\alpha\beta$. Correspondingly, the areas of the contacting segments are designated as σ_α, σ_β, etc. (per one mole of the substance or mixture), the average area of the contacting surface per 1 mole of the simple segments, such as α or β as σ_α^0 or σ_β^0, the average area of contacts per 1 mole of different types of segments as $\sigma_{\alpha\alpha}$, $\sigma_{\beta\beta}$, $\sigma_{\alpha\beta}$, etc. and the mean energies per the contact area as $\varepsilon_{\alpha\alpha}$, $\varepsilon_{\beta\beta}$, etc. Generally, the attraction forces are operative, i.e., the negative values. The surface unit is not specified because all the values are the products of the forms $\sigma\varepsilon$ or $\sigma_\beta/\sigma_\alpha$.

If the system consists of the segments of one type, i.e., only similar contacts are present, the total intermolecular energy of interaction (the negative cohesion energy) per 1 mol is written as

$$E = \delta_{\alpha\alpha}\varepsilon_{\alpha\alpha} = n_\alpha\sigma_\alpha^0\varepsilon_{\alpha\alpha}/2 \qquad [2.156]$$

where n_α is the number of segments α per one molecule. If there are two types of segments , i.e., three types of contacts, the energy is given in the following way:

$$E = \sigma_{\alpha\alpha}\varepsilon_{\alpha\alpha} + \sigma_{\beta\beta}\varepsilon_{\beta\beta} + \sigma_{\alpha\beta}\varepsilon_{\alpha\beta}. \qquad [2.157]$$

$$\sigma_\alpha = 2\sigma_{\alpha\alpha} + \sigma_{\alpha\beta} \qquad [2.158]$$

$$\sigma_\beta = 2\sigma_{\beta\beta} + \sigma_{\alpha\beta} \qquad [2.159]$$

The equilibrium constant connecting the contact areas of the three types is determined by the relation:

$$K = \frac{\sigma_{\alpha\beta}^2}{4\sigma_{\alpha\alpha}\sigma_{\beta\beta}} \qquad [2.160]$$

The numerical coefficient is included to make K equal to unity for the case of fully random contacts, i.e., when there is no preference in any contact formation between α and β segments in a hypothetical equimolar mixture. The equilibrium constant is related to the contact energies in the form:

$$K = A \exp(-\Delta\varepsilon/kT) \quad [2.161]$$

where

$$\Delta\varepsilon = 2\varepsilon_{\alpha\beta} - \varepsilon_{\alpha\alpha} - \varepsilon_{\beta\beta} \quad [2.162]$$

If there are in the system more than two types of segments, it is necessary to introduce more equilibrium constants. Values K, A, k and $\Delta\varepsilon$ in Eqs 2.160 and 2.162 should be written as $K_{\alpha\beta}$ etc., whereas value k in Eq 2.161 is not standard Boltzman constant, but dependent on the choice of the area unit. The coefficient A is determined by the steric and other restrictions. Usually Eq 2.161 is not used and there is no necessity to determine neither A nor k. From Eqs 2.158÷2.160, one can obtain:

$$\sigma_{\alpha\beta} = -2(\sigma_\alpha + \sigma_\beta)[1 - (1 + y)^{1/2}]/K' \quad [2.163]$$

$$\sigma_{\alpha\alpha} = \sigma_{\alpha/2} + (\sigma_\alpha + \sigma_\beta)[1 - (1 + y)^{1/2}]/K' \quad [2.164]$$

$$\sigma_{\beta\beta} = \sigma_{\beta/2} + (\sigma_\alpha + \sigma_\beta)[1 - (1 + y)^{1/2}]/K' \quad [2.165]$$

where

$$K' = 4(1/K - 1) \quad [2.166]$$

$$y = K'\sigma_\alpha\sigma_\beta/(\sigma_\alpha + \sigma_\beta)^2 \quad [2.167]$$

Inserting these equations to Eq 2.157 gives the molar interaction energy:

$$E = \frac{\sigma_\alpha\varepsilon_{\alpha\alpha}}{2} + \frac{\sigma_\beta\varepsilon_{\beta\beta}}{2} - \frac{(\sigma_\alpha + \sigma_\beta)\Delta\varepsilon}{K'}[1 - (1 + y)^{1/2}] \quad [2.168]$$

The value of the molar molecular interaction energy, $\sigma_\alpha\varepsilon_{\alpha\alpha}$, can be found from the standard heat of evaporation H_v^0. In another way, value E may be presented as

$$E = -H_v^0 + P\Delta V^0 - \frac{3}{2}RT \quad [2.169]$$

where ΔV^0 is the difference between the molar volumes of the liquid and vapor. The excess enthalpy per one mole or the molar heat of mixing equals to the changes of the molar energy of mixing and it is a difference between true enthalpy (or energy) of 1 mole of solution and the sum of molar enthalpies of components each of which is multiplied by its molar fraction in solution:

$$\overline{H}^E = \overline{H} - x_1\overline{H}_1 - x_2\overline{H}_2 = E - x_1E_1 - x_2E_2 \qquad [2.170]$$

From Eq 2.156 for $n_\alpha = 1$ for pure components and Eq 2.167 with $\sigma_\alpha = x_1\sigma_\beta$ and $\sigma_\beta = x_2\sigma_\alpha$, one can find:

$$\overline{H}^E = \frac{-(x_1\sigma_\alpha^0 + x_2\sigma_\beta^0)\Delta\varepsilon}{K'}\left[1 - \left(1 + \frac{K'x_1\sigma_\alpha^0 x_2\sigma_\beta^0}{x_1\sigma_\alpha^0 + x_2\sigma_\beta^0}\right)^{1/2}\right] \qquad [2.171]$$

or

$$H^E = \frac{\delta_\alpha^0\Delta\varepsilon}{K'}[1 + x_2(r-1)][1 - (1 + K'r(x_2 - x_2^2)/[1 +$$

$$+ x_2(r-1)]^2)^{1/2}] \qquad [2.172]$$

where $r = \sigma_\beta^0/\sigma_\alpha^0$. If K = 1, then K′ = 0 and equation 2.172 comes to

$$\overline{H}^E = \frac{\sigma_\alpha^0\Delta\varepsilon}{2}[r(x_2 - x_2^2)/(1 + x_2(r-1))] \qquad [2.173]$$

The heat of mixing depends on concentration, x_2, and three parameters K (or K′), r and the product $\sigma_\alpha^0\Delta\varepsilon$. These values may be found from precise measurement of H^E in relationship with solution composition. Eq 2.172 may be written in a simpler form. For this purpose, we may substitute $\Delta\varepsilon\sigma_\alpha^0/2$ by $\Delta\varepsilon$ and express value r through the fractions of contacting surfaces, z, in the following way:

$$x_2 = 1 - x_1 = \varphi_2/(nr_v\varphi_1 + \varphi_2) \qquad [2.174]$$

$$z_\beta = 1 - z_\alpha = \frac{nr_\sigma x_2}{1 + (nr_\sigma - 1)x_2} = \frac{(r_\sigma/V)\varphi_2}{1 + (r_\sigma/V - 1)\varphi_2} \qquad [2.175]$$

$$r_{\sigma/v} = r_\sigma/r_v; \quad r_\sigma = \sigma_\beta^0/\sigma_\alpha^0; \quad r_v = v_\beta^0/v_\alpha^0 \qquad [2.176]$$

$$g_k = -\frac{2}{z_\alpha z_\beta K'}[1 - (1 + z_\beta z_\alpha K')^{1/2}] \qquad [2.177]$$

Having done such substitutions, we obtain

$$\overline{N}^E = \varepsilon_\Delta x_1 z_\beta g_k \qquad [2.178]$$

Substituting ε by v in Eq 2.162 gives the expression for the excess mixing volume:

$$V^E = v_\Delta x_1 z_\beta g_k \qquad [2.179]$$

where

$$v_\Delta = \sigma_\alpha^0(2v_{\alpha\beta} - v_{\alpha\alpha} - v_{\beta\beta})/2 \qquad [2.180]$$

In principle, values of the parameters r and K′ may be applied to both H^E and V^E. In accordance with the thermodynamic properties of polymer solutions, the application of the Flory-Huggins equation allows one to estimate the interaction parameter χ_1:

$$\frac{\Delta\overline{G}_1}{RT} = \frac{\Delta\overline{H}_1 - T\Delta\overline{S}_1}{RT} = \ln\varphi_1 + (1 - V_1/V)\varphi_2 + \chi_1\varphi_2^2 \qquad [2.181]$$

Considering that

$$\chi_1 = \chi_H + \chi_S \qquad [2.182]$$

after differentiation and substitution N_1 and N_2 by x_1 and x_2, we find:

$$\chi_H = \frac{\overline{H}_1}{RT\varphi_2^2}; \quad \overline{H}_1 = \left(\frac{\delta\overline{H}_1^E}{\delta N_1}\right)_{N_2} \qquad [2.183]$$

Taking into account the parameters discussed above, we can obtain:

$$\chi_H = \frac{\varepsilon_\Delta z_\beta}{RT\varphi_2^2}\,\frac{1 - z_\alpha g_k}{1 + z_\alpha z_\beta K' g_k/2} \qquad [2.184]$$

and for K = 1:

$$\chi_H = \frac{\varepsilon_\Delta z_\beta^2}{RT\varphi_2^2} \qquad [2.185]$$

Parameter χ_S is due to the non-combinatorial contribution to the entropy of mixing, including orientation entropy, that is determined by the concentration dependence of the molecular or segmental orientation in relation to the surrounding molecules. The combinatorial entropy correction for a non-random mixing also should be included in χ_S. The last two contributions have been calculated on the basis of the same approximation that was used to derive the equation for the mixing enthalpy. The expression for the orientation contribution to the interaction parameter has been obtained in the form:

$$\chi_{s,orient} = \frac{-k'_s z_\beta}{r_\sigma \varphi_2^2}\left[\frac{z_\beta z_\alpha/(1 + K' z_\alpha z_\beta)^{1/2} + z_\alpha g_k}{1 + k'_s z_\alpha g_k}\right] \quad [2.186]$$

where:

$$k'_s = k_s/(1 - k_s) \quad [2.187]$$

For $K = g_k = 1$, we have

$$\chi_{s,orient} = -\frac{k'_s z_\beta^2}{r_\sigma \varphi_2^2 (1 + k_s z_\alpha)} \quad [2.188]$$

For the infinitely diluted solution, this value is

$$\chi_{s,orient} = -\frac{k'_s}{r_\sigma}\left(\frac{z_\beta}{\varphi_2^2}\right)^2_{\varphi_2 \to 0} \quad [2.189]$$

For the system with only one type of segments, we have

$$\chi_{s,orient} = -k_s r_\sigma / r_v \quad [2.190]$$

where k_S and k'_S are equilibrium constants. The correction for the non-randomness of mixing is given in the form:

$$\chi_{s,comb.corr} = -\frac{1}{\varphi_2^2}\left(\frac{\delta S^E_{comb.corr}}{\delta N_1}\right)_{N_2} = \frac{\sigma^0_\alpha}{4\varphi_2^2}[\ln \frac{z_\alpha}{1 - z_\beta g_k} +$$

$$+ [\frac{z_\beta(z_\beta - z_\alpha)}{(1 + K' z_\alpha z_\beta)^{1/2}} + z_\alpha z_\beta g_k] \ln \frac{1 - z_\alpha g_k)(1 - z_\beta g_k)}{z_\alpha z_\beta g_k^2}] \quad [2.191]$$

The correction given by equation 2.191 is added to the value of combinatorial entropy of mixing that is expressed usually by the Flory-Huggins equation. Parameters r_σ and K' can be found from the measurements of the excess enthalpies and excess volume of mixing. For this purpose the ratios of H^E or V^E at two concentrations are used:[38]

$$\overline{H}^E_{\varphi'/\varphi} = \frac{[1 + (nr_\sigma - 1)x'_2][1 - (1 + K' z'_\alpha z'_\beta)^{1/2}]}{[1 + (nr_\sigma - 1)x_2][1 - (1 + K' z_\alpha z_\beta)^{1/2}]} \quad [2.192]$$

The same equation is valid for $V^E_{\varphi'/\varphi}$. From two such ratios values r_σ and K may be calculated. Using mean values of these parameters

(r_σ and K′), together with appraised values V_φ^E and H_φ^E, the third term may be calculated (ε_Δ or v_Δ).

$$\varepsilon_\Delta \text{ or } (v_\Delta) = \frac{-K'\overline{H}^E(\text{or } V^E)}{2[1 + (nr_\sigma - 1)x_2][1 - (1 + K'z_\alpha z_\beta)^{1/2}]} \qquad [2.193]$$

The same equation is valid for v_A. The Eqs 2.178 and 2.179 with parameters calculated characterize the dependence of H^E and V^E and concentration. Substituting values r_σ, K′, ε_Δ and in Eq 2.184, the concentration dependence of the parameter χ_H may be found. Subtracting χ_H from χ_1 for one concentration gives the value of χ_S. If there is a great contribution of $\chi_{S\ orient}$ to χ_S, Eq 2.186 may be used to calculate k'_S and, using the value found in such a way, to calculate the concentration dependence of χ_S, χ_H, and χ_1. This procedure makes it possible to compare the theoretical and experimental values. Parameters r_σ, K, k_S, σ_α^0 and can be determined from the data obtained for two systems that contain the same type of segments and segment-segment contacts. Therefore, it is possible, using parameters for simple systems consisting of a low molecular weight components, to obtain necessary parameters for polymer solutions. It is especially important for polymer mixtures because it makes experiments easier to conduct. The application of the Huggins theory to some systems demonstrated a good agreement between the experiment and the theory.

2.7 SANCHEZ-LACOMBE THEORY

Sanchez and Lacombe[40] have proposed new equations of state for the liquid systems and polymer solutions. Their theory, called the theory of the lattice liquid, needs estimation of three parameters of the equation of state for pure components (similar to Flory's theory). The theory allows to derive the expressions for chemical potentials, heats, and volumes of mixing, critical solution temperatures, enthalpy, and entropy, contributions to chemical potentials, etc.

2.7.1 EQUATION OF STATE

For configurations available in the system consisting of N molecules with configurational energy, E, and volume, V, Sanchez and Lacombe have derived the expression:

$$\Omega \approx (1/f_0)^{N_0}(\omega/f)^N \qquad [2.194]$$

where Ω is the number of configurations available for segment in densely-packed state, $f_0 = N_0/(N_0 + r\,N)$ is the fraction of the free sites, $f = r\,N/(N_0 + r\,N)$ is the fraction of the occupied sites in the lattice. The configurational partition function is expressed as

$$z(P,T) = \sum_V \sum_E \Omega(E,V,N)\exp[-(E + PV)/kT] \qquad [2.195]$$

Because E and V are the functions of the number of the free sites in the lattice, double summation is substituted in Eq 2.195 by a single summation by N_0:

$$z(P,T) = \sum_{N_0}^{\infty} \Omega\exp[-(E + PV)/kT] \qquad [2.196]$$

In this case, the free energy of mixing can be written as

$$G = E + PV - kT\ln\Omega \qquad [2.197]$$

In the same manner as in the Flory's theory, the volume of the densely-packed segment is taken as v*, the volume of the densely-packed molecule as rv*. Parameters v* and rv* are connected with the density, ρ*, specific volume, v^*_{sp}, in the densely-packed state and molecular mass, M:

$$rv* = M/\rho* = Mv*_{sp} \qquad [2.198]$$

The total volume is

$$V = (N_0 - rN)v* \qquad [2.199]$$

The reduced values are given by the following expressions:

$$\tilde{v} = v_{sp}/v*_{sp}; \quad \tilde{\rho} = 1/\tilde{v} = \rho/\rho* \qquad [2.200]$$

$$\tilde{v} = V/V^* = v/v*; \quad V^* = rNv* \qquad [2.201]$$

The fraction of sites f can be found using the relative density as it follows from the Eqs 2.198 and 2.201 that

$$\rho = \frac{NM}{(N_0 + rN)v*} = \frac{rN\rho^*}{(N_0 + rN)} = f\rho* \qquad [2.202]$$

Therefore, $f = \rho/\rho* = \tilde{\rho}$. The full number of the pair interactions is $(N_0+rN)z/2$. If the energy of interaction between segments is ε, the mean intermolecular energy of interaction can be presented as

$$E = -(z\varepsilon/2)(N_0 + rN)f^2 \qquad [2.203]$$

$$E/rN = -\tilde{\rho}\varepsilon* \qquad [2.204]$$

where f^2 is the probability that two sites will be occupied by segments and $\varepsilon* = z\varepsilon/2$ is the full interaction energy per segment. Using these relations, the free energy of mixing can be expressed in the form of relative (reduced) values as:

$$G/rN\varepsilon* = \tilde{G} = -\tilde{\rho} + \tilde{P}\tilde{v} + \tilde{T}[(\tilde{v} - 1)\ln(1 - \tilde{\rho}) + (1/r)\ln(\tilde{\rho}/\omega)] \quad [2.205]$$

Here $\tilde{T} = T/T^*$, $T^* = \varepsilon^*/k$, $\tilde{P} = P/P^*$, $P^* = \varepsilon^*/v^*$. Differentiating $\tilde{G}$ by $\tilde{v}$ and equating the result to zero, gives the equation of state of the lattice liquid:

$$\tilde{\rho}^2 + \tilde{P} + \tilde{T}[\ln(1 - \tilde{\rho}) + (1 - 1/r)\tilde{\rho}] = 0 \quad [2.206]$$

From the equation of state, the real liquid may be completely characterized by three parameters: T^*, P^*, and ρ^*. For polymeric liquid $r \to \infty$ and equation of state (Eq 2.206) comes to

$$\tilde{\rho}^2 + \tilde{P} + \tilde{T}[\ln(1 - \tilde{\rho}) + \tilde{\rho}] = 0 \quad [2.207]$$

The parameters of the equation of state may be found from the PVT data and, in particular, from the experimental data on the density, thermal expansion coefficient, α, and compressibility, β:

$$T\alpha = \frac{1 + \tilde{\rho}\tilde{v}^2}{\tilde{T}\tilde{v}[1/(\tilde{v} - 1) + 1/r] - 2} \quad [2.208]$$

$$P\beta = \frac{\tilde{\rho}\tilde{v}^2}{\tilde{T}\tilde{v}[(1/(\tilde{v} - 1) + 1/r] - 2} \quad [2.209]$$

$$T\gamma = P*(\tilde{\rho}^2 + \tilde{\rho}) \quad [2.210]$$

At zero pressure, the following expressions are valid:

$$T\alpha = 1/[\tilde{T}(1 - \tilde{\rho}) - 2] \quad [2.211]$$

$$P* = \tilde{v}^2 T\gamma = \tilde{v}^2 T\alpha/\beta \quad [2.212]$$

2.7.2 MIXTURES OF LIQUIDS

The extension of the theory on the liquid mixtures is connected with the application of the following rules of mixing (combination rules):

1. If molecule i occupies r_i^0 sites in the pure state and has densely-packed molecular volume, $r_i v_i^*$, it should occupy r_i sites in the mixture:

$$r_i^0 v*_i = r_i v* \quad [2.213]$$

This rule guarantees the additivity of the densely-packed volumes:

$$V* = r_1^0 N_1 v*_1 + r_2^0 N_2 v*_2 = (r_1 N_1 + r_2 N_2)v* \qquad [2.214]$$

2. The total number of pair interactions in the densely-packed mixture is equal to the sum of pair interactions in the densely-packed pure state:

$$(z/2)(r_1^0 N_1 + r_2^0 N_2) = (z/2)(r_1 N_1 + r_2 N_2) = (z/2)rN \qquad [2.215]$$

where z is the coordination number of the lattice

$$r \equiv x_1 r_1^0 + x_2 r_2^0 \equiv x_1 r_1 + x_2 r_2 \qquad [2.216]$$

$$x_1 \equiv N_2/N = 1 - x_2 \qquad \text{(molar fraction)} \qquad [2.217]$$

$$N = N_1 + N_2 \qquad [2.218]$$

3. The characteristic pressures in the mixture are presented as

$$P^* = \varphi_1 P_1^* + \varphi_2 P_2^* - \varphi_1 \varphi_2 \Delta P^* \qquad [2.219]$$

$$\Delta P^* = P_1^* + P_2^* - 2P_{12}^* \qquad [2.220]$$

$$\varphi_1 = r_1 N_1 / rN = 1 - \varphi_2 \qquad [2.221]$$

The first two rules of mixing lead to the following relation for the average densely-packed segment:

$$v* = \varphi_1^0 v*_1 + \varphi_2^0 v*_2 \qquad [2.222]$$

where

$$\varphi_1^0 = r_1^0 N_1 / rN = 1 - \varphi_2^0 \qquad [2.223]$$

Concentrations φ_1 and φ_2 are expressed as:

$$\varphi_1 = \frac{\varphi_1^0}{\varphi_1^0 + \nu\varphi_2^0}; \quad \varphi_2 = \frac{\nu\varphi_1^0}{\nu\varphi_1^0 + \varphi_2^0}; \quad \nu = v*_1/v*_2 \qquad [2.224]$$

From the definition of φ_1 given by Eq 2.221 and the first mixing rule, it follows that this value is the volume fraction of the densely-packed component i, i.e.,

$$\varphi_1 = \frac{m_1/\rho*_1}{m_1/\rho*_1 + m_2/\rho*_2} \qquad [2.225]$$

where m_1 and m_2 are the mass fractions. The volume of mixing is

$$V = (N_0 + rN)v* = V*\tilde{v} \quad [2.226]$$

The reduced volume $\tilde{v}$ may be expressed through the specific volume of the mixture v_{sp} or density ρ in the following way :

$$\tilde{v} = v_{sp}/v*_{sp} = \rho*/\rho = 1/\tilde{\rho} \quad [2.227]$$

where

$$v*_{sp} = m_1/\rho*_1 + m_2/\rho*_2 = 1/\rho* \quad [2.228]$$

The interaction energy between the solvent molecule and polymer segment is

$$\varepsilon* = P^*v* = (\varphi_1 P_1^* + \varphi_2 P_2^* - \varphi_1\varphi_2\Delta P^*)(\varphi_1^0 v*_1 + \varphi_2^0 v*_2) \quad [2.229]$$

Other reduced values can be found from the relations:

$$\tilde{T} = T/T^*; \; T^* = \varepsilon*/k; \; \tilde{P} = P/P^* \quad [2.230]$$

Here ε* is given by Eq 2.229 and P^* by Eq 2.219. From Eqs 2.229 and 2.224 value T^* may be expressed as

$$T/T^* = \frac{1}{\tilde{T}} = \frac{[\varphi_1/\tilde{T}_1 + \nu\varphi_2/\tilde{T}_2]}{\varphi_1 + \nu\varphi_2} - \varphi_1\varphi_2 X \quad [2.231]$$

where $X = \Delta P^* v*/kT$.

2.7.3 FREE ENERGY AND CHEMICAL POTENTIAL

For the binary system the configurational Gibbs free energy is given as

$$G = rN\varepsilon*\tilde{G} \quad [2.232]$$

where

$$\tilde{G} = -\tilde{\rho} + \tilde{P}\tilde{v} + \tilde{T}\tilde{v}[(1-\tilde{\rho})\ln(1-\tilde{\rho}) + \frac{\tilde{\rho}}{r}\ln\tilde{\rho}] + \tilde{T}[(\varphi_1/r)\ln\varphi_1 +$$

$$+ (\varphi_2/r_2)\ln\varphi_2] \quad [2.233]$$

Differentiation $\tilde{G}$ by density or volume and equating the result to zero gives the equation of state for the mixture that is similar to the equation of state for pure liquid (Eq 2.206) but with T^*, P^*, and r found for the mixture (for the mixture $1/r = \varphi_1/r_1 + \varphi_2/r_2$). The chemical potential is expressed as

$$\mu_1 \equiv \left(\frac{\delta G}{\delta N_1}\right)_{T,P,N_2} \qquad [2.234]$$

$$\mu_1 = kT[\ln \varphi_1 + (1 - r_1/r_2)\varphi_2 + r_1^0\tilde{\rho}X_1\varphi_2^2] + r_1^0kT[-\tilde{\rho}/\tilde{T}_1 +$$

$$+ \tilde{P}_1\tilde{v}/\tilde{T}_1 + \tilde{v}[(1 - \tilde{\rho}) \ln (1 - \tilde{\rho}) + (\tilde{\rho}/r_1^0) \ln \tilde{\rho} + \frac{\tilde{\rho}}{r_1^0} \ln \tilde{\rho}]] \qquad [2.235]$$

where

$$X_1 = X(\varphi_1 - 1) = \Delta P{*}v{*}/kT \qquad [2.236]$$

The expression for μ_2 may be easily found by changing index 1 for index 2. The chemical potentials are characterized by the following properties. At $\varphi_1 = 1$ or $\varphi_2 = 1$, they transform to the corresponding molar quantities that are typical of pure components, for example,

$$\mu_1(\varphi_1 = 1) \equiv \mu_1^0 = r_1^0\varepsilon{*}_{11}\tilde{G}(\varphi_1 = 1) = r_1^0\varepsilon{*}_{11}\tilde{G}_1 \qquad [2.237]$$

At low temperatures or high pressures, the reduced densities approach their maximum values close to unity. In this limiting case (at $\tilde{\rho}$ and $\tilde{\rho}_1 \to 1$), the Flory-Huggins chemical potentials may be obtained:

$$(\mu_1 - \mu_1^0)/kT \to \ln\varphi_1 + (1 - r_1/r_2)\varphi_2 + r_1^0X_1\varphi_2^2 \qquad [2.238]$$

There is only one parameter $-\Delta P^*$ or X_1 that characterizes the binary mixture, whereas all other are known from the properties of pure components. The interaction energy is expressed in terms of dimensionless parameter, ζ, that is a measure of the deviation of P_{12}^* from the mean-geometric deviation:

$$\zeta = P_{12}^*/(P_1^*P_2^*)^{1/2} \qquad [2.239]$$

In this case, the value ΔP^* is determined by the relation:

$$\Delta P^* = P_1^* + P_2^* - 2\zeta(P_1^*P_2^*)^{1/2} \qquad [2.240]$$

2.7.4 FUNCTIONS OF MIXING

The excess volume of mixing is expressed as

$$\frac{\Delta V_{mix}}{V_0} = \frac{\tilde{v}}{\varphi_1\tilde{v}_1 + \varphi_2\tilde{v}_2} - 1 \qquad [2.241]$$

At low pressures the term $P\Delta V_{mix}$ may be neglected and the enthalpy of mixing is expressed as

$$\Delta H_{mix}/RT = r[\tilde{\rho}\varphi_1\varphi_2 X + v*[\varphi_1 P_1^*(\tilde{\rho}_1 - \tilde{\rho}) +$$

$$+ \varphi_2 P_2^*(\tilde{\rho}_2 - \tilde{\rho})]]/RT \qquad [2.242]$$

The next relation gives the entropy of mixing:

$$\Delta S_{mix}/R = -r[\frac{\varphi_1}{r_1} \ln \varphi_1 + \frac{\varphi_2}{r_2} \ln \varphi_2 + \tilde{v}[(1 - \tilde{\rho}) \ln (1 - \tilde{\rho}) + \frac{\tilde{\rho}}{r} \ln \tilde{\rho}]] +$$

$$+ \frac{r}{\varphi_1 + v\varphi_2}[\varphi_1 \tilde{v}[(1 - \tilde{\rho}_1) \ln (1 - \tilde{\rho}_1) + \frac{\tilde{\rho}_1}{r_1^0} \ln \tilde{\rho}_1] +$$

$$+ v^2 \varphi_2 \tilde{v}_2[(1 - \tilde{\rho}_2) \ln (1 - \tilde{\rho}_2) + \frac{\tilde{\rho}_2}{r_2^0} \ln \tilde{\rho}_2]] \qquad [2.243]$$

The first two terms in Eq 2.243 are the Flory -Huggins combinatorial entropy of mixing.

2.7.5 PHASE STABILITY AND SPINODAL

The negative free energy of mixing is not sufficient condition of miscibility. The necessary and sufficient conditions of miscibility in the whole range of concentration are: $d\mu_1/dx_1 > 0$ and $d\mu_2/dx_2 > 0$. For the binary mixture of lattice liquids at the constant temperature and pressure, the following condition is valid:

$$\frac{d(\mu_1/kT)}{d\varphi_1} = 2r_1\varphi_2\left[\frac{1}{2}\left(\frac{1}{r_1\varphi_1} + \frac{1}{r_2\varphi_2}\right) - \tilde{\rho}(X + \frac{1}{2}\psi^2 \tilde{T}P^*\beta)\right] \qquad [2.244]$$

where ψ is dimensionless function:

$$\psi \equiv \tilde{\rho}\frac{d(1/T)}{d\varphi_1} - \frac{d(1/r)}{d\varphi_1} + \tilde{v}\frac{d(\tilde{P}/\tilde{T})}{d\varphi_1} \qquad [2.245]$$

$$\psi = \tilde{\rho}[[v(1/\tilde{T}_1 - 1/\tilde{T}_2) + (\varphi_1^2 - v\varphi_2^2)X_1] - v(1/r_1^0 - r_2^0) +$$

$$+ \frac{\tilde{P}\tilde{V}}{\tilde{T}}(v - 1)(\varphi_1 + v\varphi_2)]/(\varphi_1 + v\varphi_2)^2 \qquad [2.246]$$

The isothermal compressibility β is defined in the following way:

$$\tilde{T}P^*\beta = \tilde{v}\left[\frac{1}{\tilde{v} - 1} + \frac{1}{r} - \frac{2}{\tilde{v}\tilde{T}}\right]^{-1} > 0 \qquad [2.247]$$

The system is stable on the condition:

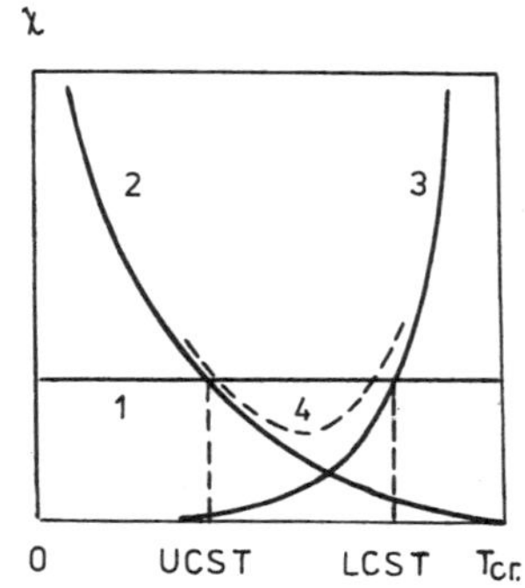

Fig. 2.4. Schematic behavior of three terms in the spinodal inequality 2 as function of temperature. The dashed curve is the sum $\tilde{\rho}X_1 + (1/2)\Psi_1^2\tilde{T}P^*\beta$. The inequality $(1/2)[1/(r_1\varphi_1) + 1/(r_2\varphi_2)] - \tilde{\rho}[X_1 + (1/2)\Psi_1^2\tilde{T}P^*\beta] > 0$ is satisfied between the UCST and LCST. Adapted, by permission, from I. Sanchez and R. H. Lacombe, *Macromolecules*, **11**, 1145 (1978).

$$\frac{1}{2}\left(\frac{1}{r_1\varphi_1} + \frac{1}{r_2\varphi_2}\right) - \tilde{\rho}(X + \frac{1}{2}\psi^2\tilde{T}P^*\beta) > 0 \qquad [2.248]$$

where $(1/r_1\varphi_1) + (1/r_2\varphi_2)$ is the contribution of combinatorial entropy, $\tilde{\rho}X$ is the energy contribution and $(1/2)\tilde{\rho}\tilde{T}P^*\beta$ is the unfavorable contribution following from the equation of state. Fig. 2.4 shows schematically the dependence of these three contributions on temperature.

For the infinitely dilute solution, the enthalpy of mixing is given by the following two equations:

$$\Delta H_{mix}^{\infty} = \lim_{\varphi_2 \to 0} \Delta H_{mix}/\varphi_2 = \left(\frac{d\Delta H_{mix}}{d\varphi_2}\right)_{\varphi_2=0} \qquad [2.249]$$

$$H_{mix}^{\infty}/RT = r_1^0(M_u/M_1)(\rho*_1/\rho*_2)[\tilde{\rho}_1X_1 + \tilde{\rho}_1^2\psi_1P_1^*\beta_1 +$$

$$+ \nu(\tilde{\rho}_2 - \tilde{\rho}_1)] \qquad [2.250]$$

where M_u is the molecular mass of the repeating unit, M_1 is the molecular mass of the solvent. When deriving Eq 2.250, the relation was used according to which

$$d\tilde{\rho}/d\varphi_1 = \tilde{\rho}^2\psi\tilde{T}P^*\beta \qquad [2.251]$$

Values of parameters χ_1 for infinitely diluted solution ($\varphi_2 \to 0$) and concentrated ($\varphi_2 \to 1$) are given by equations:

$$\chi_1(\varphi_1 \to 0) = r_1^0 \tilde{\rho}_1 (X_1 + 1/2 \psi_1^2 \tilde{T}_1 P_{*1} \beta_1) \qquad [2.252]$$

$$\chi_1^\infty(\varphi_2 \to 0) = r_1^0 [\tilde{\rho}_2 X_1 + (\tilde{\rho}_1 - \tilde{\rho}_2)/\tilde{T}_1 + (\tilde{v} - 1) \ln (1 - \tilde{\rho}_2) -$$

$$- (\tilde{v} - 1) \ln (1 - \tilde{\rho}_2) + \frac{1}{r_1^0} \ln (\tilde{\rho}_2 / \tilde{\rho}_1)] \qquad [2.253]$$

The theory considered in this part was tested for many polymer-solvent systems and a good agreement was found between theoretical predictions and experimental data.

2.8 SOME EXPERIMENTAL RESULTS OF THERMODYNAMIC PROPERTIES OF POLYMER SOLUTIONS

Presently a great deal of experimental works is available dedicated to the investigation of the thermodynamic properties of polymer solutions. They include the determination of the heat of solution and dilution, second virial coefficients, interaction parameters, Θ-temperatures, critical solution temperatures, etc. Many data are collected in the handbook[41] and the corresponding bibliography is given. In this section, we mention only the most important results showing the general regularities in the thermodynamic behavior of polymer-solvent systems.[42-49]

The most interesting results are the discovery of the existence of LCST[9,12,17,43-45] and negative entropy of mixing and enthalpy in the region of LCST.[10] These results have been predicted by the modern statistical theories.[1-7] The existence of LCST is related to the difference in the coefficients of thermal expansion between polymer and solvent, Delmas *et al.*[44] found a very good correlation between LCST and the relative density of mixing component (Fig. 2.5).

The simultaneous existence of both LCST and UCST was also observed.[17,42] With the worsening the thermodynamic quality of the solvent due to great differences in the free volumes of components, UCST and LCST draw together and form the phase diagram of the sand-glass type.[9,42] The increasing pressure leads to diminishing the difference between free volumes and as a result the solubility of a polymer in poor solvents increases.[10] It is typical that for such systems the UCST and LCST are the mirror reflections with the exception of the system with a broad molecular-mass distribution.[10] In many cases, a weak dependence of the critical concentration on the molecular mass was observed (see, for example, Fig. 1.8).

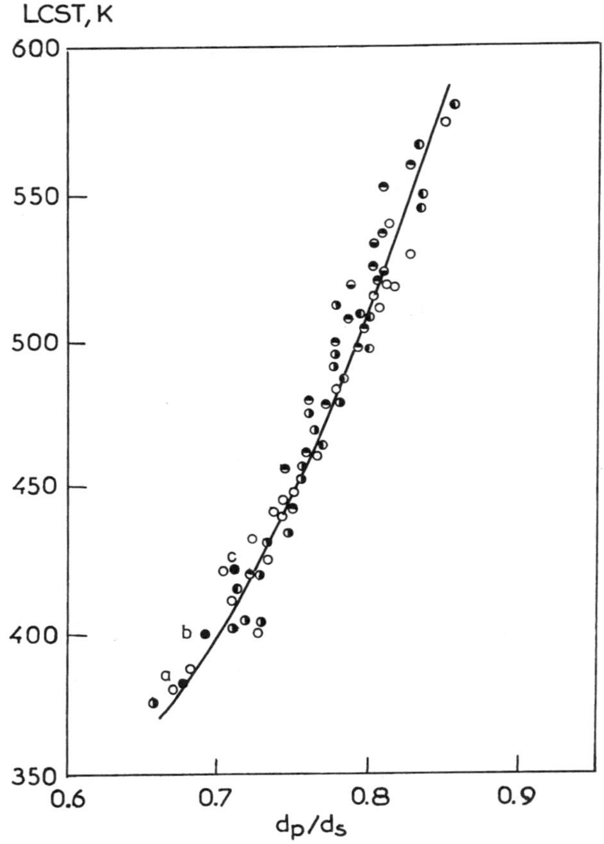

Fig. 2.5. General correlation of the polyolefines in linear and branched alkanes using the ratio of the solvent to polymer density to characterize their free volume difference: 1 - PP; 2 - PIB; 3 - PB-1; 4- P4MP-1; 5 - PP-1. Three points (●) have been added for LCST under pressure (a) PE in n-butane at 204 Atm, (b) PE in n-pentane at 80.5 Atm, and (c) PIB in pentane at 100 Atm. Adapted, by permission, from G. Charlet, R. Ducasse, and G. Delmas, *Polymer*, **22**, 1190 (1981).

Very interesting results were reported[36,45] regarding the solubility of PIB in linear and branched alkanes. It was found that the solvent quality increases with the length of the alkane molecule. This effect is related to a diminishing difference in free volumes between polymer and solvent. The substitution of linear alkane by branched alkane worsens solubility. The comparison of the results with predictions of the Prigogine-Patterson and Flory theories shows that better agreement between the theory and experiment can be obtained by using small values $S = s_2/s_1$ (0.5 – 0.8) for such chains as PS, PDMS, and others (thick chains) and high S (≈1) for thin chains such as PE or PBD.

The estimation of the interaction parameters for PIB with heptane, hexane, and octane has shown[46] that the solubility of the polymer increases with the increase in the solvent molar volume (its chain length). The dependencies of the interaction parameters and temperature are described rather well at high temperatures by the theory by Prigogine-Patterson and at low temperatures by the Flory theory. The analysis of the data on the interaction parameters[47] for the solutions of PDMS and PDES in various solvents in the temperature interval 333÷453 K shows that parameter, calculated in accordance with the Flory's theory, diminishes with temperature faster than experimental values, and has negative value at low temperatures. The interpretation of the data in the framework of the Prigogine-Patterson theory gives more reliable results.[47]

There has been also investigated the effect of branching on the polymer and deuteration of the solvent on the phase behavior of solutions.[48,50,51] It was found that branching of PS has small effect on the behavior of solution near LCST, but is rather strong in the region of UCST. The increasing degree of branching leads to a broadening the diagram of solubility of PS in cyclohexane, this effect is stronger when molecular mass of the polymer increases. In the deuterated solvent, for the ring-like polymer molecules, PS, the Θ-temperature is lower than Θ-temperature of linear chains.[50] In deuterated cyclohexane, the temperatures of the phase separation are higher as compared with protonated solvent. The phase separation curves show that effect of deuteration is connected only with the shift in Θ-temperature at which the double cluster integral in the mean-field theory disappears.[52] This shift is explained by the change in the polarizability of the solvent molecules due to deuteration.

Recently Dondos[53] has established the universal dependence between the reduced temperature, τ, ($\tau = 1 - \Theta/T$) and interactions in short and long range of the polymer (Fig. 2.6) in the form

$$\frac{\tau^2 m_s}{M_0} = 27.5a^{7.8} \qquad [2.254]$$

where m_S is the molecular mass of the statistical segments, M_0 is the molecular mass of the monomeric unit and a is the Mark-Kuhn-Houwink constant. Instead of m_S/M_0 one can take the parameter characterizing the chain rigidity $\sigma = (\overline{R_\Theta^2}/\overline{R_0^2})^{1/2}$ where R_Θ is the unperturbed

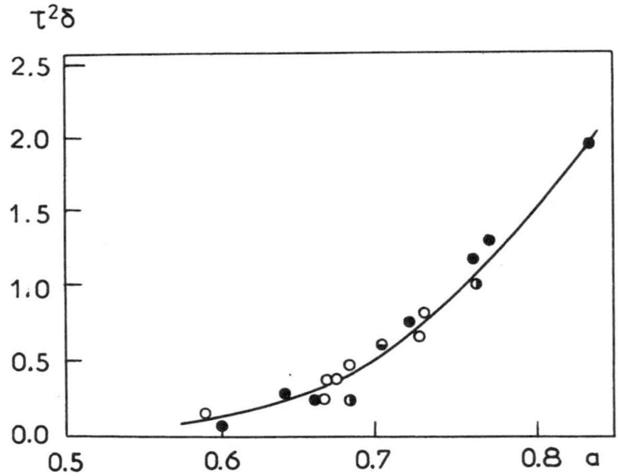

Fig. 2.6. Dependence of the reduced temperature of the polymer-solvent systems, multiplied by coefficient of the steric hindrance, of the polymers on the exponent of the MHS equation of the systems: ○ - PS, ● - PMMA, ◑ - PIB, ◒ - PmS. Adapted, by permission, from A. Dondos, *Macromolecules*, **25**, 6069 (1992).

radius of gyration, R_0 is the radius of gyration for the chain with free rotations around the bonds.

$$\tau^2\sigma = 10.6a^{8.4} \quad [2.255]$$

From this equation, it follows that if two polymers are in solution in different solvents having the same value of a, Θ-temperature for the more flexible polymer (lower σ) will be lower (higher value of τ^2) and *vice versa*. Equations 2.254 and 2.255 may be used to determine the Θ-temperature for any polymer-solvent system, if the exponents a and unperturbed dimensions of the chains in a given solvent are known. From τ^2 the rigidity parameter can also be found.

Parameter χ in Flory equation (Eq 1.43) is supposed to be independent of the concentration and molecular mass and proportional to T^{-1}. When this parameter is treated as phenomenological parameter, the comparison of the theory with experiment shows that this parameter does depend both on concentration and molecular mass and includes both energetic and entropic contributions. As a rule, the entropy contribution is larger than energetic one. The entropy corrections introduced by Huggins are very small and cannot explain the experimental results. Because of it, Koningsveld and

Kleijtiens[54-55] have introduced some additional parameters to describe the concentration dependence of χ, although introduced parameters have vague physical meaning. The Flory approach (the theory of equation of state) is a combination of a statistic mechanical model and thermodynamic phenomenology. It uses the combinatorial entropy of mixing (Eq 1.43) and a simple one-dimensional mechanical model describing the entropy contribution owing to the difference in the free volumes of components. The theory of equation of state also uses some phenomenological parameters included in the entropy contribution. In particular, this is done when considering the polymer molecules interacting with each other through these parts that are characterized by the "surface area" (fraction of sites Θ_i) having no clear physical meaning. The modification of Flory's theory (see, for example, refs. 24-26) also consists of the introduction of additional phenomenological parameters or use of some other (non-additive) rules of combination of the mixture parameters. Freed and Dudowicz[56] have noted that introduction of the phenomenological parameters of such kind in the mean-field lattice theories gives the equivalent results when describing experimental data. In a series of works, they developed the theory which, according to their opinion, is capable of explaining the molecular nature of the great entropy contribution to the phenomenological parameter χ.This theory allows also to explain the dependence of this parameter on the pressure, temperature, molecular mass, and mixture composition. In their works, they propose so-called cluster expansion of the lattice model. The monomeric units are considered to be the structures that can be disposed simultaneously on several sites of the lattice. The expression for the free energy is obtained as cluster expansion where the mean-field approximation is regenerated in the zero order. The proposed cluster theory deals with the specific structure of monomers and their interactions. Cluster expansion leads to a very simple, although very long, analytical expression that can be applied to any number of components, various molecular masses, monomer structures, compositions, temperatures, etc.. This theory is presently in its development and we refer the readers to the source[56] where the main postulates of the theory and all bibliography are given.

REFERENCES

1. I. Prigogine in *The Molecular Theory of Solutions*, Interscience, New York, 1959.
2. D. Patterson and G. Delmas, *Disc. Faraday Soc.*, **49**, 98 (1970).
3. D. Patterson, *Pure and Appl. Chem.*, **31**, 133 (1972).
4. D. Patterson, *J. Polym. Sci.*, **C16**, 3379 (1968).
5. (a) P. J. Flory, R. A. Orwoll, and A. Vrij, *J. Amer. Chem. Soc.*, **86**, 3507 (1964); (b) P. J. Flory, *J. Amer. Chem. Soc.*, **87**, 1833 (1965); 838 (1965).
6. B. E. Eichinger and P. J. Flory, *Trans. Faraday Soc.*, **64**, 2035 (1968).
7. D. Patterson, *Macromolecules*, **2**, 673 (1969).

8. W. H. Stockmayer, *J. Polym. Sci.*, **15**, 595 (1955).
9. G. Allen,C. H. Baker, *Polymer*, **6**, 181 (1965).
10. G. Delmas, D. Patterson, *Polymer*, **7**, 513 (1966).
11. P. Ehrlich and J. J. Kurpen, *J. Polym. Sci.*, **A1**, 3217 (1963).
12. D. Patterson, G. Delmas, and T. Somsinsky, *Polymer*, **8**, 503 (1967).
13. R. Simha and A. Havlik, *J. Amer. Chem. Soc.*, **86**, 197 (1964).
14. V. S. Nanda and R. Simha, *J. Phys. Chem.*, **68**, 3158 (1964).
15. V. S. Nanda, R. Simha, and T. Somsinsky, *J. Polym. Sci.*, **C12**, 277 (1966).
16. S. N. Bhattacharyya, D. Patterson, and T. Somsinsky, *Physica*, **30**,1276 (1964).
17. C. H. Baker, W. Brown, G. Gee, *et al.*, *Polymer*, **3**, 215 (1962).
18. V. M. Andreeva, A. A. Tager, I. S. Fominyh, and O. L. Zamaraeva, *Vysokomolek. soed.*, **A18**, 286 (1976).
19. Y. S. Lipatov, A. E. Nesterov, and T. D. Ignatova, *Reports Acad. Sci. USSR*, **229**, 1382 (1976).
20. A. E. Nesterov, *Ph. D. Dissertation*, Leningrad, 1966.
21. A. Bellemans and C. Naar-Colin, *J. Polym. Sci.*, **C15**, 121 (1955).
22. J. Biros, L. Zeman, and D. Patterson, *Macromolecules*, **4**, 30 (1971).
23. G. Delmas, D. Patterson, and T. Somsinsky, *J. Polym. Sci.*, **57**, 79 (1962).
24. J. A. R. Renuncio and J. M. Prausnitz, *Macromolecules*, **9**, 898 (1976).
25. V. Brandani, *Macromolecules*, **12**, 883 (1979).
26. V. Brandani, *Macromolecules*, **11**,1293 (1978).
27. K. Fujisawa, T. Shiomi, F. Hamada, and A. Nakajima, *Polymer Bull.*, **3**, 261 (1980).
28. F. Hamada, T. Shiomi, K. Fujisawa, and A. Nakajima, *Macromolecules*, **13**, 729 (1980).
29. K. K. Chee, *Polymer. J. (Japan)*, **23**, 321(1991).
30. K. Sugamiya, *Makromol. Chem.*, **178**, 565 (1977).
31. J. Dayantis, *J. Polym. Sci.*, **C39**, 35 (1972).
32. R. A. Orwoll and P. J. Flory, *J. Amer. Chem. Soc.*, **89**, 6814 (1967).
33. V. Crescenzi and G. Manzini, *J. Polym. Sci.*, **C54**; 315 (1976).
34. A. Bondi in *Physical Properties of Molecular Crystals, Liquids and Glasses*, Wiley, New York, 1968.
35. M. L. Huggins, *J. Phys. Chem.*, **74**, 371 (1970).
36. M. L. Huggins, *Polymer*, **12**, 389 (1971).
37. M. L. Huggins, *J. Phys. Chem.*, **75**, 1255 (1971).
38. M. L. Huggins, *Macromolecules*, **4**, 274 (1971).
39. M. L. Huggins, *Brit. Polymer J.*, **9**, 189 (1977).
40. I. Sanchez and R. H. Lacombe, *Macromolecules*, **11**, 1145 (1978).
41. A. E. Nesterov in *Handbook on Physical Chemistry of Polymers*, V. 1, Naukova Dumka, Kiev, 1984.
42. N. Kuwahara, S. Saeki, T. Chiba, and M. Kaneko, *Polymer*, **15**, 777 (1974).
43. G. Delmas and P. de Sant-Romain, *Eur. Polymer J.*, **10**, 1133 (1974).
44. G. Charlet, R. Ducasse, and G. Delmas, *Polymer*, **22**, 1190 (1981).
45. G. Charlet and G. Delmas, *Polymer*, **22**, 1181 (1981).
46. P. Marcille, R. Audebert, and C. Quivoron, *C. r. Acad. Sci.*, **C277**, 9 (1973).
47. M. Galin, *Macromolecules*, **10**, 1239 (1977).
48. J. M. G. Cowie, A. Horta, I. J. McEwen, and K. Prochazka, *Polymer Bull.*, **1**, 319, (1979).
49. M. Huglin, I. Idris, and M. B. Sokro, *Makromol. Chem. Rapid Commun.*, **2**, 17 (1981).
50. J. Huang, J. Chen, Ch. Li, and D. Liu, *Makromol. Chem.*, **192**, 1249 (1991).
51. I. Ichimura, K. Okano, K. Kurita, and E. Wada, *Polymer J. (Japan)*, **19**, 1101 (1987).
52. H. Yamakawa in *Modern Theory of Polymer Solutions*, Harper and Row,New York, 1971.

53. A. Dondos, *Macromolecules,* **25**, 6069 (1992).
54. R. Koningsveld and L. A. Kleintjens, *Macromolecules,* **4**, 637 (1971).
55. L. van Opatul, R. Koningsveld, and L. A. Kleintjens, *Macromolecules,* **24**, 161 (1991).
56. F. Freed and J. Dudowicz, *Theoretica Chimica Acta,* **82**, 357 (1992).

Chapter 3

THERMODYNAMICS OF POLYMER MIXING

3.1 BINARY MIXTURES

One of the main tasks of the investigation in the field of thermodynamics of mixing is to find the conditions of mixing and ranges of mutual solubility of different polymers depending on their chemical nature and molecular mass. Polymer processing proceeds as a rule at elevated temperatures, where mixed components are viscous liquids. Their mixing should be considered as mixing of two liquids. According to the general principles of thermodynamics, the formation of thermodynamically stable system should be accompanied by a decrease in Gibbs free energy, ΔG:

$$\Delta G = \Delta H - T\Delta S \qquad [3.1]$$

The necessary but not sufficient condition for the system stability is that $\Delta G < 0$, which is fulfilled if $\Delta H < 0$ and $T\Delta S > 0$, or if $\Delta H > 0$ but $|T\Delta S| > |\Delta H|$.

If there is a possibility to determine ΔG for all possible mixture compositions, temperature and composition corresponding to one-phase, stable system can be calculated. However, experimental determination of ΔG for polymer mixtures is very complicated, because many methods used for this purpose in polymer-solvent systems are not applicable to polymer mixtures. It was proposed[1] that during mixing of two polymers having a high molecular mass, the change in enthalpy is responsible for mixing considering that contribution of entropy of mixing to ΔG should be extremely small. For this case, the system was considered miscible (compatible) if $\Delta H < 0$ and immiscible (incompatible) if $\Delta H > 0$. As a rule for polymer solution in low-molecular mass liquids, the condition $\Delta H < 0$ corresponds to the condition of miscibility. However, for polymer blends, Slonimsky[2,3] showed many years ago that this condition is not true and that contribution of entropy cannot be neglected. Calculations made by Flory[4] have shown that by mixing entropy diminishes, i.e., $\Delta S < 0$. The contribution of entropy of mixing into the mutual solubility was for the first time evaluated experimentally by Tager.[5-8] It was found that entropy of

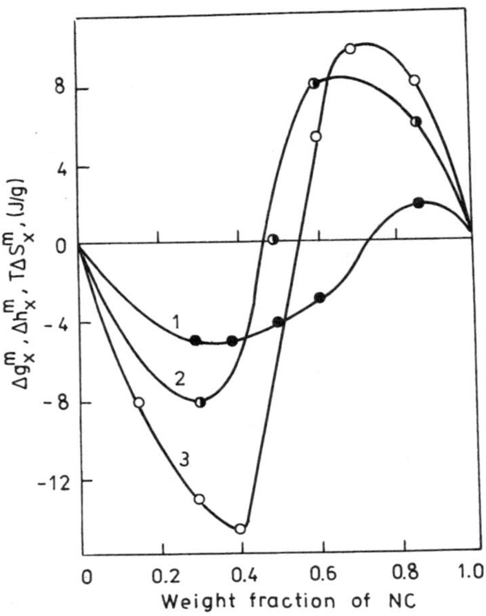

Figure 3.1. Gibbs free energy Δg_x^m (1), enthalpy Δh_x^m (2) and entropy ΔTS^m(3) of mixing (I/g) as a function of cellulose acetate-cellulose nitrate mixture composition.

mixing of two polymers is not equal to zero but may be rather high and have various signs (Fig. 3.1). For miscible polymers,[8] $\Delta S < 0$.

In many cases, to estimate miscibility of two polymers, there were used approximations of the theory of regular solutions. Presently to describe thermodynamic properties of polymers new statistical theories of polymer solutions are applied which have been developed by Prigogine, Patterson, Flory, and others.[8-11]

The main characteristics of polymer blends is thermodynamic compatibility or incompatibility of the components which is governed by the phase diagram of the system. It is necessary to emphasize from the very beginning that such consideration in principle is not correct because compatibility depends on many factors and first of all on the temperature and composition. There are almost no fully incompatible or compatible pairs of polymers but there are various conditions at which polymer pair is miscible or immiscible. However, depending on the chemical constitution of mixed components, the window of miscibility may be very broad, or, on the contrary, very narrow.

The term "compatibility" is widely used in literature. However, this term is used also to describe good adhesion between the constituents, average of mechanical properties and other properties. It is thus more correct to describe the thermodynamic behavior of polymer mixtures by using the term "miscibility". According to Olabisi, Shaw, and Robeson[12] the term "miscibility" was chosen to describe polymer-polymer mixtures with behavior similar to that expected of a single-phase system. However, it does not imply ideal molecular mixing but suggests that the level of molecular mixing is adequate to yield macroscopic properties expected of a single-phase material. Miscible polymer-polymer pair is a solution of one polymer in another. Mutual miscibility does not imply that solubility of segments is necessary related to the statistical arrangements of segments. Interactions between similar and dissimilar macromolecules can result in formation of molecular aggregates or associates (clusters) or in other non-statistical arrangements of segments.

3.1.1 APPLICATION OF THE THEORY OF REGULAR SOLUTIONS TO THERMODYNAMICS OF POLYMER MIXING

For the first time the theoretical consideration of the process of mixing of two polymers was done by Scott[13] and Tompa.[14] They used Flory-Huggins equation for the Gibbs free energy of mixing:

$$\Delta G = (RT/V_s)\left[\frac{\varphi_A}{r_A}\ln\varphi_A + \frac{\varphi_B}{r_B}\ln\varphi_B + \chi_{AB}\varphi_A\varphi_B\right] \qquad [3.2]$$

where V is the total volume of mixture, V_s is the volume of segment taken equal to the volume of repeating unit of the polymer chain (the same for both polymers), r_A and r_B are the numbers of segments of polymer A and B, $r_i = V_i/V_s$ (V_i is molar volume of component i), φ_A and φ_B are volume fractions of components, χ_{AB} is a thermodynamic interaction parameter related to the enthalpy of interaction between different segments of volume, V_S. The mixture is at a critical point when

$$\frac{\delta^2\Delta G}{\delta\varphi_A^2} = \frac{\delta^3\Delta G}{\delta\varphi_A^3} = 0 \qquad [3.3]$$

Critical values of concentration and interaction parameter are equal, correspondingly, to

$$(\varphi_A)_{crit} = 1/[1 + (r_B/r_A)^{1/2}] \qquad [3.4]$$

$$(\chi_{AB})_{crit} = (1/2)[r_A^{-1/2} + r_B^{-1/2}]^2 \qquad [3.5]$$

As follows from Eq 3.5, the value $(\chi_{AB})_{crit}$ at high molecular masses of mixing polymers should be very small (near 0) and at $M \to \infty$, polymers are immiscible at any, even very small, positive heat of mixing. The equation for spinodal has the form:

$$\frac{\delta^2 \Delta G}{\delta \varphi_A^2} = 0; \quad (\chi_{AB})_{sp} = \frac{1}{2}\left[\frac{1}{r_A \varphi_A(sp)} + \frac{1}{r_B \varphi_B(sp)}\right] \qquad [3.6]$$

where index "sp" relates to spinodal. If to suppose that chemical potentials of polymers in two coexisting phases are equal, the general equations for binodal can be obtained:

$$(\mu_A)' - (\mu_{A_0})' = (\mu_A)'' - (\mu_{A_0})''; \quad (\mu_B)' - (\mu_{B_0})' = (\mu_B)'' - (\mu_{b_0})'' \qquad [3.7]$$

$$\mu_A - \mu_{A_0} = \ln \varphi_A + (1 - r_A/r_B)\varphi_B + \chi_{AB} r_A \varphi_B^2$$

$$\mu_B - \mu_{B_0} = \ln \varphi_B + (1 - r_A/r_B)\varphi_A + \chi_{AB} r_B \varphi_A \qquad [3.8]$$

where $\mu_i - \mu_{i0}$ are chemical potentials of i-th component of the mixture. To use Eqs 3.4 and 3.5 parameter χ_{AB} is calculated from the relation:

$$\chi_{AB} = (V_s/RT)(\delta_A - \delta_B)^2 \qquad [3.9]$$

where δ_A and δ_B are solubility parameters of components. From Eq 3.9 is seen that calculated parameter is always positive. Because value $(\chi_{AB})_{crit}$ is close to 0 and contribution of combinatorial entropy (first two terms of Eq 3.2) is negligible small, the Scott's approach (Eqs 3.2 and 3.9) predicts that it is almost impossible to find a miscible polymer pair. Only when δ_A and δ_B are almost equal, the calculated value of χ_{AB} may be below the critical value. On this basis, it was proposed[15,16] to predict miscibility of two polymers in a solid state. If $\chi_{AB} < (\chi_{AB})_{crit}$, both polymers are miscible at the same component ratio which was taken for calculations of interaction parameter. Some results[16] of calculations of χ_{AB} are given for which $\Delta H < T\Delta S$.

To predict the miscibility in a solid state another approach was used[17] based on the relation proposed by Gee[18] for the heat of mixing of rubbers by swelling:

$$\Delta H_{mix} = [V_0(\delta_0 - \delta_2)(1 - \varphi_o)^2]^{1/2} \qquad [3.10]$$

where V_0 and δ_i are molar volume and solubility parameters of liquid and rubber and $(1 - \varphi_0)$ is the volume fraction of the swollen rubber. For polymer mixtures, Schneier[16] proposed the equation:

$$\Delta H_{mix} = w_1 M_1 \rho_1 (\delta_1 - \delta_2)^2 \left[\frac{w_2 M_2 \rho_2}{(1 - w_2)} - (1 - w_1) M_1 \rho_1\right] \qquad [3.11]$$

where M_i and ρ_i are molecular mass of monomeric unit and polymer density. To check this equation, the data[20] had been used. It was established that for miscible polymers, values of ΔH_{mix} are in the range from 10^{-3} to 10^{-2} J. In spite of satisfactory qualitative agreement between experimental data on miscibility with the results of theoretical calculations, some exceptions were found: for some systems values ΔH_{mix} were less than 10^{-2} J and polymers in a solid state were immiscible.[20] To describe the miscibility, Shaw[21] used two-dimensional solubility parameter. The use of multidimensional solubility parameters was found necessary to describe a phase behavior of mixtures of polymers with high dipole moment.[22] Having accepted that polymers are capable of forming regular solutions and that the main interacting element is a repeating unit of polymer chain, Shaw found the condition of equality of chemical potential of component A in phases I and II in the form:

$$\ln \varphi_{A_1} + (1 - V_1/V_2)\varphi_{A_1} + V_1(\delta_1 - \delta_2)^2/RT\varphi_{A_1}^2 = \ln \varphi_{B_1} +$$

$$+ (1 - V_1/V_2)\varphi_{B_1} + V_1[(\delta_1 - \delta_2)^2/RT\varphi_{B_1}^2 \qquad [3.12]$$

Analogous equation can be written for the component B. The first two terms of Eq 3.12 determine the entropy of mixing of regular solution, third term is connected with interactions. Eq 3.10 practically does not differ from the equation of Flory-Scott for binodal (Eq 3.8) and predicts an increase in solubility with increase in temperature. If to suppose that energy of interaction is connected only with dispersion and dipole-dipole forces, the enthalpy of interaction can be written as $(\delta_{d1}^2 - \delta_{d2}^2) - (\delta_{p1}^2 - \delta_{p2}^2)$, where subscript d relates to dispersion and p to polar interactions. In this case of high mutual solubility of two polymers, there is a good correlation between the polar and dispersion components of cohesion energy of both polymers. The analysis of this approach was demonstrated for the polymer-solvent system.[22] To calculate δ_p, the scheme is used[22] according to which

$$\delta_p^2/\delta^2 = \frac{2\mu^4/3kT}{2\mu^4/3kT + 3I\alpha^2/4 + 2\alpha v^2} \qquad [3.13]$$

where μ and α are dipole moment and polarizability, I is first ionization potential of molecule (used to be taken equal to 10 eV). Parameter δ is calculated in accordance with experimental data.[23] The evaluation of this approach has shown that with molecular mass increasing, the solubility decreases at fixed values of solubility parameters in accordance to Eq 3.12. However, temperature investigations have given contradictory results. Based on Scott approximation (Eq. 3.2) and taking interaction parameter as

$$\chi_{AB} = A + B/RT \qquad [3.14]$$

Nishi[24] proposed equation for calculation of spinodal as:

$$\frac{1}{T_S} = \frac{1}{T_{crit\infty}}\left[1 - \frac{1}{2A}\left(\frac{1}{\varphi_{BS} r_B} + \frac{1}{\varphi_{AS} r_A}\right)\right] \qquad [3.15]$$

where T_{crit} = -B/AR corresponds to the critical temperature of mixing of polymers with infinite molecular mass. T_{crit} and $T_{crit\,\infty}$ are connected at critical conditions by the relationship:

$$\frac{1}{T_{crit}} = \frac{1}{T_{crit\infty}}[1 - (1/2A)[(r_A)^{-1/2} + (r_B)^{-1/2}]] \qquad [3.16]$$

Combining Eqs 3.7, 3.8, and 3.14, one can obtain:

$$\frac{1}{T_S} = \frac{1}{T_{s,\infty}}\left[1 - \frac{1}{2A}(\varphi'_B - \varphi''_B)\right]\left[\frac{1}{r_B}\ln\frac{\varphi'_B}{\varphi''_B} - \frac{1}{r_A}\ln\frac{\varphi'_A}{\varphi''_A}\right] \qquad [3.17]$$

where φ'_i and φ''_i are calculated from

$$2[1/r_A - 1/r_B](\varphi'_B - \varphi''_B) + [(2 - \varphi'_B - \varphi''_B)(1/r_A)\ln\frac{(1 - \varphi'_B)}{(1 - \varphi''_B)} +$$

$$+ \frac{\varphi'_B + \varphi''_B}{r_B}\ln\frac{\varphi'_B}{\varphi''_B} = 0 \qquad [3.18]$$

Critical values of T_{crit} and φ_{crit} can be found from the phase diagrams, if molecular-mass distribution is rather narrow. Eqs 3.4 and 3.16 are transformed into

$$\frac{1}{T_{crit}} = \frac{1}{T_{crit,\infty}}\left[1 - \frac{1}{2A}\left(\frac{1}{\varphi_{B,crit}}\right)\right]^2 \qquad [3.19]$$

where T_{crit} and $2Ar_A$ may be found from the dependence $1/T_{crit} = 1/(1 - \varphi_{B,crit})$. The comparison of calculated spinodals, binodals, and critical points with experimental data for the polystyrene-poly(vinyl methyl ether) system, having LCST, has shown satisfactory agreement (Fig. 3.2).

The approximation of Hildebrand-Scott[25] was used[26] to find the interrelation between excess entropy of mixing at a constant pressure with excess volume of mixing:

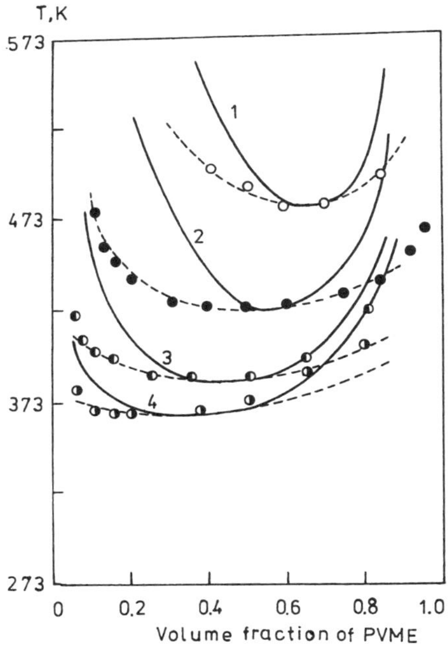

Figure 3.2. Calculated (____) spinodal and (- - - -) binodal curves for PS-PVME. $T_{c\infty}K$=343K, A m_A =14.29. Molecular weight PVME = 5.15×10^4, PS =1×10^4 (1), 2.04×10^4 (2), 5.1×10^4 (3) and 2×10^5 (4) Points are experimental. Adapted, by permission, from T. Nishi, *J. Macromol. Sci.-Phys.*, **B17**(3), 517 (1980).

$$\Delta S_p = \Delta S_v + (\delta P/\delta T)_v + \ldots \qquad [3.20]$$

Value $(\Delta S)_V$ is considered as a combinatorial entropy of mixing:

$$(\Delta S)_v = -R(n_A \ln \varphi_A + n_B \ln \varphi_B) \qquad [3.21]$$

where n_A and n_B are numbers of moles per unit volume, φ_A and φ_B are volume fractions of polymers A and B. It was found that $(\Delta S)_V$ for the 1:1 mixture of two polymers(polystyrene) and poly(2,4-phenylene-1,6-oxide) with M = 20,000 is equal to 5.85×10^{-4} J/cm^{3}K. Taking $T(\delta P/\delta T)_V \approx (\delta E/\delta V)_T \approx 347$ J/cm^3 and $\Delta V = 7\times10^{-3}$ cm^3/g, it was found that $(\delta P/\delta T)_V$ V = 8.36×10^{-3} J/cm^3 K (at 300 K). This means that one can neglect the value $(\Delta S)_V$. Then Eq 3.20 transforms to:

$$(\Delta S)_p = (\delta P/\delta T)_v \Delta V \qquad [3.22]$$

Since the value $(\delta P/\delta T)_V$ is positive, then

$$(\Delta S)_p \geq \leq 0 \text{ if } \Delta V \geq \leq 0 \qquad [3.23]$$

Eqs 3.2÷3.9 are widely used for calculations of parameters of thermodynamic interactions or Gibbs free energy, if the phase diagrams are known. Phase diagrams can be constructed if the values of interaction parameters are known. Eq 3.2 shows that because of very small values of combinatorial entropy of mixing (first two terms of equation) the miscibility of two high molecular weight polymers may be observed only when the values of χ_{AB} are close to 0. The limiting values of χ_{AB}, at which the mixing of two polymers with various molecular masses occurs, were discussed by Krause.[27,28] It was also shown[29] that to reach molecular mixing of two polymers with polymerization degrees $r_A = r_B = 1000$, values of χ_{AB} should be close to 0.002, the difference between solubility parameters (Eq 3.9) should be less than 0.42 $(J/cm^3)^{1/2}$. Therefore in systems with dispersion or weak dipole-dipole interactions, for which Eq 3.9 was derived, the molecular mixing of two polymers is a very rare case. To get miscibility of two high molecular polymers, the existence of specific intermolecular interactions such as hydrogen bonding, etc. is necessary. In this case, the unfavorable contribution of the third term in Eq 3.2 can be overcome. To take into account these favorable interactions, the third additional term was introduced[29] to modify Eq 3.2:

$$\frac{\Delta G}{RT} = \frac{\varphi_A}{r_A} \ln \varphi_A + \frac{\varphi_B}{r_B} \ln \varphi_B + \varphi_A \varphi_B \chi_{AB} + \frac{\Delta G_H}{RT} \qquad [3.24]$$

If favorable interactions are prevailing over unfavorable ones, the value of the excess Gibbs free energy of mixing becomes negative and miscibility of two polymers becomes more probable. However, in general, miscibility of two polymers should be the exception rather than the rule. Due to small entropy gains in mixing two polymers, relatively small unfavorable interactions between different monomers will generally result either in macroscopic or microphase separation. The cases of miscible pairs are discussed by us later.

3.1.2 APPLICATION OF NEW STATISTICAL THEORIES FOR DESCRIBING THERMODYNAMIC PROPERTIES OF POLYMER MIXTURES

MacMaster[30] was the first to use the new Flory theory[11] to describe thermodynamic behavior of polymer mixtures. On the basis of the modified equation-of- state, Flory derived expressions for chemical potentials, spinodals and critical points for binary mixtures of monodisperse and polydisperse polymers. In the latter case, the results by Koningsveld and collaborators[31] have been used. The equations obtained are too bulky and are not included here.

Using Flory's approach, MacMaster calculated binodals and spinodals for model systems: polystyrene-poly(vinyl methyl ether), co-polystyrene-acrylonitrile (mass fraction of acrylonitrile - 0.28)-polycaprolactone and compared calculated data with experimental results obtained from the cloud point curves. The results gave possibility to estimate the influence of various parameters of the equation-of- state on the miscibility. The main conclusions are the following:

- With increasing molecular mass of any component, the miscibility of two polymers decreases
- With decreasing the ratio of r_A/r_B, the critical composition on the phase diagram shifts stepwise to a lower concentration of polymer with higher molecular mass. This conclusions follows from the Flory-Huggins theory for polymer solutions

The change in the coefficient of thermal expansion leads to the change of all parameters of the equation of state: at T = 300K. An increase in α by 6% causes that the value of T^* diminishes by 4.6%, P^* increases by 2%, v^* decreases by 1%. To form miscible system of two polymers with molecular masses of about 200,000, the difference in their α should not exceed 4%. For polymer mixtures with molecular masses about 50,000, this difference should not exceed 10%. At very high molecular masses, the negligible change in α of one of the components may strongly influence their miscibility.

The change in the coefficient of isothermal compressibility,γ, affects only the value of P^*. In the case $\alpha_A < \alpha_B$, the differences in equation-of-state can be compensated by decreasing γ_B as compared with γ_A. Decreasing γ_B by 10% increases the critical point of mixing by 250K. The same is observed with changing γ_A: if other parameters are constant, increasing one and decreasing another value increases miscibility. Miscibility also improves due to the increased pressure.

The effect of the parameter of contact interactions, X_{AB}, is very essential. If it is small and positive, the existence of two critical solution temperatures (LCST and UCST) is possible. Increasing positive value of X_{AB}, results in merging two CSTs and in appearance of the phase diagram similar to sand-glass. The temperature at which both critical temperatures merge depends on the difference in α and increases with the decrease in $\alpha_A - \alpha_B$.

If parameter X_{AB} is negative (it may be observed when specific interactions operate, e.g., hydrogen bonding, donor-acceptor interactions, etc.), two polymers can be miscible even if their equations of state are very different. The calculations have shown that simultaneous existence of both UCST and LCST for polymer mixtures is a very rare case.

The comparison of calculated and experimental data on cloud points for the model systems revealed a very low sensitivity of cloud points to the change in temperature. MacMaster suggested that perceived insensitivity may also result from insufficient accuracy of equation-of-state for given components (i.e., parameters of state for model system), component polydispersity, and the probability of specific interactions. One of the main results, according to McMaster, is experimental verification of the existence of LCST in polymer mixtures.

The equations for spinodal obtained by McMaster were used[32] to calculate the phase behavior of the poly(caprolactone)-PVC mixture. It was shown that if only the value of X_{AB} was changed, a bimodality of phase diagram, having a large negative values, was obtained. However, if both X_{AB} and s_B/s_A are changed, it results in asymmetry of binodal. At constant X_{AB} and variable ratio of s_B/s_A, the mirror reflection of initial binodal is seen. These results show the importance of parameters of X_{AB} and s_B/s_A in describing the phase behavior of polymer mixtures (it should be noted that the value s_B/s_A is usually taken equal to 1).

The equations proposed by McMaster are too complex for practical applications. Moreover, as shown by Patterson and Robard,[33] McMaster overestimated the contribution of the term of free volume in parameter χ_{AB}. Thus, Patterson and Robard used the initial idea of the Flory-Prigogine theory in more simple approximation similar to the one used for describing the phase diagrams of polymer-solvent systems at usual and elevated pressures.[36,37] The main attention was paid to interactions favoring the mixing and stabilizing at low temperatures. It was supposed that the initial reason for the appearance of LCST is not connected with unfavorable contribution of free volume, as typical for polymer-solvent system. The essence of their approximation is the following. Using expressions of Prigogine-Flory theory[9-11] for the thermodynamic parameter of interaction, χ_{AB}, normalized to molar volume of component A:

$$\chi_{AB} = -\frac{U_A}{RT\nu}\gamma^2 + \frac{C_p}{2R}\tau^2 \qquad [3.25]$$

where:

$$-U_A = P_A^* V_A^* \tilde{V}_A^{-1} \qquad [3.26]$$

$$C_p = \frac{P_A^* M_A^* V_A^* \tilde{V}_A^{-1/3}}{T_A^*}(4/3 - \tilde{V}_A^{1/3})^{1/2} \qquad [3.27]$$

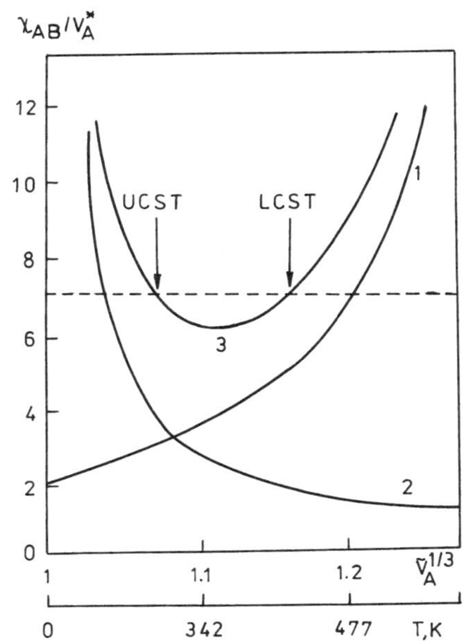

Figure 3.3. Dependence of χ_{AB}/v_A^* on T for typical polymer-solvent system with T_A^*=5000K, P_A^*=400 J/cm^3 and T_B^*=8191K (PS); χ_{AB} = 10 J/cm^3. 1 - contribution of the free volume, 2 - contribution of the interaction term, 3 - the sum of two contributions. Dashed line-critical value of parameter χ_{AB}/v_A^*, calculated for v_A^*=70 cm^3/mol and $M_{B\to\infty}$. Adapted, by permission, from D. Patterson and A. Robard, *Macromolecules*, **11**, 690 (1978).

Neglecting the concentration dependence of X_{AB}, Patterson and Robard derived the following equation for the parameter χ_{AB}:

$$\frac{\chi_{AB}}{M_A v*_A} = \frac{P_A^*}{RT_A^*}\left[\frac{\tilde{V}_A^{1/3}}{\tilde{V}_A^{1/3}-1}\left(\frac{X_{AB}}{P_A^*}\right) + \frac{\tilde{V}_A^{1/3}}{2(4/3-\tilde{V}_A^{1/3})}\tau^2\right] \qquad [3.28]$$

To satisfy the condition of symmetry, i.e., $\chi_{ij} = \chi_{ji}$, parameter χ_{AB} is related to the volume unit. If to take into account parameter Q_{AB} introduced by Flory to correct entropy, the term $-Q_{AB}/R$ should be added to Eq 3.28, or X_{AB} should be replaced by equivalent value $(X_{AB} - TQ_{AB}\tilde{V}_A^{1/3})$. Eq. 3.28 for $\chi_{AB}/M_A v_A{}^*$ shows that the term of interaction tends to attain infinity, if $\tilde{V}_A^{1/3}$ diminishes to 1 and T drops to almost 0 K. From another part, the term of free volume tends to infinity when $\tilde{V}_A^{1/3}$ attains 4/3, i.e., the temperature approaches the critical temperature of liquid-vapor for component A. The value $\tilde{V}_A^{1/3}$ term for polymers is close to 1 and therefore one should expect the prevalence of interaction term in χ_{AB}. For dilute solutions of polymers in liquids with low molecular mass in which $\tilde{V}_A^{1/3}$ may reach 4/3, the free volume term is prevailing. This means that the behavior of the systems polymer-solvent and polymer-polymer should be essentially different with changing temperature.

Fig. 3.3 shows the dependence of $\chi_{AB}/M_A v^*_A$ on $\tilde{V}_A^{1/3}$ and T for typical polymer-solvent system, calculated according to Eq 3.28. Value χ_{crit} is calculated according to Eq 3.2 at $M \to \infty$ for the polymer and $M_A v_A^* = 70$ cm^3/mol for the solvent. The system is characterized by the existence of both UCST and LCST between which the region of full miscibility exists. Although both parameters (interaction parameter and free volume) are important for χ_{AB}, the existence of UCST is determined exclusively by increasing unfavorable for mixing term of interaction (X_{AB} is positive) with decreasing temperature, whereas LCST is connected with increasing free volume term with rising temperature. If X_{AB} and/or τ are increased, S-shape curve of $\chi(T)$ should be shifted up along the temperature axis and UCST joins LCST. In this case, the diagram of sand-glass type is obtained. If molecular mass of polymer diminishes, χ_{crit} decreases along temperature axis and the region of full miscibility increases.

Dependencies of $\chi_{AB}/M_A v_A$ on $\tilde{V}^{1/3}$ and T (Fig. 3.4) have been studied for systems where both components are polymers with $\chi_{AB} = 10$ J/cm^3 = 100 atm (typical value for the systems with dispersion interactions only). The value $\tilde{V}^{1/3}$ was taken lower than in Fig. 3.3 to describe the mixtures of both polymers and oligomers. Critical values of χ_{crit} have been calculated for various molecular masses at $v^*_A = v^*_B = 1$ cm^3/g. As it should be expected, the behavior of this system is determined mainly by the interaction term, i.e., the curves for $\tau \approx 0$ are close to the curves for $\tau = 0$. For $M = 10^4$, the system is unstable in the temperature interval considered, whereas for $M = 10^3$ and $\tau = 0$, the system stabilizes with increasing temperature. With increasing τ, miscibility decreases and at a rather high temperature, LCST should appear. However, curves $\chi(T)$ are very slope, thus diminishing χ_{AB} should be observed up to very high temperatures. The contribution of the term of free volume is always positive and therefore miscibility may exist at negative values of interaction term, i.e., at $X_{AB} < 0$. Fig. 3.5 shows the dependence of $\chi_{AB}/M_A v^*_A$ on $\tilde{V}_A^{1/3}$ and T for miscible polymer pair of high molecular mass, where $\chi_{crit} = 0$ and $X_{AB} < 0$. The negative value of χ_{AB} meets condition of miscibility whereas the positive one meets condition of immiscibility. When $X_{AB} < 0$, the interaction term, favorable for mixing, may be balanced by the unfavorable free volume term, equating both terms in Eq 3.15, we obtain:

$$X_{12}/P_1^*/\tau^2 = (\tilde{V}_1^{1/3} - 1)/2(4/3 - \tilde{V}_1^{1/3}) \approx 1/7 \qquad [3.29]$$

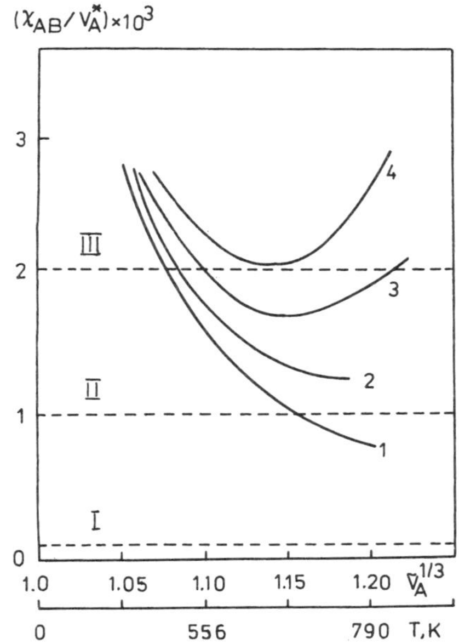

Figure 3.4. Temperature dependence of χ_{AB}/v_A^* on T for polymer-polymer mixture with $M_A = M_B$, $v_A^* = v_B^* = 1$ g/cm^3, $T_A^* = 8191$K, $P_A^* = 500$ J/cm^3, $\chi_{AB} = 10$ J/cm^3 and various τ: 0 (1), 0.1 (2), 0.15 (3), 0.2 (4). Horizontal dashed lines are critical values of χ_{AB}/v_A^* for mixtures with $M_A = M_B$: 1×10^4 (I), 2×10^3 (II), 1×10^3 (III). Adapted, by permission, from D. Patterson and A. Robard, *Macromolecules*, **11**, 690 (1978).

In the case $\tau \approx 0.15$ and $P_A^* = 400$ J/cm^2, X_{AB} has sufficiently low value (about 2 J/cm^3) for miscible system. More strictly, the value $(X_{AB} - \tilde{V}_A^{1/3}/3Q_{AB})$ should be negative.

S-shaped curve of $\chi(T)$ is typical for polymer pairs (Fig. 3.5). At low temperature, the system is miscible. However, with rise in temperature favorable interaction term decreases rapidly until its value is exceeded by unfavorable free volume for mixing term. At $\chi_{AB} = 0$, the LCST should be observed. In contrast to LCST for the polymer-solvent system, for polymer pairs, LCST is determined first of all be the interaction term which diminishes with temperature.

Differentiating Eq 3.25, we can find the ratio of temperature dependencies of both contributions to $d\chi/dT$ at LCST:

$$\frac{d\chi_{inter}}{dT} \Big/ \frac{d\chi_{free\ vol}}{dT} = \frac{3(4/3 - \tilde{V}_A^{1/3})}{4(\tilde{V}_A^{1/3} - 1)} \approx 3 \qquad [3.30]$$

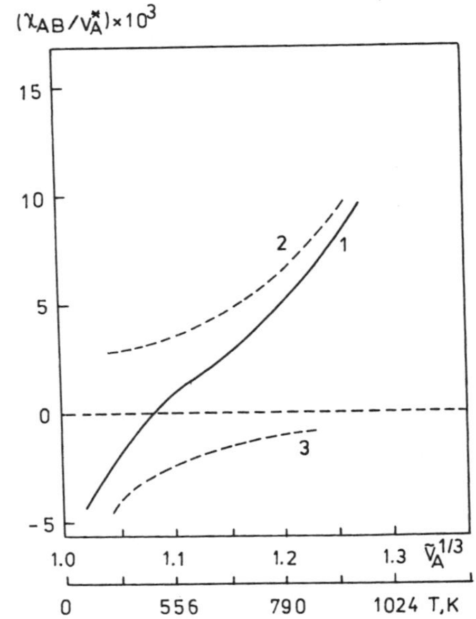

Figure 3.5. Temperature dependence of χ_{AB}/v_A^* on T for typical polymer-solvent system with $T_A^* = 8191$K, $P_A^* = 500$ J/cm^3 and $T_B^* = 7201$K (PVME); $\chi_{AB} = -1.3$ J/cm^3. 1 - contribution of the free volume, 2 - contribution of the interaction term, 3 - the sum of two contributions. Critical value of χ_{AB}/v_A^* is 0 for high molecular weight polymers. Adapted, by permission, from D. Patterson and A. Robard, *Macromolecules*, **11**, 690 (1978).

Because $\tilde{V}_A^{1/3}$ for polymers is close to 1, the interaction term changes more than free volume term. The role of the latter is less pronounced than it follows from MacMaster work[29] where he supposed that in majority of cases $\chi_{AB} = 0$.

If $Q_{AB} < 0$, the interaction is characterized by the effective value of $(X_{AB})_{eff} = (X_{AB} - TQ_{AB}\tilde{V}_A^{1/3})$. With increasing temperature this parameter changes from negative to positive, leading to more rapid, as compared with $Q_{AB} = 0$, temperature effect on χ. The role of the free volume term should be less than at $Q_{AB} = 0$. It may happen that LCST is below UCST. Such cases were observed in some polymer-solvent systems.[38,39]

Let us now consider the values τ and X_{AB} at which typical polymer systems should be miscible or immiscible with changing molecular mass. As an example, we take polystyrene with $P_A^* = 500$ J/cm^3, $T_A^* = 1891$ K, $\tilde{V}_A^{1/3} = 1.067$ at 408 K, $v^*_A = v^*_B = 1$ cm^3/g. Values τ for various M at constant X_{AB} have been obtained by equating Eq 3.25 to critical value of χ_{AB} calculated according to Eq 3.5 (Fig. 3.6).

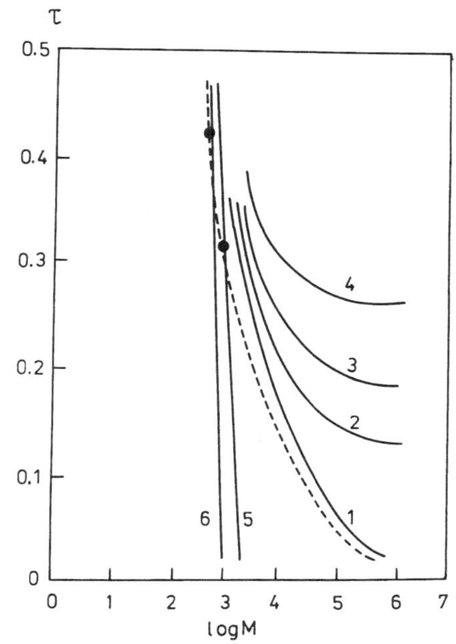

Figure 3.6. Critical values of T corresponding to phase separation for polymer-polymer systems of molecular weight M and indicated values of $X_{AB}T$=408K, $\tilde{V}_A^{1/3}$=1.067. Regions below and to the left of the full curves correspond to miscible systems. The dashed curve divides system showing UCST (to the left) from the systems showing LCST (to the right).The dots at intersections of dashed curve with curves for X_{AB} = 4 and 10 J/cm^3 thus correspond to coalescence of UCST and LCST. Values of χ_{12}: 1 - 0; 2 - (−1), 3 - (−2), 4 - (−4), 5- 4, 6 - 1. Adapted, by permission, from D. Patterson and A. Robard, *Macromolecules*, **11**, 692 (1978).

Values τ and M to the right and above any curve correspond to the phase separation, whereas values to the left and below to the region of immiscibility. The curves are critical lines and correspond to the systems at critical points. When $X_{AB} < 0$, the systems are miscible at $M \rightarrow \infty$. To reach miscibility at high molecular mass for the systems with $\tau \approx 0.15$, it is enough to have only very small negative values of X_{AB}. The points along the curves for $X_{AB} = 0$ correspond to LCST and miscibility is possible only at small molecular mass and low values of τ. If X_{AB} increases, maximum molecular mass, at which components are miscible, decreases rapidly. For $X_{AB} > 0$, the curves correspond to UCST except for very high values of τ, when they are converted into LCST. For each curve with constant X_{AB}, there exists such a value of τ at which UCST and LCST are merged. Differentiating χ_{AB} by T, we can find values of τ, (X_{AB}/P^*_A) and $\tilde{V}_A^{1/3}$ to obtain LCST when $X_{AB} > 0$:

$$\frac{\tau^2}{(X_{AB}/P_A^*)} \geq \frac{3}{2}\left(\frac{4/3 - \tilde{V}_A^{1/3}}{\tilde{V}^{1/3} - 1}\right)^2 \qquad [3.31]$$

Taking $\tilde{V}_A^{1/3} = 1.067$ and $P_A^* = 500$ J/cm^3 according to Eq 3.31, the following values can be found: $\tau = 0.77$ for $X_{AB} = 10$ J/cm^3 and $\tau = 0.31$ for $X_{AB} = 4$ J/cm^3. These values are very large for polymer-polymer systems with LCST. The largest, possible value for polymer mixture can be 1.15 which corresponds to the value T = 703 K for polystyrene. If $\tau = 0.15$ is the largest value expected for the polymer-polymer system, Eq 3.31 shows that X_{AB} should be less than 5 J/cm^3. The calculations show that the system having both UCST and LCST can hardly be observed, whereas for polymer solutions such cases were found.

The role of structural factor c/q(q = r) (contact mobility) was studied elsewhere.[40] This value is related to $T^* = (\varepsilon/k)(c/q)$, where ε/k is the potential energy of interaction expressed in the units of absolute temperature, q is the value proportional to the area of molecular surface and c is equal to 1/3 of external freedom degree. It is evident that miscibility should not be directly connected with the condition $\varepsilon_{AA} = \varepsilon_{BB}$, if c/q are different for two polymers. The analysis of the influence of c/q on the miscibility is based on the theory of rigid disturbed chains[41,42] in which the function of distribution was taken in the following form:

$$Q(N,V,T) = 1/N!)(V_f/\Lambda^3)^N(\varpi_{r,V})^N_{intr}(\varpi_{r,V})^N_{ext}\exp(N\Phi/2kT) \qquad [3.32]$$

where N is the number of molecules at temperature T and volume V, Λ is the de Broglie wavelength, V_f is the free volume, Φ is the potential field, $\omega_{r,V}$ are contributions of rotational and vibrational degrees of freedom, c is the number of external freedom degrees according to Prigogine.

$$(\omega_{r,V})_{ext} = (V_f/V)^{c-1} \qquad [3.33]$$

When considering polymer mixtures, the distribution function should obey the following border conditions:

- the second virial coefficient should be the square function of molar fractions
- at high densities the entropy of mixing should be combinatorial entropy of Flory-Huggins
- the excess values should be determined in relation to ideal gas at the same volume, temperature, and mixture composition

- the excess chemical potential of the solvent in concentrated solution should not be equal to 0 at infinite length of the solvent chain (i.e., by transition to polymer mixture)
- to account for the clusterization of molecules, when the mixture consists of segments of different potential energies, the Gibbs free energy, calculated from the distribution function, should contain the results of the disturbation theory for the mixtures of spherical molecules.[43]

At these conditions, the Gibbs free energy is expanded to temperature series:

$$G - G_{ideal\ gas} = G^{(0)} + G^1/T + G^2/T^2 + G^3/T^3 + G^4/T^4 + \dots \quad [3.34]$$

where term $G^{(0)}$ takes into account the repulsion forces and is calculated[44] as:

$$\frac{G^{(0)}}{NkT} = \frac{N\bar{c}(3\tau\tilde{\rho} - 4)\tau\tilde{\rho}}{(1 - \tau\tilde{\rho})^2} \quad [3.35]$$

($\tau = 0.7405$, $\tilde{\rho} = v^*/v$ is reduced density). Other members of Eq 3.34 reflect the attraction forces. They can be written in the form:

$$G_l = f_l(\tilde{\rho}, T_l^*, c) \quad l = 1,2,3,4 \quad [3.36]$$

The chemical potential of component j is calculated in standard form from the relationship:

$$\mu_j = (\delta G/\delta N_j)_{Nj\ (i \neq j),\ T,V} \quad [3.37]$$

The calculation of phase equilibria in polymer blends needs some parameters to be known. These are P^*, v^*_{sp}, T^*, r, ε/k, and binary constant k_{AB} which is found from:

$$\varepsilon_{AB} = (\varepsilon_A \varepsilon_B)^{1/2}(1 - k_{AB}) \quad [3.38]$$

Values ε_A, ε_B, and ε_{AB} should be known to calculate T_l^*. Some model calculations have been performed[40] for a given set parameters P^*, v^*_{sp}, T^*, and r and varying values $c/q_A/q_B$, ε_A, ε_B, ε_{AB}, and k_{AB}. It was found that:

- at constant $\varepsilon_A = \varepsilon_B$, k_{AB} and $(c/q)_A$, the mutual solubility increases at a given temperature with decreasing $(c/q)_B$
- with increasing positive values of k_{AB}, the miscibility increases whereas at negative k_{AB} and various ε_A and ε_B simultaneously both UCST and LCST are observed.

In the same work,[40] the effect of c/q of the solvent on the miscibility of two polymers in a common solvent was considered. It was shown that in the region of LCST, the mutual solubility of two polymers, miscibility increases by introducing the solvent with lower values of c/q. However, in the regions of UCST and LCST, the effect of c/q on the miscibility manifests itself in a different way. Minimum volume fractions of the solvent, which are needed for miscibility, at first decrease and then increase with growth in c/q of the solvent. The effect of difference in potential energies of polymer and solvent is not very essential. When these values are close, the dependence of minimum volume fraction of the solvent on its potential energy is characterized by a minimum.

3.1.3 THE APPLICATION OF THE NEW HUGGINS THEORY

The main concepts of the new Huggins theory have been used by Koningsveld and coworkers[31] to interpret the bimodal shape of spinodals and curves of cloud points found for some oligomer-oligomer mixtures with UCST.[45] The Huggins expression for the parameter of thermodynamic interaction g(T,φ) was applied in the following form:[31]

$$g(T,\varphi) = \varphi_A\varphi_B\beta[1-(1-\rho)-\varphi_B] - \varphi_A r_{v,A}^{-1}\ln(1+k_A z_B) -$$
$$-\varphi_B r_{v,B}\ln(1+k_B z_A) \qquad [3.39]$$

where $\rho = (\sigma_B/\sigma_A)(r_{V,A}/r_{V,B})$ (σ is the contact surface of segments in polymer), $r_{V,i} = V_i/n_iV_L$ (V_i is the molar volume of polymer; V_L is the molar volume of lattice site), n_i is the number of segments in macromolecule, $z_B = 1 - z_A = \rho\varphi_B/[1-(1-\rho)\varphi_B]$, $\beta \approx (z-2)\Delta w_{AB}/RT$. Here Δw_{AB} is the energy change by formation of one contact between two different segments. The last two terms in Eq 3.21 reflect the effect of the surrounding of polymer segment on its flexibility. By corresponding choice of parameters k_A and k_B, the bimodal shape of spinodal curves or parameter β may be obtained (Fig. 3.7). It is worth noting that bimodal phase diagram may be seen also when one uses the interaction parameter g(T,φ) in the form of its expansion into series by temperature and concentration:

$$g(T,\varphi) = g_0(T) + g_1\varphi_B + g_2\varphi_B^2 + \ldots \qquad [3.40]$$

The g value should be less 0, $g_2 > 0$ and $|g_2| > |g_1|$ (see Fig. 3.7).

An interesting interpretation of Huggins concept in relation to the effect of the surrounding molecules of one polymer on the flexibility of chains of another was given[46] based on the theories by Silberberg[47] and Huggins. The correlation between some parameters which characterize the entropy of mixing have been derived. The essence of

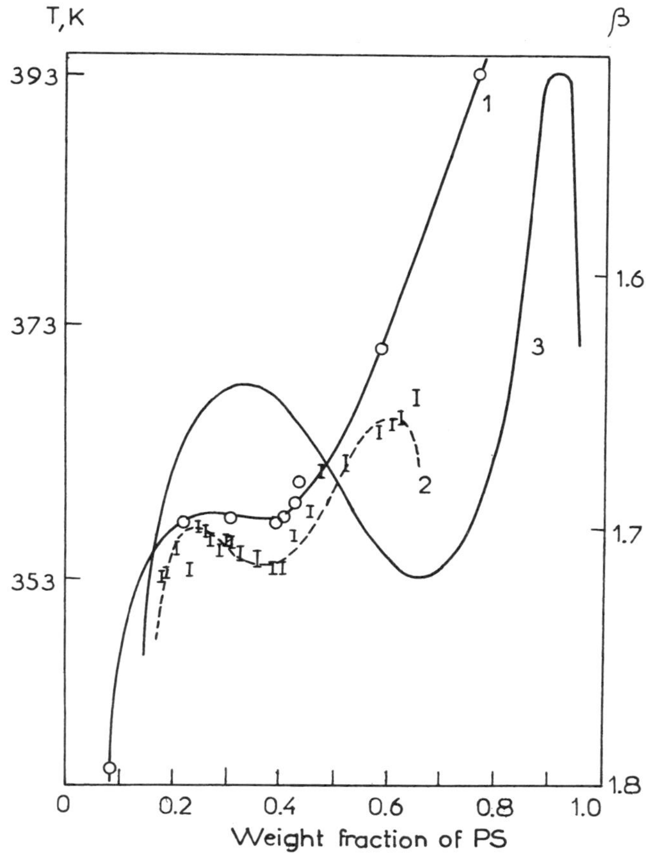

Figure 3.7. Spinodals calculated with Huggins equation (Eq 3.39) for m_A=7.2 and m_B=45; k_A=-0.69, k_B=+0.91; $r_{v,A}$=1, $r_{v,B}$=1, ρ=0.8, Δw_{AB}=RT/10. 1 - experimental, 2,3 - calculated. Adapted, by permission, from R. Koningsveld, L. A. Kleintjens, *British Polym. J.*, **9**, 212 (1977).

this approach is the following. According to Silberberg[47] the total number of configurations of the system consisting of mixture of polymolecular homopolymers A and B may be written as:

$$\Omega_{A+B} = \frac{(\sum_i N_{Ai} r_{Ai} + \sum_j N_{Bj} r_{Bj})!}{\prod_i (N_{Ai}!) \prod_j (N_{Bi}!)} \prod [V^*_{Ai} \varphi_A / V]^{N_A(r_{Ai}-1)} \times$$

$$\times \prod_j [V^*_{Bi} \varphi_B / V]^{N_{Bi}(r_{Bj}-1)} \qquad [3.41]$$

where N_{Ai} and N_{Bj} are the numbers of molecules of fractions i and j in polymer A and polymer B. The relative lengths of chains (r_{Ai} and r_{Bj}) are determined by the usual way in relation to the solvent molecule. "Parametric volume", V^*, was introduced by Silberberg for correction of overestimating the number of configurations entering the term containing factorial. The first segment of a polymer chain is disposed in the total volume, V, and then the volume available for the next r–1 segments is restricted by the parametric volume of preceding segment(s). If to suppose that V^* is independent of the mixture volume, the expression for the entropy of mixing can be derived analogous to the Flory-Huggins equation. The concentration dependence of V^* has been taken into account.[46] The expression for Gibbs free energy of mixing, ΔG, per 1 mol of segments has been derived by the usual way by adding to the entropy term van-Laar heat of mixing $g\varphi_A\varphi_B$:

$$\Delta G/RT = \sum_i \varphi_{Ai} r_{Ai}^{-1} \ln \varphi_{Ai} + \sum_j \varphi_{Bj} r_{Bj}^{-1} \ln \varphi_{Bi} + g\varphi_A\varphi_B -$$

$$- \sum_i \varphi_{ij}(1 - r_{Ai}^{-1}) \ln \tau_{Ai} - \sum_j \varphi_{Bj}(1 - r_{Bj}^{-1}) \ln \tau_{Bi} \quad [3.42]$$

where φ is the volume fraction of a polymer ($\Sigma\varphi_{ij} = \varphi_i$). Value τ_{ij} includes the term $V_{kl}^*\varphi_k/V_{ij}^*$. Huggins equation for the Gibbs free energy of mixing, which accounts for two orientation contributions into the entropy of mixing, has the following form:

$$\Delta G/RT\ bBjr_{Bj}^{-1} \ln \varphi_{Bj} + g\varphi_A\varphi_B - \varphi_A \ln(1 + k_1\rho\varphi_B Q^{-1}) -$$

$$- \varphi_B \ln(1 + k_2\varphi_A Q^{-1}) \quad [3.43]$$

where $Q = [1 - (1 - \rho)\varphi_B]$; $\rho = (\sigma_B/\sigma_A)/(v_B/v_A)$; σ_k and v_k are the contact surface and molecular volume of segment of polymer k. There exists the possibility to take into account the difference between interacting surfaces (volumes) of segments A and B. Two parameters k_{sA} and k_{sB} are, according to Huggins, empirical parameters determined as $k_{si} = k_i/(1 + k_i)$. As already noted, Koningsveld[31] has shown that equations, such as Eq 3.43, may describe the bimodal shape of spinodal. To get bimodal spinodal, the negative and positive values should be prescribed to parameters k_A and k_B. Comparing Eqs 3.42 and 3.43, neglecting r^{-1} and considering that τ is independent of the chain length, the expressions for τ_A and τ_B can be obtained:

$$\tau_A = V_A^*(\varphi_A)/V_A^*(1) \equiv 1 + k_A\rho\varphi_B Q^{-1} \quad [3.44]$$

$$\tau_B = V_B^*(\varphi_B)/V_B^*(1) \equiv 1 + k_B\varphi_A Q^{-1} \quad [3.45]$$

Silberberg[47] assumed that parametric volume may be taken equal or commensurable with the volume of polymer coil. Using this assumption, one can expect diminishing the size of polymer A coil when its concentration in the mixture decreases. For the infinitely dilute solution of polymer A in polymer B ($\varphi_A = 0$, $\varphi_B = 1$), we have $\tau_A = 1 + k_A$, whereas the polymer in bulk has $\tau_A = 1$ ($\varphi_A = 1$, $\varphi_B = 0$). On the contrary, for polymer B diluted in polymer A, $\tau_B = 1 + k_B$ ($\varphi_B = 0$, $\varphi_A = 1$). For the polymer B in bulk at $\varphi_B = 1$ and $\varphi_A = 0$, we have $\tau_B = 1$. For the polymer in bulk, the Eqs 3.44 and 3.45 give reasonable results. The neutronography data[48] show that the coil has unperturbed size in amorphous region ($\tau = 1$). Negative values of k_A indicate that τ_A should be less than 1 for the infinite dilution of polymer A in polymer B. This means the decrease of the coil sizes as compared with unperturbed size in bulk. The present data confirm such a decrease in the coil size. Various conformations of linear chain A in the melt of linear chains B have been considered[49] in the temperature interval of miscibility of two components. The essential compression of chains A should be observed analogous to the usual coiling of isolated chains in a bad solvent. However, in the melt of two polymers, sharper transition should be observed, i.e., collapse of the chain.

Using mathematic modelling[50] the analysis of the behavior of binary mixture has been done for the case when $N_A = N_B$ and $\varepsilon_{AA} = \varepsilon_{BB}$, $\varepsilon_{AB} = 0$ and concentration of one of the polymers is small (A >> B). It was discovered that with increasing dimensionless energy of interaction ε/kT, the chains of B type are compressed forming a very tight conformation whereas chains A stay practically as gaussian chains. Collapse of chains of PMMA in the mixture with PBA in a solid state was observed,[51] whereas the diminishing of coil dimension of poly-α-methyl styrene dissolved in deuterated PMMA was shown.[52] It was established[53] that by transition of parameters ψ_{AB} from negative to positive values, the coil dimensions are decreased in the miscible mixtures of asymmetric composition.

3.1.4 THE SANCHEZ-LACOMBE THEORY

To characterize the stability of the systems consisting of two polymers of high molecular mass, Sanchez has developed the theory based on the model of compressible lattice.[54-56] This theory is capable of describing the volume changes by mixing and predicts the existence of LCST. It is assumed that polymer segments are rigid cores and that four of six parameters operating in the theory can be found from PVT properties of individual components. On the basis of this theory, Sanchez has derived the expressions for chemical potentials, phase stability conditions, and interfacial tension.

3.1.4.1 Chemical Potentials

To obtain chemical potentials, the mixture is considered consisting of $N_A r_A$, $N_B r_B$,..., $N_M r_M$ segments. Their total number in the system is $rN = \Sigma r_i N_i$ where N_i is the number of molecules of i-th component and r_i is the number of segments in the molecule of component i. The Gibbs free energy of mixing for the mixture of N components is expressed as $G = rN(g_{comb} + g_{noncomb}) \equiv rNg$, where g_{comb} and $g_{noncomb}$ are combinatorial and noncombinatorial free energies per one segment of the mixture:

$$g_{comb} = \frac{kT}{v*} \sum_i \frac{\varphi_i}{r_i} \ln \varphi_i \qquad [3.46]$$

$$g_{noncomb} = -\tilde{\rho}\varepsilon* + \overline{P}\tilde{V}v* + k\tilde{T}[V(1 - \tilde{\rho}) + \frac{\tilde{\rho}}{r} \ln \tilde{\rho}] \qquad [3.47]$$

where

$$\varphi_i = r_i N_i / rN; \quad \frac{1}{r} = \sum(\varphi_i / r_i) \qquad [3.48]$$

$$\varepsilon* = \sum_i \sum_j (\varphi_i \varphi_j \sigma_{ij}^3 \varepsilon_{ij}) / V^* \qquad [3.49]$$

$$V^* = \sum_i \sum_j (\varphi_i \varphi_j \sigma_{ij}^3) \qquad [3.50]$$

$\tilde{\rho}$ and $\tilde{V}$ are reduced density and volume of the mixture ($\tilde{V} = 1/\tilde{\rho}$). Parameters ε^*_{ij} and σ_{ij} are energetic parameters characterizing the segment-segment interaction. All segments have densely-packed volume $V_1^* = \sigma^3_{ij}$ where σ_{ij} is the nearest distance between i-th and j-th segments. The attraction forces between segments are proportional to ε_{ij}. From the equilibrium condition

$$(\delta g / \delta \tilde{V})_{\rho,T,\varphi} = (\delta g_{noncomb} / \delta \tilde{V})_{\rho,T,\varphi} = 0 \qquad [3.51]$$

the equation-of-state for the mixture may be obtained:

$$\tilde{\rho}^2 + \tilde{P} + \tilde{T}[\ln (1 - \tilde{\rho}) + (1 - 1/r)\tilde{\rho}] = 0 \qquad [3.52]$$

where ρ and P are reduced temperature and pressure:

$$\tilde{T} = T / T^*; \quad T^* = \varepsilon* / k \qquad [3.53]$$

$$\tilde{P} = P / P^*; \quad P^* = \varepsilon* / V^* \qquad [3.54]$$

The chemical potential of component, μ_i, is expressed as

$$\mu_i = kT[\ln \varphi_i + (1 - 1/r)] + r_i[g_{noncomb} + (\delta g_{noncomb}/\delta\varphi_i)_{\varphi',\tilde{\rho}} - \sum_j \varphi_j(\delta g_{noncomb}/\delta\varphi_j)_{\varphi',\tilde{\rho}}] \quad [3.55]$$

Here φ' denotes that all φ_i but one should be constant. Values ε^*, 1/r, and V^* are homogeneous functions of zero, first and second order, respectively. Therefore

$$\sum_j \varphi_j(\delta g_{noncomb}/\delta\varphi_j)_{\varphi',\tilde{\rho}} \approx 2\tilde{P}V^* + (kT/r) \ln \tilde{\rho} \quad [3.56]$$

Substituting 3.56 into 3.55 we obtain

$$\mu_i = kT[\ln \varphi_i + (1 - r_i/r)] + r_j[-\tilde{\rho}[\varepsilon* + (\delta\varepsilon*/\delta\varphi)_{\varphi_i} + P\tilde{V}[(\delta V^*/\delta\varphi_i)_{\varphi'_i} + V^*] + kT\tilde{V}[(1 - \tilde{\rho}) \ln (1 - \tilde{\rho}) + (\tilde{\rho} \ln \tilde{\rho})/r_i] \quad [3.57]$$

where ε^* and V^* are given by Eqs 3.49 and 3.50. These values may be expressed in various functional form. For binary mixture

$$\varepsilon* + (\delta\varepsilon/\delta\varphi_A)_{\varphi_B} = \varepsilon*_{AA} - X_{AB}\varphi_B^2 \quad [3.58]$$

More correctly one can obtain:

$$(\delta V^*/\delta\varphi_B)_{\varphi_A} - V^* = V_A^* + (V_A^* + V_B^*)\delta\varphi_B^2 \quad [3.59]$$

where

$$X_{AB} = (P_A^* P_B^*)^{1/2} V_A^* (V_B^*/V_A^*)^2 [a + b_{AB}\delta + c_{AB}\delta^2] \quad [3.60]$$

$$a = (\tau/\nu)^{1/2} + (\nu/\tau)^{1/2} - (\nu^{1/2} + \nu^{-1/2})\xi \quad [3.61]$$

$$b_{AB} = [\varphi_A(2 + \nu\varphi_A)(\tau^{1/2} - \xi) - \varphi_B^2(\xi - \tau^{-1/2})](\nu^{1/2} + \nu^{-1/2}) \quad [3.62]$$

$$c_{AB} = \varphi_A^2(\tau^{1/2} - \xi) - (1 + \nu)(\nu^{1/2} + \nu^{-1/2}) \quad [3.63]$$

$$\tau = \varepsilon*_{AA}/\varepsilon*_{BB} = T_A^*/T_B^*; \quad \nu = \sigma_{AA}^3/\sigma_{BB}^3 = V_A^*/V_B^* \quad [3.64]$$

$$\xi = \varepsilon*_{AB}/(\varepsilon*_{AA}\varepsilon*_{BB})^{1/2} \quad [3.65]$$

$$\delta = 2\sigma*_{AB}/(\sigma_{AA}^3 + \sigma_{BB}^3) - 1 \quad [3.66]$$

Substituting Eqs 3.58 and 3.59 in 3.57, we get the value of chemical potential for one component of the mixture:

$$\mu_A/kT = \ln\varphi_A + (1 - r_A/r_B)\varphi_B + r_A\chi_{AB}\varphi_B^2 \qquad [3.67]$$

where

$$\chi_{AB} = \tilde{\rho}X_{AB}/kT + \frac{\tilde{P}_A\tilde{V}_A}{\tilde{T}_A}(1 - v^{-1})\delta + \frac{1}{\varphi_B^2}[-\tilde{\rho}/\tilde{T}_A + \tilde{P}_A\tilde{V}_A/\tilde{T}_A +$$

$$+ \tilde{V}[(1 - \tilde{\rho})\ln(1 - \tilde{\rho}) + \frac{\tilde{\rho}}{r_A}\ln\tilde{\rho}]] \qquad [3.68]$$

The chemical potential of the second component can be derived by a simple exchange of indices A and B (we may note that $X_{AB} \neq X_{BA}$). At the normal pressure, two terms related to pressure in Eq 3.68 may be omitted. From the equations given above, it follows that to calculate chemical potentials only two parameters are needed: ξ and δ (or $\varepsilon_{AB}{}^*$ and σ_{AB}). Value δ is expressed by Eq 3.65. Value ξ characterizes the deviation of volume of a real system from the ideal system (simple additive scheme):

$$\delta \le 0 \;\; V^* < V^*_{ideal}; \quad \delta = 0 \;\; V^* = V^*_{ideal}; \quad \delta > 0 \;\; V^* > V^*_{ideal} \qquad [3.69]$$

where

$$V^*_{ideal} = \varphi_A V_A^* + \varphi_B V_B^* \qquad [3.70]$$

At the temperature 0 K, the the change of sign of the mixing volume is determined by the sign of δ. Therefore Eq 3.35 supposes that

$$\delta_{AB}^3 = (\delta_{AA}^3 + \delta_{BB}^3)/2 \qquad [3.71]$$

3.1.4.2 The Stability of Phase and Spinodal

At constant T and P, the dependence of g on the composition is the necessary and sufficient condition for miscibility of binary mixture throughout the whole composition range as described by a concave curve

$$d^2g/d\varphi_B^2 > 0 \qquad [3.72]$$

Interpretation of dg/dφ_A as a function of φ_A and $\tilde{V}$ (at T = const and P = const) and application of the relation

$$d\tilde{V}/d\varphi_A = -\tilde{V}\tilde{T}\rho^*\beta[\delta^2(g/kT)/\delta\varphi_A\delta\tilde{v}] \qquad [3.73]$$

allows to show that

$$\frac{\delta^2(g/kT)}{\delta\varphi_A^2} = \left[\frac{\delta^2(g/kT)}{\delta\varphi_A^2}\right]_{\tilde{v}} - \tilde{V}\tilde{T}P^*\beta\left[\frac{\delta^2(g/kT)}{\delta\varphi_A\delta\tilde{v}}\right]^2 \qquad [3.74]$$

where β is the isothermic compressibility of the mixture determined from the expression

$$\tilde{T}P^*\beta = \tilde{V}[\tilde{\rho}/(1-\tilde{\rho}) + \frac{1}{r} - 2\tilde{\rho}/\tilde{T}]^{-1} \qquad [3.75]$$

Therefore the dependence of g on composition is concave, if in the whole concentration range the nonequality is observed:

$$\frac{1}{2}\left[\frac{1}{r_A\varphi_A} + \frac{1}{r_B\varphi_B}\right] - \tilde{\rho}[X + \frac{1}{2}\Psi^2\tilde{T}P^*\beta] > 0 \qquad [3.76]$$

where

$$-\tilde{\rho}X = \frac{1}{2}\left[\frac{\delta^2(g_{noncomb}/kT)}{\delta^2\varphi_A^2}\right]_V = \frac{1}{2}\left[-\tilde{\rho}\frac{\delta^2(1/\tilde{T})}{\delta\varphi_A^2} + \frac{\tilde{P}\tilde{V}}{\tilde{T}V^*}\frac{\delta^2V^*}{\delta\varphi_A^2}\right] \qquad [3.77]$$

$$-\tilde{\rho}\Psi \equiv \frac{1}{2}\frac{\delta^2 g_{noncomb}/kT}{\delta\varphi_A\delta\tilde{V}} = \tilde{\rho}\frac{d(1/\tilde{T})}{d\varphi_A} + \frac{\tilde{P}\tilde{V}\ln V^*}{T\varphi_A} - \frac{d(1/r)}{d\varphi_A} \qquad [3.78]$$

$$\frac{1}{r_A\varphi_A} + \frac{1}{r_B\varphi_B} = \left[\frac{\delta^2 g_{noncomb}/kT}{\delta\varphi_A^2}\right]_{\tilde{V}} \qquad [3.79]$$

Using Eq 3.58 we can obtain:

$$\frac{d(1/\tilde{T})}{d\varphi_A} = \left[\frac{d(1/\tilde{T})}{d\varphi_A}\right]_{\varphi_B} - \left[\frac{d(1/\tilde{T})}{d\varphi_B}\right]_{\varphi_A} = \frac{1}{\tilde{T}_A} - \frac{1}{\tilde{T}_B} +$$

$$+ \frac{1}{kT}(\varphi_A^2 X_{AB} - \varphi_B^2 X_{AB}) \qquad [3.80]$$

$$\frac{d^2(1/\tilde{T})}{d\varphi_A^2} = \frac{1}{2}\left[2(\varphi_A X_{BA} + \varphi_B X_{AB}) + \varphi_A^2\frac{\delta X_{BA}}{\delta\varphi_A} - \varphi_B^2\frac{\delta X_{AB}}{\delta\varphi_A}\right] \qquad [3.81]$$

Sanchez and Lacombe[56] considered also the possibility of bimodal spinodal. Their consideration turns to the following. For oligomer-oligomer mixtures, the combinatorial contribution to the nonequality 3.81 is relatively small. Therefore, the spinodal should be very sensitive to the concentration dependence of X and Ψ^2. The existence of bimodal spinodal shows that the mixture is more stable at the concentration φ_m at which the dependence T – φ is characterized by a minimum. This additional stability should be observed if X is negative near φ_m, but positive at other compositions. To fulfill this condition, it is necessary that the value $d(1/\tilde{T})/d\varphi_A$ had zero values three times in the whole concentration range. It may be shown that the zero values are determined by the square equation:

$$\tau^{1/2} - \tau^{-1/2} + (\varphi_A - \varphi_B)[(\nu^{1/2}\varphi_A \nu^{-1/2}\varphi_B)\Delta + (\nu^{1/2} + \nu^{-1/2})][(\nu\tau)^{1/2}\varphi_A + $$

$$+ (\nu\tau)^{-1/2}\varphi_B]\delta \qquad [3.82]$$

where

$$\Delta = (\nu\tau)^{1/2} + (\nu\tau)^{-1/2} - (\nu^{1/2} + \nu^{-1/2})(1 + \delta) \qquad [3.83]$$

Thus, three zero values of X throughout the whole concentration range are not possible. However, at low pressures and high r_A and r_B, the value Ψ^2 is proportional to $[d^2(1/\tilde{T})/d\varphi_A^2]$. When $d(1/\tilde{T})/d\varphi_A$ acquires zero value at φ_m, the unfavorable term of compressibility disappears. The necessary but insufficient conditions for the existence of bimodal spinodal are the following:

$$d(1/\tilde{T})/d\varphi_A]_{\varphi_A = \varphi_m} = 0 \qquad [3.84]$$

$$d^2(1/\tilde{T})/d\varphi_A^2 > 0 \qquad [3.85]$$

These conditions are satisfied even when $\nu = 1$ and $\delta = 0$. In this case, Eq 3.82 reduces to

$$\varphi_m = (r^{-1/2} - \xi)/(\tau^{1/2} + \tau^{-1/2} - 2\xi) \qquad [3.86]$$

and Eq 3.81 to

$$\frac{1}{2}\frac{d^2(1/\tilde{T})}{d\varphi_A^2} = \frac{X_{AB}}{kT} = \frac{X_{BA}}{kT} = (\tau^{1/2} + \tau^{-1/2} - 2\xi)/(\tilde{T}_A\tilde{T}_B)^{1/2} \qquad [3.87]$$

Generally, the necessary condition of bimodal spinodal at high temperatures (UCST) given by the theory is very approximate and needs that r_A and r_B were almost equal. The conditions 3.84 and 3.85 should also be fulfilled.

If $d^2(1/\tilde{T})/d\varphi_A^2 < 0$, in the concentration range (0;1) and condition 3.85 is valid, the bimodal spinodal can be expected for both oligomer-oligomer and polymer-polymer mixtures at high temperatures

(LCST). The last condition is more probable as compared with UCST. LCST is determined mainly by the compressibility term. However, there was no attempt to experimentally check the conclusions of the theory under consideration.

3.1.4.3 Criterion of Miscibility of Polymer-polymer Mixture

The combinatorial contribution into spinodal is very small for high molecular weight polymers because the compressibility terms always lead to the decrease in phase stability. The miscibility of two polymers is observed when $d^2(1/\tilde{T})/d\varphi_A^2 < 0$ throughout the whole concentration range (0;1). This condition is equivalent to the condition of negative heat of mixing at low temperatures. The heat of mixing is expressed in the following way:

$$\Delta H_{mix} = rN\,[-\tilde{\rho}\varepsilon^* + \tilde{\rho}_A\varphi_A\varepsilon^*_{AA} + \tilde{\rho}_B\varphi_B\varepsilon^*_{BB}] \qquad [3.88]$$

At low temperatures (assuming $\tilde{\rho}_A = \tilde{\rho}_B = \tilde{\rho} = 1$), we have:

$$\Delta H_{mix}(T=0) = rN\,(-\varepsilon^* + \varphi_A\varepsilon^*_{AA} + \varphi_B\varepsilon^*_{BB}) \qquad [3.89]$$

ΔH is always negative, if $d^2\varepsilon^*/d\varphi_A^2 < 0$ in the whole range of concentrations. Eq 3.89 can by written as

$$\Delta H_{mix}(T=0)/rN = \varphi_A\varphi_B(V_A^*V_B^*/V^*)[\Delta P^* + \delta[\Delta P^* + P_A^*(\nu\varphi_A - \varphi_B) -$$

$$-P_B^*(\varphi_A - \varphi_B\nu^{-1})]] \qquad [3.90]$$

where

$$\Delta P^* = P_A^* + P_B^* - \xi(\nu^{1/2} + \nu^{-1/2})(P_A^*P_B^*)^{1/2} \qquad [3.91]$$

The heat of mixing is negative at $\Delta P^* < 0$ or at

$$\xi > \xi_{min} = [(\nu/\tau)^{1/2} + (\nu/\tau)^{-1/2}]/(\nu^{1/2} + \nu^{-1/2}) \qquad [3.92]$$

A low value of ξ_{min} promotes mixing since it facilitates fulfillment of nonequality 3.92. The analysis of nonequality 3.92 shows that values of ξ_{min} at $\tau > 1$ and $\nu > 1$ are lower than at $\tau > 1$ and $\nu < 1$. The term proportional to δ in Eq 3.90 is lower than ΔP^* and changes its sign in the interval (0;1). In the concentration interval, where term gives the positive contribution, ξ should be higher in comparison with ξ_{min} to obtain the negative heat of mixing. Generally, full miscibility of high molecular polymers is possible at $\Delta H_{mix} < 0$. The properties of pure components promoting mixing, i.e., giving the negative heat of mixing, are determined by the nonequality 3.92. However, if the system is miscible at low temperatures, it may be expected to be separated into the phases at high temperatures due to the increasing contribution of compressibility term, unfavorable for mixing.

Thus, in the theory by Sanchez-Lacombe, the correlation of parameters of the mixture and parameters of pure components has been established, similar to the new theory by Huggins. For binary mixture, the necessary and sufficient condition of phase stability (miscibility) is nonequality 3.76 and condition of bimodal spinodal nonequalities 3.84. It is supposed that bimodal phase diagrams are more probable for mixtures with LCST. To observe bimodal phase diagrams, the high molecular masses of components are necessary (although such diagrams have been found[57] for low molecular weight polymers as well). Full miscibility of two polymers needs negative heat of mixing, i.e., the condition 3.92 should be realized.

The theory was tested by Sanchez[58,59] and other authors. In particular, the theory was used[60] to estimate the interaction parameters and phase diagrams for the tetramethylene Bisphenol A-polycarbonate-PS blends. For this purpose the noncombinatorial free energy of mixing was presented in the van-Laar form:

$$dg_{noncomb} = B(T)\varphi_A\varphi_B \tag{3.93}$$

where

$$B(T) = B^h - TB^S \tag{3.94}$$

The equation for spinodal was given as

$$\frac{d^2g}{d\varphi_A^2} = RT\left(\frac{1}{\varphi_A\overline{V}_A} + \frac{1}{\varphi_B\overline{V}_B}\right) - 2B_{sc} \tag{3.95}$$

where

$$B_{sc} = \frac{1}{2}\frac{d^2g}{d\varphi_A^2} \tag{3.96}$$

According to ref. 60

$$B_{sc} = B(T) + (\varphi_A - \varphi_B)\frac{dB(T)}{d\varphi_A} - \frac{1}{2}\varphi_A\varphi_B\frac{d^2B(T)}{d\varphi_A^2} \tag{3.97}$$

where parameter B(T) differs by factor kT from χ used by Sanchez. Having used the expressions for $g_{noncomb}$ and ΔH_M in the form:

$$g_{noncomb} = -\tilde{\rho}P^* + P\tilde{v} + \frac{R}{v^*}\left[\frac{1-\tilde{\rho}}{\tilde{\rho}}\ln(1-\tilde{\rho}) + \frac{1-\tilde{\rho}}{r}\right] \tag{3.98}$$

$$\frac{\Delta H_M}{V} = \tilde{\rho}^2 \Delta P^* \varphi_A \varphi_B + \tilde{\rho}[\varphi_A P_A^*(\tilde{\rho}_A - \tilde{\rho}_B) + \varphi_B P_B^*(\tilde{\rho}_B - \tilde{\rho}_A)] \qquad [3.99]$$

where

$$P^* = \varphi_A P_A^* + \varphi_B P_B^* - \varphi_A \varphi_B \Delta P^* \qquad [3.100]$$

From Eqs 3.94 and 3.95, the expression for the enthalpy part of interaction parameter can be written as follows

$$B^h = \tilde{\rho}\Delta P^* + \left[P_A^*(\tilde{\rho}_A - \tilde{\rho}_B)/\varphi_A + \frac{P_B^*(\tilde{\rho}_B - \tilde{\rho}_A)}{\varphi_B} \right] \qquad [3.101]$$

The correlation for the term of excess entropy - TB^S was found as

$$-TB^h = \frac{RT}{\varphi_A \varphi_B}\left[\frac{1}{V^*}\left(\frac{1}{\tilde{\rho}} - \tilde{\rho} \right) + \ln \frac{\tilde{\rho}}{r} \right] - \frac{\varphi_A}{V_A^*}\left[\frac{1 - \tilde{\rho}_A}{\tilde{\rho}_A} \ln(1 - \tilde{\rho}_A) + \ln \frac{\tilde{\rho}_A}{r_A} \right] -$$

$$- \frac{\varphi_B}{V_B^*}\left[\frac{1 - \tilde{\rho}_B}{\tilde{\rho}_B} \ln(1 - \tilde{\rho}_B) + \ln \frac{\tilde{\rho}_B}{r_B} \right] \qquad [3.102]$$

Using conditions of spinodal for compressible mixture[62]

$$\frac{d^2 g}{d\varphi^2} = g_{\varphi\varphi} - \frac{(g_{\tilde{\rho}\varphi})^2}{g_{\tilde{\rho}\tilde{\rho}}} = 0 \qquad [3.103]$$

where indices φ and $\tilde{\rho}$ show partial derivatives by φ and $\tilde{\rho}$. From Eqs 3.96 and 3.103, the expression for B_{sc} can be derived:

$$B_{sc} = \tilde{\rho}\Delta P^* + [[P_B^* - P_A^* + (\varphi_B - \varphi_A)\Delta P^*] + \frac{RT}{\tilde{\rho}}\left(\frac{1}{r^*_A v^*_A} - \frac{1}{r^*_B v^*_B} \right) -$$

$$- RT\left(\frac{\ln(1 - \tilde{\rho})}{\tilde{\rho}^2} + \frac{1}{\tilde{\rho}} \right)\left(\frac{1}{V_A^*} - \frac{1}{V_B^*} \right)^2 \Big/ \frac{2RT}{v^*} \times$$

$$\times \left[2\ln(1 - \tilde{\rho}) + \frac{1}{\tilde{\rho}^2(1 - \tilde{\rho})} + \frac{1 - 1/r}{\tilde{\rho}^2} \right] \qquad [3.104]$$

The values of b^h calculated from PVT data[60] are negative and become smaller with increasing temperature due to the predicted by the theory volume contraction by mixing. Negative volume of mixing for this system was found experimentally.[63] The term $-TB^S$ becomes positive with increase in temperature in accordance to Sanchez theory. The value B_{sc}, being negative, increases with temperature and diminishes by absolute value at fixed Δp^*. The use of Eq 3.103 and

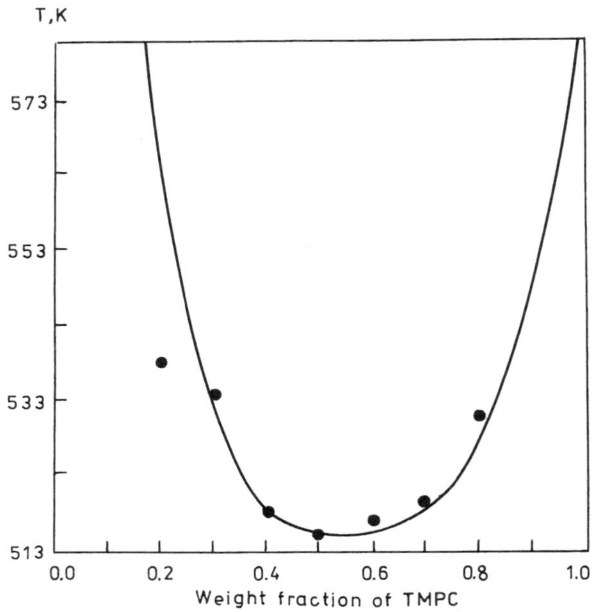

Figure 3.8. Comparison of the experimental phase separation temperatures (●) obtained by DSC method with predicted from spinodal curves (line) by using the lattice fluid theory and P^*=-0.71 J/cm^3, M_w (TMPC)=3.3×10^4, M_w (PC)=3.3×10^5, M_n (PC) =1×10^5 g/mol. Adapted, by permission, from C. K. Kim and D. R. Paul, *Polymer*, **33**, 1630 (1992).

value ΔP^* allows one to calculate the spinodal in good agreement with experimentally found temperatures of phase separation (Fig. 3.8).

3.2 EXPERIMENTAL DATA ON THERMODYNAMIC PROPERTIES OF BINARY POLYMER MIXTURES

The aim of the majority of experimental investigations of thermodynamic properties of binary blends consists of establishing their miscibility. The latter depends on the chemical nature of polymers, their molecular mass, temperature, and the mixture's composition. There are various other factors determining thermodynamic compatibility or miscibility of polymers. One of the first investigations was done by Dobry and Boyer-Kawenoki[64] who determined the compatibility of 98 polymer pairs. As a criterion of compatibility they had taken the turbidity of solutions of two polymers taken in ratio 1:1 in a common solvent or turbidity of films. If the solutions or films were transparent, the system was considered compatible. From 98 pairs studied, only 10% happen to be transparent, i.e., thermodynamically compatible. This criterion of transparency was used later by many authors[59-68] when studying films obtained either from melts or solutions. In some cases, the ternary phase diagrams have been constructed.[68-88] There were developed other criteria of compatibility and among them those which are based on dynamic mechanical measurements and DSC. If binary system has only one glass transition temperature the system is considered compatible. Depression of the melting point and the viscosity of solutions or melts was also used to estimate the compatibility. However, the best method of estimation of the phase state of polymer mixtures consists of determining their phase diagrams. Presently, there are discovered many hundreds of thermodynamically compatible (miscible) blends.[88-95] Many works were dedicated to the determination of miscibility from the data on heat effects of mixing.[2,3,96-105] First such estimations were done by Slonimsky,[2,3] S. Lipatov,[106] Voyutsky,[107] and others.[108,109] In particular, Slonimsky[2,3] discovered that only in some cases, the condition $\Delta H<0$ meets the polymer compatibility. S. Lipatov suggested[106] that in the course of polymer mixing a very important role should be played by entropy. Flory[4] has shown theoretically that at low exothermic effects, the contribution of noncombinatorial entropy of mixing (which was found to be negative) may lead to $\Delta G > 0$, showing the incompatibility of the system. Experimentally, the entropy of mixing was determined by Tager.[6-8] The concept was developed that negative entropy of mixing is the necessary condition of polymer compatibility. S. Lipatov[106] attributed the entropy contribution to the difference in packing density of individual polymers and their mixtures.

The conclusion according to which the entropy of mixing should be negative for compatible pairs follows from the experimental observation of negative values of excess volume of mixing.[97,110-115] The excess volume of mixing is connected with the entropy of mixing by equation such as Eq 3.23. If $\Delta V^E < 0$ then $\Delta S^E{}_p < 0$. It is worth noting that the comparison of the value of ΔV^E with miscibility was done mainly for mixtures with UCST where the theories really predict ΔV^E and $\Delta S^E < 0$ (see refs. 9-11, 54-56). If the mixtures have $\Delta V^E > 0$, ΔS^E should also be >0, as follows from Eq 3. 23. The theories of the equation of state predict the diminishing of the volume by mixing because one of the polymers expands more than the other, i.e., there is observed asymmetry of characteristic parameters of components T_i^* and P_i^*. However, for miscible pairs of PMMA with PVF and PEA, the positive values of excess volume of mixing have been found[116] in spite of strong difference in reduced volumes of components. Positive value of excess volume of mixing was also found[117] for the system PMMA/PVDF. In this work, the positive value of ΔV^E was discussed in the framework of the theory of equation of state. Varying characteristic parameters

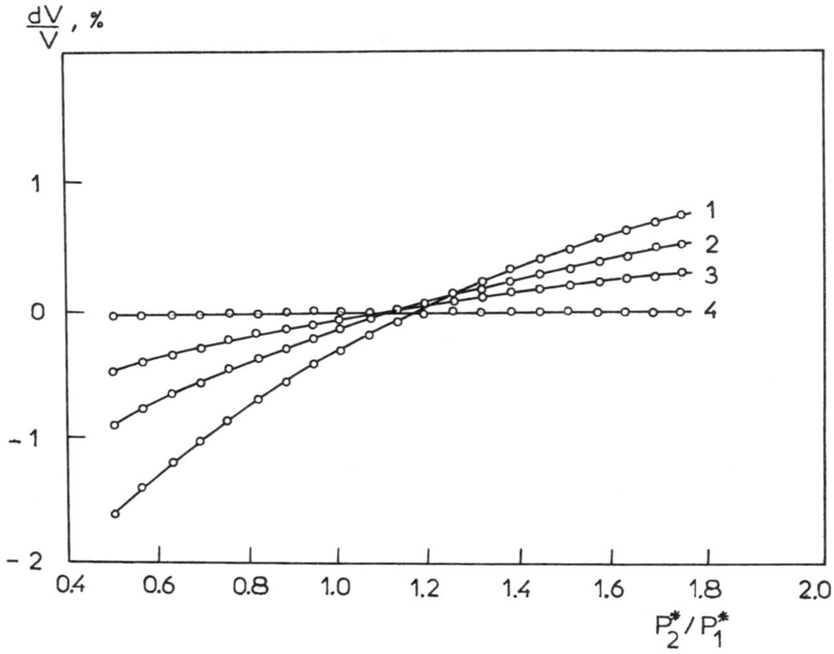

Figure 3.9. Dependence of the excess volume dV/V of mixing (50 wt% composition) on the ratios of pressures P_2^*/P_1^* for different ratios of the characteristic temperatures T_2^*/T_1^*: T_1^*=8,000K (1), 7,000K (2), 6,500K (3), 6,000K (4). Adapted, by permission, from W. Wolf and J. H. Wendom, *Polymer*, **30**, 1524 (1989).

for mixture components, it was found that ΔV^E may be either positive or negative, the transition between these values proceeding by increasing asymmetry of characteristic temperatures of components (Fig. 3.9). The conclusion has been drawn that the necessary conditions for positive values ΔV^E by mixing are strong asymmetry of thermal expansion coefficients $\alpha_A >> \alpha_B$ which leads to $T_B^* >> T_A^*$ and asymmetry of isothermal compressibility coefficient $\gamma_A >> \gamma_B$ leading to $P_A^* >> P_B^*$ (Figs. 3.9 and 3.10). Possible change of asymmetry of characteristic parameters by mixing was explained by the existence of strong specific interactions[117] enhancing local variations of structure. This effect may influence chain configuration through deformation of potentials of bonds rotation (a typical example is the formation

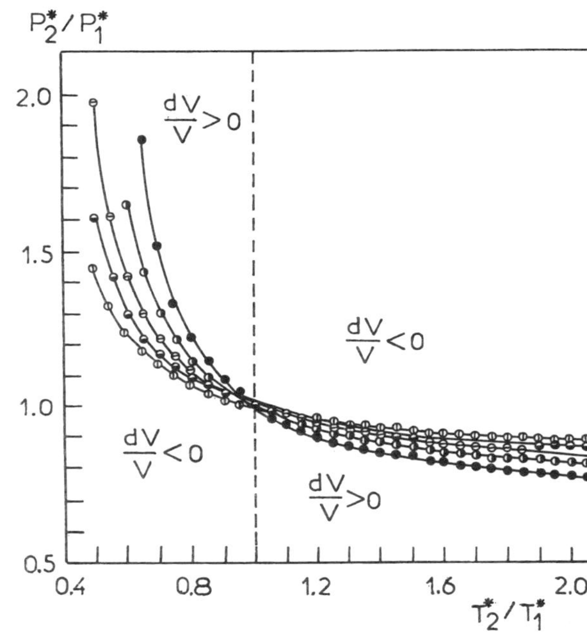

Figure 3.10. Dependence of the sign of the excess volume dV/V of mixing on the asymmetry of the ratios P_2^*/P_1^* and T_2^*/T_1^*. The curves represent the transition from positive to negative values and *vice versa*: T_1^*=6,000K (1, ●), 7,000K (2), 8,000K (3), 9,000K (4), 10,000K (5, O). Adapted, by permission, from W. Wolf and J. H. Wendom, *Polymer*, **30**, 1524 (1989) .

of ice from water where strong interactions determine specific configurations with higher free volume).

A great deal of work was dedicated to the estimation of the parameter of thermodynamic interaction, χ_{AB}, using different methods (small-angle scattering of heat neutrons,[118-132] melting point depression,[133-149] heats of mixing,[100-102] cloud points,[150-152] light scattering,[153] osmotic pressure,[154] inverse gas chromatography,[155-173] and interdiffusion via forward recoil spectroscopy.[174-176] The analysis of numerous data on parameters of thermodynamic interaction shows that their values, determined, for example, by using common solvent, are very small. To explain this fact, it was supposed[177] that in the ternary polymer-polymer-solvent systems two polymers do not fully mix, i.e., the number of contacts A-B is much smaller than it should be for random mixing. The strive to avoid the coil overlapping should lead to their repulsion and to segregation of two different macromolecular species. Yamakava and Fujii[178] suppose that interaction of polymer segments in dilute solution differs from that in the case of small molecules. This difference is connected with lower number of degrees of freedom for small molecules not bound into a chain. At any case, for polymer pairs, value χ_{AB} is very small even at the critical conditions and diminishes with growth of molecular mass of polymers mixed. These facts meet Flory conclusions[179] according to which the polymers with high molecular mass may be miscible only at the negative values of χ_{AB} when the interactions of the type A-B dominate over interactions A-A and B-B. This case may be realized only for specific interactions between two polymers. Therefore the results[150] showing the miscibility of atactic poly(vinyl ethylene) and *cis*-1,4-polyisoprene, are unique example, as in this system, according to the data from Fourier IR spectroscopy, there are no specific interactions but only van-der-Waals interactions. The reason for miscibility seems to be extremely low parameter χ_{AB} (0.0004). In this case, even a small gain in combinatorial entropy leads to the mutual miscibility. For the same system, it was also found[181] that $\chi_{AB} < 0$ and diminished by absolute value with increasing temperature, leads to the appearance of LCST. The negative energy of interaction may be the result of enhancement of local dispersion forces engendered by subtle alterations of the short range structure upon blending.

The majority of data on interaction parameters was derived from the measurements of melting point depression according to the equation:

$$-\frac{\Delta H_A V_A}{R V_B}\left(\frac{1}{T_{m,A}} - \frac{1}{T^0_{m,A}}\right) + \ln \varphi_B / N_B\left(\frac{1}{N_B} - \frac{1}{N_A}\right) = \chi_{AB}\varphi_A^2 \qquad [3.105]$$

where V_i and H_i are molar volume of monomeric unit of i-th component and enthalpy of melting of one mole of segments correspondingly, $T_{m,A}$ is the melting point of crystallisable component in the mixture. Eq 3.105 is not quite convenient because it does not provide the temperature and concentration dependence of parameter χ_{AB} for polymers with high molecular mass. This deficiency may be avoided by using inverse gas chromatography, IGC, that allows to determine parameters χ_{AB} in the whole composition range at various temperatures in the region for equilibrium sorption of a solvent[156,182] (sorbate). The only shortcoming of this method consists of the dependence of the measured values of parameter on chemical nature of sorbate.[168,169,173,184-191] To account for this circumstance, the concepts of Prigogine-Flory-Patterson theory should be used[158,192] (for more detail see part 3.1.2). In spite of this shortcoming, the IGC method is widely used for estimating the temperature and composition dependence of χ_{AB} for polymer mixtures.

From the studies of a set of polar and nonpolar sorbates, values of thermodynamic interaction parameter χ_{ij} between sorbate and PVA and PMMA have been calculated[187] and parameters χ_{AB} between the same for polymers with different molecular mass. It was found that for various sorbates including polar and nonpolar ones value $\Delta\chi = \chi_{SA} - \chi_{SB}$ is very small (no more than 0.1). Therefore these polymers have similar PVT properties. Value χ_{AB} was found independent of the type of sorbate used. Since in a given case both polymers are similar, it is difficult to explain this difference by non-uniform distribution of sorbate due to any preferential affinity to one of the polymers.[32] The parameter of contact interactions X_{AB} is small and positive, indicating the lack of specific interactions. The contribution of the term of free volume to χ_{AB} is small as well. Authors have not found the change in miscibility of PVA and PMMA with changing of molecular mass of PMMA by 25 times. On this basis the conclusion has been drawn that given polymers have similar chemical nature and PVT properties. As a result, the unfavorable noncombinatorial contribution to free energy of mixing (parameter of contact interactions and free volume term) is small. It was also observed[168] that for PVA-PBMA mixtures, parameter χ_{AB} practically does not depend on the mixture composition. For the mixtures of PS (molecular mass 110,000) with PBMA (molecular mass 320,000), it was found that the mixture is not miscible throughout all composition range.[193] However, for mixtures of PS with molecular mass 1,790 with the same PBMA, the miscibility was observed at any composition, although the concentration dependence of interaction parameter is very complicated (Fig. 3.11). The increasing miscibility with diminishing molecular mass was explained by entropy, not enthalpy, contribution to free energy of mixing. For the mixtures of PVDF and PMMA, interaction parameter was found to be negative and dependent on the mixture composition.

In Table 3.1, the data on parameters χ_{AB} are presented for the mixtures of PVC with various polyacrylates and polymethyl acrylates.[173] Low values of parameter show the existence of specific interactions between mixture components enhancing their miscibility.

Very interesting results have been obtained for the mixtures of PS with poly-2-dimethyl-1,4-phenyleneoxide.[191] Values χ_{Si}, $\chi_{S(A+B)}$, and χ_{AB} were determined and the dependence of $\chi_{S(AB)}$ on concentration was constructed. It was found that this dependence is characterized by a break point at 60-70% of PS. At 60%, the transition was discovered from positive to negative values of χ_{AB}. Despite of this transition, the system has only one glass transition point. According to the data on scattering of heat neutrons,[194] parameter χ_{AB} for the

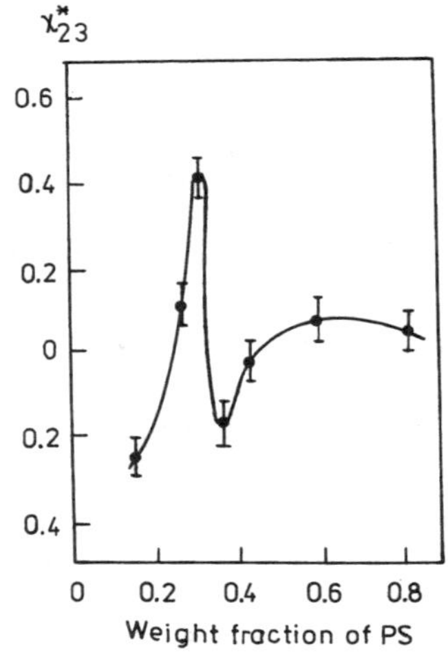

Figure 3.11. Composition dependence of the polymer-polymer interaction parameter for PS-PBMA blends. Adapted, by permission, from G. Di Paola, P. Degre, *Macromolecules,* **14**, 1456 (1981).

Table 3.1: Values of parameters χ_1 and χ^*_{23} per PEG segment for the mixtures PEG-40,000-PEG-2,000 at φ_3 = 0.3 and T = 353K by IGC[348]

Sorbate	χ_{12}	χ_{13}	χ^*_{23}
hexane	1.812	1.726	-0.0090
heptane	1.972	1.870	-0.0080
CCl_4	0.386	0.330	-0.0070
benzene	0.102	0.032	-0.0084
methyl ethyl ketone	0.328	0.285	-0.0092

same system is negative and $\Theta \approx 618$ K. To explain the convergence of data on χ_{AB} and glass transition, authors[191] refer to the data[195] where using DSC method, the broadening of the glass transition region with diminishing concentration of PS was discovered. This effect was explained by increase of microheterogeneity of the mixtures. The same conclusion follows from dielectric data.[196] Data of IGC are in a good agreement with postulated microheterogeneity of blends containing small amounts of PS.

When using values χ_{AB} and ΔG to characterize the miscibility, one has to bear in mind the following. If the mixture of crystalline and amorphous polymers is characterized by depression of melting point, according to equation 3.105, parameter χ_{AB} should be negative indicating the miscibility of components. However, the determinations of the same values using IGC or measurement of equilibrium sorption give the values characterizing the system only at a given temperature and composition. The system may be miscible at one composition and temperature and immiscible at another. So in this case one has to determine full concentration dependence of χ_{AB} and ΔG to judge the miscibility of two polymers.

3.2.1 PHASE BEHAVIOR OF BINARY POLYMER MIXTURES

To estimate thermodynamic compatibility (miscibility) of two polymers using various methods, it is necessary to study all concentration range and to have coinciding characteristics determining miscibility. Binodals and spinodals provide important information because, together with characteristics of the phase state, they give data for analysis of the mechanism of phase separation.

Presently, there are many reported phase diagrams of binary polymer mixtures.[197-201] There were discovered the systems with LCST,[60,63, 203-224] with UCST,[31,127, 225-253] and systems which have both LCST and UCST.[254-261] The diagrams of "sand-glass" type[256,259] and closed-loop immiscibility gaps[254, 262-264] were also described.

According to the phase behavior all the polymer mixtures can be classified in the following way:[265]

1. systems with specific interactions between mixture components
2. mixtures of oligomers
3. systems composed of chemically similar polymers
4. mixtures of homopolymers with copolymers and mixtures of copolymers.

Following this classification, the first type of the system is typical of miscible mixtures that have only LCST, whereas the second type has UCST. Types 2 and 3 may have both UCST and LCST. Miscible mixtures formed by high molecular polymers are characterized by the existence of LCST,[60,63,203-224,266-277] although both CST[254-261] and "sand-glass" diagrams were also discovered (Fig. 3.12). For oligomer mixtures UCST is observed as a rule.[31,127,225-234,243-253] Phase diagrams of closed type (Fig. 3.13) were discovered for polymer-oligomer mixtures.[262-265]

The dependencies of the phase behavior on the composition and molecular mass are rather well described by the current theories of polymer solutions and blends. At the same time, some peculiarities of such systems have been found that cannot presently be explained. For

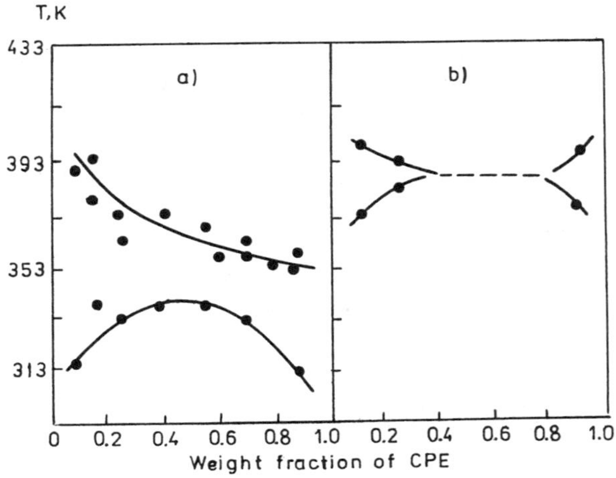

Figure 3.12. Phase diagram for blends of CPE ($M_w = 1.08\times10^5$, 27.4 wt% Cl) with PMMA (a)($M_w = 1.17\times10^5$) and CPE ($M_w = 1.2\times10^5$, 47.8 wt% Cl) with the same PMMA (b). Adapted, by permission, from D. J. Wolf, S. Laingle, C. Zhikuan, *Polymer*, **22**, 1005 (1981).

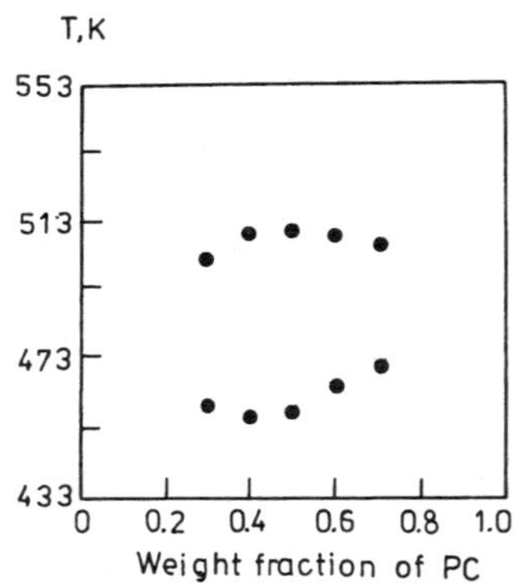

Figure 3.13. An miscibility loop cloud point phase diagram of PC/PMMA blends at the heating rate 0.5°C/min. Adapted, by permission, from T. Kyu, D.-S. Lim, *J. Polymer Sci.-Lett.*, **27**, 423 (1989).

example, for the mixture PVDF-PEMA (molecular masses 100,000 and 4,000, correspondingly), it was found[274] that the critical concentration of mixing is situated in the composition regions enriched in PVDF despite of the fact that its molecular mass is much higher (Fig. 3.14). This behavior differs from predicted by Flory-Huggins theory and by new statistical theories. It was shown[274] that such behavior can be described in the framework of modified Flory's theory. For this

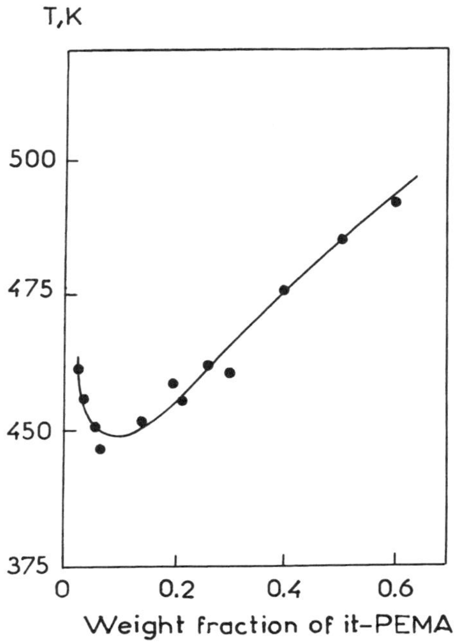

Figure 3.14. Cloud point curve of blends of PVF with low molecular weight it-PEMA recorded by turbidimetry. Adapted, by permission, from G. ten Brinke, A. Eshuis, E. Roerdink, and G. Challa, *Macromolecules*, **14**, 867 (1981).

purpose the number of freedom degrees in equation of state is accepted to be non-additive and expressed as

$$c = c_A\varphi_A + (1 - \varphi_A)c_B - \varphi_A(1 - \varphi_A)c_{AB} \qquad [3.106]$$

It is then possible to apply the original Flory's equation-of-state, introducing the following designations:

$$P^* = \varphi_A P_A^* + \varphi_B P_B^* - \varphi_A \Theta_B X_{AB} \qquad [3.107]$$

$$\tilde{T} = T/T^* = (1/P^*)[\varphi_A \tilde{T}_A P_A^* + \varphi_B \tilde{T}_B P_B^* - (c_{AB}\varphi_A\varphi_B kT/V^*)] \qquad [3.108]$$

Equations for spinodal have been calculated[174] for PEMA-PVDF by varying parameters X_{AB} at $c_{AB} > 0$. Choice of this condition is possible due to the existence of specific interactions in this mixture leading to volume decrease by mixing as compared with additive values. Thus, the number of outer freedom degrees diminishes as well. Fig. 3.15

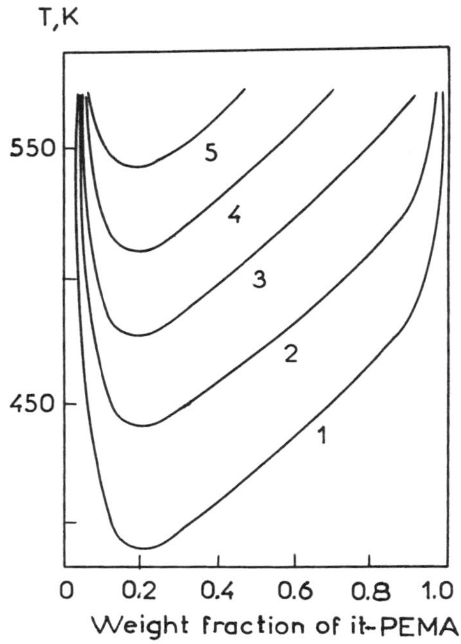

Figure 3.15. Simulated spinodals for it-PEMA/PVF mixtures (m_A/m_B=0.66, S_A/S_B=1, c_{AB}=0.02). X_{AB} - 2.09 (curve 1), - 4.18 (2), - 6.17 (3), - 8.36 (4), - 10.45 J/cm^3 (5). Adapted, by permission, from A. Eshuis, E. Roerdink, and G. Challa, *Macromolecules,* **14**, 867 (1981).

shows calculated spinodals for two values of S_B/S_A and c_{AB}. Fig. 3.16 demonstrates that because of negative values of X_{AB} and at $S_A/S_B = 1$ and $c_{AB} = 0$, phase diagrams are obtained having minimum in the composition range enriched in PEMA, calculated diagrams are situated above experimental ones. Hence to obtain the agreement between calculated and experimental phase diagrams, it is necessary to use the positive values of c_{AB}. Better coincidence may be found if to take the following values: $r_A/r_B = 0.66$, $S_A/S_B = 0.55$ and $c_{AB} = 0.02$ (Fig. 3.16). In this case, the deviation of the number of freedom degrees from additivity is almost the same as found for the system PDMS-solvent. The best result was found at $X_{AB} = -6.27$J/cm^3. The shift of the phase diagram to the composition region enriched with polymer of higher molecular mass was also observed for mixtures of PDMS (mol. mass 850) with PIB (mol. mass 250) with UCST.[225] For these mixtures

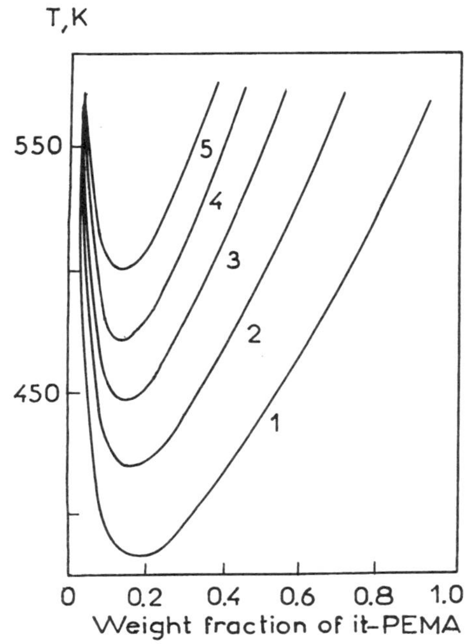

Figure 3.16. Simulated spinodals for it-PEMA/PVF mixtures (m_A/m_B=0.66, S_A/S_B=0.55, c_{AB}=0.02). X_{AB}: - 2.09 (1), - 4.18 (2), - 6.17 (3), - 8.36 (4), - 10.45 J/cm^3 (5). Adapted, by permission, from A. Eshuis, E. Roerdink, and G. Challa, *Macromolecules*, **14**, 867 (1981).

$r_{PIB}/r_{PDMS} < 1$ and $c_{AB} > 0$ and the shape of phase diagram may be attributed to non-additivity of the outer freedom degrees.

For some oligomeric mixtures the bimodality of phase diagrams has been discovered.[31,225,230,231,247] For example, for mixtures of PDMS with n-alkanes of various chain length, it was found[231] that the temperature of phase separation first increases with molecular mass of alkane and then changes very slowly. The bimodal shape of phase diagram with two plateau regions or with two sharp maxima is typical of these systems. For relatively long chains of alkane and PDMS, the curve with two plateau and for unfractionated PDMS, the curves with two maxima were observed. For the mixtures of PPG and PEG at some ratio of molecular masses the "shoulder" and bimodality on phase diagrams also are described.[230] According to section 3.1.2, the bimodality of phase diagrams can be described within a framework of a new theory by Huggins[31] and is predicted by the Sanchez-Lacombe theory.[54-56]

The majority of miscible polymer pairs described up to now belong to the systems with LCST, i.e., they are characterized by the negative enthalpy of mixing. To describe the exothermal processes of mixing, the concept of "complementary dissimilarity" was proposed.[12] Its essence consists of assumption that macromolecules of mixed polymer should have some functional groups which supplement each other and are capable of formation of H-bonds or donor-acceptor complexes. The same concept was suggested as a basis for the analysis of behavior of dilute polymer solutions.[278]

As a typical example, PVC can be mixed with polymers containing acceptor groups, including polylactones,[32,279-282] polyethyl acrylates, polymethylacrylates,[103] and other polymers.[283-286] A characteristic feature of these mixtures consists of formation of H-bonds and donor-acceptor bonds.[32,287] For the mixtures of PVC with poly(alkyl acrylate) and poly(alkyl methacrylate), the dependence of χ_{AB} on the alkyl group length was found to pass through a minimum,[103] i.e., there is some minimal concentration of ester groups at which these polymers can be mixed with PVC. This can be due to a definite ratio between specific and dispersion interactions.[103] To evaluate this assumption, the heats of mixing were determined for oligomers modelling acrylates (mol. mass 1,000-3,000) and chlorinated paraffin (mol. mass 437, Cl content 53.1%) modelling PVC. In all cases, the heats of mixing were negative except for the mixture with oligomethacrylate. The contribution of dispersion interactions to the heat of mixing was estimated on the basis of the theory of solubility parameters. The contribution of specific interaction was estimated as a difference between total and calculated heat of mixing. The miscibility of components requires that contribution of specific interactions is larger than dispersion interactions. It is already modeled by the additional term (characterizing specific interactions) in the expression for the free energy of mixing (Eq 3.24). The heat of mixing of model compounds correlates with miscibility of corresponding high molecular compounds. However, there was no reliable correlation between heats of mixing and parameters χ_{AB}. This can be explained by contribution of noncombinatorial mixing entropy to χ_{AB}. Specific interactions explain also the miscibility of chlorinated PVC with polylactones,[281,282] where the interactions of the type -CH...OOC- between CH(D) groups of PVC and COO- groups of polylactone occur. The miscibility of some copolymers of PVC with polylactones also depends on the interactions of dipoles C=O and C=Cl.[288] It is interesting that polylactones are not miscible with poly(vinyl fluoride) and poly(vinylidene fluoride).[289] It is probably due to the essential difference in electronegativity of atomic radii and molar attraction between atoms of Cl and F (F has higher electronegativity than Cl and forms stronger bonds with other atoms).[290] The miscibility of polylactones with poly(vinyl bromide) is improved compared with PVC. Poly(vinylidene fluoride) is miscible with many poly(alkyl acrylates) and poly(alkyl methacrylates)[134,268] containing short alkyl radicals. Specific interactions can explain many cases of miscibility of chlorinated polymers with various polyesters and other polymers.[283-294]

To estimate the presence of specific interactions in a mixture, the Fourier transform IR spectra are widely used.[292,295-305] It is possible to estimate quantitatively the strength and number of intermolecular bonds in the miscible mixture poly(-4-vinyl phenol) -poly(ethylene-co-vinyl acetate).[297] The fraction of carbonyl groups participating in H-bonding and the heat of interaction can be found from intensities of adsorption bands at 1738 and 1708 cm^{-1} (frequencies of free and bound C=0 groups). The calculations were performed for two samples: one obtained by evaporation of a solution at room temperature, the second preliminary heated to 463 K. The spectra of the first sample were taken after heating to the demixing temperature and for the second sample after cooling from demixing temperature. In the first case, carbonyl groups participate in the formation of H-bonds. If to denote the intensities (areas) of the absorption bands near 1738 and 1708 cm^{-1} for the first case as I_1 and I_2 and for the second case as I'_1 and I'_2 then the following equations can be written:

$$a_2 = (I_1 - I'_1) = a_1(I'_2 - I_2) \qquad [3.109]$$

$$a_1/a_2 = (I_1 - I'_1)/(I'_2 - I_2) \qquad [3.110]$$

where a_1 and a_2 are absorption coefficients for the bands at 1738 and 1708 cm^{-1}, respectively. From the ratio 3.110, one can calculate the fraction of carbonyl groups participating in the formation of H-bonds.

To calculate the strength of intermolecular interactions, the mixture of poly(vinyl phenol) and copolymer is considered as a complex with molar concentrations designated as [-OH] and [-C=O]. The concentration of carbonyl groups participating in the H-bonding is designated as x. Then in a simplified form

$$[-OH] + [-C{=}O] \leftrightarrow K \leftrightarrow [-OH...O{=}C]; \quad K = x/([-OH] - x)([-C{=}O] - x)$$

$$\ln K = (-\Delta H/RT) + C$$

where K is the equilibrium constant, ΔH is the enthalpy of complex formation, and C is a constant. The dependence of lnK on 1/T is presented in Fig. 3.17. Below glass transition temperature of miscible system (T_g = 360 K), this dependence becomes nonlinear due to possible non-equilibrium state of the system. From the initial part of

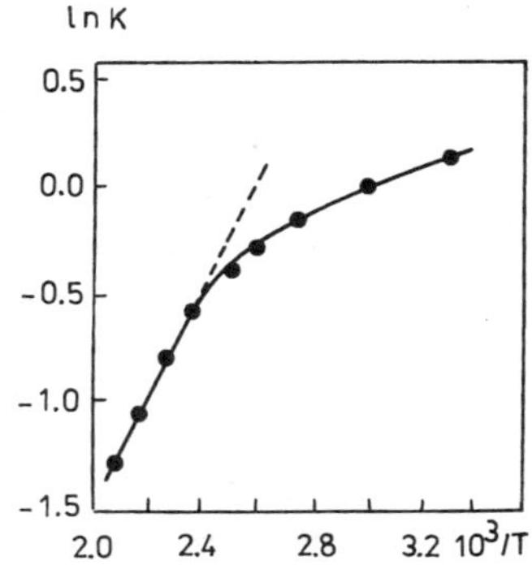

Figure 3.17. Graphs of lnK vs. T^{-1} based on the data obtained for 80:20 (by wt) PVPh-EVA (45) blend. Adapted, by permission, from E. J. Moskala, S. E. Howe, P. C. Painter, and M. M. Coleman, *Macromolecules*, **17**, 1671 (1984).

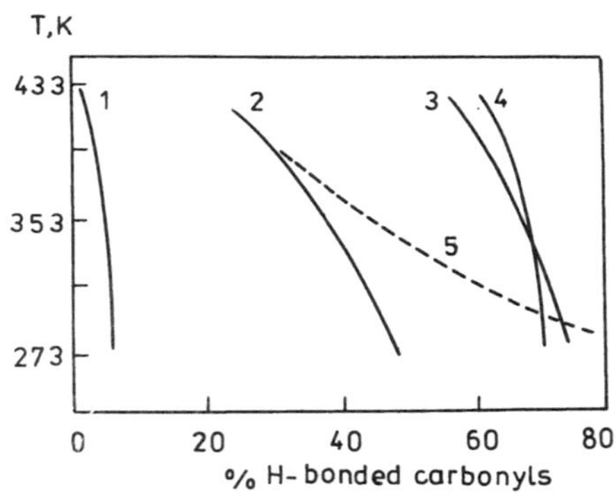

Figure 3.18. Graph of temperature vs.. fraction of hydrogen-bonded carbonyls for samples of 80:20 wt% blends heated to 463K and measured as a function of decreasing temperature: 1 - EVA(25), 2 - EVA(45), 3 - EVA(70), 4 - PVAc (curves 1-4 - experimental data, curve 5-calculation). Adapted, by permission, from E. J. Moskala, S. E. Howe, P. C. Painter, and M. M. Coleman, *Macromolecules*, **17**, 1671 (1984).

the dependence, there were determined ΔH = 21.4 kJ/mol and C = –6.55. From these values the fraction of carbonyl groups forming H-bonds was calculated. This calculated fraction exceeds experimental results below glass transition temperature which indicates the non-equilibrium state of the system caused by small mobility of macromolecules. With increasing amount of vinylacetate in copolymer, the temperature of demixing increases, i.e., the dissociation of H-bonds occurs at higher temperature (Fig. 3.18).

It is worth noting that, from the temperature dependence of the shift of frequencies of valent vibrations of C=0 group, it is possible to approximate LCST.[306] Spectroscopic method was used to describe the changes in free energy of complex formation in binary mixtures where strong and directed interactions are present.[307-309] The associative model was used for analysis. The associated complexes are considered as linear chains of small molecules bound at their ends by hydrogen bonding forming new, distinguishable molecular species. The H-bonding interactions are considered independently of the forces involved in mixing. The number and distribution of species was estimated through the equilibrium constants related to the changes in free energy. This approximation was used[307] to describe mixing of polyurethanes and polyethers under assumption that the formation of H-bonds is not influenced by covalent bonding. Considering polymer chains as flexible and assuming that functional groups are separated by other species,[307] it was concluded that covalent bonding did not prevent formation of H-bonds. Eq 3.14 was used to calculate the additional term $\Delta G_H/RT$ that is expressed as[311]

$$\Delta G_H/RT = \varphi_B \ln \frac{\varphi_{B1}}{\varphi_B^0 \varphi_B^{1/n_H^0}} + \frac{\varphi_A}{r} \ln \frac{\varphi_{0A}}{\varphi_A} + \varphi_B \left[\left(\frac{\Gamma_1^0}{\Gamma_2^0} \right) - \left(\frac{\Gamma_1}{\Gamma_2} \right) \right] +$$

$$+ \varphi_B \frac{\Gamma_1}{\Gamma_2} \frac{X}{1+X} \qquad [3.111]$$

where $n_H^0 = \Gamma_2^0/\Gamma_1^0$ is the mean number length of hydrogen-bound chain in pure component where the self-associated species are formed:

$$\Gamma_1 = (1 - K_2/K_B) + (K_2/K_B) \frac{1}{1 - K_B \varphi_{B1}} \qquad [3.112]$$

$$\Gamma_2 = (1 - K_2/K_B) + (K_2/K_B) \frac{1}{(1 - K_B \varphi_{B1})^2} \qquad [3.113]$$

Here $X = K_A \varphi_{0A}/r$, K_A is association equilibrium constant for formation of hydrogen bond between B and A units, K_B is self-association equilibrium constant between B units, K_2 is equilibrium constant for formation of dimers from B units, φ_A and φ_B are volume fractions of non-self-associated species A and self-associated species B, φ_{0A} and φ_{B1} are volume fractions of total and free monomers, $r = v_A/v_B$. The following equations have been derived to estimate the equilibrium constant K_A:

$$\varphi_B = \varphi_{B1}[(1-K_2/K_B)+(K_2/K_B)]\frac{1}{(1-K_B\varphi_{B1})^2}\left[1+\frac{K_A\varphi_{0A}}{r}\right] \quad [3.114]$$

$$\varphi_A = \varphi_{0A} + K_A\varphi_{0A}\varphi_{B1}\left[(1-K_2/K_B) + (K_2/K_B)\frac{1}{1-K_B\varphi_{B1}}\right] \quad [3.115]$$

where $\varphi_{0A} = (1-\varphi_A)[1 + K_A\varphi_{B1}/K_B\varphi_B]^{-1}$ for stoichiometric ratio.

The equations for spinodal have also been derived:[311]

$$2\chi - \left[\frac{1}{\varphi_{B1}}\frac{\delta\varphi_{B1}}{\delta\varphi_B} - \frac{1}{r\varphi_{0A}}\frac{\delta\varphi_{0A}}{\delta\varphi_B}\right] + \left[\frac{1}{\bar{n}_H^0\varphi_B} + \frac{1}{r\varphi_{0A}}\right] = \frac{1}{\varphi_B N_B} + $$

$$+\frac{1}{\varphi_A N_A} \quad [3.116]$$

where

$$\frac{\delta\varphi_{B1}}{\delta\varphi_B} = \frac{1 + [K_A\varphi_{0A}\varphi_{B1}/r\varphi_A]}{1 + 4K_B\varphi_{B1} + [K_A\varphi_{0A}^2/r\varphi_A]} \quad [3.117]$$

$$\frac{\delta\varphi_{0A}}{\delta\varphi_B} = \frac{\varphi_{0A}(1 + K_A\varphi_{0A})\frac{\delta\varphi_{B1}}{\delta\varphi_B}}{\varphi_A} \quad [3.118]$$

under supposition that interacting units and chemically repeating units are identical. Parameter χ is calculated from Eq 3.9 and is considered as a linear function of 1/T.

The equilibrium constants K_A and K_B, which describe B-A and B-B hydrogen bonds, depend on the enthalpy of these bonds, h_B and h_A, via usual thermodynamic relations:

$$\delta \ln K_B/\delta(1/T) = -h_B/R \quad [3.119]$$

$$\delta \ln K_A/\delta(1/T) = -h_{BA}/R \quad [3.120]$$

$$K_B = K_B^0 \exp[-h_B/R(1/T - 1/T^0)] \quad [3.121]$$

$$K_A = K_A^0 \exp[-h_{BA}/R(1/T - 1/T^0)] \quad [3.122]$$

where K_A^0 and K_B^0 are equilibrium constants at absolute temperature. In some works[311-322] this theory was applied to calculate phase diagrams and a good agreement between calculated and experimental data was found.

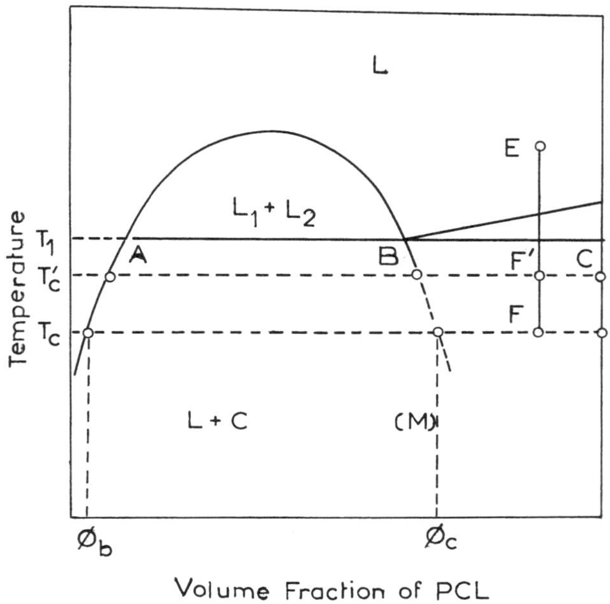

Figure 3.19. Speculative illustration of the phase diagram of PCL-PSO system. L and C denote liquid and perfect crystalline phases, respectively, and the dashed curve represents the lower limit of φ_{PCL} in the amorphous layer between lamellae. Adapted, by permission, from S. Nojima, K. Satoh, and T. Ashida, *Macromolecules*, **24**, 942 (1991).

Besides hydrogen bonds, the miscibility of binary polymer mixtures may depend on ion-ion,[323-333] ion-dipole,[221,324] and other[334] interactions. Among mixtures of crystallisable polymers with no specific interactions, there were discovered systems where components are capable of co-crystallization: PE-deuterated PE,[335,336] poly(isopropylene)-deuterated PP,[337] poly(isopropylene)-polybutene,[338] and PEHD-PELD.[339]

For the mixtures of amorphous and crystalline polymers, phase diagrams show both processes - crystallization and phase separation.[340,341] These two transitions determine very complicated morphology. Typical phase diagram of the mixture oligostyrene-poly(caprolactone), PCL, is presented in Fig. 3.19. In this diagram, L and C represent disordered (amorphous) and ordered (crystalline) phases. In the region L_1+L_2 two disordered phases coexist, whereas in the region L + C there coexist a disordered phase and perfect crystal. At the temperature T_1, three phases coexist: A, B, and C (eutectic temperature). By quenching the system in the region L+C, outside the coexistence curve, two phases are always formed. One of them is enriched with oligostyrene and the other is a perfect crystal of PCL. By crystallization, the amorphous layer of PCL is always formed between lamellas. This layer is enriched with oligostyrene. As a result, the third phase, M, appears which is shown by the dotted line and represents the amorphous phase between lamellas. In this phase, the limiting miscibility of two components occurs. The amorphous layer can be enriched with oligostyrene along the composition axis as shown by the dotted line. However, the following increase in the oligostyrene content initiates the phase separation in the layer. This phase always exists but is not in thermodynamic equilibrium and may change its composition in accordance with thermodynamic and kinetic factors (or T_g and φ_{PCL}) during crystallization. When the mixture is transferred from homogeneous state to the two-phase region designated as L + C, both phases are in the state of thermodynamic equilibrium. One of them is enriched with oligostyrene and another consists of pure PCL. Hence the crystalline domain structure (alternating lamellar structure and amorphous layer) favors diminishing the free energy of the system. Amorphous component is chucked out the lamellas during crystallization and partially transferred to the amorphous layer between lamellas. The structure is then formed consisting of two domains: one fully amorphous with composition φ_B and another crystalline consisting of lamellas (perfect crystal). Between them the amorphous layer is situated which contains PCL and has a composition $\varphi_A >> \varphi_C$.

3.2.2 CONCENTRATION AND TEMPERATURE DEPENDENCIES OF PARAMETERS OF THERMODYNAMIC INTERACTIONS IN THE MIXTURES OF OLIGOMERS AND POLYMER HOMOLOGUES

During last years many reactive oligomers were developed, such as oligoesters, oligoethers, oligoesteracrylates, etc. They are widely used for polymer production. The main peculiarity of these oligomers consists of strong intermolecular interaction determining the formation of associates in the reaction system. The latter influences reaction mechanism and structure of cured polymer.[342-345] In many cases, the mixtures of oligomers are used. Knowledge of thermodynamic behavior of such systems is important to understand the properties and morphology of networks formed on the base of these oligomers.[345,346] Using the IGC method the thermodynamic reaction parameters were determined for mixtures of poly(ethylene glycol) ($M = 1.5\times10^4$) with poly(propylene glycol) (M = 2,000), PPG, and mixtures of poly(ethylene glycol adipate), PEGA, with PPG.[200.347] The composition and temperature dependencies of the interaction parameter, χ_{AB}, were measured (Fig. 3.20). The curve of the dependence of χ_{AB} for the mixture of

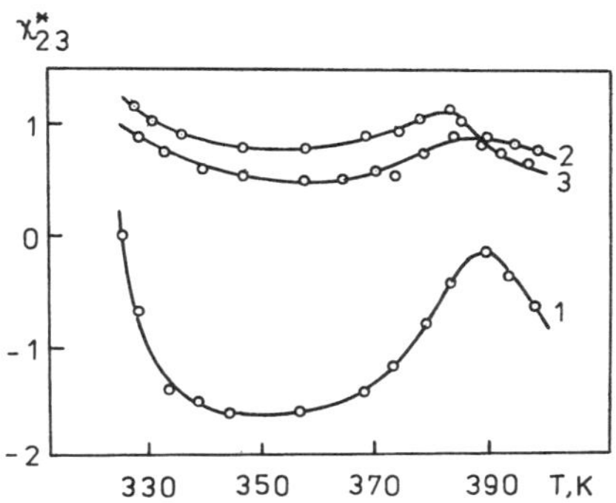

Figure 3.20. Dependence of $\chi^{+}_{AB}=(V_1/V_2)\chi_{AB}$ on T for PPG-PEGA: 1 - 88.0, 2 - 78.0, 3 - 65.0 wt% of PEGA. Adapted, by permission, from Y. S. Lipatov, A. E. Nesterov, and T. D. Ignatova, *Eur. Polym. J.*, **15**, 775 (1979).

PPG-PEGA passes through a minimum in the temperature range of 343-363 K and reaches its maximum at 385-390 K. The presence of a minimum on the temperature dependence makes the behavior of the system similar to binary mixtures with both UCST and LCST. In the temperature region of 328-343 K (Fig. 3.20), $\chi_H = -T(\delta\chi/\delta T)_P > 0$ (typical of UCST) and at the same time the maximum at 370-390K shows that $\chi_H = -T(\delta\chi/\delta T) < 0$ (typical of LCST). Simultaneously, the maximum at 385-390 K indicates that it is not LCST. The analogous maxima of the dependencies of χ_{AB} and T for these systems are characteristic also for the mixtures of PEG-PPG (Fig. 3.21). This type of the dependence can be explained in the following way. Specific interactions favor miscibility. The weakening of these interactions, for example, H-bonds, should lead to the demixing and formation of two phases. Therefore, the appearance of LCST in such systems may be the result of the weakening of H-bonds. This effect in PPG-PEGA and PEG-PPG mixtures is induced by increasing temperature. The presence of a maximum in the temperature dependence $\chi_{AB} = f(T)$ can be a sign of the system with closed phase diagrams, i.e., when UCST is situated above LCST.[262,264] The weakening of hydrogen bonds for these systems follows from the data obtained by IGC method, in particular, from the temperature dependence of logarithm of the retention volume on inverse temperature for polar (methanol) and non-polar (heptane) sorbates. There are some break points for methanol that are not present in the curves for heptane. The lack of the breaks for heptane shows that by heating the mixture of PPG-PEGA, there are no phase transitions. Increasing retention volume shows that, with increasing temperature, the hydrogen bonds are weakened.

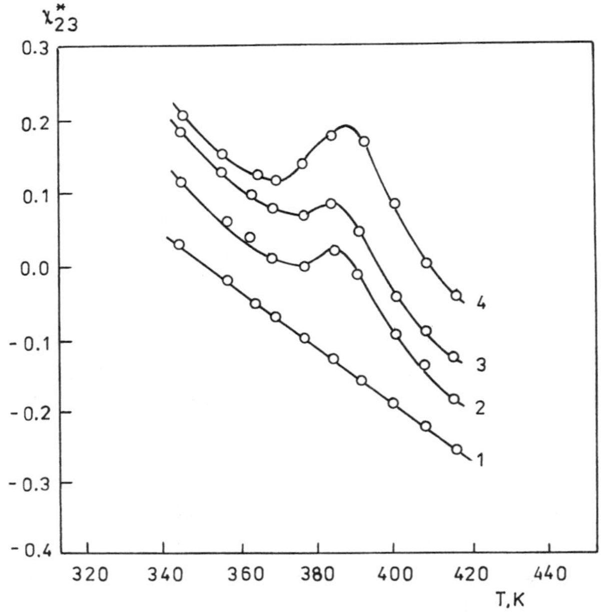

Figure 3.21. Temperature dependence of parameters χ_{AB} (for segment PEG) for PPG-PEG: 1 - 12, 2 - 31, 3 - 41, 4 - 69 vol% PPG. Adapted, by permission, from Y. S. Lipatov, A. E. Nesterov, and T. D. Ignatova, *Eur. Polym. J.*, **15**, 775 (1979).

With the weakening of the H-bonds, the number of functional groups of stationary phase, capable of specific interaction with methanol, increases. As a result, the relative retention volume $V_{g,rel} = V_{g(methanol)}/V_{g(heptane)}$ increases in the region of maximum of $\chi_{AB} = f(T)$ (Fig. 3.22). Similar break points were observed for mixtures of macrodiisocyanates and PPG.[346] The increase of χ_{AB} with temperature decreasing can be related to phase separation of the liquid crystalline type.

Minimum in the dependence of χ_{AB} on T was observed also for the mixture PEG-PEGA (Fig. 3.23). χ_{AB} was negative. It can be expected for system to demix on heating. Fig. 3.24 shows the dependence of the specific retention volume of sorbate on inverse temperature. For the mixtures PEG-PEGA above the melting point (point C), there are observed some break points (designated by arrows) which can be related to the temperature of phase separation.[349] The values of χ_{AB} above and below the temperature of additional break points on the dependence of $\ln V_g$ on 1/T indicate phase separation (Fig. 3.25).

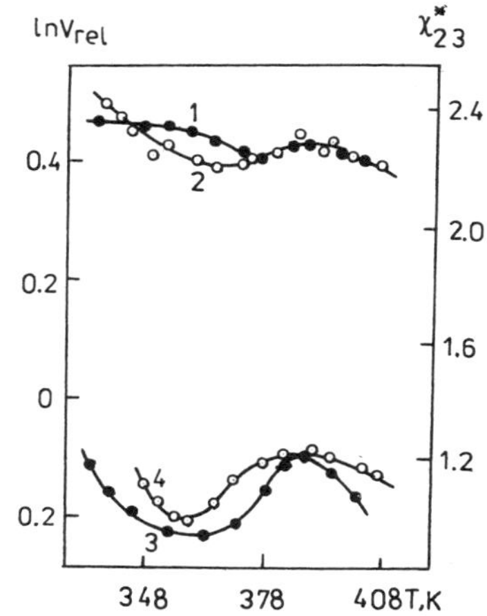

Figure 3.22. Dependence of log V_{rel} (1,2) and χ^*_{23} (3,4) on temperature for PPG-PEGA at 30 wt% of PPG: 1,3 - initial samples, 2,4 - heat-treated samples. Adapted, by permission, from Y. Lipatov, A. E. Nesterov, T. D. Ignatova, *Eur. Polym. J.*, **15**, 775 (1979).

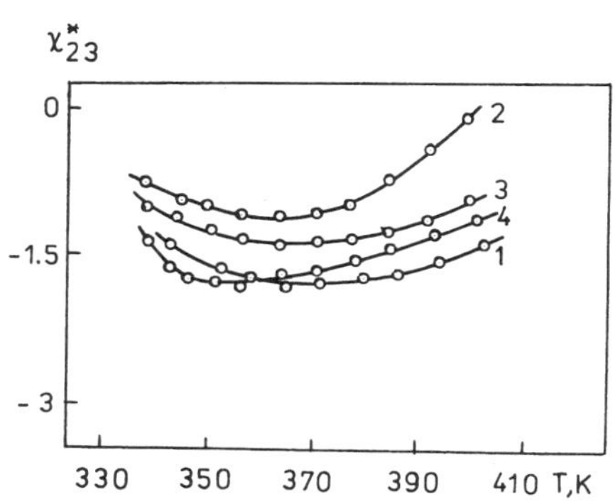

Figure 3.23. Dependence of χ_{23} on T for PEGA-PEG: 1 - 20, 2 - 30, 3 - 70, 4 - 80 wt% PEGA. Adapted, by permission, from Y. S. Lipatov, A. E. Nesterov, T. D. Ignatova, *Eur. Polym. J.*, **15**,775 (1979).

The values of χ_{AB} below the break temperature on the curves $\ln V_g - 1/T$ (Fig. 3.25) are negative and above the break temperature – positive. Measurements of the temperature dependence of the light scattering intensity show that above the same break points a sharp

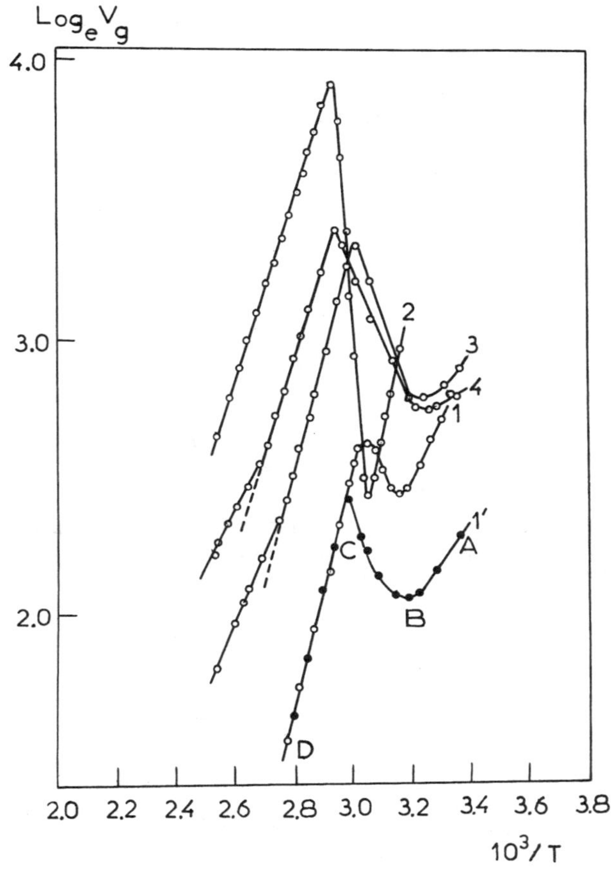

Figure 3.24. Dependence of logV$_G$ on 1/T for PEGA (1, 1′), PEG (2) and their mixtures in the ratio 10/90 (3), 30/70 (4), 70/30 (5). Curve 1′ - after quenching the melt PEGA from 378 K.

increase in the intensity occurs. This is a criterion of the phase separation for the systems with LCST.[350] Fig. 3.26 shows the concentration dependence of the phase separation temperature constructed from the break points on the curves $\ln V_g = f(1/T)$. This dependence has an unusual shape (convex upwards). In other sources, only closed region of immiscibility was observed.[262-264] This region of miscibility is restricted underneath by crystallisation. For this system a depression of melting points of both components and eutectic points are found

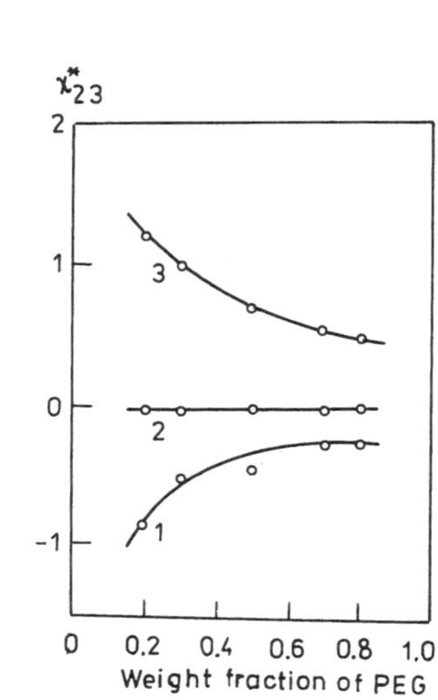

Figure 3.25. Concentration dependence of χ_{AB} for PEG/PEGA mixtures at 345K (1) 10^0 higher of break point on $\lg V_G = f(1/T)$ dependence (2) and 385K (3).

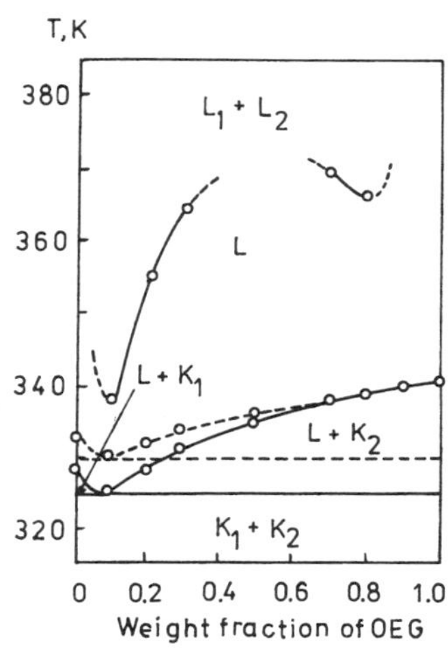

Figure 3.26. Phase diagram for PEG-PEGA mixtures. Solid lines - initial system, dashed lines - after quenching the melts from 378K.

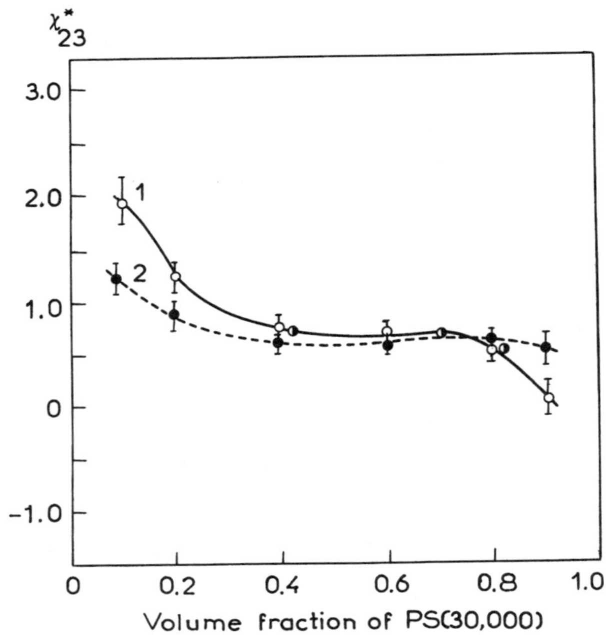

Figure 3.27. Temperature and concentration dependence of χ_{AB} for mixtures PS (2,000)-PS (30,000). 1 - 415, 2 - 454K. Vertical lines - scatter of experimental data of parameter, determined by using various sorbates).

(Fig. 3.26). These data confirm the suggestions by Nishi[24] that two crystallizable polymers miscible in melt crystallize by quenching with the formation of eutectics.

According to the theories of the equation of state for solutions and mixtures of polymers, the difference in parameters of the equation and, in particular, in free volumes is one of the essential reasons for immiscibility in binary mixtures. Even at the same values of contact interaction energies, the difference in free volumes may lead to the phase separation. To account for the contribution of these factors to the thermodynamic behavior of polymeric mixtures, it is important to have such binary systems where one parameter may be varied. From this point of view, the mixtures of polymer homologues give the best possibilities. Polymer homologues are traditionally considered as thermodynamically stable systems. For the mixture of polymer homologues with considerably different molecular masses, the difference in the coefficients of thermal expansion can be rather high. For example,

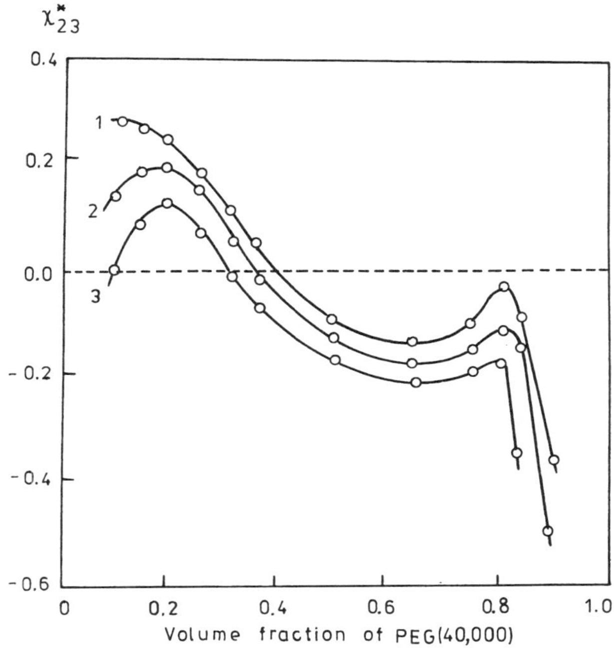

Figure 3.28. Temperature and concentration dependence of χ_{AB} for PEG (2,000)-PEG (40,000): 1 - 345, 2 - 370, 3 - 409K.

for PS with molecular masses 2,000 and 30,000 expansion coefficients are equal to 0.821×10^{-3} and 0.515×10^{-3}, respectively.[351] For poly(ethylene glycol) with molecular masses of 2,000 and 40,000, the difference in expansion coefficients is small but, due to different specific content of OH-groups, the energy of interactions differs substantially. These two polymers were taken to estimate the role of two contributions in thermodynamic properties of mixtures.[200,347] The interaction parameters for the mixtures of PS-PS and PEG-PEG were found in a broad range of composition and temperatures. For the mixtures of PEG 2,000 and PEG 40,000 parameter χ_{AB} is a function of both temperature and composition. Concentration dependence is bimodal (Figs. 3.27 and 3.28).

Parameter χ_{AB} increases with temperature due to the weakening of hydrogen bonding. Fig. 3.28 shows that at some compositions of mixtures (10, 35, and 80 wt% PEG 40,000), with increasing temperature, χ_{AB} changes from negative to positive values. Therefore, it could

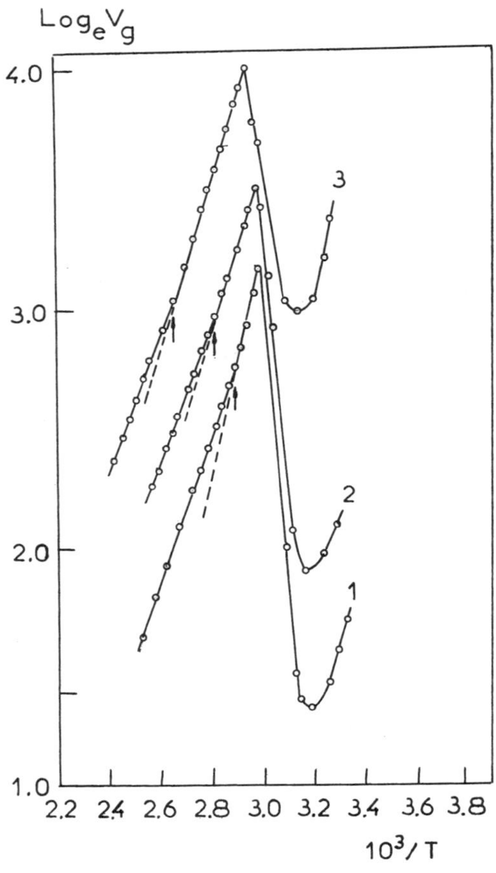

Figure 3.29. Dependence of $\ln V_G$ on 1/T for the mixtures PEG (2,000)-PEG (40,000): 1 - 10, 2 - 35, 3 - 80 wt% PEG (40,000).

be expected that the phase separation should occur. To establish this effect from the measurements of temperature dependence of light scattering is not possible because the refractive indices are similar. IGC method can be used more effectively. Fig. 3.29 shows that the dependencies of $\ln V_g$ and 1/T have two additional break points above the melting point of the mixture. This effect may be the evidence of the phase separation in the system[349] and may serve as a criterion of the phase separation. However, because of the dependence of χ_{AB} on the sorbate nature the strong correlation between the temperatures of phase separation and break points on dependence ln V_g = f(T) can hardly be expected.

For the mixtures without specific interactions (PS-PS), a very high positive values of χ_{AB} are observed (Fig. 3.28) that are larger than χ_{crit}. Interaction parameter in this system diminishes with increasing temperature as it should for the system with dispersion interactions between components. In principle, for the system of different thermal expansion coefficients and having close values of the solubility parameter, there should be observed the LCST and χ_{AB} should increase with temperature. However, Patterson[33] has shown that effect of volume changes is not very essential for polymer mixtures. Based on the theory of equation of state, value χ_{AB} was calculated according to Eqs 3.25-3.28. From the data on expansion coefficients[351] the characteristic and reduced temperatures and volumes of components were determined. Taking $P_B^* \approx 500$ J/cm^3 (typical for PS), and selecting values $\tau = 1 - T_B^*/T_A^*$, the magnitude of χ_{AB}/V_B^* was found in dependence on temperature for the system PS (2,000)–PS (30,000) and for mixtures of various fractions of PS with PS (30,000). The value of contact interactions, X_{AB}, was taken one order lower (1 J/cm^3) than usually observed for dispersion interactions because the solubility parameters for various PS should be very close. Fig. 3.30 shows that even at such small value of X_{AB}, the magnitude of χ_{AB}/V_B^* for PS (3,000)–PS (30,000) is much higher than the critical value calculated from the equation

$$\frac{(\chi_{AB})_{crit}}{V_B^*} = \frac{1}{2}\left[\frac{1}{(V_B^*)^{1/2}} + \frac{1}{(V_A^*)^{1/2}}\right]^2 \qquad [3.123]$$

To fulfill the symmetry condition, i.e., $\chi_{ij} = \chi_{ji}$, parameter χ_{AB} is related to the volume unit. The magnitude of χ_{AB}/V_B^* becomes less critical, only for the system PS (10,000)–PS (30,000) at very high temperatures. With increasing molecular mass of the second component, the temperature was found at which $\chi_{AB} < (\chi_{AB})_{crit}$. For the mixtures PS (5,000)–PS (30,000) parameter χ_{AB}/V_B^* does not reach a critical value and these mixtures are immiscible. Fig. 3.30 shows that in all cases with temperature decreasing, parameter χ_{AB}/V_B^* increases in accordance with experimental data for the mixture PS (2,000)–PS (30,000) (Fig. 3.28).

The change of χ_{AB}/V_B^* for the mixtures of PS (30,000) with fractions of growing molecular mass was also studied. Fig. 3.31 shows that, when molecular mass of the second component reaches 10,000, we have $\chi_{AB}/V_B^* > (\chi_{AB})_{crit}/V_B^*$. If the energy term is equal to zero, (in principle, it may be expected for the mixtures of polymer homologues), the term of free volume may determine the immiscibility of compo-

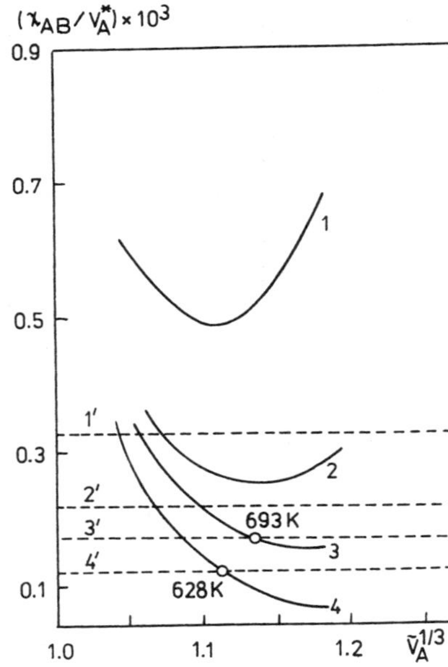

Figure 3.30. Temperature and concentration dependencies of parameters χ_{AB}/v_B^* calculated from Eq 3.28 (solid lines) at χ_{AB}=1 J/cm^3 and critical values of the same parameters calculated from Eq 3.123 (dashed lines) for the mixtures of PS (30,000) with PS fractions: 1,1′ - M=3,000, T=0.13, 2,2′- M=5,000, T=0.068, 3,3′ - M=7,000, T=0.055, 4,4′ - M=10,000, T= 0.034. 1′-4′ - critical values of χ_{AB}.

nents (curve 1, Fig. 3.32). At small values of τ, it is practically impossible to reach lower solution temperatures (see Fig. 3.32, curve 5). For the system with specific interactions and, in particular, for PEG-PEG mixtures, the energy term becomes positive with increasing temperature (usually this term is negative) (Fig. 3.33). Together with the growth of the free volume this term leads to increasing χ_{AB} with temperature. The value τ was taken constant because the difference in free volumes between PEG (2,000) and PEG (40,000) is very small. χ_{AB} should increase with diminishing molecular mass of one of the mixture components at a constant value of molecular mass of another. This agrees with experimental data for PEG (2,000)-PEG (40,000) in spite of a slight difference in expansion coefficients of the components.[352] The experimental data show that the mixture of polymer homologues can be immiscible even at high difference in expansion

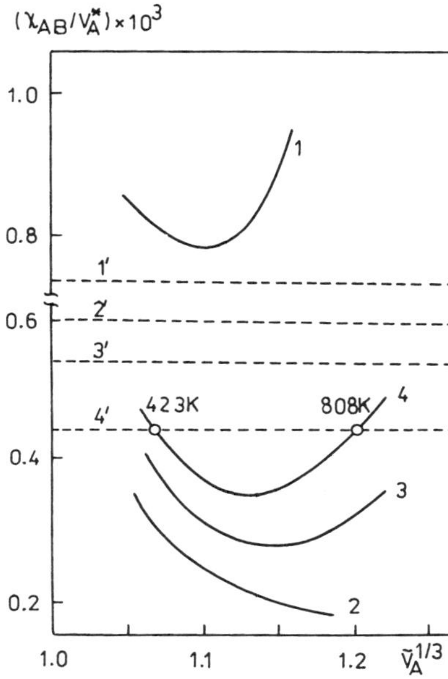

Figure 3.31. Temperature dependencies of parameters χ_{AB}/v_B^* (solid lines) and their critical values (dashed lines) from mixtures of PS fractions with molecular mass: 2,000-5,000 (1,1'), 3,000-5,000 (2,2'), 3,000-7,000 (3,3'), 3,000-10,000 (4,4'). χ_{AB} =1 J/cm^3.

coefficient and small difference in the contact interaction parameters (or solubility parameters), if there are no specific interactions. On the other hand, for polar systems the main reason of immiscibility consists of essential weakening of specific interactions even when the difference in expansion coefficients is small. One of the characteristic features of the PEG-PEG system is bimodal concentration dependence of χ_{AB}. Bimodal character of phase diagrams was predicted theoretically.[46,56] However, the physical meaning of this phenomenon is not always clear. For the mixtures of fractions of poly(butadiene), PB, parameters χ_{AB} were found to be negative, showing their miscibility. The possibility of the negative interaction parameter for the mixtures without specific interactions was discussed earlier.[181] However, the concentration relationship of χ_{AB} depends on the combination of molecular masses of components. For the mixture PB (20,000) with PB (390,000), the bimodal concentration dependence of χ_{AB} was observed (Fig. 3.34). With increasing molecular mass of the second component, bimodality disappears (Fig. 3.35) and after reaching mo-

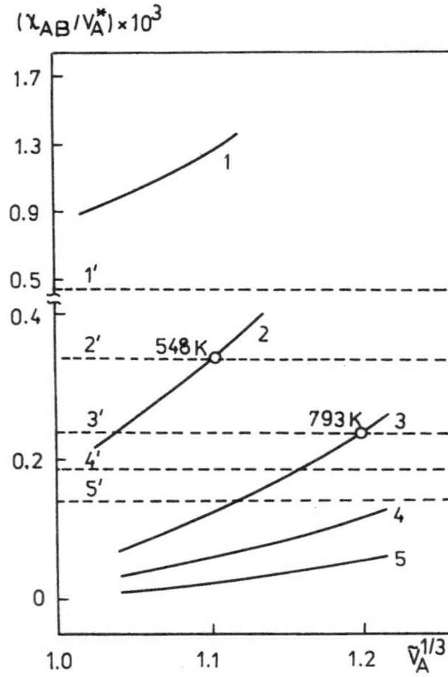

Figure 3.32. Temperature dependence of the free volume term of parameter χ_{AB}/v_B^* (solid lines) at $\chi_{AB}=0$ and critical values (dashed lines) for mixtures of PS (30,000) with fractions: 2,000 (1,1′), 3,000 (2,2′), 5,000 (3,3′), 7,000(4,4′), 120,000 (5,5′).

lecular mass of 70,000, there is no more dependence of the interaction parameter on concentration. For all mixtures parameter diminishes with temperature (becomes "more" negative) but for the mixture of PB (390,000) with PB (20,000) and PB (390,000) with PB (40,000), where χ_{AB} does not change. From the temperature dependence of χ_{AB} for the latter mixtures, the enthalpy part of the interaction parameter, $\chi_H = -T(d\chi_{AB}/dT)$ was calculated and noncombinatorial entropy contribution, $\chi_S = \chi_{AB} - \chi_H$, estimated. In all cases the enthalpy contribution to parameter χ_{AB} is positive, i.e., the system has UCST. Value χ_S happens to be negative, showing that noncombinatorial contribution in the entropy of mixing is positive ($\Delta S_{noncomb} = -R\chi\varphi_B^2$).

The concentration dependencies of χ_H and especially of χ_S (Figs. 3.35 and 3.36) are unusual. At the content of PB (390,000) 30-70%, both χ_H and χ_S are close to zero, i.e., the behavior of the system in this concentration range is similar to an ideal solution. This feature distinguishes the behavior of the system under consideration from

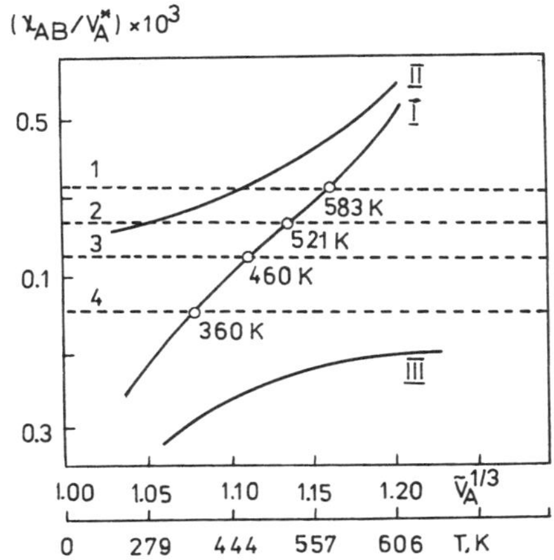

Figure 3.33. Temperature dependence of the total parameter χ_{AB}/v_B^* (I), free volume term (II) and interaction term (III) for the mixture PEG-PEG at χ_{AB}=-0.87 J/cm^3 and T=0.1. Dashed lines - critical values for mixtures PEG with molecular masses 2,000-40,000 (1), 3,000-40,000 (2), 5,000-40,000 (3), >40,000 (4).

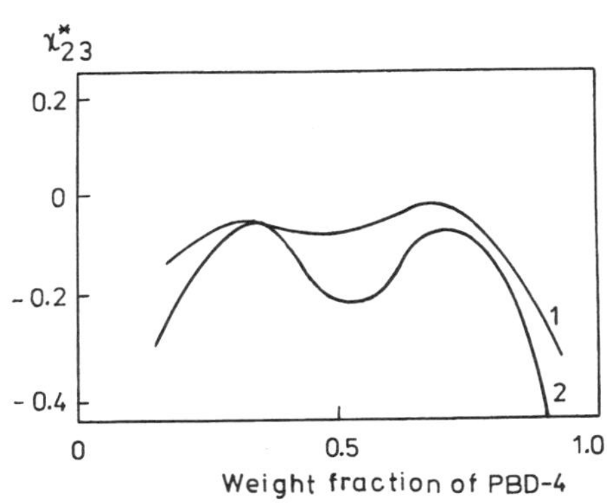

Figure 3.34. Concentration dependence of parameter χ_{23} for PBD (20,000)-PBD (500,000) mixtures at temperatures (K): 1 - 323, 2 - 457. Parameters calculated for PBD segment. Adapted, by permission, from Y. S. Lipatov, A. E. Nesterov, V. F. Shumsky, T. D. Ignatova, and A. N. Gorbatenko, *Eur. Polym. J.*, **18**, 981 (1982).

usual polymer solutions in low molecular solvents (see ref. 11). Diminishing positive noncombinatorial contribution to the mixing entropy can be associated with the processes of molecular aggregation in solutions. At least for the more strict explanation of these data, it is

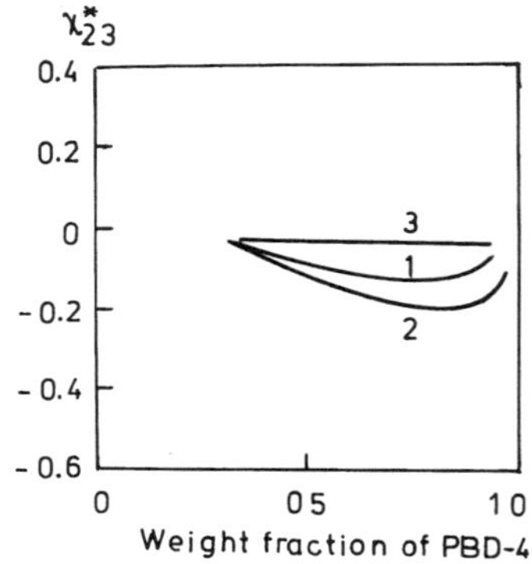

Figure 3.35. Concentration dependence of χ_{23} for PBD (40,000)-PBD (500,000) (1,2) and PBD (70,000) -PBD (500,000) (3) at temperatures (K): 1,3 - 323, 2 - 357. Parameter calculated for PDB segment. Adapted, by permission, from Y. S. Lipatov, A. E. Nesterov, V. F. Shumsky, T. D. Ignatova, and A. N. Gorbatenko, *Eur. Polym. J.*, **18**, 981 (1982).

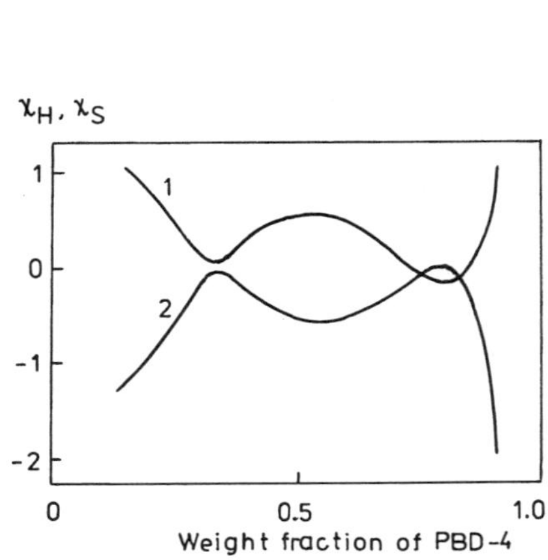

Figure 3.36. Concentration dependence of enthalpy (1) and entropy ((2) parameters for PBD (20,000)-PBD (500,000) (a) and PBD (40,000)-PBD (500, 000) (b). Adapted, by permission, from Y. S. Lipatov, A. E. Nesterov, V. F. Shumsky, and T. D. Ignatova, A. N. Gorbatenko, *Eur. Polym. J.*, **18**, 981 (1982).

necessary to know the molecular sizes of chains in the mixture. It is worth noting that the concept of the formation of molecular aggregates of fractions of different molecular mass was put forward by S. M. Lipatov[353] as early as in 1930.

3.2.3 INFLUENCE OF VARIOUS FACTORS ON THE MISCIBILITY

3.2.3.1 Deuteration and Tacticity of Polymers

In many works[119,125,235,240,276,354-357] the deuteration of one of the mixture components shifted the temperature region of miscibility. For deuterated PS, the curve of phase separation of its mixture with PVME is situated 40^0 higher than for protonated PS.[276] The mixtures of poly(α-methyl styrene) with PVME affect the cloud point curve only when deuteration proceeded in the benzene ring.[240] In the mixtures of oligomeric PS and PB, the deuteration of the latter shifts the miscibility interval to the higher temperatures by 14°C.[354,355] This shift corresponds to the difference in the heat of mixing 0.2 kJ/mol of styrene units in 1:1 blends. For the mixtures of protonated and deuterated PS with molecular masses 15,400 and 11,500, there was observed UCST ($\chi_{AB} = 1.7\times10^{-4}$ at 433 K).[119] The immiscibility in the mixtures of protonated and deuterated shift of the cloud point curve.[357] Phase separation in the mixture of protonated and deuterated poly(1,4-butadiene) was discovered[358] when there is a great difference in their molecular masses. The positive interaction parameter was observed for the mixtures of protonated and deuterated poly(vinyl ethylene), PVE, with poly(ethyl ethylene)[126] containing 50 vol% of deuterated chains.

To explain these effects, it was supposed[126] that the excess of free energy of mixing for isotopic mixtures is connected with the difference in the polarizabilities and free volumes of segments of protonated and deuterated polymers: $(V_H - V_D)/V = 0.0019$ and 0.0039 for PVE and poly(ethyl ethylene), correspondingly. The miscibility of two polymers depends also on the stereoregularity.[137,359-363] The optimum miscibility was observed for poly(styrene-*co*-allyl alcohol)-PMMA mixture when PMMA was syndiotactic and the miscibility was lower when PMMA was isotactic.[137] Isotactic PMMA is not miscible with PVC in the whole range of compositions whereas syndiotactic PMMA forms miscible mixture when the content of syndiotactic PMMA does not exceed 60%. At higher content, two phases are formed, one contains pure syndiotactic PMMA and the other is a mixture of syndiotactic PMMA with PVD in the ratio of 60:40.[359] LCST in the mixtures of PMMA with PVC, chlorinated PVC, CPVC, and poly(vinylchloride-*co*-vinylidene chloride) increases with increase in the syndiotactic part of PMMA and decreases in its mixtures with the same copolymer.[360] PEO shows miscibility with atactic and syndiotactic PMMA and is not miscible with isotactic PMMA.[361] The investigation of the mixtures of poly(o-chlorostyrene-*co*-p-chlorostyrene) with atactic and isotactic PS shows the depression of the cloud point curve when polymer has isotactic structure, i.e., regular structure of chain prevents miscibility.[362] The values of the interaction parameters for the mixtures of PMMA with PVDF are equal to –0.13, –0.10, and –0.06 for isotactic, atactic, and syndiotactic PMMA, correspondingly.[137]

3.2.3.2 Branching of Chains

In some works,[364-366] the effect of the chain branching on the miscibility (cloud point curves) was studied. It was discovered that for the mixtures of branched PS and linear PVME, the shift of cloud point curves was observed to the side of phase diagram enriched with branched polymer as compared with the mixture of linear polymer. This effect was attributed to exclusion of some part of the branch from the interaction with another polymer. To estimate the fraction of the star macromolecule excluded from interaction with surrounding molecules in the mixture, the cloud point curves were compared with the same curves for the mixtures of linear polymers with the same molecular mass as branches. For this purpose, the concentration axis on the cloud point curves were multiplied by a factor 1 – K, which was selected that both curves overlap. Factor 1 – K represents the fraction of the star molecule which interacts with other molecules in the mixture, K is the fraction of non-interacting molecules. The "excluded" part of PS was estimated for the mixtures PVME-PS with 14, 18, and 22 branches, selecting values K by putting over cloud point curves. When the temperatures of minima on the curves did not coincide, the vertical shift of curves was done initially considering the difference of minima temperatures. It was found that for the numbers of the branches 14, 18, and 22 the K values were 0.21, 0.26, and 0.27, correspondingly. As expected, the fraction of the "core" excluded from the interaction diminishes with decreasing the number of branches. The observed shift of the cloud point curves to the lower temperatures was explained by kinetic effects of phase separation due to a greater number of entanglements of the chains as compared with usual branched macromolecules.[364]

3.2.3.3 Effect of Pressure

Many works have been dedicated to the investigation of the pressure influence on the temperature of the phase separation.[233,367-370] For example, in mixtures PEA/PVDF (this pair has LCST), the cloud points decrease with increase of pressure, reach the minimum, and increase again, if the content of the second component does not exceed 30%. At 40% PVDF, only monotonous growth of cloud point proceeds. In mixtures of poly(2,6-dimethyl-1,4-phenylene oxide)-poly(styrene-*co*-fluorostyrene), LCST increases with growth in pressure in the whole composition range.[368] For oligomeric mixtures with UCST (mixtures of oligomers of styrene and butadiene, styrene and propylene), the increase in pressure increases UCST.[233,370]

To describe the effect of pressure on the phase behavior of polymer mixtures, the modified theory of the equation of state was used.[190] Chemical potential of component A in the mixture is given by the expression:

$$\Delta\mu_A/R = \ln\varphi_A + (1 - r_A/r_B)\varphi_B + P_A^*V_A^*/RT[3\tilde{T}_A \ln(\tilde{V}_A^{1/3} - 1)/(\tilde{V}^{1/3} - 1) + 1/\tilde{V}_A - 1/\tilde{V} + \tilde{P}_A(\tilde{V} - \tilde{V}_A)] + V_A^*X_{AB}\Theta_B^2/RT\tilde{V} - V_A^*Q_{AB}\Theta_B^2/R \quad [3.124]$$

Applying the spinodal condition $\delta(\Delta\mu/RT)/\delta\varphi = 0$, it was found[24]

$$-1/\varphi_A + (1 - r_A/r_B) + (P_A^*V_A^*/RT)(-D/(\tilde{v} - \tilde{v}^{2/3} + P_A^*V_A^*D/RT\tilde{v}^2 + P_A^*V_A^*D/RT + V_A^*X_{AB}2\Theta_B^2\Theta_A/RT\tilde{v}\varphi_A\varphi_B - V_A^*X_{AB}D\Theta_B^2/RT\tilde{v}^2 - V_A^*Q_{AB}2\Theta_B^2\Theta_A/R\varphi_A\varphi_B = 0 \quad [3.125]$$

where $D = \delta\tilde{v}/\delta\varphi^2$ and

$$\frac{\delta\tilde{v}}{\delta\varphi_B} = [\delta\tilde{P}/\delta\varphi_B - \delta\tilde{T}/\delta\varphi_B(\tilde{P}/\tilde{T} + 1/\tilde{T}\tilde{v}^2)] \Big/ \left[\frac{2}{\tilde{v}^{5/3}} - \frac{\tilde{T}(3\tilde{v}^{1/3} - 2)}{3\tilde{v}^{5/3}(\tilde{v}^{1/3} - 1)^2}\right] \quad [3.126]$$

$$\delta\tilde{P}/\delta\varphi_B = P/P^{2*}[P_A^* - P_B^* - \Theta_B X_{AB}(1 - \Theta_A/\Theta_B)] \quad [3.127]$$

$$\delta\tilde{T}/\delta\varphi_B = (\tilde{T}/\tilde{P})(\delta\tilde{P}/\delta\varphi_A) + \frac{P_B^*\tilde{T}_B - P_A^*\tilde{T}_A}{P^*} \quad [3.128]$$

From known values of the volume expansion coefficients and coefficients of isothermal compression, the reduced values of $\tilde{P}_i$, τ_i, V_i^*, and parameters P^* and T^* can be calculated.[10,11] Value $\tilde{v}$ at $P = 0$ is calculated according to $\tilde{v}^{1/3} = 1/(1 - \tilde{T}\tilde{v})$. When the pressure is applied then

$$\tilde{v} = 1/[1 - \tilde{T}\tilde{v}/(\tilde{P}\tilde{v}^2 + 1)] \quad [3.129]$$

The procedure of calculation can be applied at any pressures and temperatures, if $\tilde{T}$ is calculated. The only approximation in this approach is that the theory of the equation of state is used in a

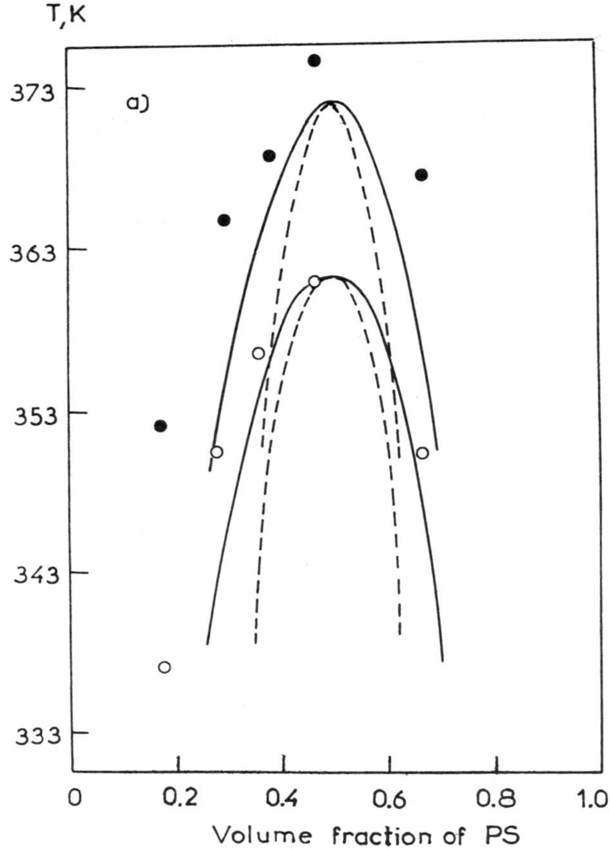

Figure 3.37a. Simulated (___) binodals and(- - -) spinodals of PBD (2,350)-PS (1,520) mixtures at 1 (A) and 1,000 (B) atm. (a) and EVA (45 wt% VA) - CPE (93 wt% Cl). Points are experimental data for a) O - 1 atm, ● - 1,000 atm. Adapted, by permission, from S. Rostami and D. J. Walsh, *Macromolecules,* **18**, 1228 (1985).

simplified form with constant values of P^*, T^*, and V^*, whereas effects of pressure on X_{12} and Q_{12} are neglected.

The theory was applied for comparison with experimental data for liquid mixtures of poly(ethylene-*co*-vinyl acetate) (45 wt% of vinyl acetate) with chlorinated PE (57 and 45 wt% Cl) and for mixtures of polyethersulfone and PEO. A good agreement was found. As a rule,the increase in pressure increases the temperature of phase separation for the systems with LCST.[190]

There was also investigated the effect of pressure on the phase behavior of the mixtures having UCST:[371] oligomeric PB-oligomeric

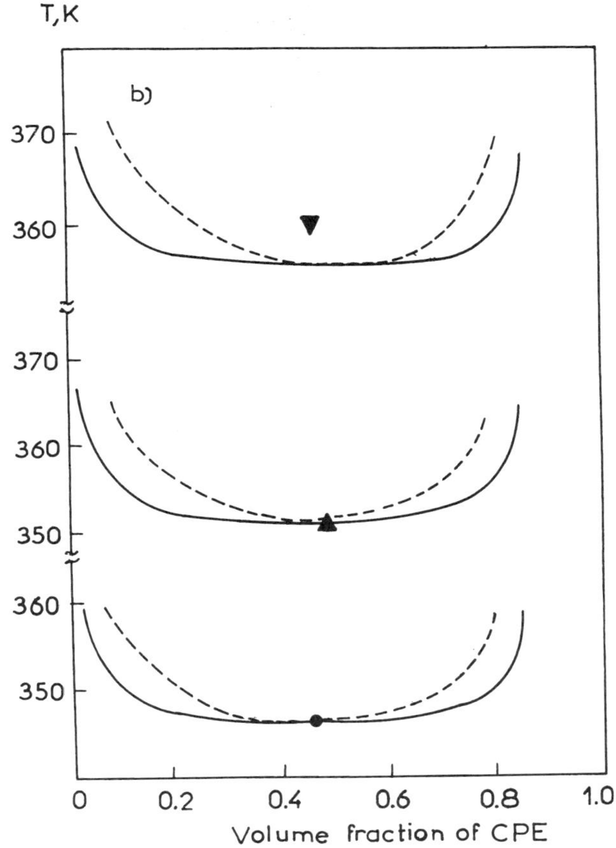

Figure 3.37b. Simulated (___) binodals and(- - -) spinodals of PBD (2,350)-PS (1,520) mixtures at 1 (A) and 1,000 (B) atm. (b) at 1(A) , 300 (B) and 858 (C) atm. Points are experimental data for b) ● -1 atm, ▲ - 300 atm, ▼ - 858 atm. Adapted, by permission, from S. Rostami and D. J. Walsh, *Macromolecules,* **18**, 1228 (1985).

PS and methoxylated oligopropylene oxide-oligoethylene oxide. The general conclusions are the following. Mixtures with UCST and negative heats of mixing increase the miscibility region along with increase in pressure, whereas for positive heats of mixing this region narrows. Mixtures with LCST and with high molecular masses increase their miscibility with pressure. The calculation of binodal and spinodal according to the proposed relationships has shown a satisfactory agreement with experimental data at various pressures for oligomeric and polymeric mixtures[371] (Fig. 3.37).

Table 3.2: Parameters χ_{23} per PEG segment for the mixture of PPG-PEG found using various methods

Mixture	Cloud point[252]	Phase equilibrium[252]	IGC[347]
PEG-500-PPG-1500	0.106-0.086	-	-
PEG-600-PPG-2025	0.104-0.084	0.091	-
PEG-600-PPG-1000	-	0.130	-
PEG-600-PPG-1500	-	0.160	-
PEG-600-PPG-2025	-	0.170	-
PEG-15000-PPG-1050	-	-	0.121-0.114

3.2.3.4 Asymmetry of the Length of Statistical Segment

A correlation between parameter of thermodynamic interaction and difference in the length of statistical segments, a, of mixed polymers was discovered[372] for some polyolefines. Some data on the correlation between χ and $\Delta a/\tilde{a}$ (where $\tilde{a} = a_1\varphi_1 + a_2\varphi_2$) are given in the Table 3.2. The conclusion has been drawn that the difference in the length of statistical segment may affect phase separation. 1,4-PI and 1,2-PB form single phase system at all temperatures. At $a_{1,4\ PI}$=5.9Å and $a_{1,2PB}$=6.0Å, we have $\Delta a/\tilde{a}$=0.16. For this system $\chi = 0.901T^{-1} + 4.17\times10^{-3}$ and system should have UCST.

However, because for this system $\delta\ \ln a_{1,4\ PI}/\delta T \approx 0.5\times10^{-3}$ and $\delta\ \ln a_{1,2PB}/\delta T \approx 0$, with increasing temperature, value Δa should increase probability of appearance of UCST.

3.2.4 POSSIBLE ERRORS IN ESTIMATING THE MISCIBILITY OF MIXTURES OBTAINED FROM SOLUTIONS IN A COMMON SOLVENT

The samples for estimating the miscibility of two polymers are prepared by evaporation of the solutions of polymers in common solvent.[64,219,270,373-384] Miscible polymers in solvents with a great difference in the interaction parameter between each polymer and solvent give "closed" regions of immiscibility.[75,383,384] The mixtures obtained from such solutions may be two-phase mixtures because during evaporation, the system passes through the region of immiscibility. Also, rapid increase in viscosity causes the two-phase structure of the mixture to be frozen.[385] Investigation of the phase behavior of such mixtures may lead to detection of false UCST.[386,387] For immiscible polymers with high molecular masses, the rapid evaporation may lead to the formation of pseudo-homogeneous system with false LCST.[388] A typical example is the mixture of Bisphenol A polycarbonate with PMMA that was considered to be immiscible.[389] Thorough investigation of this system at various conditions was performed.[219] If the mixture was obtained from solutions in methylene chloride or tetrahydrofuran by evaporation or precipitation using methanol, the mixture behaved as two-phase system. If it was formed from the same solution, using precipitation by heptane, the single-phase behavior with LCST is typical in the whole composition range. Another example is the mixture of PS with PMMA.[290] By precipitation of the mixture from toluene, a pseudo-homogeneous system was obtained capable of demixing by annealing.

The mixture of miscible PMMA and SAN formed from a common solvent gives two-phase system with modulated structure.[391] By rapid heating, this system was transferred from a room temperatures to temperatures above LCST after which the time dependence of the intensity of light scattering was measured. During annealing the intensity of light scattering decreases, the position of the maximum intensity is observed at the same angle. The observed effect corresponds to "phase dissolution" with preservation of permanent periodic structure. This apparent dissolution in the two-phase region above LCST of the films obtained from solution was interpreted as a change in composition of two phases along the binodal curve. The mixture PS/PVME with LCST was chosen as a model. For this mixture, there are available parameters needed to construct the phase diagram (Fig. 3.38). If to transfer the single-phase mixture with composition, C_a, to the region of the phase diagram with temperature, T_1, the intensity of light scattering increases until equilibrium compositions C_1 and C_2 are reached (Fig. 3.38).

After the transition of the system to temperature $T_2 < T_1$, the amplitude of the periodic fluctuation concentration wave and, therefore, the light scattering intensity maxima diminish with time until the equilibrium concentrations C_1' and C_2' are established. The light scattering intensity of two-phase system is proportional to the square difference of compositions of two phases, $I \sim |C_1 - C_2|^2$. The change of this difference up to $|C'_1 - C'_2|^2$ leads to diminishing the scattering intensity. Therefore, if the dried film consists of two phases, for example with composition C_1 and C_2, then by decreasing temperature from T_1 to T_2, the dissolution of components will occur that leads to diminishing light scattering. If the same film is kept at higher temperature than T_1, phase separation gives a greater composition difference and increases the scattering intensity. The effect of "apparent dissolution"[391] in the two-phase region is very important for determination of LCST for films dried from solutions and having two-phase structure. Indeed, when such two-phase system is transferred to higher temperatures, during the annealing, the scattering intensity

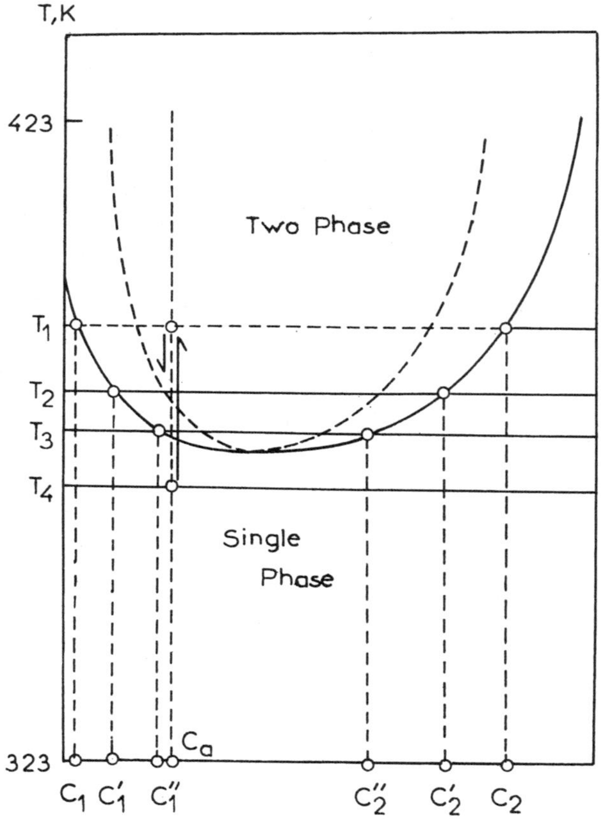

Figure 3.38. LCST-type phase diagram calculated and temperature jump ($T_4 \rightarrow T_1$) and drop (T_1, T_2 or T_3) manipulation. v^*_A= 151,000 cm^3/mol, v^*_B=46,000 cm^3/ mol, T^*_A=8,191K, T^*_B=7,201K, P^*_A=500 J/cm^3, X_{AB}=-0.922 J/cm^3. Adapted, by permission, from T. Ohnaga and J. Maruta, T. Inoue, *Polymer*, **30**, 1845(1989).

begins to drop. One may think that this effect can be explained by dissolution of two phases and by transition of the system to homogeneous state. On the contrary, if the system is transferred to higher temperature and the scattering intensity increases, one may believe that this system transits into two-phase state. This transition boundary (the temperature above which the growth and below which the drop of intensity begins) can be thought to be LCST. But it is not the case. To obtain correct results, a prolonged thermal treatment is necessary.[391]

The pseudo-homogeneous, non-equilibrium systems are more frequently encountered when system undergoes phase separation with very low rate.[391,393-396] Slow phase separation by heating may lead to the overestimation of the length of miscibility region, especially when LCST is situated near T_g. The effect of low rate of phase separation on the estimation of the cloud point curves was investigated in detail for the mixtures of Bisphenol A polycarbonate with PMMA.[396] The phase equilibrium was reached only after prolonged annealing above T_g.

3.3 MISCIBILITY OF POLYMERS AT THE INTERFACE WITH SOLID

3.3.1 PHASE DIAGRAMS OF FILLED POLYMER MIXTURES

Reinforcement of polymer blends is a new principle of creating polymeric composite materials (polymer alloys). The phase state of reinforced or filled blends determines their morphology and physical and mechanical properties.[397] From the thermodynamic point of view the filled polymer mixture can be characterized if the phase diagrams are known for filled and unfilled polymers. Many experimental data[398,399] show that introduction of filler changes the shape of phase diagram. This effect can be explained by consideration of the changes in matrix composition due to formation of the boundary surface layer near the filler interface. This layer has different composition than bulk. The enrichment of the surface layer of the polymer mixture with one of the components should shift a figurative point on the phase diagram along the composition axis. Let component A be adsorbed preferentially and surface layer be enriched with this component, i.e., have different composition than bulk. Then, the phase separation of the system with UCST caused by decrease in temperature will begin with delay in the surface layers (at lower temperature) as compared with the bulk. For the systems with LCST, the phase separation due to heating begins later (at higher temperatures). The reason for this effect is the difference in temperature of phase transition from one to two-phases for various compositions. Thus, the reliability of experimentally found phase diagrams depends on the volume ratio between the parts of the mixture in the surface layer and in bulk. In real systems, these effects may not be so obvious because of the composition gradient between the surface layer and bulk. The shift of the phase transition temperature is larger when the selectivity of interaction increases. These processes cause some experimental difficulties in determination of phase diagrams of filled polymer mixtures. For the systems with UCST, the turbidity due to cooling and for LCST the clearing due to heating appear earlier in the bulk. It is unclear in what way the

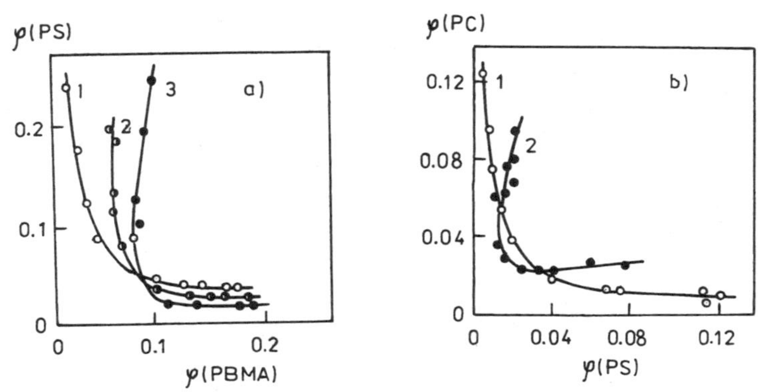

Figure 3.39. Binodals for the mixtures PS-PBMA-benzene (a) and PS-PC-chloroform (b): 1 - without filler, 2 - in the presence of filler, 0.52 g/100 ml, 3 - 1,0 g/100 ml at 295.5K. Adapted, by permission, from Y. S. Lipatov, A. E. Nesterov, T. D. Ignatova, N. P. Gudima, and O. T. Gritsenko, *Eur. Polym. J.*, **22**, 83 (1986).

coexistence of these two regions reflects upon the experimentally found cloud points curves.

Let us now consider some experimental data on phase diagrams of filled polymer alloys. The effect of solid surface on the equilibria in ternary systems polymer-polymer-solvent was studied elsewhere.[400] This effect is connected with adsorption of one or both components of the polymer mixture at the interface with solid. Fig. 3.39 shows that introduction of silica into ternary system PS-PBMA-benzene and PS-polycarbonate-chloroform results in the displacement of the binodal and changes its shape. For both polymer pairs, the region of miscibility broadens. Increase in the amount of silica leads to a broadening of the miscibility window. It is evident that the effect cannot be explained by a simple change in solution composition after adsorption. If it were so, the position and shape of binodal would not be changed. The points on the binodal corresponding to the definite volume fractions of components would only be displaced along the initial binodal. The simple explanation of the broadening of the region of mutual miscibility of two polymers is based on their polydispersity. It is known that high molecular weight fractions are primary adsorbed on solid surface.[401] As a result of such preferential adsorption, the solution becomes enriched with low molecular weight fractions of the polymer which has been selectively adsorbed and the mutual miscibility increases. It should be noted that effect depends on the ratio of components in solution. To understand the phenomenon, it is also

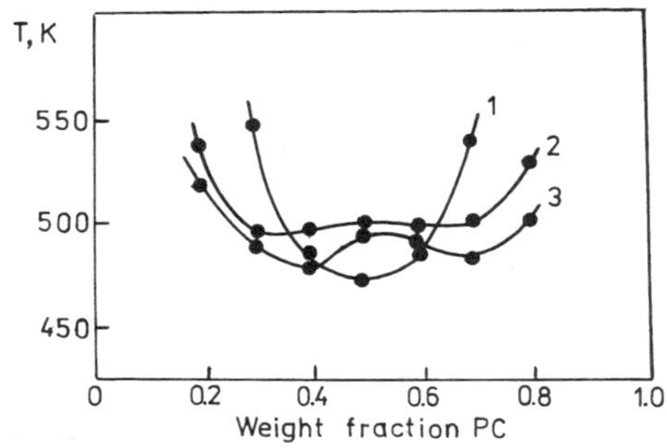

Figure 3.40. Cloud point curves for the mixtures PS with PC unfilled (1), filled with 10 wt% of fume silica modified by ethylene glycol (2), and modified by butanol (3).

necessary to take into account different thermodynamic quality of solvent for each component, which determines adsorption from the mixture. Fig. 3.40 shows the phase diagram for polystyrene-polycarbonate alloy with various fillers. The diagrams were constructed from the cloud point results.[402] The presence of filler changes the shape of phase diagram. In the region of the mean compositions, the phase separation temperature is higher, i.e., the system has higher thermodynamic stability. However, the excess of one of the components decreases the phase separation temperature. Fig. 3.41 represents the phase diagram of the PS-PBMA mixture. Even 10% filler decreases thermodynamic stability. It is the change in matrix composition which causes formation of border layer near the filler surface and composition differs in interface and bulk. Two regions coexist in the system of a given composition, characterized by their own phase separation temperatures. These regions are the bulk of a matrix and the matrix border layer. Their compositions are not the same as in the initial mixture. It is unclear, however, in what way may the coexistence of these two regions reflect upon the experimentally found cloud points.

The cloud point curves for unfilled poly(vinyl acetate) - poly(methyl methacrylate) mixtures and filled with 10% of non-modified and silanized fumed silica show that curves for filled mixtures are situated at much lower temperature than typical for unfilled blend.[403] It should be noted that the temperature of phase separation decreases sharply in the region where one of the mixture components has a predominant content. The addition of silane-treated fumed silica also decreases temperature of phase separation. These effects were attributed to the difference in adsorption of both components at the interface

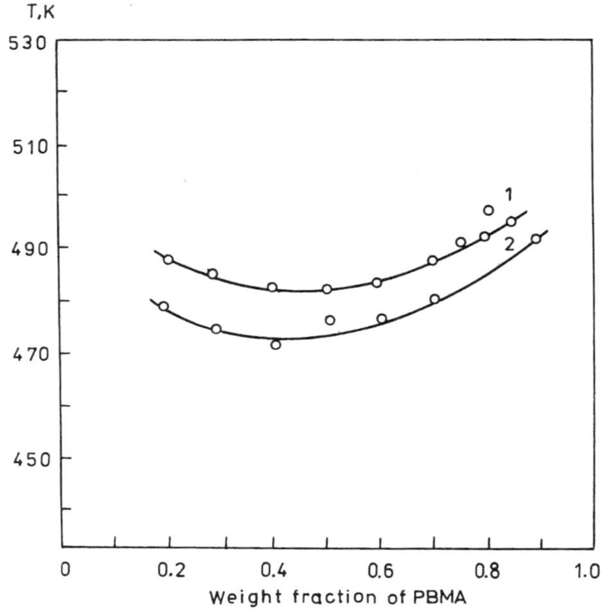

Figure 3.41. Cloud point curves for the mixtures PS-PBMA unfilled (1) and filled with 10 wt% of fumed silica (2).

with solid[404] and to the asymmetry of interactions in the system. The phase separation temperature in this alloy also depends on the filler content (Fig. 3.42). The phase separation temperature depends on the ratio of components. This fact may be interpreted as indirect confirmation of the role of asymmetry of interactions. Increased filler content should affect the asymmetry of the interactions and therefore affect temperature of phase separation. The selective adsorption leads to the change in composition of the border layers and in the matrix bulk and, simultaneously, to the changes in miscibility of components because the miscibility is a function of composition.

The limited experimental data on phase diagrams of filled polymer alloys and other factors involved in the processes of phase separation give no background to present any general concept regarding the solid surface effect on the character of phase diagrams. Only some general considerations may be presented. In the region of metastable states between the binodal and spinodal, the border layers with preferred amount of one of the alloy's component may be formed. These layers can be considered as a nuclei of a new phase obtained by

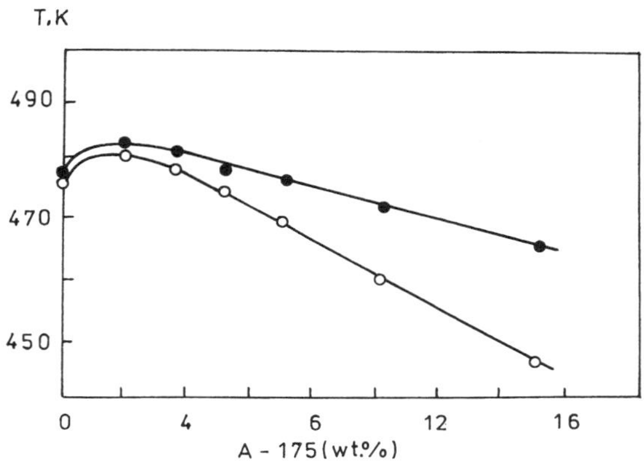

Figure 3.42. Phase separation temperature as a function of concentration of fume silica for the mixtures PVA/PMMA 0.8/0.2 (o) and PVA/PMMA 0.6/0.4 (●).

separation according to the nucleation and growth mechanism, initiating the phase separation. If the system happens to be inside the spinodal, these layers are capable to initiate the appearance of the selective fluctuations of the composition with the formation of periodic modulated structure of interconnected regions of spinodal decomposition. However, when the distance between filler particles approaches by the order of magnitude the radius of the intermolecular interaction and all the matrix is transferred into the state of the surface layer, there is no composition redistribution. The filler influence manifests itself by the change in conditions of interaction between different macromolecules, which may lead to the changes in the phase diagram and total retardation of the phase separation.[397]

These considerations show that filler introduced into polymer mixture changes conditions of the phase structure formation in addition to its traditional role of reinforcement.

3.3.2 PHASE SEPARATION IN THIN FILMS OF POLYMER MIXTURES AT THE INTERFACE WITH SOLID

The selectivity of interaction of the polymer mixture components with the solid plays an important role in the thermodynamic behavior of filled polymer mixtures.

Let us consider some simple thermodynamic relations.[405] The free energy of mixing two different polymers, ΔG_{AB}, can be approximated as

$$\Delta G_{AB} = RT\chi_{AB}\varphi_A\varphi_B \quad [3.130]$$

where φ_A and φ_B are volume fractions of polymers A and B. For a binary mixture in contact with solid surface, we can express the change in the free energy of the system due to interaction with a solid as

$$\Delta G_{mix} = \Delta G_{AS} + \Delta G_{BS} - \Delta G_{AB} \quad [3.131]$$

where $\Delta G_{AS} = RT\chi_{AS}\varphi_A\varphi_{S(A)}$ and $\Delta G_{BS} = RT\chi_{BS}\varphi_B\varphi_{S(B)}$ are free energies of interaction of components with the surface, φ_A and φ_B are volume fractions of components and $\varphi_{S(i)}$ is the volume fraction of active points occupied by the segments A and B at the interface. The thermodynamic stable system is formed when $\Delta G_{mix} < 0$. Two cases can be distinguished.

Case 1. The interaction of each component with the surface is symmetric, i.e., the energies of pair interactions are approximately equal. In this case, $\Delta G_{AS} \approx \Delta G_{BS}$. If both values are negative and the components are immiscible ($\Delta G_{AB} > 0$), then ΔG_{mix} is negative. For miscible polymers, free energy of their mixing is negative and adsorption proceeds only when $|(\Delta G_{AS} + \Delta G_{BS}) > |\Delta G_{AB}|$. Thus, the interactions with the surface are preferential when polymers are immiscible. These relations are valid independent of the fraction of polymers A and B interacting with the surface. For symmetric interactions there is no selective adsorption and the composition of the mixture in the interphase is the same as in bulk.

Case 2. The interactions are non-symmetric, ΔG_{AS} is larger or smaller than ΔG_{BS}. In this case, the sum $(\Delta G_{AS} + \Delta G_{BS})$ is negative and components are immiscible, there is a selective adsorption of one of the components. The composition of the interphase differs from the bulk. Preferential adsorption is equal to increase in the motive force of phase separation of two immiscible polymers. The interphase is enriched with component A if $-\Delta G_{AS} > -\Delta G_{BS}$ and *vice versa*. For miscible pairs with non-symmetrical interactions, the following relationship should be valid for adsorption: $|\Delta G_{AS} + \Delta G_{BS}| > \Delta G_{AB}$.

These considerations are consistent with experimental data which show the enrichment or depletion of the surface layer of one of the components of the binary mixture (see below). As a result of changing the composition of the system, their miscibility in the interphase may increase or decrease. The same is valid for the bulk. One of important factors here is also the redistribution of polydispersed components according to their molecular mass between the surface and the bulk. The second reason for changes in miscibility is the difference in conditions of interaction between two polymers due to the conformational restrictions in the interphase and in the bulk.

In comparison with theoretical and experimental works in the field of polymer alloys and blends, the studies of filler influence on the phase behavior of these systems is at its very early stage and there are only few data available and some attempts to describe theoretically the thermodynamic behavior of polymer mixtures near the interface with solid. The only review which can be mentioned was written by Dolinny and Ogarev,[406] where the results of theoretical and experimental investigations were analyzed with regard to the phase transitions in semi-infinite two-component systems (capable of mixing liquids and polymers). In the framework of the mean field theory, which accounts for short range interactions between the solid surface and mixture components, the conditions of the formation of wetting layer of one of the equilibrium phases on the substrate surface were discussed. Phase transitions in polymer solutions were explained on the basis of Cahn theory and the data on the concentration profile at the interface with solid. The rapid increase of the thickness of the layer in contact with substrate can be considered as a phase transition of the first order. It was shown[407,408] that the phase diagrams of thin layers of polymer solutions are essentially changed as indicated by comparison with phase diagrams of the bulk solution. The transitions at the interface between solid and solid mixture have not been considered.

The two-parameter model of polymer-polymer excluded volume in conjunction with the renormalization group was developed to include an interacting plane surface.[409,410] It was theoretically shown that such system has two types of volume exclusion, the usual polymer-polymer excluded volume interactions and the "excluded volume" constraint of the boundary. There is a competition between the gain of internal energy with the formation of attractive surface contacts and the loss of configurational entropy due to the proximity of the surface.[411] Reich and Cohen[412] were the first who proposed the thermodynamic description of the effect of interface on the phase separation in polymer mixtures. They considered the lattice-cell model in which the coordination number of lattice points is uniform throughout the lattice, but the number of lattice points in the surface layer was smaller than in the bulk. The number of pair interactions in the total sample of N lattice sites is

$$N = Z/2 - (1/2)[(2H_b/H)N(Z - Z_b)] \qquad [3.132]$$

where: N is total number of lattice sites in entire system, H is sample thickness, H_B is the boundary layer thickness, Z and Z_b are coordination numbers of lattice points in the bulk and in the surface layer. At $Z_b < Z$, thermodynamic interaction parameter is equal to:

$$\overline{\chi}_{AB}(T,H) = \chi^0_{AB}(T)F(H) \qquad [3.133]$$

where χ_{AB} is interaction parameter corrected for the surface effects and χ^0_{AB} is the interaction parameter neglecting surface effects, i.e., for $H_b/H = 0$. Function F(H) is obtained as

$$F(H) = 1 - 2(1 - Z_b/Z)(H_b/H) \qquad [3.134]$$

Since the value of $\overline{\chi}_{AB}$ at the spinodal is invariant at fixed composition, it follows that

$$\frac{dT_{sp}}{dH} = -\frac{(\delta\overline{\chi}_{AB}/\delta H)_T}{(\delta\overline{\chi}_{AB}/\delta T)_H} = \frac{F(H)(\delta\chi^0_{AB}/\delta T)}{\chi^0_{AB}(T)(\delta F/\delta H)} \qquad [3.135]$$

From this equation it follows that for a system with LCST ($d\chi^0_A/T > 0$), dT_{sp}/dH is negative, i.e., the system stability should increase with decreasing thickness of the surface layer. The opposite effect should hold for mixtures with UCST, where $d\chi_{AB}/dT < 0$ and dT_{sp}/dH is positive.

The effects of upgrading the stability of a binary system can also be observed for systems characterized by UCST. To predict the dependence of the spinodal temperature on the film thickness, parameter χ_{AB} at T_{sp} is expressed as

$$\chi_{AB}(T_{sp}) = (1/2)[(1/\varphi_A) - (V_A/V_B\varphi_A)] \qquad [3.136]$$

where φ_A is the volume fraction and V is molecular volume. From this equation, it may be obtained that

$$\chi^0_{AB}(T_{sp})\left(1 - \frac{2(1 - Z_B/Z)H_B}{H}\right) = \frac{1}{2}\left(\frac{1}{\varphi_A} + \frac{v_A}{v_B}\frac{1}{\varphi_B}\right) \qquad [3.137]$$

Expressing $\chi^0_{AB} = \chi^0 - \chi_1/T$, χ^o and $\chi_1 > 0$, and using this expression for $H_b/H = 0$ and for finite small H, we find:

$$1/T^0_{sp} - 1/T_{sp}(H) = \frac{1}{2\chi_{AB}}\left(\frac{1}{\varphi_A} + \frac{V_A}{V_B\varphi_B}\right)\left(\frac{1}{F(H)} - 1\right) \qquad [3.138]$$

where T_{sp} is the temperature on the spinodal at $H_b/H = 0$. If $T_{sp}(H)$ does not differ much from unity, $1/T_{sp}$ is a linear function of 1/H:

$$\frac{1}{T_{sp}(H)} = \frac{1}{T^0_{sp}} - \frac{1}{\chi_{AB}}\left(\frac{1}{\varphi_A} + \frac{V_A}{V_B\varphi_B}\right)\left(1 - \frac{Z_b}{Z}\right)\frac{H_b}{H} \qquad [3.139]$$

This equation allows one to predict the film thickness at which T_{sp} differs from $T_{sp}(H)$. Calculations have demonstrated the effect of the

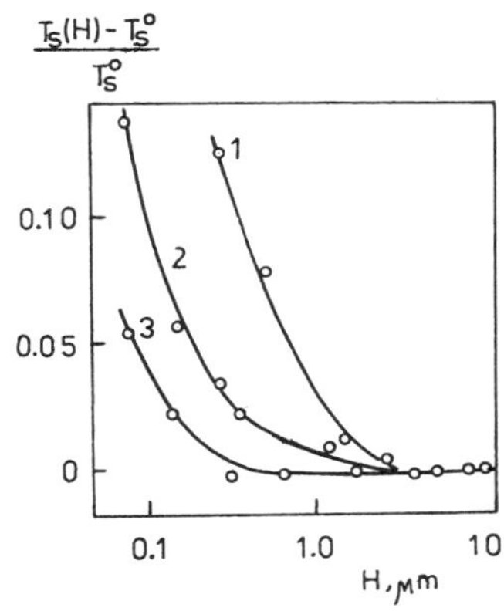

Figure 3.43. Effect of film thickness on spinodal temperature of films formed on a gold substrate, T°s, spinodal temperature of thick films: 1 - 15% PS, 2 - 50 % PS, 3 - 67% PS. Adapted, by permission, from S. Reich and Y. Cohen, *J. Polym. Sci.-Phys.*, **19**, 1255 (1981).

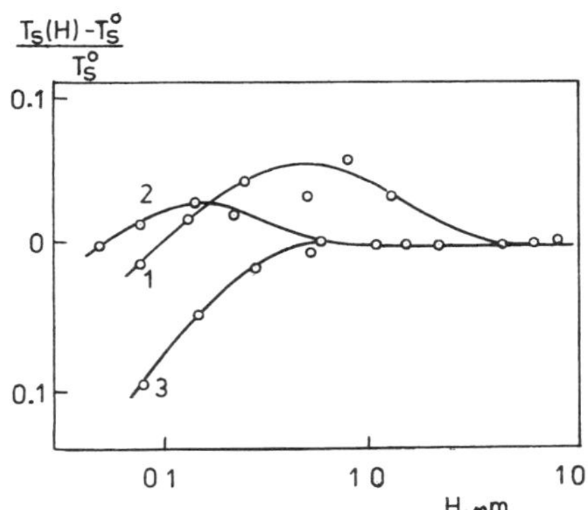

Figure 3.44. Effect of film thickness on spinodal temperature of films formed on a glass substrate. T°s, spinodal temperature of thick films: 1 - 15% PS, 2 - 50% PS, 3 - 67% PS. Adapted, by permission, from S. Reich and Y. Cohen, *J. Polym. Sci.-Phys.*, **19**, 1255 (1981).

layer thickness on the spinodal decomposition in PS-poly(vinyl methyl ether) mixture, beginning with thickness of 0.1-0.8 mm. In Fig. 3.43, T_{sp} increases in films on gold support with film thickness decreasing. For the films on the glass support, the complicated dependence of T_{sp} on the film thickness was found for various mixture compositions (Fig. 3.44). The reasons for such differences are explained by the contribution of destabilizing influence of electrostatic interactions between component and support and by selective adsorption of one of the

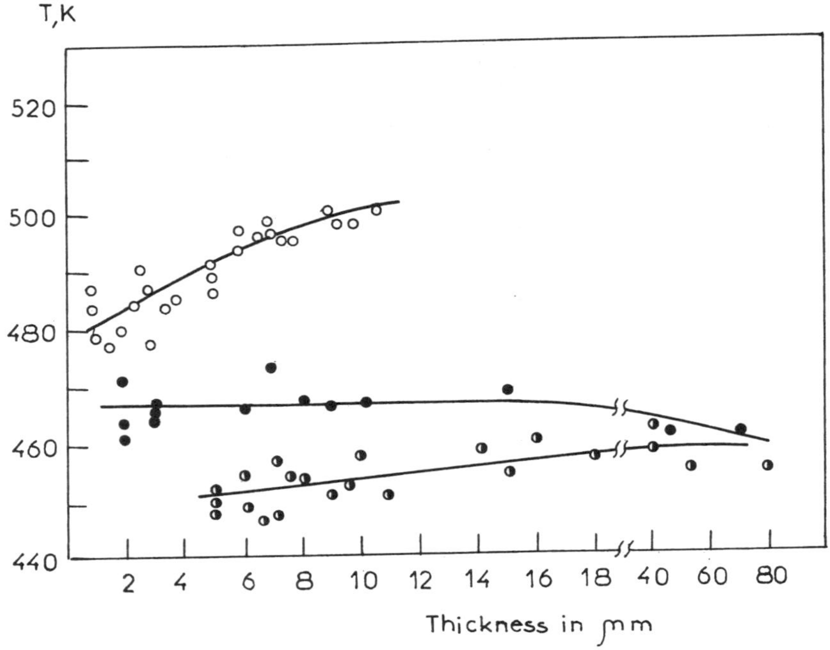

Figure 3.45. Temperature of phase separation of 0.7/0.3 (by weight) PVA/PMMA mixture on non-silanized (o, ◐) and silanized (●) glass as a function of film thickness. Adapted, by permission, from A. Nesterov, V. Gorichko, and Y. Lipatov, *Makromol. Chem., Rapid Commun.*, **12**, 571 (1991).

components leading to the change in composition in the boundary layer.

There should be various contributions to the phase separation temperature due to the existence of the border layer having its own composition and the bulk with still another composition. Thus, the film thickness at the interface with solid affects the phase separation temperature. Fig. 3.45 shows this dependence for the PMMA-PVA system on various supports. On the glass surface, there were observed two phase separation temperatures for film thickness of 10-12 μm.[413] Films casted on silanized glass had only one phase separation temperature regardless of film thickness. The appearance of the second high temperature region of phase separation may be due to the formation of adsorption layer on non-silanized glass. At small thicknesses (5-12 μm), the components of the mixture may be already affected by the surface force field (i.e., form surface layer) and only one temperature of phase separation is observed which corresponds to this layer. With increasing film thickness, the fraction of the film which was not subjected to the action of the surface increases and another transition appears. On the silanized glass, the probability of formation of surface layer is very small and only one temperature of phase separation is detected.

The increase in miscibility of polymers in capillaries was predicted theoretically.[414,415] The miscibility in a capillary may increase if the ratio of a capillary diameter to the chain dimension is very small.

3.3.3 THERMODYNAMIC INTERACTION PARAMETERS IN FILLED POLYMER MIXTURES

Even if we have a correctly drawn phase diagrams, they only give answer regarding the range of composition and temperature at which the system is one- or two-phase system. The phase diagrams do not characterize the system in the regions where the incomplete phase separation already took place. Such characteristics are only obtained when calculation of changes in the free energy of mixing is possible in the presence of a filler or when the thermodynamic interaction parameter can be determined.

In some our works,[416-418] the values of interaction parameter, χ_{AB}, have been determined for unfilled and filled polymer alloys in the melt state, using the inverse gas chromatography method, in spite of many experimental difficulties and uncertainties. This parameter characterizes how far is the system from its equilibrium state. Depending on the composition, amount of filler, and its chemical nature, the parameter χ_{AB} may increase or decrease, i.e., the change of thermodynamic stability has a very complicated character. By introduction of fillers into binary mixtures of crystallizing polymers, the crystallinity degree may also be changed due to the changes in the thermodynamic stability. For the system with a lower thermodynamic stability in the melt, the crystallinity degree is higher in a solid state, whereas for more stable systems crystallinity degree decreases on cooling.[419-421]

Consider the data on χ_{AB} for filled PS-PBMA, determined using inverse gas chromatography. Because of the method involved all the data relate to the melted state of mixture. Fig. 3.46 shows the increase in thermodynamic stability of PS-PBMA mixture by introducing silica (parameter χ_{AB} diminishes). For PS-PC mixture, the dependence of χ_{AB} on composition is rather complex. For low and high PC content in the mixture, χ_{AB} diminishes because preferential adsorption of PC proceeds. In the composition range, where preferential adsorption of PS takes place, χ_{AB} becomes greater than for the mixture without filler.

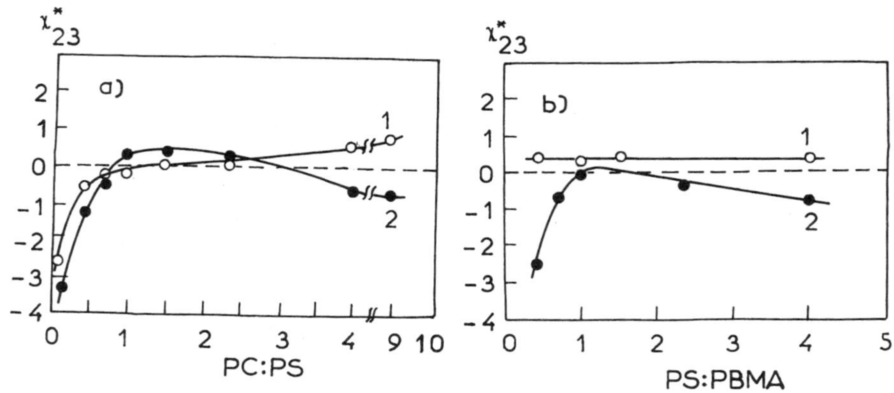

Figure 3.46. Dependence of χ^*_{23} on the PS-PC composition (a) and PS-PBMA (b). Filled with 10 wt% of fume silica - 2, unfilled - 1. Adapted, by permission, from Y. S. Lipatov, A. E. Nesterov, and T. D. Ignatova, *Eur. Polym. J.*, **22**, 83 (1986).

The experimental data can be treated on the basis of statistical theories of polymer solutions. Parameter χ_{12} (polymer-solvent) may be presented in the form:[422]

$$\chi_{23} = -\frac{U_1 \nu^2}{RT} + \frac{C_p \tau^2}{2R} \qquad [3.140]$$

where $\tau = 1 - T^*_1/T^*_2$. In Eq 3.140, the first term is due to the interchange energy required to form contacts between polymer segment and solvent. The second term gives structural contribution arising from changes of free volume of mixing. The quantities T^*_i are the temperature reduction parameter of the solvent (component 1) and the polymer (component 2), τ is the temperature-dependent molecular parameter, U_1 and C_P are the configurational energy and its temperature derivative. The second term contributes unfavorably to the miscibility due to the differences in free volumes of components. It is evident that the unfavorable contribution to mixing is diminished when the difference between T^*_1 and T^*_2 becomes smaller and *vice versa* (τ is always positive).

What happens when we introduce filler into the mixture of two polymers? It was shown[423] that in polymer layer at the interface with solid there is a decrease of the thermodynamic interaction parameter χ_{12} (solvent-polymer). Such behavior was explained by diminishing molecular packing density in the surface layer and by increasing fractional free volume. These data indicate that the selective adsorption also leads to increase of the free volume of adsorbed polymer. Calculations[400] have shown that the difference between T^* for two pairs of polymers in a definite range of composition decreases due to a looser packing. As a result both τ^2 and χ_{AB} diminish. Thus, there is an increase in thermodynamic stability of the systems. This phenomenon is related to the adsorptional interaction of components with the surface and with the selective adsorption of one of them. The transition of macromolecules on the interface into the surface layer, changes the conditions of interaction between macromolecules. The interaction parameter χ_{AB} decreases. The difference between free volumes of components in the surface layer and in the bulk contributes markedly to increase in thermodynamic stability.

For the system PBMA-polyethylene filled with various amounts of fumed silica, the difference $\Delta\chi_{AB} = \chi_{AB} - \chi_{AB(filled)}$ have been estimated.[424] This difference is a measure of increasing stability of the system caused by filler presence. The experimental results show that for a given system, the increasing amount of the filler contributes to the increased interaction between different segments in the surface layer.[399] The most general interpretation of these results may include dependence of χ_{AB} on the lattice coordination number in the surface layer. It was assumed that coordination number depends on the conformational state of macromolecules -- different in the surface layer and in the bulk. Fig. 3.47 shows concentration dependencies of the enthalpy and entropy contributions into the total interaction parameter, χ_H, and χ_S which characterize the change of the interaction parameter in the surface layers of polymer alloy. These contributions were calculated from the temperature dependencies of χ_{AB}. With increasing the fraction of PBMA in the system, χ_H essentially drops, whereas χ_S goes through a maximum. These results show the important role of non-combinatorial entropy contribution to the interaction in the interphase which depends on the composition. Using traditional model of filled polymers, it was assumed that the interaction parameter found experimentally is a sum of parameters, characterizing the surface layer and the bulk:

$$\chi_{AB(filled)} = \chi_{AB}(1 - \varphi) + \chi_{AB(surface\ layer)}\,\varphi \qquad [3.141]$$

where φ is a volume fraction of the mixture in the surface layer. The value $\chi_{AB(surface\ layer)}$ can be calculated from the experimental data on the volume fractions of the surface layers given in ref. 425. The results of calculation show that the interaction parameter in the surface layer, $\chi_{AB(surface\ layer)}$, is considerably lower than in the bulk. This allows the conclusion that the phase interface with solid increases the thermodynamic stability of immiscible polymer pair even in the melt. The

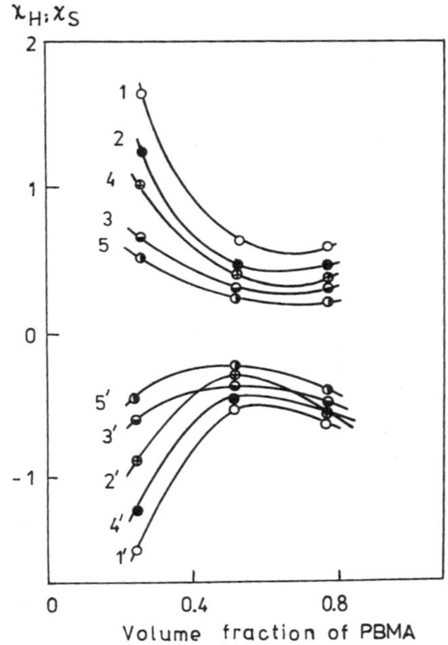

Figure 3.47. Concentration dependence of parameters χ_H (1-5) and χ_S (1′-5′) for PE-PBMA: without filler (1,1′) with 5 (2,2′) and 15 (3,3′) wt% of fumed silica and 5 (4,4′) and 15 (5,5′) wt% of fumed silica modified by diethylene glycol at 417K.

effect is probably connected with the changing conditions of interaction of different chains. It means that by increasing the total surface area of filler, it is possible to transfer system from thermodynamically unstable state to the stable state (effect of increasing compatibility of components in the surface layer). The additive relation 3.141 may be transformed to

$$\Delta\chi_{AB}/\varphi = (\Delta\chi_{AB})_{(surface\ layer)} \qquad [3.142]$$

where $(\Delta\chi_{AB})_{(surface\ layer)} = \chi_{AB(surface\ layer)} - \chi_{AB}$. Eq 3.142 shows changes of the interaction parameter for the hypothetical system where both components are fully in the state of the surface layer. Value $(\chi_{AB})_{(surface\ layer)}$ can be considered as a measure of thermodynamic activity of a filler.

The value of interaction parameter in the surface layers depends on the filler nature (its activity) at comparable specific surfaces. Not all fillers increase thermodynamic stability of the filled alloy. The effect also depends on the conditions of the blend production.[426] Theoretically the effect of surface on the behavior of polymer mixtures was considered by Noolandi.[427]

The same effects of decreasing the interaction parameters in the melt due to the filler presence were observed for binary mixtures of crystallizing polymers (polyethylene-polypropylene-kaolin, polyethylene-polyoxymethylene-kaolin).[420] The crystallinity degree in solid state (all measurements of χ_{AB} were performed in the state of a melt due to the requirements of the inverse gas chromatography) correlates with χ_{AB}. Maximum χ_{AB} in the filled melt corresponds to the maximum in crystallinity degree. Therefore, if the filled system is less stable, its crystallinity degree increases and *vice versa*. These data indicate that, during crystallization from melt containing miscible components, the phase separation is hindered and crystallization degree is lower at the same cooling rate. If the system is not stable, the phase separation is more complete and crystallinity increases.

Different observations were recorded for filling compatible polymers.[428] By introducing filler (kaolin) to the miscible mixture poly(vinyl acetate)- ethylene-vinyl acetate copolymer, the interaction parameters increase in the surface layers in the whole range of compositions. This indicates the diminishing thermodynamic stability of filled mixture. The effect is explained by independent adsorption of both components at the interface decreasing the interaction between component macromolecules.

The method of determination of interaction parameter (inverse gas chromatography from the retention volume) leads to some uncertainties. According to the theoretical relationships, interaction parameter is calculated from the retention volumes of pure components. For filled polymer mixtures, it is not clear, what volume to use for calculations, namely, the retention volume of each pure component, or the same volume of each component in the presence of filler. The comparison of calculation methods was done[429] and both values of retention volume were used: "unperturbed" retention volume, for pure components, and "perturbed retention volume" for each component with a corresponding amount of filler. Such approach is important because introduction of filler leads to the loosening of molecular packing and to the change in the retention volume. The calculations of χ_{AB} based on "perturbed" retention volume give in the majority of cases a lower value than these obtained for the unfilled mixture. This can be considered as a sign of increasing miscibility in filler presence. Correspondingly, the calculations of χ_{AB} based on the "unperturbed" retention volumes result in lower changes in parameter in the first method. In some cases, the calculated values of χ_{AB} for the filled systems are higher than for unfilled system. Fig. 3.48 shows that the

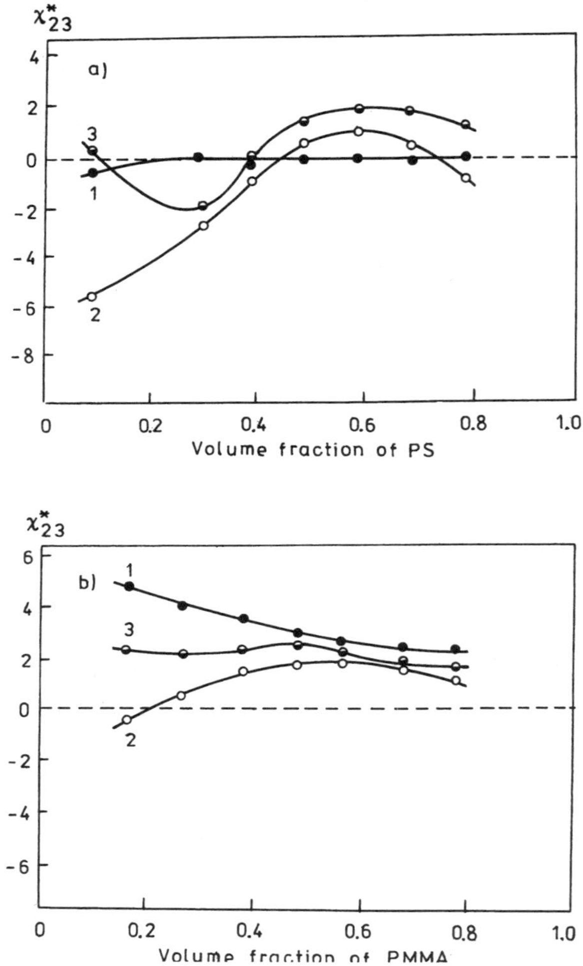

Figure 3.48. Dependence of the thermodynamic interaction parameter calculated from IGC using two methods: a) PBMA-PS and b) PBMA-PMMA. 1 - without filler, 2 - calculation by the second method, 3 - calculation by the first method.

concentration dependence of χ_{AB} determined by two methods is the same. The result may be explained considering that the transition into the state with a lower packing density means the transition to the thermodynamically less stable state.[430] By decreasing polymer packing density of the filled system, its entropy diminishes and enthalpy increases. The polymer matrix in the presence of filler is always less stable thermodynamically, regardless if it has one polymer or a polymer mixture. Therefore, by analysis of the stability of filled system one can distinguish the contributions of decreasing stability of each component to the total stability. The calculations according to the first method account for perturbing surface action. They indicate that interaction of perturbed chains of each component increases the stability of binary mixture. The filled mixture is more stable than mixture without filler. At the same time, the filled system can be considered less stable if its stability estimation is done without accounting for perturbing surface action. Thus, the comparison of two calculation methods allows one to distinguish between the effects related to instability caused by surface action of each polymer component and the effect of change of interaction between perturbed chains of both components. The difference in χ_{AB} calculated by the two methods shows the contribution of perturbing action of the surface to the component interaction in the interphase region near the phase border.

The perturbing action of the surface cannot be accounted for by comparing the phase diagrams for filled and unfilled mixtures. For comparison of the phase diagram with the data on χ_{AB} we have to use the unperturbed χ_{AB} values. The comparison of data calculated by two methods allows estimation of various contributions to the system stability -- due to the changes in the perturbed component interaction and the decreasing stability of each component in the filled system. The calculation by the first method gives only changes in the interphase interaction between two polymers, whereas the second method represents the total effect of filler action on the thermodynamic stability of the system.

Thus, the introduction of a filler changes the thermodynamic stability of the polymer mixture. The effect is determined by two opposite contributions to the stability, one determined by decreasing packing density and another by changing the system volume by mixing. Taking into account the relative character of χ_{AB} estimation, the analysis of the experimental data can be done within the framework of either method of calculation. It is important to consider that perturbing action of the surface favors the component mixing in the border layer near the filler surface. The role of both contributions to the stability changes with mixture composition and filler content. This explains the complex character of χ_{AB} changes with composition.

3.3.4 ON THE EQUILIBRIUM AND NON-EQUILIBRIUM COMPATIBILIZATION OF POLYMER MIXTURES

The data presented in preceding paragraphs considered the filler effect on the miscibility of two polymers. There are two main methods of polymer mixture compatibilization using particulate fillers.[431] The first may be called equilibrium compatibilization. Its essence is based on equilibrium increase of thermodynamic stability of binary system as a result of symmetric pair interactions of each component with the filler surface. When the energy of interactions polymer-filler pair are approximately the same, the interphase layers are formed near the interface. These layers have different properties but the same composition as the initial mixture. This is possible if filler does not have essential selectivity of adsorption of one of the components. The thermodynamic stability in the interphase increases due to conformational changes. Depending on the surface nature and interaction energy, the effects under consideration can be manifested in various ways. It is known that for ternary polymer systems the increase in asymmetry of pair interactions decreases thermodynamic stability.[432-434] However, even if there is some asymmetry or selectivity of adsorption, the increase in thermodynamic stability can be encountered if there is a preferential adsorption of high molecular mass fractions from the bulk at the interface or enrichment of the bulk with low molecular fractions which causes better miscibility in accordance with general thermodynamic principles.

This effect depends on the alloy composition, choice of components, and their ratio. By an appropriate choice of components, the polymer-solid interaction parameter of a pair, χ_S, may be selected for which introduction of particulate filler increases compatibilization. There exists some analogy to improvement of miscibility of two polymers by introducing third polymer miscible with each component of the mixture. For a miscible polymer pair, the filler decreases the compatibility, this effect is most likely result of interaction asymmetry.

The second method of compatibilization is not a true increase in miscibility of components. It occurs when strong interactions of each component with the surface take place independently of interaction symmetry. The phase separation, which should proceed through the change of temperature conditions during blend production, is hindered due to pure kinetic reasons causing a sharp decrease in molecular mobility of adsorbed macromolecules of both polymers. The polymer mixture is in the non-equilibrium state. The structure corresponding to more equilibrium states. Such system has a high kinetic stability. For real systems both suggested mechanisms of compatibilization take place simultaneously.

3.4 TERNARY POLYMER SYSTEMS

3.4.1 TERNARY SYSTEMS POLYMER-POLYMER-SOLVENT

The theory of ternary systems was developed by Scott[13] and Tompa.[14] They analyzed so-called symmetric case when the interaction of each polymer with solvent is the same $\chi_{SA} = \chi_{SB}$ and $r_A = r_B = r$. The equation for the free energy of mixing in the ternary solution is given by:

$$\Delta G = \frac{RTV}{V_S}[\varphi_S \ln \varphi_S + \frac{\varphi_A}{r_A} \ln \varphi_A + \frac{\varphi_B}{r_B} \ln \varphi_B + \chi_{AS}\varphi_S\varphi_A + + \chi_{BS}\varphi_S\varphi_B + \chi_{AB}\varphi_A\varphi_B] \quad [3.143]$$

where V_S is the molar volume of a solvent. The corresponding equation for binodal is written in the following form:

$$\left(\frac{\delta^2\Delta G}{\delta\varphi_A^2}\right)\left(\frac{\delta^2\Delta G}{\delta\varphi_B^2}\right) = \frac{\delta^2\Delta G}{\delta\varphi_A\delta\varphi_B} = 0 \quad [3.144]$$

For the symmetric case, Scott has derived the binodal equation as:

$$\ln \Theta'_A + \varphi\chi_{AB}(1 - \varphi_s)(\Theta'_B)^2 = \ln \Theta'_B + \varphi\chi_{AB}(1 - \varphi_s)(\Theta'_A)^2 \quad [3.145]$$

where $\Theta_A = \varphi_A/(\varphi_A + \varphi_B)$ and $\Theta_B = \varphi_B/(\varphi_A + \varphi_B)$. The shape of binodal according to Eq 3.145 does not depend on χ_{SB} and χ_{SA}. This means that the phase separation in ternary systems is determined only by the unfavorable interactions between segments of different chains ($\chi_{AB} > 0$). The role of a solvent is to dilute the mixture and decrease both the number of effective contacts between segments of various type and interaction parameter $\chi_{AB}(1 - \varphi_S)$. At $\varphi_S \rightarrow 1$, the effective interaction parameter approaches zero and the system becomes single-phase system. The corresponding critical value of the interaction parameter in the ternary system is determined by the relation:

$$(\chi_{AB})_{crit} = (1/2)[(1/r_A)^{1/2} + (1/r_B)^{1/2}]^2[1/(1 - \varphi_s)] \quad [3.146]$$

From Eq 3.146, it follows that the critical magnitude of the interaction parameters is reached by a given polymer pair independent of the solvent nature (at the same φ_S) for different solvents. From this consideration, it follows that if two polymers are not miscible in a given solvent, they will not be miscible in any other solvent. This conclusion and its apparent confirmation by the data by Dobry and Boyer-Kawenoki[64] have slowed down the search for miscible polymer pairs by almost ten years. In 1956, Kern[65] reported data on critical concentrations of mixing for PS and PMMA in numerous solvents. It

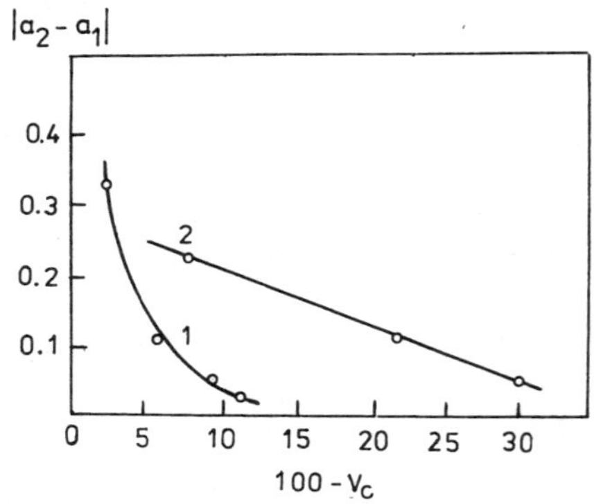

Figure 3.49. Relation between V_C and difference of affinities of polymers with solvent: 1 - PV2P-PMMA, 2 - PS-PMMA. Adapted, by permission, from C. Hugelin and A. Dondes, *Makromol. Chem.*, **126**, 206 (1969).

was discovered that the critical concentration of mixing changes approximately from 13 vol% in dimethyl aniline to 2.5 vol% in p-xylol. According to Scott theory, this should correspond almost to five-fold change in χ_{AB}. The essential influence of solvent on the miscibility was found[385,66,67] for films of blends of various polymers. DSC data show that films obtained from the solvents in which they are not miscible, stay immiscible by removing solvent, as well and *vice versa*. For immiscible films, two glass transition temperatures T_g by the DSC data are usually observed. Even when only one T_g is detected (an intermediate position between T_gs of pure components), the dielectric data show the microheterogeneity of such mixtures. It was also shown[66] that the critical concentration of mixing correlates with the value of $|a_2 - a_1|$ where a_i is the constant of Mark-Houwink equation connecting the intrinsic viscosity with molecular mass (Fig. 3.49). Because a_i are interrelated with the parameter of polymer-solvent interactions, their asymmetry increases the immiscibility of polymers in a solution.

The available data on the influence of solvent on the miscibility of polymers in solution were the stimulating incentive to reconsider the conclusion of the Scott theory concerning the role of a solvent in the miscibility of ternary systems. Zeman and Patterson[383] have used for this purpose the following equations for spinodals:[435]

$$\sum r_i\varphi_i - 2\sum r_i r_j(\chi_i + \chi_j)\varphi_i\varphi_j + \varphi r_1 r_2(\chi_1\chi_2 + \chi_1\chi_3 + \chi_2\chi_3)\varphi_1\varphi_2\varphi_3 = 0 \quad [3.147]$$

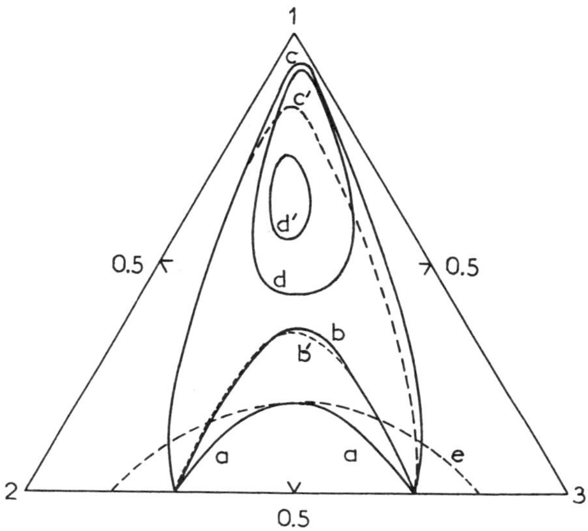

Figure 3.50. Calculated spinodals for a ternary system: solvent (1), polymer (2) and polymer (3). Values P_1=1, P_2=P_3 =1000 are taken for all curves. For spinodals 1,2,2′,3,3′ χ_{23}=0.0025 and the pairs of interaction parameters χ_{12}, χ_{13} are as follows: a) 0.40, 0.40, b) 0.40, 0.45, b′) 0.40, 0.35, c) 0.40, 0.50, c′) 0.40, 0.30. For spinodals d and d′ χ_{23}=0.0 and χ_{12} and χ_{13} are the same as for 3 and 3′. Curve e is a binodal calculated with the same choice of parameters as for spinodal 1. Adapted, by permission, from L. Zeman and D. Patterson, *Macromolecules*, **5**, 513 (1972).

where

$$2\chi_1 = \chi_{12} + \chi_{13} - \chi_{23} \qquad [3.148]$$

and other values of χ_i are found by the usual cyclic change of indices. In Eq 3.148, φ_i are the volume fractions, r_i are the numbers of segments in components (Patterson applies another designations of the system components, namely: 1 for solvent, 2 and 3 for two mixed polymers. In this section, this designations are followed for convenience). In Eqs 3.148 and 3.149, parameters χ_{ij} correspond to three energies of the formation of contacts between segments -- contrary to Flory parameters where contacts relate to the molecule of the first polymer, i.e., χ_{ij} (Flory) = $r_i\chi_{ij}$ (Tompa).

On the basis of Eq 3.148, Zeman and Patterson have constructed spinodals for changes of parameters χ_{1i} and χ_{23}. The analysis of the data allows for the following conclusions:

1. For symmetric case, i.e., at $\chi_{12} = \chi_{13}$ and for constant χ_{23}, the spinodals do not depend on χ_{1i}. Almost the same spinodals are obtained at a small difference in parameters χ_{1i} showing only greater region of miscibility (Fig. 3.50). It is interesting that improvement of the thermodynamic quality of the solvent for polymer 3 leads to the same effect as its worsening (curves 2 and 2′). Higher difference in χ_{ij} provokes the greater broadening of the immiscibility region (curves 3 and 3′). At $\chi_{12} = \chi_{13}$ and $\chi_{23} = 0$, the spinodals disappear, i.e., the polymers become miscible both in the presence and absence of solvent. However, at higher difference in χ_{1i} at $\chi_{23} = 0$, the region of immiscibility shifts to the lower concentrations (curves 4 and 4′), i.e., the immiscibility region remains at $\chi_{23} = 0$ and even at $\chi_{23} < 0$, but spinodals have shape of closed loops that do not intersect the axis of triangle diagram (binary mixtures are fully miscible).

In a model and real systems,[75] the restriction of miscibility in the region of small concentrations is connected with an inequality of polymer-solvent interactions, whereas in the region of high concentrations the immiscibility is the result of unfavorable polymer-polymer interactions (see discussion on the concentration dependence of parameter χ_{23}).

2. The effect of inequality of polymer-solvent interactions and unfavorable polymer-polymer interactions are typical of polymers with high molecular masses. When molecular mass of mixed polymers decreases, even at an essential difference in χ_{1i} (for example, at $\chi_{23}=0$) (Fig. 3.51), the closed loops of immiscibility rapidly narrow and at low molecular masses they disappear. Increasing molecular mass of one of the components broadens the miscibility region shifting it mainly to higher concentrations of low molecular weight polymer.

3. For the case of partial miscibility in the polymer-solvent system by simultaneous change of χ_{12} and χ_{13}, the critical points at high concentrations stay at the very bottom of phase diagram, whereas at low concentrations are shifted on spinodal to the end region of phase diagram, i.e., to the region of pure mixtures 1-2 or 1-3. As a result, the critical temperature appears with $(\chi_{1i})_{crit}=(1/2)[1+(r_1/r_2)^{1/2}]^2$ and $(\varphi_{1i})_{crit} =(1/2)[1 + (r_1/r_2)^{1/2}]^{-1}$ (Fig. 3.51). Subsequent increase of χ_{12} is accompanied by the appearance of a critical point characteristic of binary mixture 1-3 (Fig. 3.48, curve 3), this point, because of a further growth of χ_{1i}, comes out the region restricted by the phase diagram.

4. Phase diagrams of closed type should be very rare because special conditions must be fulfilled for their appearance. In particular, the difference between χ_{1i} and χ_{23} should be close to zero or negative. According to authors opinion,[383] such situation may occur for the ternary mixtures of polymers, having similar chemical nature (where

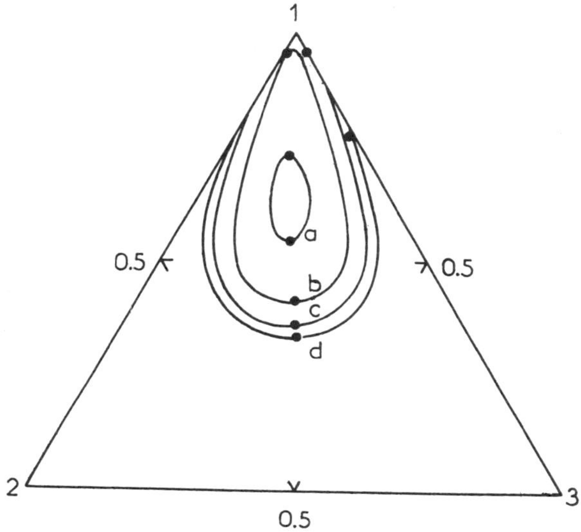

Figure 3.51. Calculated spinodals and critical points for the ternary system: solvent (1), polymer (2) and polymer (3) with $\chi_{23}=0$ and the following pairs of interaction parameters χ_{12} and χ_{13}: a) 0.40, 0.30, b) 0.532, 0.432, c) 0.632, 0.532, d) 0.70, 0.60. $P_1=1$ and $P_2=P_3=1000$ for all curves. The critical points which lie within the phase diagram are shown as dots. Adapted, by permission, from L. Zeman and D. Patterson, *Macromolecules*, **5**, 513 (1972).

$\chi_{23} = 0$) but differing in the chain flexibility. The latter determines the difference in free volumes of components or the contributions of the term of equation of state to χ_{12} and χ_{13}. Polymer homologues with essentially different molecular masses and thermal expansion coefficients may serve as an example of such systems.[351] For the system oligoethylene glycol (2,000)-benzene at 353 K, we have $\chi_{12} = 0.102$, whereas for oligoethylene glycol (40,000)-benzene, $\chi_{13} = 0.032$. Here the interaction parameter for the binary oligomeric mixture is negative and therefore for the ternary mixture of polymer homologues, the phase diagram of closed type may be obtained.

Presently there were discovered many systems with LCST,[90,91] i.e., with $\chi_{23} < 0$. Therefore, the diagrams of closed type can be typical of such miscible mixtures.[75,384] The determining role of inequality of interaction parameters in the mutual miscibility of two polymers in a common solvent was confirmed[433] by measurement of parameters χ_{23} for solutions of PS and PVME in a wide range of temperatures. It was found that if the difference $\Delta\chi = \chi_{12} - \chi_{13}$ diminishes with decreasing temperature, the immiscibility region narrows. The PS-PVME mixture has LCST[24,30] and parameter χ_{23} should increase with temperature. If the miscibility in solution was determined by parameter χ_{23}, as follows from the Scott's theory, then the opposite effect would be observed, namely, the broadening the immiscibility region. These experimental data fully confirm the theoretical conclusions by Patterson and stress the role of inequality of χ_{1j} in the immiscibility of two polymers in a common solvent. The calculations of spinodals[383] for parameters χ_{1i} and χ_{23} and various molecular masses of mixed polymers give the same conclusions.

The data discussed show that the role of a solvent in the ternary polymer-polymer-solvent systems is much more important than predicted by the Scott's theory. Moreover, these data confirm that polymers immiscible in solution are immiscible in the absence of a solvent and *vice versa.*

By studying the thermodynamic behavior of the ternary systems, a very narrow region of single-phase state was discovered[69,435] between LCST and UCST. This phenomenon may occur only when the critical temperatures of solutions of two binary polymer-solvent mixtures are close. The narrow interval of miscibility appears near the curves of cloud points of one of binary mixtures (Fig. 3.52). To find out the reasons of such behavior, the spinodal have been calculated theoretically[69] at variation of parameters χ_{12} and χ_{13} and their temperature dependence. The narrow region between LCST and UCST was observed if χ_{12} and χ_{13} were close; it diminished with increasing temperature. The value of χ_{23} should be very small (0.02) to get a very narrow region of miscibility, whereas the temperature dependence should by positive and strong. However, the model system that was used (oligostyrene and dimethyl siloxane mixture) had UCST (308 K) and therefore χ_{23} should decrease with temperature increasing. On the basis of these data, it was concluded that introduction of solvent should rapidly decrease parameter χ_{23} and change its temperature dependence from the negative to positive. The formation of a cloud of segments around gravity centre of macromolecule at small concentrations[69] is a possible reason. It prevents intermolecular segment-segment contacts and affects apparent value of χ_{23} in a dilute ternary system (it becomes lower than its true magnitude for the binary polymer-polymer mixture). With increasing total concentration of polymers in a mixture, the conformational effects begin to play very important role in changing χ_{23}. Okazawa[69] analyzed the data by Koningsveld[435] who has discovered a narrow miscibility region for the mixture of isotactic PP with poly(ethylene diphenyl ether). At the same time, for another pair: atactic PP and the same polyether such region was not discovered. Okazawa supposed that these effects are connected with the difference in gyration radii for isotactic and atactic

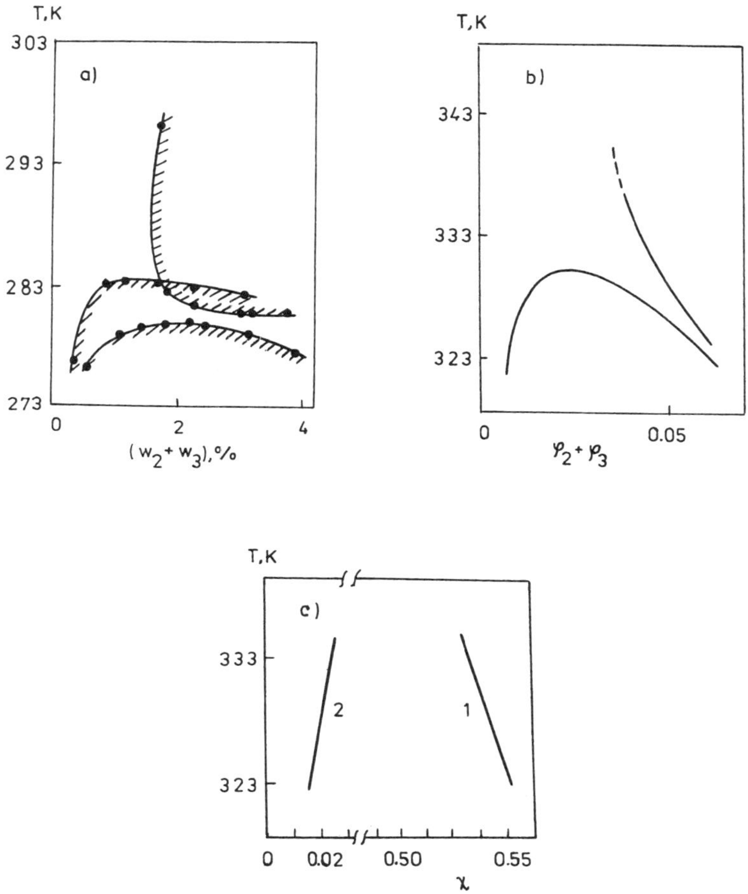

Figure 3.52. a - Experimental cloud point curves for the PS-PDMS: a-(1:1)-ethyl acetate ternary system (●) and the PDMS-ethyl acetate binary system (o) b-computed spinodal curve using the values of χ_{ij} shown in (c) and $r_1=r_2=2{,}000$. For (a) M_w (PS)=6.7×10^5, M_w (PDMS)=3.9×10^4. Adapted, by permission, from T. Okazawa, *Macromolecules*, **8**, 371 (1975).

PP. In atactic PP, the probability of the formation of intersegment contacts is lower than in isotactic PP.

Aharoni[72] observed the bimodality of closed cloud point curves for three miscible mixtures. The molecular attraction between polymers was greater than between any polymer and solvent. Thus, the reason for bimodality becomes clear considering that the same effects were observed for immiscible polymers, as well.[70]

Many works on the ternary polymer-polymer-solvent systems are dedicated to the determination of the polymer-polymer interaction parameter and analysis of the correctness of the data obtained by various methods (by composition of coexisting phases,[177] thermodynamic cycles,[8] inverse gas chromatography).[155,156] Long since the first work on the application of inverse gas chromatography for estimating χ_{23} in the mixtures of low molecular weight components,[155] the effect of the sorbate nature on the magnitude of parameter was discovered. Olabisi[32] found that by using a non-polar sorbate, the values of parameters χ_{23} for ε-caprolactone-PVC mixtures were positive and with polar sorbate they were negative. The negative values were attributed to the specific interactions between the components of a stationary phase and to complex formation. As a consequence, some part of interacting groups was excluded from the interactions. Olabisi's supposition that polar sorbate may not interact with that part of polar groups of the stationary phase which participates in the complex formation was confirmed for various systems.[156,346] For the mixture of PPG (2,000) and macrodiisocyanate, based on PPG (2,000) and 4,4'-diphenyl diisocyanate with blocked end-groups, it was shown[346] that methanol used as sorbate gives two sharp transitions on the curves of the dependence of ln V_g on 1/T due to its ability to form H-bonds with the mixture components.

These transitions are not connected with phase transformations because they do not occur when non-polar sorbate (heptane) is used (Fig. 3.53). The temperature transitions in the methanol presence are due to the weakening of hydrogen bonds by heating the mixture during experiment. The polar groups formed during process interact with methanol and increase the retention volume. Similar effect was observed for segmented polyurethanes.[436,437] These data show that some polar groups, interacting with sorbate, may be excluded from the sphere of interactions. These conditions may lead to negative values of parameters χ_{23} calculated from the following equations:[349]

$$\chi_{(1(2+3)} = \chi_{12}\varphi_2 + \chi_{13}\varphi_3 - \chi_{23}\varphi_2\varphi_3 \quad [3.149]$$

$$\ln V_{g(2+3)} = \varphi_2 \ln V_{g(2)} + \varphi_3 \ln V_{g(3)} + \varphi_2\varphi_3\chi_{23} \quad [3.150]$$

where $V_{g(2+3)}$ is the specific retention volume of sorbate by the polymer mixture, $V_{g(i)}$ is specific retention volume by individual components and φ_i are the volume fractions of components.

The dependence of the parameter χ_{23} on the sorbate nature was evaluated by many authors.[32,169,173,184] For example, for the PVC-dioctylphtalate very high negative values of χ_{23} were obtained[169] showing high miscibility of components up to the volume fraction of dioctylphtalate equal to 0.25. By increasing concentration, parameter becomes positive at the volume fraction equal to 0.55. The transition

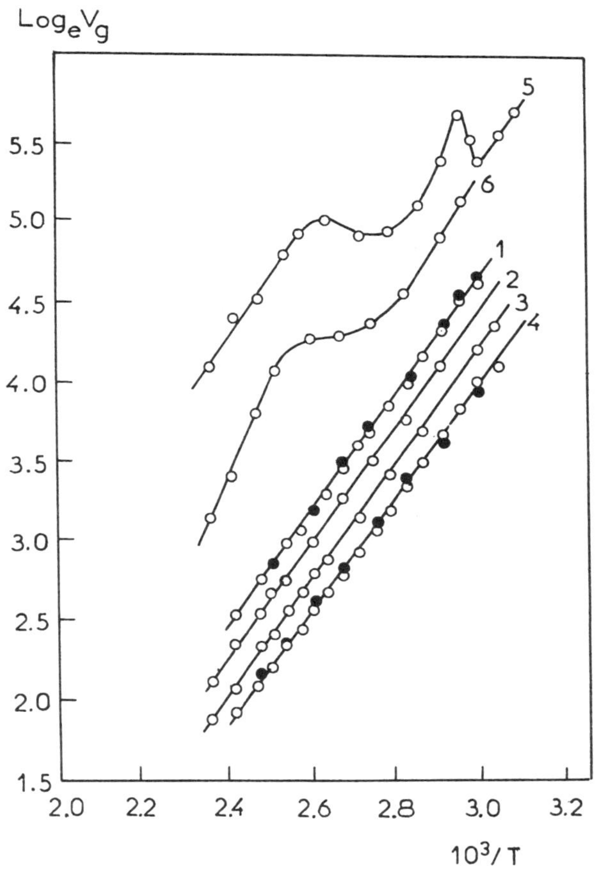

Figure 3.53. Dependence of lnV_G on 1/T of n-heptane (curves 1-4) and methanol (curves 5,6): 1 - poly(propylene glycol), 2 and 5 - mixtures of 19 wt% macrodiisocyanate and 81% poly(propylene glycol), 3 - mixture of 57 wt% of macrodiisocyanate and 43 % poly(propylene glycol), 4 - macrodiisocyanate, 6 - mixture of 29 wt% macrodiisocyanate and 71% poly(propylene glycol). Adapted, by permission, from A. E. Nesterov, T. E. Lipatova, K. Dusek, Z. Peltzbauer, M. Houska, J. Hradil, and Y. S. Lipatov, *Angew. Makromol. Chem.*, **52**, 39 (1976).

from negative to positive values was explained by a high mobility of dioctylphtalate molecules. At its high concentration, molecules preferentially contact each other with less contacts with PVC. When the probe, such as n-alkane, enters the stationary phase, the contacts between dioctylphtalate molecules are easier destroyed than contacts with PVC molecules. The miscibility of sorbate with dioctylphtalate is higher than with PVC or PVC-dioctylphtalate mixture because of the selective interaction of the sorbate molecules with dioctylphthalate. The parameter χ_{23} found from Eq 3.150 is smaller than parameter for all the mixture components freely interacted with probe molecule. Eq 3.150 gives the value χ_{23} by subtraction of $\chi_{1(23)}$ from the sum $(\varphi_2\chi_{12} + \varphi_3\chi_{13})$. The diminishing χ_{123} (increasing $V_{g(2+3)}$) leads to positive values of χ_{23}. The dependence of χ_{23} on the sorbate nature was explained[75] by the preferential interaction of sorbate with the polymer having lower χ_{1i}. This effect was called "$\Delta\chi$-effect". It could be expected that if the values of χ_{1i} are very close, then the experimentally found values of χ_{23} should be close, independent of the sorbate nature. Such experiments were done for the mixtures of polymer homologues of PEG and confirmed[348] the validity of this supposition (Table 3.1). The only necessary condition is to avoid the influence of support nature, coating thickness, the rate of gas-carrier, etc. In some cases, the comparable data on χ_{23} may be obtained even when χ_{1i} are different. This is the case of mixtures PEG-PPG for which parameters χ_{23} were determined by inverse gas chromatography[347] and by composition of coexisting phases[252] (Table 3.2). The analogous result was obtained for the mixtures of poly(epichlorohydrin) with poly(methyl acrylate).[167] It was established that the "true" values of interaction parameters can be determined for poor solvent interacting with both polymers.

Tager[438] concluded that to determine ΔG_{mix} and χ_{23} it is necessary to use thermodynamically good solvents. According to the data,[438] Gibbs free energy of mixing from thermodynamic cycles does not depend on solvent nature. For graft copolymers (considered as a mixture of two homopolymers), the parameters found from the same cycles were dependent on the solvent nature.[439]

Recently, it was noted[440] that the usual methods of the estimating χ_{23} by measurements of the properties of the ternary solutions (IGC and vapor sorption) do not give true value of interaction parameter, if the Scott's equation is used in the following form:

$$\chi_{1p} = \xi_2\chi_{12} + \xi_3\chi_{13} - \xi_2\xi_3\chi'_{23} \qquad [3.151]$$

where $\xi_i = \varphi_i/(\varphi_2 + \varphi_3)$, i=2,3, φ_i is the segment fraction of i-th component in the mixture, $\varphi_p=\varphi_2+\varphi_3$, $\chi_{1\,p}=\Delta\mu/RT\varphi_p^2$, $\chi'_{23}=\chi_{23}(V_1^*/V_2^*)$. To have a true value, the Flory theory of equation of state was used and the following equation for the true interaction parameter was derived:

$$\chi_{1p} = \chi_{12}\frac{\tilde{v}_{12}}{\tilde{v}}\frac{\Theta_2\Theta_p}{\Theta_{2B}^2} + \chi_{13}\frac{\tilde{v}_{13}}{\tilde{v}}\frac{\Theta_3\Theta_p}{\Theta_{3B}^2} - \chi'_{23}\frac{\tilde{v}_{23}}{\tilde{v}}\frac{\varphi_2\varphi_3}{\varphi_p^2}\frac{\Theta_p^2}{\varphi_p^2}\frac{S_1}{S_3} + \Gamma \qquad [3.152]$$

where Γ is the term of the free volume expressed as

$$\Gamma = \frac{P_1^* 1 V_1^*}{RT\varphi_p^2}\left[\Gamma_{12} - \frac{\tilde{v}_{12}\Theta_2\Theta_p}{\tilde{v}\Theta_{2B}^2}\Gamma_{12} - \frac{\tilde{v}_{13}\Theta_3\Theta_p}{\tilde{v}\Theta_{3B}^2}\Gamma_{13}\right] +$$

$$+ \frac{P_1^* V_1^*}{RT\varphi_p^2}\frac{\tilde{v}_{23}}{\tilde{v}}\frac{S_1}{S_2}\frac{\Theta_2}{\Theta_3}\Theta_p^2\Gamma_{23} \qquad [3.153]$$

where

$$\Gamma_{ij} = \tilde{v}_i^{-1} - \tilde{v}_j^{-1} - 3\tilde{T}_i \ln\left[(\tilde{v}_{ij}^{1/3} - 1)/(\tilde{v}_i^{1/3} - 1)\right] \qquad [3.154]$$

$$\Theta_{iB} = S\varphi_p/(S_1\varphi_1 + S_i\varphi_p) \qquad [3.155]$$

$$\Theta_p = \Theta_2 + \Theta_3 \quad (i = 2,3) \qquad [3.156]$$

This equation has no simplification. Eq 3.152 differs from Eq 3.151 because binary interaction parameters are multiplied by the surface, not the segment, fractions and contain the free volume term. Supposing $\tilde{v}_2 = \tilde{v}_3 = \tilde{v}_{23}$, $\tilde{v}_{12} = \tilde{v}_{13} = \tilde{v}_{1p}$ and $S_2 = S_3$, and comparing Eqs 3.152 and 3.151, we obtain:

$$\chi_{23}^A = \chi_{23}^T \frac{\tilde{v}_3 S_1 \Theta_p^2}{\tilde{v} S_3 \varphi_p^2} \qquad [3.157]$$

This equation shows that the difference between $\chi^A{}_{23}$ and $\chi^T{}_{23}$ depends on the solvent nature and polymer concentration in solution. Therefore, the error in estimating χ_{23} depends both on the method of determination and on the solvent nature. This difference has the following limits:

$$(\chi_{23}^A/\chi_{23}^T)_{\varphi_p \to 0} = (S_3/S_1)\tilde{v}_3/\tilde{v}_1 \qquad \text{(vapor sorption)} \qquad [3.158a]$$

$$(\chi_{23}^A/\chi_{23}^T)_{\varphi_p \to 1} = (S_1/S_2) \quad \text{(IGC)} \qquad [3.158b]$$

Because $S_1/S_3 \approx 2$ and the ratio $\tilde{v}_1/\tilde{v}_3 \approx 1.2$, the parameters χ_{23} calculated in approximation of additivity differ from the true values in the limits 1/2÷2. For the case $\varphi_p \to 1$ (IGC), we have:

$$\chi_{23}^A = \chi_{23}^T(S_1/S_2) + (\chi_{12} - \chi_{13})\frac{(S_3 - S_2)v*_2}{(\varphi_2 S_2 + \varphi_3 S_3)v*_1} - k \qquad [3.159]$$

Parameter k includes value at $\varphi_p \to 1$ according to the equation:

$$k = (v*_2/v*_1)[\Gamma/\xi_2\xi_3 + \chi_{12}(\tilde{v}_2/\tilde{v}_{23} - 1)[\xi_3^{-1} + (S_2 - S_3)/(\xi_2 S_2 +$$

$$+ \xi_3 S_3)] + \chi_{13}(\tilde{v}_3/\tilde{v}_{23} - 1)[\xi_2^{-1} - (S_2 - S_3)/(\xi_2 S_2 + \xi_3 S_3)] \quad [3.160]$$

Eq 3.159 explains the dependence of χ_{23} on the sorbate nature. These values depend on the affinity of a solvent to each polymer in the mixture ($\Delta\chi = \chi_{12} - \chi_{13}$) and on the number of contact points per solvent segment, S1. If $\chi_{12} = \chi_{13}$ or $S_2 = S_3$ (as observed for the mixtures of polymer homologues), then

$$\chi_{23}^{A} = \chi_{23}^{T} \frac{S_1}{S_3} \quad [3.161]$$

To obtain the true value of parameter, it is necessary to calculate values Γ and k. In a simplified form:

$$\chi_{23}^{A} = \frac{\chi_{23}^{T}}{v*_2 S_{23}} + \frac{(\chi_{12} - \chi_{13})}{v*_1 S_1} \frac{(S_3 - S_2)}{(\varphi_2 S_2 + \varphi_3 S_3)} \quad [3.162]$$

The analysis of the data of IGC for various sorbates, in the framework of the relations discussed above, was performed for a miscible pair poly(4-hydroxystyrene)-PVA and resulted in the value $\chi_{23} < 0$, indicating the miscibility of the pair and suitability of equations.[441] The effect of sorbate nature on IGC results was also discussed elsewhere.[442] The free energy of mixing, proportional to fractions of molar surfaces of mixed components not to their volume fraction, is the assumption of the equation for χ_{23}:

$$\chi'_{23} = X_{23}(S_1/S_2)\left(\varphi_2 \frac{S_2}{S_3} + \varphi_3 \frac{S_3}{S_2}\right) + X_{12}\left(1 - \frac{S_3}{S_2}\right) + X_{13}\left(1 - \frac{S_2}{S_3}\right) \quad [3.163]$$

where S_i is the molar surface of i-th component. The first term in Eq 3.163 is proportional to the parameter of contact interactions X_{23}/S_2 and depends on the sorbate nature through molar surface S_1 in analogy to the dependence of χ_{23} on V_1. Two last terms show that χ'_{23} depends also on the individual parameters X_{12} and X_{13}. The character of this dependence changes with some parameters of the mixture. If the ratios of the surface to the volume for both components are close, these changes are small, but become essential if contact parameters are considerably different. If individual values of X_{1i} are found experimentally for various sorbates and values of S_1 are correctly calculated, then the correct values of χ'_{23} should be obtained for other sorbates by using only one connecting parameter: S_2/S_3. Another equation for the determination of the interaction parameter was proposed[443] based on the properties of ternary solutions:

$$\chi = \xi_1^2 \chi_{11}^{b}(\varphi_1) + \xi_2^2 \chi_{22}^{b}(\varphi_2) + 2\xi_1 \xi_2 \chi*_{12}(\varphi_1, \varphi_2) \quad [3.\ 164]$$

where $\chi^b_{ii}(\varphi_i)$ are parameters of interaction in binary solution (polymer i and solvent 0) of polymer volume fraction φ_i and $\chi^*_{12}(\varphi_1\varphi_2)$ are the contributions to χ of all types of interactions between molecules of immiscible polymers 1 and 2 in a solution. Values φ_i and ξ_i are calculated from the relations: $\varphi_i = v_i n_i/V = P_i n_i/(n_0 + P_1 n_1 + P_2 n_2)$, where n_i and P_i are the number of moles and relative length of the chain of component i. Because $\varphi_0 + \varphi_1 + \varphi_2 = 1$, the composition of a solution is determined by variables φ_1 and φ_2. It may also be determined by variables φ and ξ_1: $\xi = 1 - \xi_2$, $\xi_1 = \varphi_1/\varphi$, where φ is the total volume fraction of polymer 1 and 2 in the mixture, ξ_1 is the volume fraction of polymer 1 in the mixture. All terms in Eq 3.108 are considered at constant P and T. This approximation was used[444] to calculate binodals for polymer-polymer-solvent mixtures to show how the affinity of solvent to each polymer influences the ternary phase diagram. The systems PS-PIB-cyclohexane and PS-PIB-benzene were chosen. The difference between these two system is based on fact that cyclohexane is a poor solvent for PS and a good solvent for PIB, whereas benzene is a good solvent for both polymers. The comparison of theoretically calculated and experimentally found binodals led to conclusion that, in the case of immiscible polymers, the affinity of the common solvent to one of the polymers influences the shape of the phase diagram. In particular, it was found that for a solvent with different affinity to the mixture components, the region of immiscibility shifts rapidly diagram towards the polymer-solvent axis whereas the tie-lines are parallel to the same axis. For a good solvent for both polymers ($\Delta\chi$ is small), the binodal curves are symmetric and tie-lines are parallel to the polymer-polymer axis of the ternary diagram (Fig. 3.54). The same regularities operate for the ternary polymer-polymer-polymer mixtures (see below), i.e., by adding the third polymeric component to two immiscible polymers.

In some works,[445-449] the concentration dependence of χ_{23} for immiscible polymers in a common solvent has been investigated. For the mixtures of PS and PMMA in toluene[448] and bromobenzene,[449] it was shown that the interaction parameter in a dilute solution does not change, whereas in a semi-dilute solution it increases with concentration. The growth of χ_{23} begins at concentrations exceeding the cross-over concentration, C^*, (Fig. 3.55).[448,449] In all cases, the linear increase of χ_{23} was observed. The extrapolation of the parameter value to $\varphi \to 1$ gives χ_{23} corresponding to polymer pair without solvent.[449]

The change of macromolecular coil dimensions in the ternary polymer-polymer-solvent systems was also considered.[450-457] de Gennes[450] and Joanny[451,452] extended the scaling approach to the ternary polymer-polymer-good solvent systems where polymers differ only by molecular mass. They found the additional regime of high-concentra-

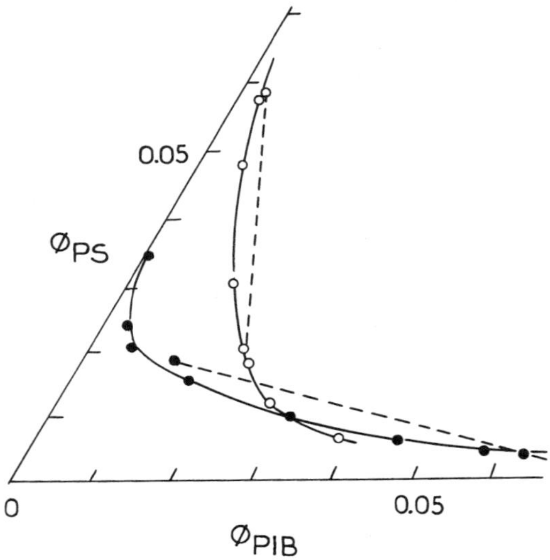

Figure 3.54. Binodal curves (solid lines) and tie lines (dashed lines) for PS-PIB$_1$ + chloroform (o) and PS-PIB$_2$ + benzene systems (●) at 303K. $M_w(PIB_1) = 1.54\times10^5$, $M_w(PIB_2) = 1.52\times10^5$, $M_w(PS) = 5.36\times10^5$. Adapted, by permission, from Z. Tong, Y. Einaga, T. Kitagawa, and H. Fujita, *Macromolecules*, **22**, 450 (1989).

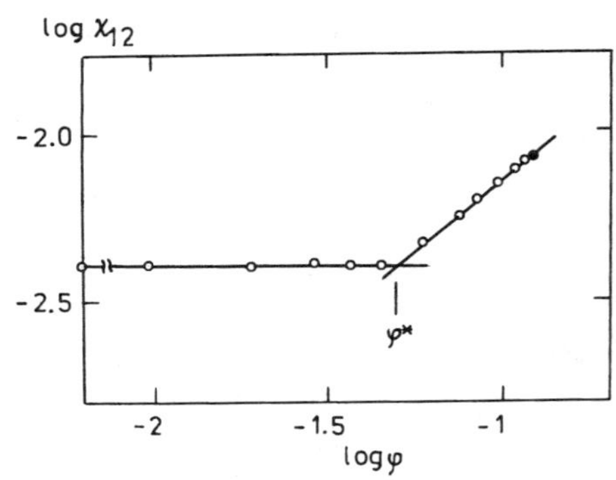

Figure 3.55. Plot of logχ_{23} vs.. logφ for PS/PMMA/bromobenzene ternary solution: the filled and half-filled circles represent the values at the spinodal and dilute limit, correspondingly; M_w (PS)$=2.83\times10^5$, M_w (PMMA)$=2.11\times10^5$. Adapted, by permission, from T. Fukuda, M. Nagata, and H. Inagaki, *Macromolecules*, **19**, 1411 (1986).

tion dependence of the coil dimensions with an exponent −1/4 when N>P and $1 < C_p/C_p^* < (N/P)^{4/5}$, where N and P are the polymerization degrees of the guest and host polymers (polymer with a higher concentration is solution is called "host" and with a lower concentration is called "guest") and C^* is crossover concentration. At the concentration $(C_p/C_p^*) > (N/P)^{4/5}$, the contraction of coil dimensions is described by the exponent −1/8 in dependence on the concentration of the host-polymer.

Nose[453] expanded the scaling analysis to include the effects of thermodynamic interaction between two types of polymers and established that collapse of the guest chain was predicted at the concentration close to C^* at high χ for N/P >> 1. When N = P the effect of the interaction parameter becomes essential at $C_p/C_p^* > 1$ and $\chi > (C_p/C_p^*)^{-5/8}$. At a given χ, the exponential law is predicted for the dimension changes with exponent −1/3.

The dependence of the gyration radius of the guest polymer on the concentration of the host-polymer also was studied experimentally.[454-457] Two mixtures were investigated: (1) poly-β-vinylnaphtalene with $M_w = 7.8\times10^5$ (guest polymer) in solution of PIB ($M_w=6.1\times10^5$) (host polymer) in mixed solvent (toluene-decalin = 1:1.3 by weight, poor solvent for poly(vinyl naphthalene) at 303 K, (2) poly(naphthyl methacrylate) ($M_w=5.2\times10^6$) (guest) and PMMA ($M_w=4.4\times10^6$) (host) in benzene (poor solvent for poly(naphthyl methacrylate) at 293 K).[454] The data are given in Fig. 3.56. With increasing polymer-host concentration, the dimensions of the coils of host-polymer decrease in both cases. There exists some region of polymer-host concentration where the dimensions of coils do not change. This region is narrower for polymers with higher immiscibility. The dependence of the guest polymer dimensions on the host polymer concentration is described by a power law:

$$(\overline{R}^2) = KC^{-1/3}$$

where K is a constant.

The dependence of the gyration radius of PS (guest polymer) on the concentration of PMMA (host polymer) in benzene – a good solvent for both polymers – was studied[456] in the concentration region of host polymer from dilute to semi-dilute regime. It was found that in dilute regime there was no change in coil dimensions for guest polymer, whereas at the concentrations of host polymer above C^*, the chains of the guest polymers contract by power law with the exponent −0.127. The diminishing of the gyration radius of PS (guest) vs. concentration of PMMA (host) in benzene[456] and toluene[457] follow the power law with different exponents for various combinations of molecular masses of polymers. It was discovered that when the Θ-conditions are reached by the addition of PMMA to toluene, the gyration radius of PS diminishes only by 15% which is much less than for Θ-conditions

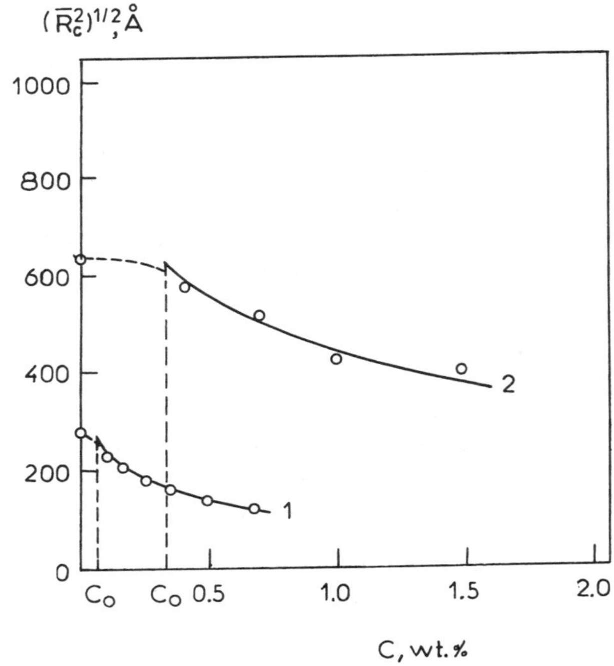

Figure 3.56. Dependence of the coil dimensions of poly(vinyl naphthalene) (1) and poly(naphthyl methacrylate) (2) on the second polymer (host) concentration in solution.

reached by adding a poor solvent. The reason of this effect is the exclusion of PMMA from PS coils, surrounded by pure toluene, adsorbed on PS coils. In both cases[455,456] the dependence of $(R^2/R^2_0)^{1/2}$ on scaled concentration C/C^* is given by a straight line for various combinations of molecular masses with the slope −0.266[455] and −1.[457] For the mixture PS-PMMA-ethylbenzoate (good solvent for both polymers), it was established that at smaller molecular mass of host polymer in dilute solution its chains penetrate the host-polymer domains.[455,458,459] In such solutions, due to the repulsion of coils, the second virial coefficient and interdiffusion coefficient diminish. In dilute solutions this effect prevails over the change in coil dimensions and mobility of macromolecules of host polymer. These results are important because they may explain some kinetic effects during bimolecular reactions between macromolecules with functional end-groups.

3.4.2 TERNARY POLYMER-POLYMER-POLYMER SYSTEMS

To describe the thermodynamic behavior of the ternary polymer mixtures, the same approaches can be used as for the ternary polymer-polymer-solvent systems. The interaction parameter can be written as[13]

$$\chi = \chi_{12}\varphi_1\varphi_2 + \chi_{13}\varphi_1\varphi_3 + \chi_{23}\varphi_2\varphi_3 \qquad [3.165]$$

where χ_{ij} are parameters of binary interactions of polymers i and j and φ is the volume fraction of corresponding polymer. For the polymer mixtures, if the critical values of $\chi_{crit}\rightarrow 0$ from Eq 3.109, it follows that at $\chi_{ij} > 0$ (unfavorable interactions), the ternary mixture is immiscible. To realize the miscibility, it is necessary that one of the components is miscible with two other (negative value of χ). For the first time, it was demonstrated for the mixtures of immiscible PMMA and PEMA to which the third component (PVDF) was added.[460] PVDF is miscible with both PMMA and PEMA. The ternary system is also miscible. It was confirmed later that the addition of the third component to two immiscible polymers makes all system miscible.[461-474]

The general picture of the effect of the third component on the thermodynamic behavior of the ternary mixture was developed on the basis of the Flory theory of equation-of-state.[475] To derive the equation of state of multicomponent system, there were introduced some additional parameters that are similar to the classical interaction parameter of the Flory-Huggins theory:

$$\sum_{j=2}^{N} \sum_{i=1}^{j-1} (X_{ij}/\tilde{v} - T\tilde{v}Q_{ij}) \qquad [3.166]$$

where X_{ij} and Q_{ij} are the exchange enthalpy and entropy of mixing, correspondingly. In principle, the exchange parameters are functions of concentration (neglected to simplify the calculations). According to Flory

$$X_{ij} = S_i\Delta\eta_{ij}/2v*_i^2 \qquad [3.167]$$

where

$$\Delta\eta_{ij} = (\eta_{ii} + \eta_{ij})/2 - \eta_{ij} \qquad [3.168]$$

where η_{ii}, η_{jj} and η_{ij} are characteristic energetic parameters describing intermolecular homo- and hetero-contacts. The characteristic pressure for the mixture is given as

$$P* = \sum_{i=1}^{N} \varphi_i P*_i - \sum_{j=2}^{N} \sum_{i=1}^{j-2} \varphi_i \Theta_j X_{ij} \qquad [3.169]$$

The reduced temperature is presented as

$$\tilde{T} = T/T^* = (1/P^*) \sum_{i=1}^{N} P_i^* \tilde{T}_i \varphi_i \qquad [3.170]$$

Eqs 3.169 and 3.170 give the usual equation of state:

$$\tilde{P}\tilde{v}/\tilde{T} = (\tilde{v}^{1/3} - 1)^{-1} - \tilde{T}\tilde{v} \qquad [3.171]$$

If the values of $\tilde{P}$ and $\tilde{T}$ are known from the corresponding properties of the individual components, using Eqs 3.169 and 3.170 value $\tilde{v}$ for the mixture can be found from Eq 3.171. The free energy of mixing can be found from the standard equation of statistical thermodynamics:

$$\Delta F^M = -kT \ln \left(z / \sum_{i=1}^{N} z_i \right) \qquad [3.172]$$

Considering each term of distribution function, one can obtain the generalized Helmholz free energy for multicomponent system in the form

$$\Delta F_M / N_\varphi kT = \sum_{i=1}^{N} (\varphi_i / r_i) \ln \varphi_i + 3 \sum_{i=1}^{N} \varphi_i (c_i - c) \ln [(2\pi m_i kT)^{1/2}/h] +$$

$$+ 3 \sum_{i=1}^{N} c_i \varphi_i \ln [(\tilde{v}_i^{1/3} - 1)/(\tilde{v}^{1/3} - 1)] + (v*/kT) [\sum_{i=1}^{N} \varphi_i P_i^* (1/\tilde{v}_i -$$

$$- 1/\tilde{v}) + \sum_{j=2}^{N} \sum_{i=1}^{j-1} \varphi_i \Theta_j (X_{ij}/\tilde{v} - T\tilde{v}_i \Theta_{ij})] \qquad [3.173]$$

In this equation, the average number of translation degrees of freedom for one segment of the mixture is

$$c = \sum_{i=1}^{N} \varphi_i c_i; \quad c_i = P_i^* v*_i / kT_i^* \qquad [3.174]$$

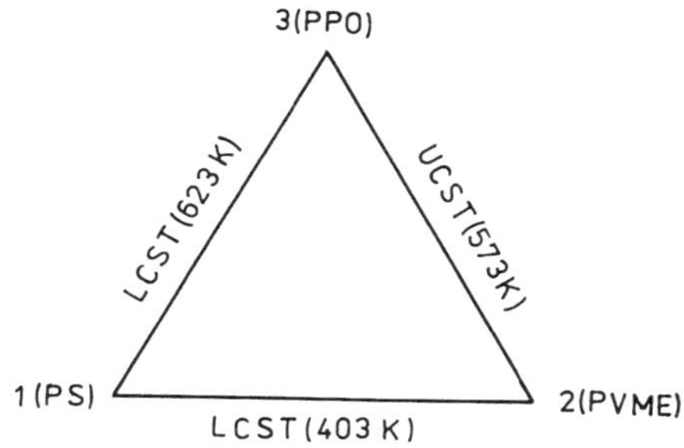

Figure 3.57. Binary blend phase behavior calculated from Flory's equation-of-state (EOS) theory for the systems: PS(1)/PVME (2), PS(1)/poly(2,6-dimethyl-1,4-phenyleneoxide) (PPO) (3) and PVME(2)/PPO(3). X_{12}=-2.0 J/cm^3, X_{13}=-5.7 J/cm^3, X_{23}=2.0 J/cm^3, Q_{12}=Q_{13}=-0.00277 J/deg×cm^3, Q_{23}=0.00277 J/deg×cm^3, S_1/S_2 =1.56, S_1/S_3 =0.75, S_2/S_3=0.48, r_i=r_2=r_3=400, P^*_i(J/cm^3): 480 (PS), 420 (PVME), 540 (PPO); $\alpha_i.10^{-4}$ (K^{-1}): 5.80 (PS), 6.73 (PVME), 7.23 (PPO); v_i (cm^{3}g^{-1}) 0.934 (PS), 0.950 (PVME), 0.93 (PPO). Adapted, by permission, from S. Klotz and H. J. Cantow, *Polymer*, **31**, 315 (1990).

The fraction of a surface is $\Theta_i = (S_i/S)\varphi_i$, where S_i is a segment surface area. The total surface area in a mixture is

$$S = \sum_{i=1}^{N} S_i\varphi_i \qquad [3.175]$$

The parameters of the equation of state can be calculated using Eqs 3.169-3.171 and 3.174-3.175. Eq 3.173 can be used, in principle, only for monodispersed polymers. For thermodynamically stable ternary system, there are considered only two independent variables φ_1 and φ_2 ($\varphi_3 = 1 - \varphi_1 - \varphi_2$). Thus the stability condition at constant temperature and pressure can be written as[475]

$$\begin{vmatrix} \delta^2 G/\delta\varphi_1^2 & \delta^2 G/\delta\varphi_1\delta\varphi_2 \\ \delta^2 G/\delta\varphi_1\delta\varphi_2 & \delta^2 G/\delta\varphi_2^2 \end{vmatrix} > \phi \qquad [3.176]$$

$$\delta^2 G/\delta\varphi_i^2 > 0 \qquad [3.177]$$

The solutions of Eqs 3.176 and 3.177 can be found from Eq 3.173 with corresponding parameters if to calculate derivatives. Because these equations are very bulky, we do not give them. There is no analytical solution of Eqs 3.176 and 3.177, but they can be solved numerically using iteration process by substituting the corresponding parameters obtained from the investigation of the mixture properties. In order to

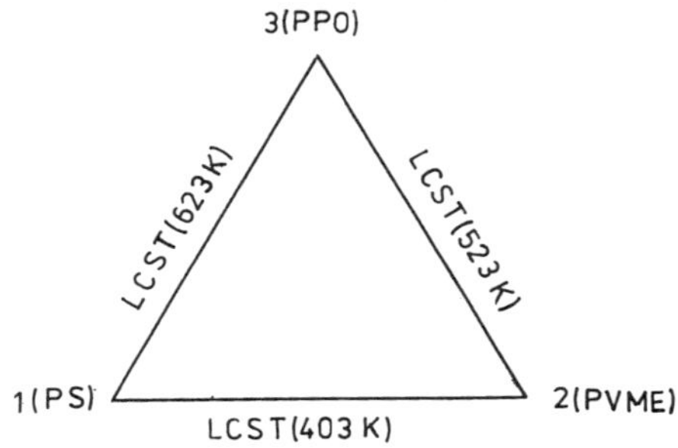

Figure 3.58. Binary blend phase behavior calculated from Flory's equation-of-state theory for the system shown in Figure 3.57, but X_{23} =-3.5 J / cm^3. PVME (2)/PPO (3) is assumed to exhibit LCST behavior. Adapted, by permission, from S. Klotz and H. -J. Cantow, *Polymer*, **31**, 315 (1990).

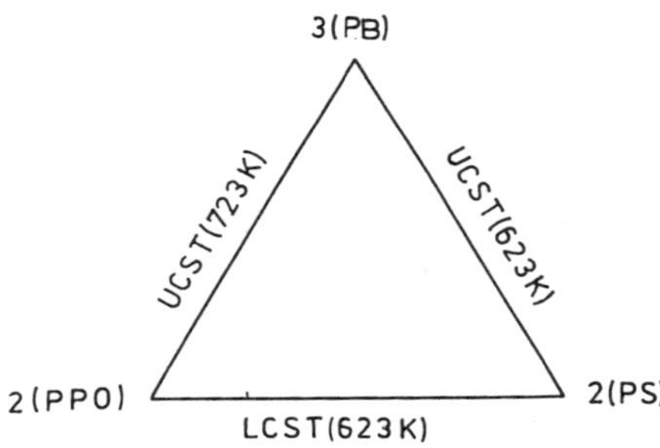

Figure 3.59. Binary blend phase behavior calculated from Flory EOS theory for the systems PPO (1)/PS (2), PPO (1)/PB (3) and PS (2)/PB (3). X_{12} =-5.7 J/cm^3, X_{13}=X_{23}=7.0 J/cm^3, Q_{12}=-0.0027 J/deg×cm^3, Q_{13}=Q_{23}= 0.0065 J/cm^3, S_1/S_2=1.34, S_1/S_3=1.98, S_2/S_3=1.48, r_i=r_1=r_2=400, P^* (PB)=420 J/cm^3, (αPB)=7×10^{-4}K^{-1}, v(PB)=1.176 cm^3/g. Adapted, by permission, from S. Klotz and H. -J. Cantow, *Polymer*, **31**, 315 (1990).

calculate the partial derivatives, the equation of the following type can be applied:

$$(\delta^2 G/\delta\varphi_i^2)_{P,T} = (\delta^2 F/\delta\varphi_i^2)_{T,v} + (\delta^2 F/\delta\varphi\delta v)_T\,(\delta v/\delta\varphi_i)_{T,P} = 0 \quad [3.178]$$

It is evident that these equations are suitable only for systems with weak specific interactions between components.

Using the developed equations, the thermodynamic behavior of ternary mixtures has been studied[475] with asymmetry of pair interactions. The following pair systems have been chosen:

- two with LCST and one with UCST (Fig. 3.57)
- three with LCST (Fig. 3.58)
- two with UCST and one with LCST (Fig. 3.59).

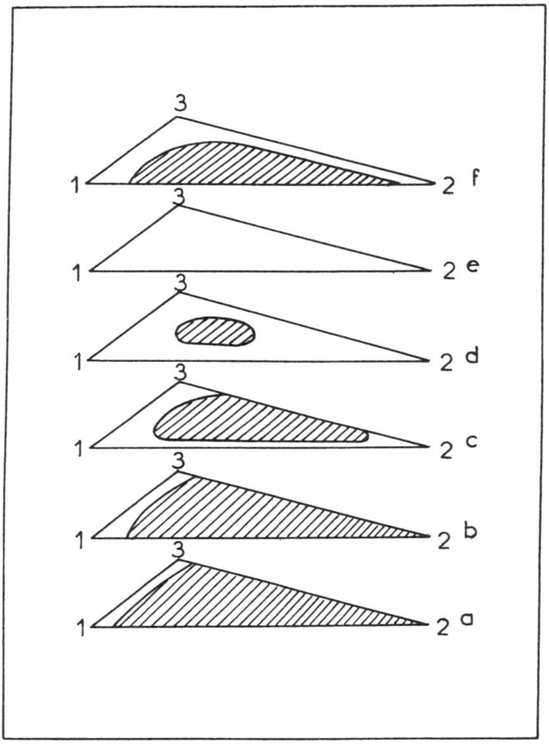

Figure 3.60. Ternary blend phase behavior calculated from Flory EOS theory. The EOS parameters are taken from PS (1), PVME (2), and PPO (3). The binary interaction parameters used in this model are given in Figure 3.57 a) T= 373K, b) T=493K, c) T=573K, d) T=673K, e) T=773K, T=873K. Adapted, by permission, from S. Klotz and H. -J. Cantow, *Polymer*, **31**, 315 (1990).

For these systems, the literature data on phase diagrams and parameters of the equation-of-state are available. These data allow one to use equations for estimating the effect of adding the third polymer component to a binary mixture of miscible or immiscible polymers on the phase behavior of the ternary mixture. For example, by adding the component 2 immiscible with one of the components of binary mixture (Fig. 3.60) to miscible mixture 1-3, the phase separation proceeds as demixing of the single-phase regime. With increasing temperature the region of miscibility broadens. The system formed by polymers 2 and 3 has UCST and, with increasing temperature, it becomes single-phase system. This means that the immiscibility region decreases with temperature increasing as well as the appearance of the closed region of immiscibility (Fig. 3.60). Further increase in temperature

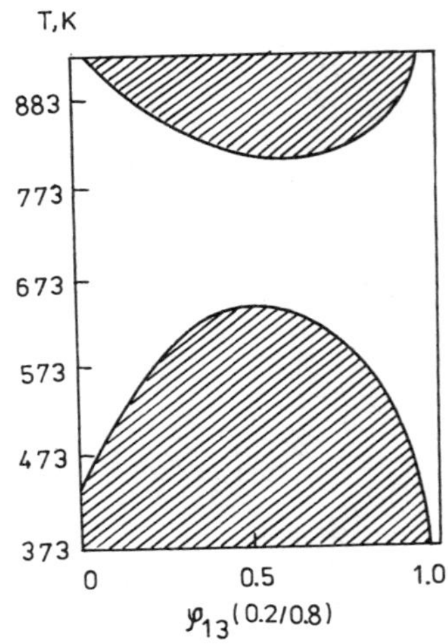

Figure 3.61. Concentration-temperature phase diagram of the system shown in Figure 3.60. The blend ratio φ_1/φ_3 is taken constant at φ_1/φ_3=0.25 and the concentrations of the homopolymer φ_2 of the blend φ_{i3} are varied (φ_2+φ_{13})=1. Adapted, by permission, from S. Klotz and H. -J. Cantow, *Polymer*, **31**, 315 (1990).

makes the ternary mixture fully miscible (Fig. 3.60) and the region of immiscibility appears on the side of 1-2 mixture on triangle diagrams. This region is determined by LCST for 1-2 mixture. Fig. 3.61 shows that the phase behavior at lower temperatures is controlled by UCST-behavior of the 2-3 system. At higher temperatures, the LCST-behavior of the 1-2 system prevails. At a definite ratio of parameters of exchange interaction, fully miscible ternary system may be designed (Fig. 3.60e).

Another model system consists of three pairs with LCST (Figs. 3.58 and 3.62). It is unexpected that the closed region of immiscibility appears (Fig. 3.62). This system is artificial because the parameters of exchange interactions are negative. The presence of closed region of immiscibility indicates miscibility of all three pairs. With increasing temperature, the region of immiscibility broadens and shifts to the region of higher concentrations of the component that has the lowest LCST with each other as compared with LCST of binary mixture of two components. Phase diagram in the region of low content of this component practically does not change.

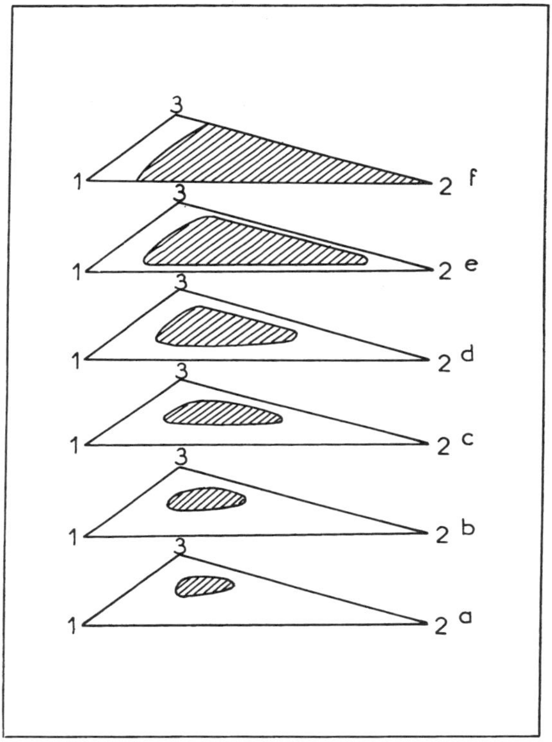

Figure 3.62. Ternary blend phase behavior calculated from Flory's EOS theory. Parameters are taken from PS (1), PVME (2), and PPO (3). The binary parameters used in this model are given in Figure 3.58. a) T=273K, b) T=373K, c) T=423K, d) T=473K, e) T=573K, f) T=673K. Adapted, by permission, from S. Klotz, H. -J. Cantow, *Polymer*, 31, 315 (1990).

For the ternary polymer-polymer-solvent system, Patterson[33] explained the immiscibility by the asymmetry of pair interactions. Therefore, the region of immiscibility for the ternary mixtures can be explained in the same way. It was found[476] that, in the ternary system PCL/SAN/CMA, all the binary pairs are miscible even though the ternary mixture reveals a broad immiscibility region.

For the third model system (Fig. 3.59), there was a broadening of the miscibility regions with temperature increasing due to improvement of miscibility of 1-3 and 2-3 systems (Fig. 3.63). Further temperature increase leads to the formation of fully miscible system (Fig. 3.63e). With increasing temperature, LCST is reached for 1-2 mixture and the region of immiscibility arises adjoining the composition axis

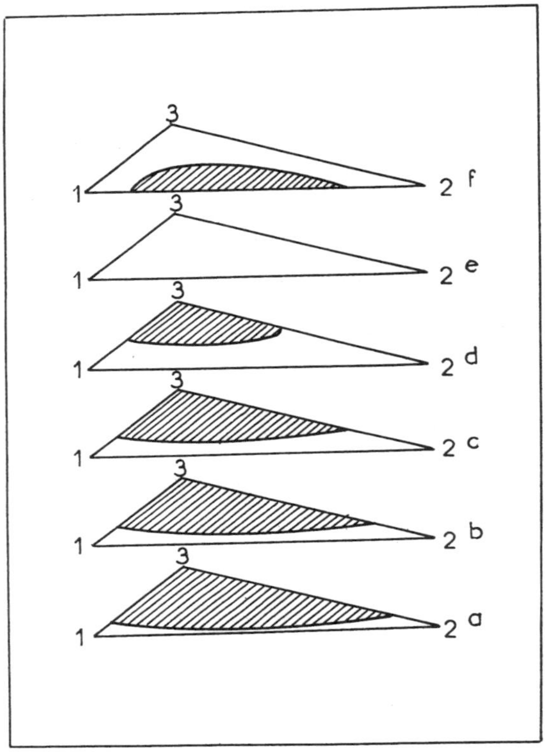

Figure 3.63. Ternary blend phase behavior calculated from Flory's EOS theory. Parameters are taken from PPO (1), PS (2), and PB (3) (Figures 3.57 and 3.59) and the binary interaction parameters are taken from Figure 3.59. a) T=273K, b) T=473K, c) T=573K, d) T=673K, e) T=773K, f) T=873K. Adapted, by permission, from S. Klotz and H. -J. Cantow, *Polymer*, **31**, 315 (1990).

1-2. It may be explained as follows. The third component is added to the miscible system 1-2. This component is immiscible with polymers 1 and 2 at low temperature. It is miscible at the higher temperature. LCST of the system 1-2 shifts to higher temperatures.

There are only a few data on the phase diagrams of ternary mixtures.[207,465,470,477-481,483] These data confirm the model.[475] The data[33] for the PS/PVME/chloroform mixture (a model system of the third type) give example of the close region of immiscibility. The ternary system having two LCST and one UCST (mixtures PS/PVME/PCL where PS/PVME and PVME/PCL have LCST[477] and mixtures PVC/PCL/CPVC where PVC/PCL and CPVC/PCL have

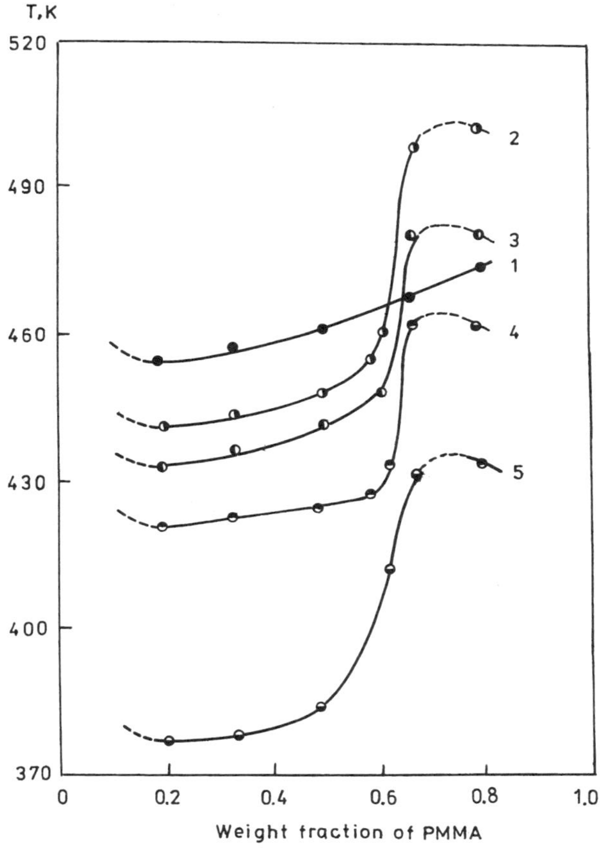

Figure 3.64. Cloud point curves for the mixtures PVA-PMMA (1) and PVA-PMMA-PEG with 5 (2), 10 (3), 20 (4), and 40 (5) wt% PEG.

LCST[465]) and the system with two UCST and one LCST (PS/PCV/PCL where PC/PS and PCL/PS have UCST).[477] For the last system a good agreement exists between experimental and calculated data.

For some systems, there are only incomplete data on the phase behavior. It was found[467] that PC is miscible with poly(3,4-cyclohexane-dimethyl-succinate), PCDS, or poly(1,4-butylene adipate), PBA. However PC is not miscible with SAN, i.e., the situation is similar to presented in Fig. 3.55. The addition of 1% polyester leads to miscibility at the ratio SAN:PC 1:1 and 3:1. On the phase diagram, the region of immiscibility adjoins the PC/SAN axis of the ternary diagram, i.e., is

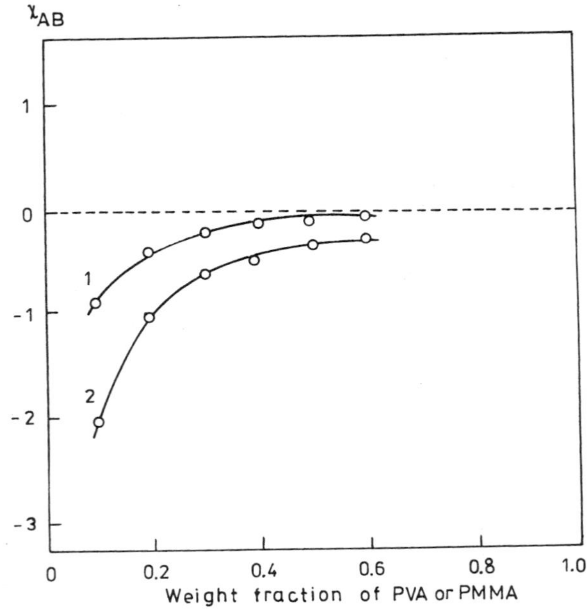

Figure 3.65. Concentration dependence of parameters χ_{AB} for PVA-PEG (1) and PMMA-PEG (2) mixtures, found from depression of the melting point.

shifted to higher concentration of PC. This is so because PC/PEF and PC/SAN have LCST, whereas PC/SAN - UCST. The mixture PVC/PCL/CP corresponds to the scheme in Fig. 3.55. Addition of PVCL leads to the improvement of miscibility.[465] Among other mixtures, the following were studied p-MSAN/PEMA/ PMMA,[470] (two LCST and one UCST) and PVC/PMMA/SAN.[472]

Su and Fried[478] attempted to adapt principles of Flory-Higgins theory to ternary mixtures. The phase diagrams were constructed from the binodals and spinodals calculated for various ratios of parameters χ_{ij}. Mutual solubility is favored when the ternary mixtures are composed of partially miscible systems and all pair interactions are athermic. The miscibility is restricted in symmetric ternary systems especially if one of the polymer pairs is either more miscible or immiscible. For symmetric mixtures of partially miscible polymer pairs both two and three-phase windows of miscibility are predicted. Some systems were investigated with all three pairs miscible.[207,480,481] These were: PMMA/PVA/poly(epichlorohydrin), PECH,[481] phenoxy/PVME/PCL,[480] and PMMA/PEO/PECH.[207] The ternary phase diagrams were obtained for some other systems.[483,484-486]

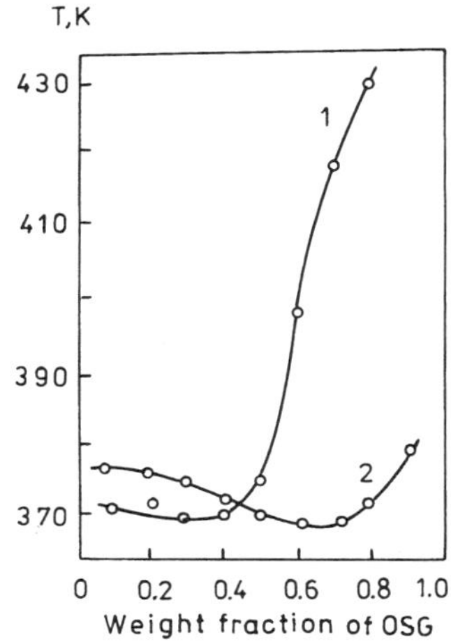

Figure 3.66. Cloud point curves for the mixtures PEG-OSG (1), PEG-OSG-PCL (20 wt PCL) (2).

All pairs of PVA/PMMA/PEG system[484] are here miscible. PVA/PMMA has LCST and PEG/PMMA is not separated up to temperature of degradation. Addition of PEG (Fig. 3.64) leads to the change in the phase separation temperatures. Also, the shape of the cloud point curves changes (they acquire S-shape). The temperature of phase separation at small additions of PEG (5-10 wt%) is lower than for PVA/PMMA mixtures without PEG. In the region of the preferential content of PMMA, these temperatures are higher for the same mixtures without PEG. Increasing PEG content, the phase separation decreases temperature of phase separation but S-shape of the cloud point curve is preserved. The asymmetry of the cloud point curves can be determined from the difference in parameters of interaction χ_{ij} of PEG with PVA and PMMA (Fig. 3.65). The parameters were measured by melting point depression of PEG in mixtures with PVA and PMMA.[484] The data in Fig. 3.65 show that ternary mixture should have a closed immiscibility region, which broadens with increasing temperature and shifts to the side of the component having smaller interaction parameter with PEG (i.e., PMMA). Such behavior was predicted for the mixtures where all the pairs have LCST.[475] Another system with three miscible pairs is oligostyrene glycol OSG/PEG/PCL[485] (Fig. 3.66). Interaction parameters for this system were measured from melting point depression of crystallizing components PEG and PCL (Fig. 3.67). It is seen that for the mixtures

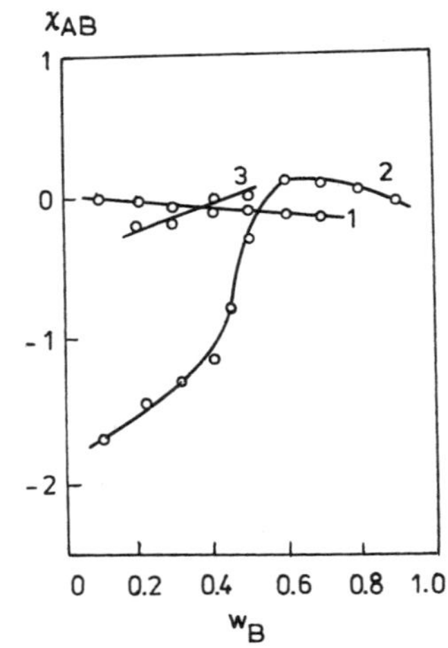

Figure 3.67. Concentration dependence of χ_{ij} for PEG-OSG (1), PEG-PCL (2), and PCL-OSG (3) found from depression of the melting point.

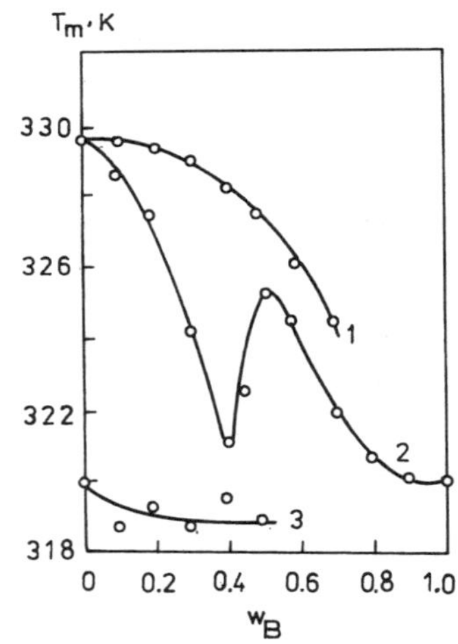

Figure 3.68. Dependence of the melting points on composition for the blends PEG-OSG (1), PEG-PCL (2), and PCL-OSG (3).

PEG/OSG and PCL/OSG, the concentration dependence of parameters χ_{ij} is linear. For PEG/PCL a very complicated dependence is observed. It depends on the effect of concentration on the melting point

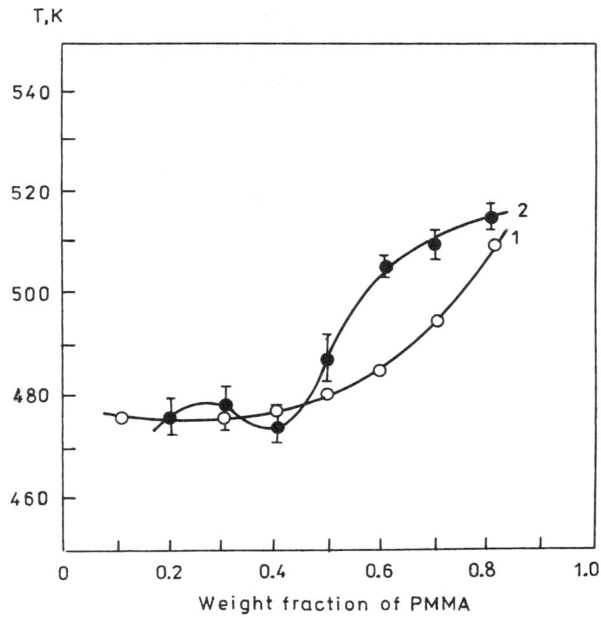

Figure 3.69. Cloud point curves for PVA-PMMA (1) and PVA-PMMA blends with addition of 10-40 wt% of SAN (2). Heating rate 1.3K/min.

of this mixture (Fig. 3.68). Fig. 3.66 shows that the addition of PCL to PEG/OSG mixture leads to the increase in the phase separation temperature in the region of primary content of PEG. This temperature decreases when high OSG concentrations are present in the mixture. The effect is caused by higher affinity of PCL to PEG than to OSG (parameters χ_{ij} for the PEG/PCL mixture at small concentration of PCL are smaller than for OSG/PCL) (Fig. 3.67).

The ternary mixture PVA/PMMA/SAN with 28 mol% of acrylonitrile has different behavior.[486] In this system, two pairs PVA/PMMA and SAN/PMMA have LCST and one (SAN/PVA) has UCST (SAN is not miscible with PVA at room temperature). The addition of SAN to PVA/PMMA mixture gives a complicated shape of concentration dependence of the cloud point curve (Fig. 3.69). In the region of preferential content of PVMA, the phase separation temperatures are higher than temperatures for the same mixtures without SAN. The behavior of this system is close to predicted for the mixtures where two pairs have LCST and one UCST.[475] By introduction of the third component into miscible system PVA/PMMA, the demixing of single-phase region may occur. Temperature increase causes the transition

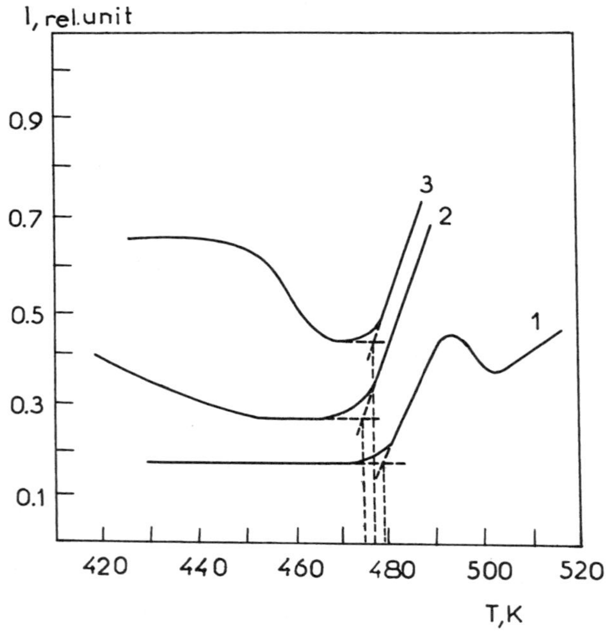

Figure 3.70. Temperature dependence of the light scattering intensity of films PVA-PMMA 60/40 (1) and PVA-PMMA-SAN 54.5/36.4 / 9.1 (2) and 48.8 / 32.5 /1.87 (3).

to single-phase and system becomes more transparent (curves 2 and 3, Fig. 3.70). Further temperature increase leads again to immiscibility caused by LCST in PVA/PMMA mixture.

The available data on the ternary polymer-polymer-polymer systems confirm in a qualitative form the theoretical predictions. The same principles are valid as for mixtures of homopolymers with copolymers having monomeric units of both homopolymers. For example, in the system PEO/PMMA/PEG all pair interaction parameters are negative and the system should have the closed region of immiscibility.[475] Substitution of PEG and PEO for poly(epichlorohydrin-*co*-ethylene oxide) results in the closed region of immiscibility for PMMA/copolymer system (all three interaction parameters are negative).

Table 3.3: The miscible polymers

Component 1	Component 2	Criteria of miscibility or determination method	Refs.
Cellulose	Poly(ethylene terephthalate)	FTIR	491
Nitrocellulose (12,62 and 13,42%N)	Poly-ε-caprolactone Poly(ethylene adipate) Poly(butylene adipate) Polyvalerolactone	FTIR	492
Cyclic oligomers of Bisphenol-A polycarbonate	Bisphenol chloral polycarbonate Bisphenol-A-polyformal Bisphenol-A-polysulfone Poly(ether sulfone) Poly(methyl methacrylate) Poly(phenyl sulfone) Poly(xylylene ether)	The films obtained from solutions are transparent	493
Oligodimethylsiloxane	Oligomethylphenylsiloxane	LCST	232
Oligobutadieneurethane	n-Ethylenedimethacrylate with n=1-5 n-Oxyethylenedimethacrylate with n=1-13	UCST	494
Hydrogenated oligocyclopentadiene	Isotactic polypropylene	UCST and LCST	197
Hydrogenated oligocyclobutadiene	Isotactic poly(butene) Isotactic polypropylene	$\chi_{AB}<0$ T_g,UCST and LCST	495 496
Polyamide-6	Trogamide T (product of condensation of dimethylterephthalate and 2,2,4'-three-methyl-1,6-hexadiamine)	DSC,T_g	497
Poly(aryl ether ether ketone)	Poly(aryl ethersulfone)	$\chi_{AB}<0$,LCST	498
Polybenzimidazole	Polyimide on basis of poly(2,2 (m-phenylene)-5,5-dibenzimidazole), Polyimide on basis of 3,3',4,4'-benzophenone tetracarbonyl anhydride and 4/1 moles 2,4-toluilenediisocyanate and 4,4 '-diphenylmethanediisocyanate	LCST	305
Polybutadiene	Deuterated polybutadiene Polyisoprene (M_η=35,550)	UCST LCST	499 502
Protonated polybutadiene	Deuterated polybutadiene	Transition from UCST to LCST with decreasing of 1,2-unit in p-PBD or d-PBD	500
1,4-Polybutadiene (from 3 to 92% 1,4-units)	cis-1,4-Polyisoprene	LCST	501
Chlorinated polybutadiene	Chlorinated polyethylene	LCST	209
Poly(butylene terephthalate)	Polyarylate	$\chi_{AB}<0$	503
Poly(2,6-dimethyl-1,4-phenylene oxide)	Deuterated polystyrene Ionated polystyrene	Diffusion LCST (6-47% iode)	504 505

Component 1	Component 2	Criteria of miscibility or determination method	Refs.
Partially sulfonated poly(2,6-dimethyl-1,4-phenylene oxide)	Poly(p-fluorostyrene) Poly(o-fluorostyrene) Poly(p-chlorostyrene) Poly(p-bromostyrene)	DSC,T_g	506
Brominated poly(2,4-di-methyl-1,4-phenylene oxide)(50-76%Br)	Polystyrene	LCST	205
	Poly(butylene terephthalate)	DSC,DMA	508
	Poly(hydroxy ether of phenylphthalein)	T_g, $\chi_{AB}<0$	509
	Poly(methyl methacrylate)	Immiscibility loop	262
	Poly(methyl methacrylate)	LCST	219, 374, 378, 510
	Atactic PMMA	LCST	512
	Syndiotactic PMMA	UCST and LCST	513
	Syndiotactic PMMA (M from 8 300 to 55 000)	Immiscibility loop	513
Hexafluoro-Bisphenol-A-polycarbonate	Poly(ethyl methacrylate) Poly(n-propyl methacrylate)	T_g	199
Poly(hydroxyether of Bisphenol-A)	Poly-ε-caprolactone	IR, LCST	514
Poly(epichlorohydrine)	Poly(methyl acrylate)	LCST	515
	Poly(ethyl acrylate)	LCST	515
	Poly(n-propyl acrylate)	LCST	515
	Poly(methyl methacrylate)	LCST	515
	Poly(ethyl methacrylate)	LCST	515
	Poly-n-butyl methacrylate)	LCST	515
	Poly(propyl methacrylate)	T_g	209
	Poly(isopropyl methacrylate)	T_g	209
	Poly(N-vinyl-2-pyrrolidone)	LCST	212, 516
Poly(ether ether ketone)	Poly(ether imide)	$\chi_{AB}<0$	517
Poly(ether sulfone)	Poly(ether imide)	T_g, LCST	518
Poly(ethylene glycol)	Poly(propylene glycol)	χ_{AB}, UCST	530
Poly(ethylene glycol) maleate	Epoxy resins (IPN)	T_g	519
	Epoxy resins cured by phthaleic anhydride		520
Poly(ethylene oxide)	Cellulose	$\chi_{AB}<0$	521
	Amorphous polyurethane on basis ester and 4,4-diphenylmethanediisocyanate	DSC	522
	Poly(ethyl acrylate)	$\chi_{AB}<0$	507
	Poly(epichlorohydrine)	LCST	207
	Poly(carboxylic acid)	Complex	523
	Poly(hydroxy ether Bisphenol-A)	B<0, NMR	524
	Poly(D(-)-(3-hydroxybutyrate)	χ_{AB}, LCST	432
	Polyisophthalamide (from 2 to 10 methylene groups)	UCST, LCST and hour-glass vs. number of methylene groups	316
	Poly(hydroxy ether of bisphenolphthalein)	T_g, $\chi_{AB}<0$	509
	Poly(methyl methacrylate)	LCST	207

Component 1	Component 2	Criteria of miscibility or determination method	Refs.
Poly(ethylene oxide)	Atactic PMMA	LCST	526
	Atactic,syndiotactic PMMA	T_m, T_g	361
	Deuterated PMMA	$\chi_{AB}<0$	124
	Poly(ethyl methacrylate)	T_m	527
Poly(hydroxy ether phenolphthalein)	Poly(butylene terephthalate)	DSC,T_g	528
cis-Polyisobutylene	*trans*-Polyisobutylene	DSC	529
Poly(methyl methacrylate)	Poly(α-methylstyrene)	LCST	530
	Poly(diglycidyl ether of Bisphenol-A)	FTIR	531
	Poly(hydroxy ether of Bisphenol-A)	T_g	532
	Poly(3-hydroxybutyrate)	DSC(to 20% PHB)	533
	Poly(1-hydroxy-2,6-phenylene methylene)	T_g	534
	Chlorinated polypropylene	LCST	277
	Poly(vinyl chloroacetate)	LCST	535
	Poly(vinyl phenol)	T_g	534
	Deuterated PMMA	UCST	588
Poly(chloromethyl methacrylate)	Poly-ε-caprolactone	$\chi_{AB}<0$	536
	Poly(methyl methacrylate)	T_g	537
	Poly(ethyl methacrylate)	T_g	537
	Poly(n-propyl methacrylate)	T_g	537, 238
	Poly(isopropyl methacrylate)	T_g	537, 238
	Poly(acetonyl methacrylate)	T_g	539
	Poly(2-chloromethyl methacrylate)	$\chi_{AB}<0$	536
	Poly(tetrahydrofurfuryl methacrylate)	T_g	537, 238
	Poly(ethylene adipate)	$\chi_{AB}<0$	540
	Poly(butylene adipate)	$\chi_{AB}<0$	540
	Poly(hexamethylene sebacate) Poly(2,2' -dimethyl-1,3-propylene succinate) Poly(2,2' -dimethyl-1,3-propylene adipate)	$\chi_{AB}<0$	540
Poly(3-chloropropyl methacrylate)	Poly(butylene adipate) Poly(2,2-dimethyl-1,3-propylene adipate) Poly(hexamethylene sebacate) Poly-ε-caprolactone	$\chi_{AB}<0$	541
Poly(1,1,1,3,3,3-hexafluoro-isopropyl methacrylate)	Poly(methyl methacrylate) Poly(ethyl methacrylate) Poly(tetrahydrofurfuryl methacrylate) Poly(2-fluoroethyl methacrylate)	LCST	542

Component 1	Component 2	Criteria of miscibility or determination method	Refs.
Poly(1-chloroethyl methacrylate)	Poly(methyl methacrylate)	LCST, T_g	543
	Poly(ethyl methacrylate)	LCST, T_g	543
	Poly(n-propyl methacrylate)	LCST, T_g	543
	Poly(isopropyl methacrylate)	LCST, T_g	543
	Poly(n-butyl methacrylate)	LCST, T_g	543
	Poly(tetrahydrofurfuryl methacrylate)	LCST, T_g	543
Poly(3-chloroethyl methacrylate)	Poly(tetrahydrofurfuryl methacrylate)	LCST	544
Poly(2-fluoroethyl methacrylate)	Poly(tetrahydrofurfuryl methacrylate)	LCST	542
Poly(2-iodoethyl methacrylate)	Poly(butylene adipate)	LCST	542
	Poly(2,2'-dimethyl-1,1,3-propylene adipate)	LCST	542
	Poly(hexamethylene sebacate)	LCST	542
	Poly-ε-caprolactone	LCST	542
Poly(2,2'-dichloroethyl methacrylate)	Poly(ethyl methacrylate)	LCST	545
	Poly(butyl methacrylate)	LCST	272
	Poly(n-propyl methacrylate)	LCST	542
	Poly(isopropyl methacrylate)	LCST	542
	Poly(isoamyl methacrylate)	LCST	272
Poly(2,2,2-trichlorooctyl methacrylate)	Poly(methyl methacrylate)	LCST	545
	Poly(ethyl methacrylate)	LCST	545
	Poly(n-propyl methacrylate)	LCST	545
	Poly(isopropyl methacrylate)	LCST	545
	Poly(tetrahydrofurfuryl methacrylate)	LCST	545
Poly(vinyl acetate)	Cyclic oligomer Bisphenol-A polycarbonate (M=1530)	LCST	493
	Poly(2-bromoethyl methacrylate)	LCST, DSC	576
	Poly(2-chloroethyl methacrylate)	LCST, DSC	576
	Poly(D(-)-(3-hydroxybutyrate)	$\chi_{AB}<0$	144
	Poly(4-hydroxystyrene)	$\chi_{AB}<0$	577
	Poly(methyl acrylate)	T_g	578
	Poly(methyl methacrylate)	LCST	579
Poly(vinyl alcohol)	Cellulose	DSC, DMA	567,568
	Poly(acrylic acid)	LCST	569
	Poly(methacrylic acid)	LCST	569
	Poly(vinyl acetate)	DMA, T_g	570
	Poly(vinyl pyrrolidone)	NMR	572
	ibid.	FTIR, $\chi_{AB}<0$	573
	ibid.	T_g, $\chi_{AB}<0$	574
	ibid.	T_g, IR	575
Poly(vinyl butyral)	Polyurethane on the basis of methylenebisdicyclohexyl diisocyanate	DSC, DMA	580

Component 1	Component 2	Criteria of miscibility or determination method	Refs.
Poly(vinyl chloride)	Nitrile rubber	$\chi_{AB}<0$	581
	Poly(butyl acrylate)	LCST	574
	Poly(butyl methacrylate)	LCST	320
	Poly(methyl methacrylate)	LCST, T_g	114,582, 583,584
	Poly-ε-caprolactone	$\chi_{AB}<0$	161
	Poly(methylene oxide)	$\chi_{AB}<0$	585
	Poly(ethylene oxide)	$\chi_{AB}<0$, T_g	525,586
	Poly(acetonyl methacrylate)	T_g	110
	Poly(isopropyl methacrylate)	T_g	582
	Poly(N-vinyl-2-pyridine)	T_g	212
	Poly(tetramethylene oxide)	$\chi_{AB}<0$	587
	Poly(propylene oxide)	$\chi_{AB}<0$	587
Chlorinated poly(vinyl chloride)	Poly-ε-caprolactone	LCST (67.2% Cl)	465
	Poly(methyl methacrylate)	LCST	360
	Polyurethane	T_g	546
	Atactic PMMA	LCST	491
Poly(vinyl methyl ether)	Poly(2-chlorostyrene)	LCST	560
	Poly(ethyl acrylate)	LCST, T_g	561
	Poly(propyl acrylate)	LCST, T_g	561
	Poly(butyl acrylate)	LCST, T_g	561
	Poly(hydroxyether of Bisphenol-A)	LCST	102
	Poly(hydroxyether of Bisphenol-A)	$\chi_{AB}<0$	509
	Polisophthalamide (2-10 methylene groups)	UCST, LCST	316
Poly(vinyl ethylene)	cis-1,4-polyisoprene	UCST, $\chi_{AB}<0$	180, 181
	Poly(ethyl ethylene)	UCST	126
Poly(vinyl ethyl ethylene)	cis-1,4-polyisoprene	DSC	299
Poly(vinylidene fluoride)	Poly(acetonyl methacrylate)	$\chi_{AB}<0$	142
	Poly(methyl methacrylate)	UCST, LCST	254,260
	Poly(methyl methacrylate)	$\chi_{AB}<0$	161
	Poly(ethyl acrylate)	LCST	547
	Poly(ethyl methacrylate)	LCST	548
	Poly(vinyl pyrrolidone)	DMA	549
Poly(vinyl fluoride)	Poly(vinyl phenol)	LCST	550
Poly(vinyl phenol)	Poly(butylene terephthalate)	DSC, T_g	556
	Poly(ethylene terephthalate)	DSC, T_g	556
	Poly(2,2'-dimethylpropylene terephthalate)	DSC, T_g	556
	Poly(ethylene-2,6-naphthalene dicarboxylate)	DSC, T_g	556
Poly(2-vinyl pyridine)	Poly(2-acrylamido-2-methyl-propanesulfonic acid)	IR, TA	551
	Poly(diglycidyl ether of Bisphenol-A)	FTIR	551

Component 1	Component 2	Criteria of miscibility or determination method	Refs.
Poly(4-vinyl pyridine)	Poly(2-acrylamido-2-methyl-propanesulfonic acid)	IR, TA	551
	Poly(diglycidyl ether of Bisphenol-A)	FTIR	551
	Cellulose	NMR	552
Poly(vinyl pyrrolidone)	Cellulose	NMR	553
	Poly(vinyl formal)	DSC, DTA	554
	Poly(vinyl butyral)	IR, complex	554
Poly(vinyl propionate)	Poly(ethyl acrylate)	DSC, T_g	555
Poly(2,2'-m-phenylene-5,5'-benzimidazol)	Polyimide based on 3,3',4,4'-benzophenonetetracarboxydianhydride and 3,3'-diaminobenzophenone	DSC, IR	557
Poly(4,4'-bisphenylene-dicarboxamide)	Polyamide 46	T_g	558
	Polyamide 6	T_g	558
	Polyamide 66	T_g	558
	Polyamide 69	T_g	558
	Polyamide 610	T_g	558
	Polyamide 11	T_g	558
Poly(2,2'-bis(trifluoro-methyl)-4,4'-bisphenylene-dicarboxamide)	Polyamide 46	T_g	558
	Polyamide 6	T_g	558
	Polyamide 66	T_g	558
	Polyamide 69	T_g	558
	Polyamide 610	T_g	558
	Polyamide 11	T_g	558
Poly(p-phenylene benzothiazole)	Polyamide 66	LCST	208
Sulfonated poly(oxy-1,4-phenyleneoxy-1,4-phenylenecarbonyl-1,4-phenylene)	Poly(ether imide)	T_g	559
	Poly(amide imide)	T_g	559
Polystyrene	Poly(α-methyl styrene)	UCST, T_g	239,240
	Poly(α-methyl styrene)	$\Delta H^M < 0$	530,562
	Poly(2-chlorostyrene)	LCST	563
Deuterated polystyrene	Poly(α-methyl styrene)	LCST	127,356
	Poly(vinyl methyl ether)	LCST	121
Sulfonated polystyrene	Poly(alkyl oxide) with NH groups	FTIR	299
Na-sulfonated polystyrene ionomer	Polyamide 6	$\chi_{AB}<0$	148
Zn-sulfonated polystyrene ionomer	Poly(2,6-dimethyl-1,4-phenylene oxide)	DMTA	564
(Li, Mg, Zn, Mn)-sulfonated polystyrene ionomers (sulfonation degree is 6.5-9.0)	Polyamide 6	LCST which increases in order Mn<Zn<Li<Mg	565
(K, Na, Li) polystyrene ionomers	Poly(propylene oxide)	LCST which increases in order Li<Na<K	221
Polystyrene containing n(1,1,1,3,3,3-hexafluoro-2-hydroxyisopropyl)-α-methyl-styrene	Poly(butyl methacrylate)	LCST	566

3.5 THERMODYNAMICS OF FORMATION OF STABLE DISPERSION IN BINARY POLYMER MIXTURES

The condition $\Delta G < 0$ characterizes the process of formation of thermodynamically stable and equilibrium system.[487,488] $d^2\Delta G/d\varphi^2 > 0$ is the criterion of stability of the system. Compatibility is considered to be a result of mixing two polymers on molecular level. However, condition $\Delta G < 0$ describes also another processes -- the process of spontaneous emulsification of one polymer in the medium of another. It leads to the formation of a stable colloid systems. In this case, thermodynamic parameters of the system are changed due to dispersion of one polymer in another but also because of formation of an interface between two polymers. By dispersion, parts of dispersion medium and dispersed phase transit into interfacial layers (interphase). The change in entropy of such transition can by found from a dependence of interfacial tension, γ, on temperature and composition of the mixture considered in thermodynamics of surface phenomena:[489]

$$a d\gamma = \left(S_{AB}\frac{v_{A\gamma}}{v_{AB}} - S_{A\gamma}\right)dT + \sum_{i,k=1}^{n}\left[(N_A^{(2)} - N_A^{(1)})\frac{v_{A\gamma}}{v_{AB}} - \right.$$

$$\left. - (N_i^{(\gamma)} - N_i^{(1)})\right]\frac{\delta^2 G}{\delta N_i^{(1)}\delta N_k^{(1)}}\, dN_k^{(1)} \qquad [3.179]$$

a is the molar surface area (surface area per mole of matter in the interfacial layer), S_{AB} and v_{AB} are changes in entropy and volume during the transition of 1 mole of matter from phase 1 to phase 2, correspondingly; $S_{A\gamma}$ and $v_{A\gamma}$ are the same during the transition from phase 1 to the interfacial layer, and N_i is the molar fraction of the i-th component. Upper indices denote: (1) - phase 1, (γ) - interfacial layer. Let phase 1 be dispersion medium and phase 2 - dispersed phase. For a binary mixture Eq 3.122 transits into expression:

$$a\left(\frac{d\gamma}{dT}\right)_p = -S_{A\gamma} + \frac{N_A^{(\gamma)} - N_A^{(1)}}{N_A^{(2)} - N_A^{(1)}} S_{AB} \qquad [3.180]$$

Let us assume that the first component forms the dispersion medium and the second the dispersed phase. For immiscible binary polymer mixtures, where $N_A^{(1)} = 1$ in phase 2 and its surface layer, the first component is virtually absent (i.e., $N_1^{(2)} = N_1(\gamma) = 0$) and both components are non-volatile, Eq 3.180 has a very simple form:

$$a_A(d\gamma/dT) = S^{(1)} - S_A^{(\gamma)} \qquad [3.181]$$

where a_A and S_A^1 are the molar surface and molar entropy of interfacial layer on the side of phase 1. A similar equation is also obtained for the phase 2 and therefore

$$\Delta S^{(\gamma)} = -d\gamma/dT \qquad [3.182]$$

where

$$\Delta S^{(\gamma)} = [(S^{(1)} + S^{(2)}) - (S_A^{(\gamma)} + S_A^{(\gamma)})]/(a_A + a_B) \qquad [3.183]$$

The latter value is the change in the entropy of the two-phase polymer system during the formation of unit surface area of the interfacial layer (surface density of entropy change). Experimental measurements are necessary to determine the entropy change. From equation describing the change in free energy, ΔG, of the system during dispersion:

$$\Delta G = -T\Delta S_c + \gamma \sum_{i=1}^{n} A_i \qquad [3.184]$$

Then, accounting for the entropy change due to the formation of the interphase layers, we obtain

$$\Delta G = -T\Delta S_c - T\Delta S^{(\gamma)} \sum_{i=1}^{n} A_i + \gamma \sum_{i=1}^{n} A_i = -T\Delta S_c +$$

$$+ (\gamma - T\Delta S^{(\gamma)}) \sum_{i=1}^{n} A_i \qquad [3.185]$$

In Eqs 3.184 and 3.185, ΔS_k is the change in combinatorial entropy during dispersion and A_i is the surface area of the i-th particle of the dispersed phase. Eq. 3.185 fundamentally changes the concept of the processes leading to the formation of stable dispersion systems. As for all pairs of polymers investigated to date, value $d\gamma/dT$ is negative.[490] Even without the contribution of combinatorial entropy in accordance with Eqs 3.182 and 3.185, ΔG decreases from the beginning of dispersion, provided that

$$T\Delta S^{(\gamma)} > \gamma \qquad [3.186]$$

For the studied pairs of polymers,[490] $d\gamma/dT$ is within the range 0.002-0.02×10^{-3} J/m^{2}K in the temperature range of 400-500 K. The calculated values $T\Delta S^{(\gamma)}$ are $(0.8\text{-}10)\times10^{-3}$m^2 which are very close to experimental values of interfacial tension (taking into account the

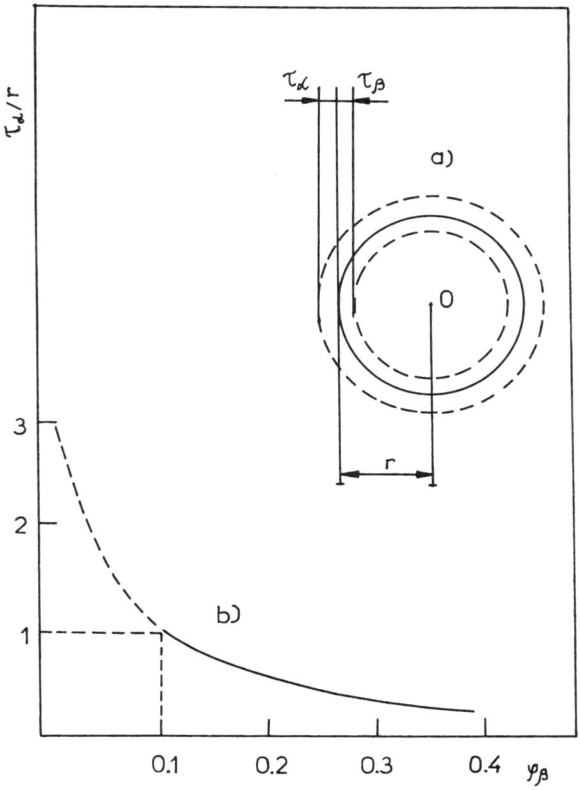

Figure 3.71. The model of a spherical particle of radius, r, with the surface layer from the side of dispersion medium (δ_A) and particle δ_B (a) and the change of possible dispersion degree with increasing the content of dispersed phase (b). Adapted, by permission, from Y. S. Lipatov and A. E. Feinerman, *J. Colloid & Interface Sci.*, **98**, 126 (1984).

scatter of experimental data). From this consideration, it follows that the process of spontaneous emulsification (dispersion) cannot be considered as a seldom case of formation of binary polymer mixtures. From condition 3.186, it follows that dispersion is simplified with increase in temperature because the left-hand of the inequality increases, whereas the right-hand side (interfacial tension) decreases.

Now let us discuss causes that inhibit spontaneous dispersion to the level of a single macromolecule. Let us consider the simplest model of a spherical particle of the dispersed phase (Fig. 3.71a). The thickness of the interfacial layer is determined by the characteristics of interfacial interaction. It does not depend on the particle radius, r. Therefore, during dispersion, the fraction of interfacial layers in the total volume of the system increases, i.e., the ratio δ/r increases. At a specific degree of dispersion, the polymers of the dispersion medium and dispersed phase pass completely into the interfacial layer and, as a result, the entropy gain of the transition into the interfacial layer diminishes or disappears. For a polymer of the dispersed phase, the condition of its transformation into the interfacial layer in accordance with Fig. 3. 71 is determined by the relationship:

$$\delta_B/r = 1$$

For a polymer of the dispersion medium, an analogous condition can be obtained when applying a model in the form of a cubic lattice whose sites accommodate particles of the dispersed phase of the same radius. The number of particles, N_{dp}, of dispersed phase (number of lattice sites)within the volume unit is determined from the expression:

$$N_{dp} = \varphi_B/[(4/3)\pi r^3] \quad [3.187]$$

For this case, the volume of the dispersion medium polymer, found in the interfacial layer, can be evaluated as

$$V_A^{(\gamma)} = \varphi_B[(4/3)\pi(r + \delta_A)-4/3\pi r^3]/(4/3)\pi r^3 = \varphi_B[(r + \delta_A)^3 - r^3]/r^3 \quad [3.188]$$

The transition of the dispersion medium polymer into the interfacial layer ceases when these layers touch neighboring particles, thus, for the densest packing, we obtain:

$$[1 + (\delta_A/r)]^3 = [(1/\varphi_B) - 1]0.74 + 1 \quad [3.189]$$

From Eqs 3.188 and 3.189, it can be seen that, if the volume fraction of a component in the dispersed phase is of no importance, then the content of dispersion medium directly determines the maximum degree of dispersion of dispersed phase.

Fig. 3.71b shows how the maximum degree of dispersion varies with increase in the dispersed phase content. Condition 3.188, marked by dashed line, allows one to evaluate the maximum concentration of the dispersed phase above which a sharp increase in the size of its particles occurs. This conclusion explains why the formation of dispersion systems is often observed in the region of small content of one of the components when the stability of dispersion system is determined only by entropy effects accompanying the dispersion process. The thickness of the interfacial layer, based on the data on polymer adsorption, is larger than 0.1 mcm and hence the minimum size of the dispersed phase particles for binary polymer systems greatly exceeds the size of a single macromolecule.

Another cause affecting the conditions of dispersion is the change in the boundary surface curvature. For the evaluation of the influence of this curvature, the modified Eq 3.180 can be used, which accounts for the shift of two-phase equilibrium with the presence of a curved surface:

$$d\gamma/dT = -\frac{\Delta S^{(\gamma)}}{(a + \Delta a)} + f(r)dr/dT \quad [3.190]$$

where

$$\Delta a = (2/r)(v_A^{(\gamma)} - v_A); \quad f(r) = \frac{2\gamma(v_A^{(\gamma)} - v_A)}{ar^2 + 2r(v_A^{(\gamma)} - v_A)} \quad [3.191]$$

In these equations, v_A and $v_A^{(\gamma)}$ are the molar volumes of matter inside the i-th phase and in the interfacial layer respectively (incompatible polymers are under consideration only). The expected effect depends on the sign of the difference $(v_A^{(\gamma)} - v_A)$ and the derivative dr/dT (Eqs 3.190 and 3. 191). On the basis of the conclusion made during discussion of Eq 3.186, we can say that the sign of dr/dT is negative. For the sign of the volume difference, here one can rely on many data which indicate that for liquids the increase in entropy in the surface layer is accompanied by a decrease in density within it. If this is true for the polymer-polymer interface, then with an appreciable decrease in r, $\Delta S^{(\gamma)}$ decreases owing to the decrease of its surface density and the derivative dγ/dT. The reduction in the entropy gain causes the appearance of a minimum in the dependence $\Delta G = f(r)$, i.e., cessation of further dispersion.

The conclusions made above are in a good agreement with experimental data on thermodynamic parameters of mixtures of polymers and oligomers and, in particular, with the results of measurements of temperature dependence of parameter χ_{AB}. From these data the enthalpy and entropy contributions χ_H and χ_S have been calculated. As seen from Fig. 3.36a,b, by diminishing the content of one oof the components, there is observed a sharp increase of χ_S and χ_H, leading to $\chi_{AB} < 0$ and $\Delta G < 0$. Sharp increase of noncombinatorial entropy of mixing $(\Delta S = -RT\chi_S\varphi_B^2)$ is hence the result of the process of spontaneous emulsification. As a result the fraction of the interfacial layer in the system increases.

REFERENCES

1. G. Gee, *Rev. Chem. Soc.*, **1**, 265 (1947).
2. G. L. Slonimsky and G. V. Struminsky, *J. Phys. Chem. (Russ.)*, **30**, 2144 (1956).
3. G. V. Struminsky and G. L. Slonimsky, *J. Phys. Chem.(Russ.)*, **30**, 1941 (1956).
4. P. J. Flory, B. E. Eichinger, and R. A. Orwoll, *Macromolecules*, **1**, 287 (1968).
5. A. A. Tager, T. I. Scholokhovich, and I. M. Scharova, *Vysokomolek. soed.*, **A17**, 2766 (1975).
6. A. A. Tager, *Vysokomolek. soed.*, **A19**, 1659 (1977).
7. A. A. Tager and T. I. Scholokhovich, *Vysokomolek. soed.*, **A18**, 1175 (1976).
8. A. A. Tager, T. I. Scholokhovich, and Y. S. Bessonov, *Eur. Polym. J.*, **11**, 321 (1975).
9. I. Prigogine in *The Molecular Theory of Solutions*, Interscience, New York, 1959.
10. D. Patterson, *J. Polym. Sci.*, **C16**, 3379 (1968).
11. B. E. Eichinger and P. J. Flory, *Trans. Faraday Soc.*, **64**, 2035 (1968).
12. O. Olabisi, L. M. Robeson, and M. T. Shaw in *Polymer-Polymer Miscibility*, Head Press, New York, 1989.
13. R. L. Scott, *J. Chem. Phys.*, **17**, 268 (1949).
14. H. Tompa in *Polymer Solutions*, Butterworths, London, 1956.
15. D. J. David and T. F. Sincock, *Polymer*, **33**, 4505 (1992).
16. T. Pazonyi and M. Dimitrov, *Rubber Chem. Technol.*, **40**, 1119 (1967).
17. B. Schneier, *J. Appl. Polym. Sci.*, **17**, 3175 (1973).
18. G. Gee, *Trans. Faraday Soc.*, **38**, 276 (1942).
19. P. A. Small, *J. Appl. Polym. Sci.*, **3**, 71 (1953).
20. L. Bohn, *Rubber Chem. Technol.*, **41**, 495 (1968).
21. M. T. Shaw, *J. Appl. Polym. Sci.*, **18**, 449 (1974).
22. C. M. Hansen, *J. Paint. Technol.*, **39**, 104, 505 (1967).
23. C. W. Bunn, *J. Polym. Sci.*, **16**, 323 (1955).
24. T. Nishi, *J. Macromol. Sci.*, **B17**, 517 (1980).
25. J. H. Hildebrand and R. L. Scott in *Regular Solutions*, Prentice-Hall, New York, 1962.
26. A. F.Yee and M. A. Maxwell, *J. Macromol. Sci.*, **B17**, 513 (1980).
27. S. Krause, *J. Macromol. Sci.*, **C7**, 251 (1972).
28. S. Krause in *Polymer Blends*, Eds. D. R. Paul and S. Newman, Vol. 1., Academic Press, New York, 1978.
29. M. M. Coleman, C. J. Serman, D. E. Bhagwagar, and P. C. Painter, *Polymer*, **31**, 1187 (1990).
30. L. P. McMaster, *Macromolecules*, **6**, 760 (1973).
31. R. Koningsveld and L. A. Kleintjens, *J. Polym. Sci.*, **C61**, 221 (1977).
32. O. Olabisi, *Macromolecules*, **8**, 316 (1975).
33. D. Patterson and A. Robard, *Macromolecules*, **11**, 692 (1978).
34. K. S. Siow, G. Delmas, and D. Patterson, *Macromolecules*, **5**, 29 (1972).
35. S. Saeki, N. Kuwahara, S. Konno, and M. Kaneko, *Macromolecules*, **6**, (1973).
36. L. Zeman, J. Biros, G. Delmas, and D. Patterson, *J. Phys. Chem.*, **76**, 1206 (1972).
37. B. A. Wolf and G. Blaum, *J. Polym. Sci.*, **C61**, 589 (1977).
38. S. Saeki, N. Kuwahara, M. Nakata, and M. Kaneko, *Polymer*, **17**, 685 (1976).
39. T. Shiomi, K. Imai, C. Watanabe, and M. Miya, *J. Polym. Sci.-Phys. Ed.*, **22**, 1305 (1984).
40. D. Liu and J. M. Prausnitz, *Macromolecules*, **12**, 454 (1979).
41. S. Beret and J. M. Prausnitz, *Macromolecules*, **8**, 878 (1975).
42. M. D. Donohue and J. M. Prausnitz, *AIChE Journal*, **24**, 849 (1978).
43. D. Henderson, *J. Chem. Phys.*, **61**, 926 (1974).
44. N. F. Carnahan and K. E. Starling, *J. Chem. Phys.*, **51**, 635 (1969).
45. R. Koningsveld and L. A. Kleintjens, *Brit. Polym. J.*, **9**, 212 (1977).

46. R. Koningsveld and R. F. Stepto, *Macromolecules*, **10**, 1166 (1977).
47. A. Silberberg, *Discuss. Faraday Soc.*, **49**, 162 (1970).
48. G. Allen and A. Maconnachie, *Brit. Polym. J.*, **9**, 184 (1977).
49. J. E. Joanny and F. Brochard, *J. Phys. (France)*, **42**, 1145 (1981).
50. A. Sariban and K. Binder, *Makromol. Chem.*, **189**, 2357 (1988).
51. K. A. Peterson, A. D. Stein, and M. D. Fayer, *Macromolecules*, **23**, 111 (1990).
52. W. A. Kruse, R. G. Kirste, J. Haas, B. J. Schmitt, and D. J. Stein, *Makromol. Chem.*, **177**, 1145 (1976).
53. P. Cifra, F. E. Karasz, and W. J. MacKnight, *Macromolecules*, **25**, 192 (1992).
54. I. C. Sanchez and R. H. Lacombe, *J. Phys. Chem.*, **80**, 2352 (1976).
55. I. C. Sanchez and R. H. Lacombe, *Macromolecules*, **11**, 1145 (1978).
56. I. C. Sanchez and R. H. Lacombe, *J. Macromol. Sci.*, **B17**, 565 (1980).
57. J. J. Ziska, J. W. Barlow, and D. R. Paul, *Polymer*, **22**, 918 (1981).
58. I. C. Sanchez in *Polymer Blends*, Eds. D. R. Paul and S. Newman), Vol. 1, Academic Press, New York, 1978.
59. I. C. Sanchez in *Polymer Compatibility and Incompatibility. Principles and Practices*, Ed. K. Solc, MMI Press Symposium. Series 2, Harwood Academic Press, New York, 1982, p. 59.
60. C. K. Kim and D. R. Paul, *Polymer*, **33**,1630 (1992).
61. I. C. Sanchez, *Polymer*, **30**, 471 (1989).
62. I. C. Sanchez, *Macromolecules*, **24**, 908 (1991).
63. A. C. Fernandes, J. W. Barlow, and D. R. Paul, *Polymer*, **27**, 1789 (1986).
64. A. Dobry and F. Boyer-Kawenoki, *J. Polym. Sci.*, **2**, 90 (1947).
65. R. J. Kern, *J. Polym. Sci.*, **21**, 19 (1956).
66. Ch. Huglin and A Dondos, *Makromol. Chem.*, **126**, 206 (1968).
67. G. M. Bristow, *J. Appl. Polym. Sci.*, **2**, 120 (1959).
68. D. Berek, D. Lath, and V Durdovic, *J. Polym. Sci.*, **C16**, 659 (1967).
69. T. Okazawa, *Macromolecules*, **8**, 371 (1975).
70. D. G. Weligan and C. M. Burns, *J. Appl. Polym. Sci.*, **18**, 521 (1974).
71. B. H. Clampitt, *Polym. Eng. Sci.*, **19**, 827 (1974).
72. S. M. Aharoni, *Macromolecules*, **11**, 277 (1978).
73. S. Djadoun, R. N. Goldberg, and H. Morawetz, *Macromolecules*, **10**, 1015 (1977).
74. V. N. Kuleznev, L. S. Krokhina, Y. G. Oganesov, and A. M. Zlatsen, *Kolloid Zh. (Russ.)*, **33**, 98 (1971).
75. A. Robard, D. Patterson, and N. G. Delmas, *Macromolecules*, **10**, 706 (1977).
76. R. J. Hefford, *Polymer*, **25**, 979 (1984).
77. K. L. Petrak, *J. Appl. Polym. Sci.*, **29**, 855 (1984).
78. W. Narasimhan, D. Lloid, and C. M. Burns, *J. Appl. Polym. Sci.*, **23**, 749 (1979).
79. W. Narasimhan, R. Y. M. Huang, and C. M. Burns, *J. Polymer, Sci.-Chem. Ed.*, **22**, 3895 (1984).
80. D. Patterson, *Polym. Eng. Sci.*, **22**, 64 (1982).
81. W . Lau, C. M. Burns, and R. Y. M. Huang, *J. Appl. Polym. Sci.*, **29**, 1531 (1984).
82. D. R. Lloyd and C. M. Burns, *J. Appl. Polym. Sci.*, **22**, 593 (1978).
83. J. L. White and R. D. Patel, *J. Appl. Polym. Sci.*, **19**, 1775 (1975).
84. J. L. White, D. G. Salladay, D. O. Quisenberry, and D. L. MacLeon, *J. Appl. Polym. Sci.*, **16**, 2811 (1972).
85. S. A. Vshivkov and N. A. Komolova,*Vysokomolek. soed.*, **A23**, 2780 (1981).
86. L. S. Krokhina and V. N. Kuleznev,*Vysokomolek. soed.*, **A20**, 1981 (1978).
87. J. Hashizume, A. Teramoto, and M. Fujita, *J. Polym. Sci.-Phys. Ed.*,**19**, 1405 (1981).
88. M. W. J. Van den Esker and A. Vrij, *J. Polymer Sci.-Phys. Ed.*, **14**, 1943 (1976).
89. L. M. Robeson in *Polymer Compatibility and Incompatibility. Principles and Practices*, Ed. K. Solc, MMI. Press Symposium Series 2, Harwood Academic Press, New York, 1982, N2, p. 177.

90. S. Krause, *Pure and Appl. Chem.*, **58**, 1553 (1986).
91. A. A. Tager and V. S. Blinov, *Adv. Chem. (Russ.)*, **56**, 1004 (1987).
92. A. E. Nesterov and Y. S. Lipatov in *Phase State of Solutions and Mixtures of Polymers. Handbook*, Naukova Dumka, Kiev, 1987.
93. V. N. Kuleznev and L. S. Krokhina, *Adv. Chem. (Russ.)*, **42**, 1278 (1973).
94. L. A. Utracki in *Polymer Alloys and Blends*, Hanser Publishers, New York, 1989.
95. K. Fujita, T. Kyu, and R. St. J. Manley, *Macromolecules*, **29**, 91 (1996).
96. R. S. Barrum, S. H. Goh, J. W. Barlow, and D. R. Paul, *J. Polym. Sci.-Letters Ed.*, **23**, 395 (1985).
97. T. Shiomi, F. Hamada, T. Nasako, K. Yoneda, K. Imai, and A. Nakajima, *Macromolecules*, **23**, 229 (1990).
98. P. Novakov, Ch. Konstantinov, and P. Mitanov, *J. Appl. Polym. Sci.*, **16**, 1827(1972).
99. S. Akajima and K. Miasa, *Polym. J. (Japan)*, **11**, 157 (1979).
100. N. F. Weeks, F. E. Karasz, and W. J. MacKnight, *J. Appl. Polym. Sci.*, **48**,4068 (1974).
101. P. Pedrosa, J. A. Pomposo, E. Calahorra, and M. Cortazar, *Macromolecules*, **27**, 102 (1994).
102. C. Uriarte, J. I. Equiazabal, M. Llanos, J. I. Iribarren, and J. J. Iruin, *Macromolecules*, **20**, 3038 (1987).
103. D. J. Walsh and G. L. Chang, *Polymer*, **25**, 499 (1984).
104. A. Nakajima, *Macromolecules*, **23**, 229 (1990).
105. J. I. Equiazabal, M. Cortazar, J. J. Iruin, and G. M. Guzman, *J. Macromol. Sci.*, **B27**, 19 (1988).
106. S. M. Lipatov and G. V. Lipatova, *Kolloid Zh., (Russ.)*, **21**, 517 (1959).
107. S. S. Voyutsky in *Solutions of High Molecular Compounds*, M. Goskimizdat, 1960.
108. S. Ichihara, A. Komatsu, and T. Hata, *Polymer J. (Japan)*, **2**, 640 (1971).
109. J. E. Harris, D. R. Paul, and J. W. Barlow, *Polym. Eng. Sci.*, **23**, 676 (1983).
110. V. N. Kuleznev and K. M. Igosheva,*Vysokomolek. soed.*, **4**, 1858 (1962).
111. J. R. Fried, F. E. Karasz, and W. J. MacKnight, *Macromolecules*, **11**, 150 (1978).
112. B. G. Rancy, *J. Polym. Sci.*, **C51**, 89 (1975).
113. J. J. Hickman and R. M. Ikeda, *J. Polym. Sci.-Phys. Ed.*, **11**, 1713 (1973).
114. T. Sato, Y. Tsujita, A. Takizawa, and T. Kinoshita, *Macromolecules*, **24**, 158 (1991).
115. S. Savard, D. Levesque, and R. E. Prud'homme, *J. Appl. Polym. Sci.*, **23**, 1943 (1979).
116. D. J. Walsh and S. Rostami, *Macromolecules*, **17**, 216 (1984); **18**, 3 15 (1985).
117. M. Wolf and J. H. Wendorf, *Polymer*, **30**, 1524 (1989).
118. C. C. Han, M. Okada, Y. Muroga, F. L. McCrackin, B. J. Baurer, and Q. Tran-Cong, *Polym. Eng. Sci.*, **26**, 2 (1986).
119. F. S. Bates and G. D. Wignall, *Macromolecules*, **19**, 932 (1986).
120. F. S. Bates, S. B. Dierker, and G. D. Wignall, *Macromolecules*, **19**, 1938 (1986).
121. M. Shibayama, H. Yang, R. S. Stein, and C. C. Han, *Macromolecules*, **18**, 2179 (1985).
122. H. Yang, M. Shibayama, R. S. Stein, and C. C. Han, *Polymer Bull.*, **12**, 7 (1984).
123. C. Herkt-Maetzky and J. Schelton, *Phys. Rev. Lett.*, **51**, 896 (1983).
124. H. Ito, T. P. Russell, and G. D. Wignall, *Macromolecules*, **20**, 2213 (1987).
125. G. Hadziioannou and R. S. Stein, *Macromolecules*, **17**, 767 (1984).
126. F. S. Bates, L. J. Fetters, and G. D. Wignall, *Macromolecules*, **21**, 1086 (1988).
127. A. Rameau, Y. Gallot, P. Marie, and B. Farnoux, *Polymer*, **30**, 386 (1989).
128. M. Warner, J. S. Higgins, and A. J. Carter, *Macromolecules*, **16**, 1931 (1983).

129. P. E. Tomlins and J. S. Higgins, *Macromolecules*, **21**, 425 (1988).
130. T. P. Russell and R. S. Stein, *Amer. Chem. Soc. Polym. Prepr.*, **20**, 24 (1979).
131. R. P. Kambour, R. C. Bopp, A. Maconnachie, and W. J. MacKnight, *Polymer*, **21**, 133 (1980).
132. J. Jelenic, R. G. Kirste, B. J. Schmitt, and S. Schmitt-Steker, *Makromol. Chem.*, **18**, 2057 (1979).
133. T. K. Kwei and H. L. Frisch, *Macromolecules*, **11**, 1267 (1978).
134. T. Nishi and T. T. Wang, *Macromolecules*, **8**, 909 (1975).
135. T. K. Kwei, G. D. Patterson, and T. T. Wang, *Macromolecules*, **9**, 780 (1986).
136. H. Bergmans and N. Overbergh, *J. Polym. Sci. Phys. Ed.*, **15**, 1757 (1977).
137. E. Roerdink and G. Challa, *Polymer*, **19**, 173 (1978).
138. E. Martuscelli, G. Demma, E. Drioli, L. Nicolas, and S. Spina, *Polymer*, **20**, 571 (1979).
139. T. K. Kwei, T. Nishi, and R. F. Roberts, *Macromolecules*, **7**, 667 (1974).
140. Seo Seung Won, Jo Won Ho, and Ha Wan Shik, *J. Appl. Polym. Sci.*, **29**, 567 (1984).
141. S. A. Liberman, A. S. Gomes, and E. M. Macchi, *J. Polym. Sci. Chem. Ed.*, **22**, 2809 (1984).
142. Y. F. Chong, S. Y. Lee, and S. H. Goh, *Eur. Polym. J.*, **26**, 1145 (1990).
143. L. C. Cesteros, J. R. Quintana, J .A. Fernandez, and I. A. Katime, *J. Polym. Sci. Phys. Ed.*, **27**, 2567 (1989).
144. P. Greco and E. Martuscelli, *Polymer*, **30**, 1475 (1989).
145. R. M. Briber and F. Khoury, *Am. Chem. Soc. Polym. Prepr.*, **26**, 310 (1985).
146. A. C. Fernandes, J. W. Barlow, and D. R. Paul, *J. Appl. Polym. Sci.*, **29**, 1971 (1984).
147. N. Hirose and T. Takanashi, *Polymer J. (Japan)*, **21**, 347 (1989).
148. X. Lu and R. A. Weiss, *Macromolecules*, **24**, 4381 (1991).
149. E. Martuscelli, M. Pracella, and Yue Wang Ping, *Polymer*, **25**, 1097 (1984).
150. H. Ueda and F. E. Karasz, *Macromolecules*, **18**, 2719 (1985).
151. A. C. Balazs, F. E. Karasz, W. J. MacKnight, H. Ueda, and I. C. Sanchez, *Macromolecules*, **18**, 2784 (1985).
152. G. Ronca and T. P. Russell, *Macromolecules*, **18**, 665 (1985).
153. T. Fukuda, M. Nasata, and H. Inagaki, *Macromolecules*, **17**, 548 (1984).
154. T. Shiomi, K. Konno, K. Yoneda, T. Tomita, M. Miya, and K. Imai, *Macromolecules*, **18**, 414 (1985).
155. D. D. Deshpande, D. Patterson, H. P. Schreiber, and C. S. Su, *Macromolecules*, **7**, 530 (1974).
156. A. E. Nesterov in *Inverse Gas Chromatography of Polymers*, Naukova Dumka, Kiev, 1987.
157. S. Klotz, R. H. Schuster, and H. J. Cantow, *Makromol. Chem.*, **187**, 1491 (1986).
158. M. Iriarte, E. Epsi, A. Etxeberria, M. Valero, M. Fernandez-Berridi, and J . J. Iruin, *Macromolecules*, **20**, 5546 (1991).
159. B. M. Mandal, C. Bhattacharyya, and S. N. Bhattacharyya, *J. Macromol. Sci.*, **A26**, 175 (1989).
160. A. Edelman and A. Fradet, *Polymer*, **30**, 317 (1989).
161. G. Di Paola-Baranyi, S. J. Fletcher, and P. Degre, *Macromolecules*, **15**, 885 (1982).
162. B. Riedl and R. E. Prud'homme, *Am. Chem. Soc. Polym. Prepr.*, **28**, 138 (1987).
163. P. Munk, P. Hattam, A. A. Abdel-Azim, and Du Quangguo, *Makromol. Chem. Macromol. Symp.*, **38**, 205 (1990).
164. K. S. Schikhaliev, *Reports of Acad. Sci. USSR*, **45**, 33 (1989).
165. J. M. Elorza, M. J. Edel-Berrid, J. J. Iruin, and C. Uriarte, *Makromol. Chem.*, **189**, 1855 (1988).

166. G. Bhattacharyya, N. Maiti, B. M. Mandal, and S. N. Bhattacharyya, *Macromolecules*, **22**, 4062 (1989).
167. P. Munk, P. Hattam, and Q. Du, *J. Appl. Polym. Symp.*, **33**, 373 (1989).
168. C. S. Tyagi and S. Husain, *Indian J. Technol.*, **22**, 435 (1984).
169. C. S. Su, D. Patterson, and H. P. Schreiber, *J. Appl. Polym. Sci.*, **20**, 1025 (1976).
170. G. Di Paola-Baranyi, *ACS Polym. Mater. Sci. Eng.*, **58**, 735 (1988).
171. M. J. El-Hibri, W. Cheng, and P. Munk, *ACS Polym. Mater. Sci. Eng.*, **58**, 741 (1988).
172. T. I. Kirillova and A. A. Tager, *Vysokomolek. soed.*, **A26**, 1584 (1984).
173. D. J. Walsh and J. G. McKeown, *Polymer*, **21**, 1335 (1980).
174. P. F. Green and B. L. Doyle, *Phys. Rev. Lett.*, **57**, 2407 (1986).
175. R. J. Composto, E. J. Kramer, and D. M. White, *Macromolecules*, **21**, 2580 (1988).
176. R. J. Composto and E. J. Kramer, *Am. Chem. Soc. Polym. Prepr.*, **29**, 401 (1988).
177. G. Allen, G. Gee, and J. P. Nicholson, *Polymer*, **1**, 56 (1960).
178. H. Yamakawa and M. Fujii, *J. Chem. Phys.*, **58**, 1523 (1973).
179. P. J. Flory in *Principles of Polymer Chemistry*, Cornell Univ. Press, New York, 1953.
180. C. M. Roland, *Rubber Chem. and Technol.*, **61**, 866 (1988).
181. D. W. Tomlin and C. M. Roland, *Macromolecules*, **25**, 1992 (1992).
182. D. G. Gray, *Progr. Polym. Sci.*, **5**, 60 (1977).
183. R. N. Novitskaya, and M. S. Vigdergauz, *Izvestiya Acad Nauk USSR, Ser. Chim.*, **12**, 2798 (1970).
184. C. S. Su and D. Patterson, *Macromolecules*, **10**, 708 (1977).
185. T. C. Walsh, D. P. Sheehy, J. S. Riffle, and J. E. Mc Grath, *Macromolecules*, **14**,1791 (1981).
186. V. S. Blinov, A. A. Tager, and V. N. Yarunina, *Vysokomolek. soed.*, **A26**, 477 (1984).
187. A. K. Nandi, B. M. Mandal, and S. N. Bhattacharyya, *Macromolecules*, **18**, 1454 (1985).
188. J. M. Elorza, M. J. Fernandez-Berridi, J. J. Iruin, and G. M. Guzman, *Polym. Eng. Sci.*, **24**, 287 (1984).
189. D. P. Sheehy, T. C. Ward, and J. E. McGrath, *Am. Chem. Soc. Polym. Prepr.*, **24**, 106 (1983).
190. D. J. Walsh, J. S. Higgins, S. Rostami, and K. Weeraperuma, *Macromolecules*, **16**, 391 (1983).
191. G. Di Paola-Baranyi, J. Richer, and W. M. Prest, *Can. J. Chem.*, **63**, 223 (1985).
192. D. Patterson, *Macromolecules*, **2**, 673 (1969).
193. G. Di Paola-Baranyi and P. Degre, *Macromolecules*, **14**, 1456 (1986).
194. A. Maconnachie, K. P. Kambour, D. M. Walsh, *Macromolecules*, **17**, 2645 (1984).
195. W. M. Prest, D. J. Luca, and F. J. Roberts in *Thermal Analysis in Polymer Characterization*, Heyden, Philadelphia, 1981, p. 24.
196. W. J. McKnight, J. Stoelting, and F. E. Karasz in *Multicomponent Polymer Systems*, Interscience, Washington, 1971, p. 29.
197. S. Cimmino, E. Di Pace, F. E. Karasz, E. Martuscelli, and L. J. Silvestre, *Polymer*, **34**, 972 (1993).
198. J. M. Gomez-Elvira, J. L. Halary, L. Monnerie, and L. J. Fetters, *Macromolecules*, **27**, 3370 (1994).
199. C. K. Kim and D. R. Paul, *Polymer*, **33**, 4929 (1992).
200. Y. S. Lipatov, A. E. Nesterov, and T. D. Ignatova, *Eur. Polym. J.*, **15**, 775 (1979).

201. W. Li and R. E. Prud'homme, *J. Polymer Sci. Phys. Ed.*, **31**, 719 (1993).
202. Y. S. Lipatov, A. E. Nesterov, V. F. Shumsky, T. D. Ignatova, and S. N. Gorbatenko, *Eur. Polym. J.*, **18**, 981 (1982).
203. C. P. Doube and D. J. Walsh, *Polymer*, **20**, 1115 (1979).
204. E. M. Pearce, T. K. Kwei, and B. Y. Min, *J. Macromol. Sci.*, **A21**, 1181 (1984).
205. S. H. Goh and S. Y. Lee, *Am. Chem. Soc. Polym. Prepr.*, **28**, 148 (1987).
206. R. Orwoll, L. St. Clair, and K. D. Dobbs, *J. Polym. Sci. Phys. Ed.*, **19**, 1385 (1981).
207. K. E. Min, J. S. Chiou, J. W. Barlow, and D. R. Paul, *Polymer*, **28**, 1721 (1987).
208. H. H. Chuah, T. Kyu, and T. E. Helminiak, *Polymer*, **28**, 2130 (1987).
209. M. A. Masse, H. Ueda, and F. E. Karasz, *Macromolecules*, **21**, 3438 (1988).
210. A. C. Fernandes, J. W. Barlow, and D. R. Paul, *J. Appl. Polym. Sci.*, **32**, 5481 (1986).
211. K.- H. Illers, W. Heckmann, J. Hambrecht, and J. C. V. Washington: *Colloid, Polym. Sci.*, **262**, 557 (1984).
212. Q. Guo, *Makromol. Chem. Rapid Commun.*, **11**, 279 (1990).
213. H.- W. Kammer and C. Kummerloewe, *Acta Polymerica*, **41**, 269 (1990).
214. S. H. Goh, *Polym. Bull.*, **17**, 221 (1987).
215. Y. D. Shibanov and T. M. Radjatov, *Vysokomol. Soed.*, **B30**, 281 (1988).
216. J. Piglowski and M. Kozlowski, *Polym.-tworz. wielkoczasteczk (Poland)*, **34**, 611 (1989).
217. J. M. G. Cowie and S. Saeki, *Polym. Bull.*, **6**, 75 (1981).
218. S. H. Goh and S. Y. Lee, *Eur. Polym. J.*, **24**, 923 (1988).
219. J. S. Chiou, J. W. Barlow, and D. R. Paul, *J. Polymer Sci. Phys. Ed.*, **25**, 1459 (1987).
220. T. Kyu and J. M. Saldanha, *J. Polymer Sci. Lett. Ed.*, **26**, 33 (1988).
221. M. Hara and A. Eisenberg, *Macromolecules*, **20**, 2160 (1987).
222. D. R. Paul and J. W. Barlow, *J. Macromol. Sci.*, **C18**, 109 (1980).
223. E. Roerdink and G. Challa, *Polymer*, **21**, 1161 (1980).
224. J. S. Chiou and D. R. Paul, *J. Appl. Polym. Sci.*, **33**, 2935 (1987).
225. G. Allen, G. Gee, and J. P. Nicholson, *Polymer*, **2**, 8 (1961).
226. G. Allen, Z . Chait, C. L. Chong *et al.*, *Polymer*, **25**, 239 (1984).
227. B. A. Wolf and W. Schuch, *Makromol. Chem.*, **182**, 1801 (1981).
228. B. A. Wolf and G. Blaum, *Makromol. Chem.*, **180**, 2591 (1979).
229. C. M. Roland, *Macromolecules*, **20**, 2557 (1987).
230. L. I. Komaricheva and Y. Y. Shibanov, *Vysokomolek. soed.*, **A30**, 1436 (1988).
231. M. Huglin, I. Idris, and M. B. Sokro, *Makromol. Chem. Rapid Commun.*, **2**, 17 (1981).
232. C. M. Kuo and C. J. Clarson, *Macromolecules*, **25**, 2192 (1992).
233. B. A. Wolf and G. Blaum, *J. Polym. Sci.*, **C61**, 251 (1977).
234. A. E. Chalykh and I. N. Sapojnikova, *Compos. Polym. Mater.*, (Kiev) **7**, 46 (1980).
235. B. Hammouda, R. Briber, and B. Bauer, *Nist. Techn. Note.*, **96**, 1292, (1992).
236. J. S. Higgins and A. J. Carter, *Macromolecules*, **17**, 2197 (1984).
237. S. Lau, J. Pathak, and B. Wunderlich, *Macromolecules*, **15**, 1278 (1982).
238. J. M. G. Cowie and I. J. McEven, *Polymer*, **26**, 1662 (1985).
239. J. M. Wigmaier and G. Mignard, *Eur. Polym. J.*, **23**, 989 (1987).
240. J.- L. Lim and R.- J. Roe, *Polymer*, **29**, 1227 (1988).
241. J.- L. Lim and R.- J. Roe, *Macromolecules*, **20**, 2168 (1987).
242. S. Saeki, J. M. G. Cowie, and I. J. McEwen, *Polymer*, **24**, 60 (1983).
243. B. A. Wolf and R. Jend, *Makromol. Chem.*, **178**, 1811 (1977).
244. A. E. Chalykh, N. I. Druz, and A. E. Krejtus, *Vysokomolek. soed.*, **B24**, 747 (1982).
245. A. E. Chalykh, N. N. Avdeev, and S. M. Meznikowski, *Vysokomolek. soed.*, **B22**, 464 (1980).

246. M. V. Huglin and I. M. Idris, *Polymer*, **24**, 1434 (1983).
247. R. Koningsveld, L. A. Kleintjens, and M. H. Onclin, *J. Polym. Sci.*, **C18**, 363 (1980).
248. A. A. Berlin and S. M. Meznikowski, *Pure and Appl. Chem.*, **51**, 2345 (1979).
249. A. E. Chalykh, N. M. Avdeev, A. A. Berlin, and S. M. Mezhikowski, *Reports Acad. Sci. USSR*, **238**, 893 (1978).
250. R.- J. Roe and W.- Ch. Zin, *Macromolecules*, **13**, 1221 (1980).
251. T. Russel, G. Hadziioannou, and W. K. Warburton, *Macromolecules*, **18**, 78 (1985).
252. A. Friday, D. R. Cooper, and C. Booth, *Polymer*, **18**, 171 (1977).
253. G. F. Roginskaya, V. P. Volkov, A. E. Chalykh *et al.*,*Vysokomolek. soed.*, **A21**, 2111 (1979).
254. H. Saito, Y. Fujita, and T. Inoue, *Polymer J. (Japan)*, **19**, 405 (1987).
255. J. M. G. Cowie and S. Saeki, *Polym. Bull.*, **6**, 75 (1981).
256. S. L. Zacharias, G. ten Brinke, W. J. MacKnight, and F. E. Karasz, *Macromolecules*, **16**, 381 (1983).
257. G. Cong, Y. Huang, W. J. MacKnight, and F. E. Karasz, *Macromolecules*, **19**, 2765 (1986).
258. X. Yang, Y. Huang, S. Zhao, and G. Cong, *Annual Reports Polymer Physics Laboratory*, Academia sinica, **1990**, p. 109.
259. D. J. Walsh, S. Laingle, and C. Zhikuan, *Polymer*, **22**, 1005 (1981).
260. H. Tomura, H. Saito, and T. Inoue, *Macromolecules*, **25**, 1611 (1992).
261. S. Shimada, Y. Hori, and H. Kashiwabara, *Macromolecules*, **21**, 3454 (1988).
262. T. Kyu and D.- S. Lim, *J. Polymer Sci. Lett. Ed.*, **27**, 424 (1989).
263. M. J. Hill and P. J. Barham, *Polymer*, **33**, 2530; 4099 (1992).
264. A. E. Nesterov, V. V. Horichko, and Y. S. Lipatov, *Reports Acad. Sci. Ukraine*, **B12**, 49 (1987).
265. T. Shiomi, F. Hamada, T. Nasako, K. Yoneda, K. Imai, and A. Nakajima, *Macromolecules*, **23**, 229 (1990).
266. L. I. Khomutov, N. A. Lashek, and G. F. Mikulski, *Vysokomolek. soed.*, **B25**, 249 (1983).
267. S. Akajima, I. Matsumoto, M. Nakata, and R. Kaneko, *Kobunshi Ronbunshu (Japan)*, **33**, 238 (1976).
268. R. E. Bernstein, C. A. Cruz, D. R. Paul, and J. W. Barlow, *Macromolecules*, **10**, 681 (1977).
269. R. Casper and L. Morbitzer, *Angew. Makromol. Chem.*, **58/59**, 1 (1977).
270. J. L. Habary, J. M. Ubrich, J. M. Nunzi *et al.*, *Polymer*, **25**, 956 (1984).
271. J. E. Harris, S. H. Goh, D. R. Paul, and J. W. Barlow, *J. Appl. Polym. Sci.*, **27**, 839 (1982).
272. S. M. Low, S. H. Goh, and S. Y. Lee, *Macromolecules*, **27**, 3764 (1994).
273. E. Nolley, D. R. Paul, and J. W. Barlow, *J. Appl. Polym. Sci.*, **23**, 623 (1979).
274. G. ten Brinke, A. Eshuis, E. Roerdink, and G. Challa, *Macromolecules*, **14**, 867 (1981).
275. C. Trembley and R. E. Prud'homme, *J. Polymer Sci. Phys. Ed.*, **22**, 1857 (1984).
276. H. Yang, G. Hadziioannou, and R. Stein, *J. Polymer Sci. Phys. Ed.*, **21**, 159 (1983).
277. C. Zhikuan, S. Ruona, D. J. Walsh, and J. S. Higgins, *Polymer*, **24**, 263 (1983).
278. A. A. Tager, *Vysokomolek. soed.*, **A26**, 659 (1984).
279. R. E. Prud'homme, *Polym. Eng. Sci.*, **22**, 90 (1982).
280. D. Allard and R. E. Prud'homme, *J. Appl. Polym. Sci.*, **27**, 559 (1982).
281. G. Belorgey, M. Aubin, and R. E. Prud'homme, *Polymer*, **23**, 1051 (1982).
282. H. Jager, E. J. Vorenkamp, and G. Challa, *Polym. Commun.*, **24**, 290 (1983).
283. E. M. Woo, J. W. Barlow, and D. R. Paul, *J. Appl. Polym. Sci.*, **28**, 1347 (1983).

284. M. Aubin and R. E. Prud'homme, *Am. Chem. Soc. Polym. Prepr.*, **24**, 182 (1983).
285. M. Aubin and R. E. Prud'homme, *Polym. Eng. Sci.*, **24**, 350 (1984).
286. J. S. Ziska, J. W. Barlow, and D. R. Paul, *Polymer*, **22**, 918 (1981).
287. B. N. Shtarkman in *Plastification of Polyvinylchlorid (in Russ.),* M. Khimia, 1975.
288. M. Aubin, Y. Bedard, M.- F. Morrisette, and R. E. Prud'homme, *J. Polymer Sci. Phys. Ed.*, **21**, 233 (1983).
289. M. Aubin and R. E. Prud'homme, *J. Polymer Sci. Phys. Ed.*, **19**, 1245 (1981).
290. U. Sheppard and S. Scharts in *Organic Chemistry of Fluor,* M. Mir, 1972.
291. P. Alexandrovich, F. E. Karasz, and W. J. MacKnight, *Polymer*, **18**, 1022 (1977).
292. M. Iskander, C. Tran, L. M. Robeson, and J. E. McGrath, *Polym. Eng. Sci.*, **23**, 682 (1983).
293. L. M. Robeson and J. E. McGrath, *AIChE Symp. Ser.*, **73**, 151 (1977).
294. L. M. Robeson and J. E. McGrath, *Polym. Eng. Sci.*, **17**, 300 (1977).
295. D. F. Varnell and D. F. Coleman, *Polymer*, **22**, 1324 (1981).
296. R. A. Weiss, *J. Polymer Sci. Phys. Ed.*, **26**, 1545 (1988).
297. E. J. Moskala, S. E. Howe, P. C. Painter, and M. M. Coleman, *Macromolecules*, **17**, 1671 (1984).
298. D. F. Varnell, E. F. Moskala, P. C. Painter, and M. M. Coleman, *Polym. Eng. Sci.*, **23**, 658 (1983).
299. A. Garton, S. Wang, and R. A. Weiss, *J. Polymer Sci. Phys. Ed.*, **26**, 1545 (1988).
300. S. P. Ting, E. M. Pearce, and T. K. Kwei, *J. Polym. Sci., Lett. Ed.*, **18**, 201 (1980).
301. E. J. Moskala and M. M. Coleman, *Polym. Commun.*, **24**, 206 (1983).
302. D. E. Bhagwagar, P. C. Painter, and M. M. Coleman, *Macromolecules*, **25**, 1361 (1992).
303. D. F. Varnell, M. Runt, and R. E. Prud'homme, *Polymer*, **24**, 37 (1983).
304. A. Garton, M. Aubin, and R. E. Prud'homme, *J. Polymer Sci. Lett. Ed.*, **21**, 45 (1983).
305. S. Stankovic, G. Guena, D. J. Williams *et al.*, *Polym. Commun.*, **29**, 14 (1988).
306. D. F. Coleman, E. J. Moskala, P. C. Painter *et al.*, *Polymer*, **24**, 1410 (1983).
307. P. C. Painter, Y. Park, and M. M. Coleman, *Macromolecules*, **21**, 66 (1988).
308. P. C. Painter, Y. Park, and M. M. Coleman, *Macromolecules*, **22**, 570 (1989).
309. P. C. Painter, J. F. Graf, and M. M. Coleman, *J. Chem. Phys.*, **92**, 6166 (1990).
310. M. M. Coleman, J. F. Graf, and P. C. Painter in *Specific Interactions and the Miscibility of Polymer Blends,* Technomic Publishing, Inc., Lancaster, USA, 1991.
311. P. C. Painter, Y. Park, and M. M. Coleman, *Macromolecules*, **22**, 580 (1989).
312. M. M. Coleman, M. Lichkus, and P. C. Painter, *Macromolecules*, **22**, 586 (1989).
313. Y. Xu, J. F. Graf, P. C. Painter, and M. M. Coleman, *Polymer*, **32**, 3103 (1991).
314. M. M. Coleman, J. Hu, Y. Park, and P. C. Painter, *Polymer*, **29**, 1659 (1988).
315. M. M. Coleman, J. Y. Lee, C. J. Serman, Z. Wang, and P. C. Painter, *Polymer*, **30**, 1298 (1989).
316. J. Hu, P. C. Painter, M. M. Coleman, and T. D. Krizan, *J. Polymer Sci. Phys. Ed.*, **28**, 149 (1990).
317. C. J. Serman, Y. Xu, P. C. Painter, and M. M. Coleman, *Polymer*, **32**, 516 (1991).
318. C. J. Serman, Y. Xu, P. C. Painter, and M. M. Coleman, *Macromolecules*, **22**, 2015 (1989).
319. C. J. Serman, P. C. Painter, and M. M. Coleman, *Polymer*, **32**, 1049 (1991).

320. M. M. Coleman, X. Yang, P. C. Painter, and J. F. Graf, *Macromolecules*, **25**, 4414 (1992).
321 Y. Xu, P. C. Painter, and M. M. Coleman, *Makromol. Chem. Macromol. Symp.*, **52,** 61; 91 (1991).
322. M. M. Coleman, Y. Xu, and P. C. Painter, *Makromol. Chem. Macromol. Symp.*, **52**, 75 (1991).
323. P. Smith and A. Eisenberg, *J. Polymer Sci. Lett. Ed.*, **21** 233 (1983).
324. L. Natansohn, R. Murali, and A. Eisenberg, *Makromol. Sci. Macromol. Symp.*, **16,** 175 (1988).
325. A. Eisenberg, P. Smith, and Z. L. Zhou, *Polym. Eng. Sci.*, **22**, 1117 (1982).
326. Z. L. Zhou and A. Eisenberg, *J. Polymer Sci. Phys. Ed.*, **21**, 595 (1983).
327. M. Rutkowska and A. Eisenberg, *Macromolecules*, **17**, 821 (1984).
328. M. Rutkowska and A. Eisenberg, *J. Appl. Polym. Sci.*, **30**, 3317 (1985).
329. M. Rutkowska and A. Eisenberg, *J. Appl. Polym. Sci.*, **29**,755 (1984).
330. M. Rutkowska and A. Eisenberg, *J. Polymer Sci. Phys. Ed.*, **33**, 2833 (1987).
331. L. Natansohn and A. Eisenberg, *Macromolecules*, **30**, 323 (1987).
332. X. Zhang and A. Eisenberg, *Polym. Adv. Technol.*, **1**, 7 (1989).
333. X. Zhang and A. Eisenberg, *J. Polymer Sci. Phys. Ed.*, **28**, 1841 (1990).
334. N. Ohno and J. Kumanitani, *Polym. J. (Japan)*, **11**, 947 (1975).
335. D. G. H. Ballard and A. Cunningham, *Polymer*, **18**, 259 (1977).
336. M. I. Bank and S. J. Krimm, *J. Polymer Sci. Phys. Ed.*, **8**, 193 (1970).
337. D. G. H. Ballard, P. Cheshire, and G. W. Lougman, *Polymer*, **19**, 379 (1978).
338. R. M. Gohil and J. Peterman, *J. Macromol. Sci.*, **B18**, 217 (1980).
339. S.-R. Hu, T. Kyu, and R. S. Stein, *J. Polymer Sci. Phys. Ed.*, **25**, 71 (1987).
340. W. R. Burghardt, *Macromolecules*, **22**, 2482 (1989).
341. S. Nojima, K. Satoh, and T. Ashida, *Macromolecules*, **24**, 942 (1991).
342. A. S. Zubko, A. E. Nesterov, Sh. G. Vengerovskaya, and L. S. Schejnina, *Vysokomolek. soed.*, **A17**, 1054 (1975).
343. M. Sh. Yagfarov, A. A. Mukhutdinov, and N. S. Igonina in *Sintez and Physikokhimiya Polymerov (in Russ.),* **12**, 94 (1973).
344. S. S. Gudzera, Y. L. Spirin, and V. V. Magdinets, *Sintez and Physikokhimiya Polymerov (in Russ.),* **14**, 101 (1975).
345. T. E. Lipatova in *Catalytic Polymerization of Oligomers and The Formation of Polymer Networks (Russ.),* Naukova Dumka, Kiev, 1974.
346. A. E. Nesterov, T. E. Lipatova, K. Dusek, Z. Peltzbauer, M. Houska, J. Hradil, and Y. S. Lipatov, *Angew. Makromol. Chem.*, **52**, 39 (1976).
347. Y. S. Lipatov, A. E. Nesterov, and T. D. Ignatova, *Vysokomolek. soed.*, **A22**, 2284 (1980).
348. Y. S. Lipatov and A. E. Nesterov, *Int. Symp. on Macromol. Chem.*, Tashkent, 17-21 Oct., 1978; Moskva, *Nauka*, 1978, p. 23.
349. A. E. Nesterov and Y. S. Lipatov, *Polym. Networks Blends*, **4**, 213 (1994).
350. T. G. Scholte, *Eur. Polym. J.*, **6**, 1063 (1970).
351. A. Rudin, R. A. Wagner, K. K. Chee *et al.*, *Polymer*, **18**, 124 (1977).
352. A. E. Nesterov in *Handbook on Physical Chemistry of Polymers*, Vol. 1, Naukova Dumka, 1984.
353. S. M. Lipatov in *High Molecular Compounds (Liophilic colloids) (in Russ),* Acad. Sci. Belor. SSR, Tashkent, 1943.
354. E. L. Atkin, L. A. Kleintjens, R. Koningsveld, and L. J. Fetters, *Makromol. Chem.*, **185**, 377 (1984).
355. F. Cangelosi and M. T. Shaw, *Am. Chem. Soc. Polym. Prepr.*, **24**, 258 (1983).
356. M. V. Rei, M. Scott, J. J. Pireaux, C. Cregoire, R. Caudano, A. Lapp, J. Lopes da Silva, and A. M. Roteglo Do Rego, *Surface Sci.*, **211**, 782; **212**, 789 (1989).
357. J.- L. Lin, D. Rigby, and R.- J. Roe, *Macromolecules*, **18**, 1609 (1985).
358. F. S. Bates, L. J. Fetters, and G. D. Wignall, *Phys. Rev. Lett.*, **55**, 2425 (1985).

359. J. W. Schurer, A. De Boer, and G. Challa, *Polymer*, **16**, 1086 (1975).
360. E. Lemieux, R. E. Prud'homme, R. Forte, R. Jerome, and Ph. Teyssie, *Macromolecules*, **21**, 2148 (1988).
361. C. Silvestre, S. Cimmino, E. Martuscelli *et al.*, *Polymer*, **28**,1190 (1987).
362. C. Silvestre, S. Cimmino, F. E. Karasz, and W. J. MacKnight, *J. Polymer Sci. Phys. Ed.*, **25**, 2531 (1987).
363. J. Vanderschueren, A. Janssens, M. Ladang, and J. Niezette, *Polymer*, **23**, 395 (1982).
364. P. S. Sremcich, A. B. Faust, J. W. Gilmer, and J. W. Mays, *Polym. Commun.*, **30**, 146 (1989).
365. A. G. Bringuier, J. W. Gilmer, and J. W. Mays, *Am. Chem. Soc. Polym. Prepr.*, **22**, 118 (1987).
366. A. B. Faust, P. S. Sremcich, J. W. Gilmer, and J. W. Mays, *Macromolecules*, **22**, 1250 (1989).
367. Y. Suzuki, Y. Miyamoto, H. Miyaji, and K. Asai, *J. Polymer Sci. Lett. Ed.*, **20**, 563 (1982).
368. Y. Maeda, F. E. Karasz, W. J. MacKnight, and R. Vukovic, *J. Polymer Sci. Phys. Ed.*, **24**, 2345 (1986).
369. V. P. Privalko, K. N. Petrenko, and Y. S. Lipatov, *Reports Acad. Sci. Ukraine*, **B8**, 50 (1986).
370. L. Van Opatul and R. Koningsveld, *Polymer*, **33**, 3445 (1992).
371. D. J. Walsh and S. Rostami, *Macromolecules*, **18**, 1228 (1985).
372. F. S. Bates, M. F. Schulz, J. M. Rosedall, and K. Almdal, *Macromolecules*, **25**, 5547 (1992).
373. R. P. Kambour, P. E. Gundlach,T. C. W. Wange, D. M. White, and G. W. Yeager, *Am. Chem. Soc. Polym. Prepr.*, **28**, 140 (1987).
374. T. Kyu and J. M. Saldanha, *Am. Chem. Soc. Polym. Prepr.*, **28**, 124 (1987).
375. T. Kyu, J. M. Saldanha, and D.- S. Lim, *Am. Chem. Soc. Polym. Prepr.*, **29**, 454 (1988).
376. T. Kyu, D.- S. Lim, *Am. Chem. Soc. Polym. Prepr.*, **29**, 354 (1988).
377. J. M. Saldanha and T. Kyu, *Macromolecules*, **20**, 2840 (1987).
378. T. Kyu and J. M. Saldanha, *Macromolecules*, **21**, 1021 (1988).
379. C. J. T. Landry and P. M. Hendichs, *Macromolecules*, **22**, 2157 (1989).
380. J. Pilar, A. Sikora, J. Labsky, and S. Schlick, *Macromolecules*, **26**, 137 (1993).
381. T. Kyu and P. Zhuang, *Polym. Commun.*, **29**, 99 (1988).
382. A. E. Nesterov, Y. S. Lipatov, and T. D. Ignatova, *Reports Acad. Sci. Ukraine*, **B7**, 629 (1975).
383. L. Zeman and D. Patterson, *Macromolecules*, **5**, 513 (1972).
384. E. H. Hellmann, G. P. Hellmann, and A. R. Rennie, *Colloid Polym. Sci.*, **269**, 343 (1991).
385. M. Bank, J. Leffingwell, and C. Thies, *Macromolecules*, **4**, 43 (1971).
386. M. E. Fowler, J. W. Barlow, and D. R. Paul, *Polymer*, **28**, 1177 (1987).
387. A. E. Nesterov, Y. S. Lipatov, N. P. Gudima, and T. D. Ignatova, *Vysokomolek. soed.*, **B31**, 781 (1989).
388. I. Inoue, T. Ougizawa, O. Yasuda, and K. Miyasaka, *Macromolecules*, **18**, 56 (1985).
389. V. A. Izvozchikova, I. N. Razinskaya, L. V. Adamova *et al.*, *1 Vsesoyuznaya konf. Smesi Polymerov*, Ivanovo, 1986, p. 26.
390. D. Braun, P. R. Kohl, and G. P. Hellmann, *Makromol. Chem.*, **189**, 1671 (1988).
391. J. Maruta, T. Ougizawa, and T. Inoue, *Polymer*, **29**, 2056 (1988).
392. T. Ohnaga, J. Maruta, and T. Inoue, *Polymer*, **30**, 1845 (1989).
393. M. A. Gashgari and C. W. Frank, *Macromolecules*, **21**, 2782 (1988).
394. T. Izumitani and T. Hashimoto, *J. Chem. Phys.*, **83**, 3694 (1985).

395. M. Takenaka, T. Izumitani, and T. Hashimoto, *Macromolecules*, **20**, 2257 (1987).
396. M. Nishimoto, H. Kekkula, and D. R. Paul, *Polymer*, **32**, 272 (1991).
397. Y. S. Lipatov in *Polymer Reinforcement*, ChemTec Publishing, Toronto, 1995.
398. A. E. Nesterov, Y. S. Lipatov, V. V. Horichko, and O. T. Gritsenko, *Ukr. Polym. J.*, **1**, 132 (1992).
399. Y. S. Lipatov in *Controlled Interfaces in Composite Materials*, Ed. H. Ishida, Elsevier, Amsterdam, 1990, p. 599.
400. Y. S. Lipatov, A. E. Nesterov, T. D. Ignatova, N. P. Gudima, and O. T. Gritsenko, *Eur. Polym. J.*, **22**, 83 (1986).
401. Y. S. Lipatov and L. M. Sergeeva in *Adsorption of Polymers*, John Wiley, New York, 1974.
402. A. E. Nesterov, V. V. Horichko, and Y. S. Lipatov, *Reports Acad. Sci. USSR*, **299**, 655 (1988).
403. A. E. Nesterov, Y. S. Lipatov, V. V. Horichko, and O. T. Gritsenko, *Polymer*, **33**, 619 (1992).
404. Y. S. Lipatov, *Polym. J. (Japan)*, **23**, 659 (1991).
405. Y. S. Lipatov, *J. Adhesion*, **37**, 181 (1992).
406. A. N. Dolinny and V. A. Ogarev, *Adv. Chemistry (in Russ)*, **57**, 1769 (1988).
407. A. N. Dolinny, *Vysokomolek. soed.*, **A32**, 1929 (1990).
408. A. N. Dolinny, *Vysokomolek. soed.*, **A32**, 1938 (1990).
409. A. M. Nermirovsky and K. F. Fried, *J. Chem. Phys.*, **83**, 4166 (1985).
410. J. F. Douglass, A. M. Nemirovsky, and K. F. Fried, *Macromolecules*, **19**, 2041 (1986).
411. G. H. Weiss and R. J. Rubin, *Adv. Chem. Phys.*, **52**, 363 (1983).
412. S. Reich and Y. Cohen, *J. Polymer Sci. Phys. Ed.*, **19**,1255 (1981).
413. A. E. Nesterov, V. V. Horichko, and Y. S. Lipatov, *Makromol. Chem. Rapid Commun.*, **12**, 571 (1991).
414. E. Raphael, *J. de Physique*, **50**, 803 (1989).
415. P. Cifra, F. E. Karasz, W. J. MacKnight, *J. Polymer Sci. Phys. Ed.*, **30**, 1401 (1992).
416. Y. S. Lipatov, *Colloid and Polymer Sci.*, **264**, 377 (1986).
417. V.V. Shifrin, Y. S. Lipatov, and A. E. Nesterov, *Vysokomolek. soed.*, **A27**, 369 (1986).
418. A. E. Nesterov, Y. S. Lipatov, and N. P. Gudima, *Reports Acad. Sci. Ukraine*, **B11**, 44 (1985).
419. Y. S. Lipatov, V. V. Shifrin, O. I. Vasilenko, and M. S. Krivko, *Vysokomolek . soed.*, **B28**, 869 (1986).
420. A. E. Nesterov, Y. S. Lipatov, and T. D. Ignatova, *Vysokomolek. soed.*, **A22**, 2665 (1980).
421. Y. S. Lipatov, E. G. Gudova, and G. V. Titov, *Vysokomolek. soed.*, **A31**, 1962 (1989).
422. D. Patterson, *Pure and Appl. Chem.*, **31**, 133 (1972).
423. Y. S. Lipatov, A. E. Nesterov, and V. V. Shifrin, *Reports Acad. Sci. USSR*, **276**, 405 (1984).
424. Y. S. Lipatov, V. V. Shifrin, E. G. Gudova, and O. I. Vadilenko, *Vysokomolek. soed.*, **A31**, 1464 (1989).
425. Y. S. Lipatov, T. S. Khramova, T. T. Todosijchuk, and E. G. Gudova, *J. Colloid Interface Sci.*, **123**, 143 (1988).
426. Y. S. Lipatov, V. V. Shifrin, and O. I. Vasilenko, *Vysokomol. soed.*, **A29**, 1400 (1987).
427. J. Noolandi, *Abstr. Pap. 194th Meet. Amer. Chem. Soc.*, New Orlean-Washington, 1987, p. 998.

428. Y. S. Lipatov, V. V. Shifrin, and O. I. Vasilenko, *Ukrainian Chem. J.*, **55**, 306 (1989).
429. Y. S. Lipatov, O. I. Vasilenko, and N. P. Gudima, *Reports Acad. Sci. USSR*, **311**, 649 (1990).
430. Y. S. Lipatov, V. A. Kargin, and G. L. Slonimsky, *Zh. Phys. Chem. (Russ)*, **30**, 1076, 1202. (1956).
431. Y. S. Lipatov, *Science and Engn. Composite Mater.*, **2**, 69 (1991).
432. J. S. Yoon, C. S. Choi, S. J. Maing, H. J. Choi, H.- S. Lee, and S. J. Choi, *Eur. Polym. J.*, **29**, 1359 (1993).
433. G. Defiew, G. Groeninckx, and H. Reynayers, *Contemp. Top. Polym. Sci.*, **6**, 423 (1989).
434. J. A. Pomposo, E. Calahorra, I. Equiazabal, M. Cortazar, *Macromolecules*, **26**, 2104 (1993).
435. R. Koningsveld, H. A. Chermin, and M. Gordon, *Proc. Roy. Soc. London.*, **A**, 1970, p. 319.
436. A. E. Nesterov, S. Y. Lipatov, and T. D. Ignatova, *Vysokomolek. soed.*, **B20**, 93 (1978).
437. Y. S. Lipatov, A. E. Nesterov, and T. D. Ignatova, *Ukrainian Chem. J.* , **41**, 939 (1975).
438. A. A. Tager and V. S. Blinov, *Vysokomolek. soed.*, **A23**, 2122 (1981).
439. B. A. Golender, Z. G. Sagdieva, Sh. Fuzailov, and S. A. Tashmukhammedov, *Vysokomolek. soed.*, **B18**, 688 (1976).
440. M. G. Prolongo, R. M. Masegosa, and A. Horta, *Macromolecules*, **22**, 4346 (1989).
441. E. G. Lescano, C. C. Salom, and M. G. Prolongo, *Macromolecules*, **25**, 6849 (1992).
442. Z. Y. Al-Saigh and P. Munk, *Macromolecules*, **10**, 803 (1984).
443. Y. Einaga, Z. Tong, and H. Fujita, *Macromolecules*, **20**, 2027 (1987).
444. Z. Tong, Y. Einaga, T. Kitagawa, and H. Fujita, *Macromolecules*, **22**, 450 (1989).
445. L. Ould-Kaddour and C Strazielle, *Polymer*, **28**, 459 (1987).
446. L. Ould-Kaddour and C. Strazielle, *Eur. Polymer J.*, **24**, 117 (1988).
447. M. C. Anasagasti, I. Katime, and C. Strazielle, *Makromol. Chem.*, **188**, 201 (1987).
448. L. Ould-Kaddour and C. Strazielle, *Polymer*, **33**, 889 (1992).
449. T. Fukuda, M. Nagata, and H. Inagaki, *Macromolecules*, **19**, 1411 (1986).
450. P. G. De Gennes, *J. Polymer Sci.-Symp.*, **61**, 313 (1977).
451. J. F. Joanny, P. Grant, L. A. Turkevich, and P. J. Pincus, *J. Appl. Phys.*, **52**, 5943 (1981).
452. J. F. Joanny, P. Grant, and L. A. Turkevich, *J. Physique*, **42**, 1045 (1981).
453. T. Nose, *J. Physique*, **47**, 517 (1986).
454. V. E. Eskin and A. E. Nesterov, *Kolloid J. (Russ)*, **28**, 904 (1966).
455. M. S. Kent, M. Tirrell, and T. P. Lodge, *Polymer*, **32**, 314 (1991).
456. R. Kuhn and H.- J. Cantow, *Makromol. Chem.*, **122**, 65 (1969).
457. C.- J. Lin and S. L. Rosen, *J. Polymer Sci. Phys. Ed.*, **20**, 1497 (1982).
458. M. S. Kent, M. Tirrell, and T. P. Lodge, *Macromolecules*, **25**, 5383 (1992).
459. M. S. Kent, A. F. Faldi, M. Tirrell, and T. P. Lodge, *Macromolecules*, **25**, 4501 (1992).
460. T. K. Kwei, H. L. Frisch, W. Radigan, and S. Vogel, *Macromolecules*, **10**, 157 (1977).
461. Y. Y. Wang and S. A. Chen, *Polym. Eng. Sci.*, **21**, 47 (1981).
462. M. Rink, L. Montrini, P. Gasparini, and A. Pavan, *Plast. Rubber Process. Appl.*, **3**, 45 (1983).
463. S. H. Goh, D. R. Paul, and J. W. Barlow, *Polym. Eng. Sci.*, **22**, 34 (1982).
464. S. H. Goh, K. S. Siow, and K. S. Yap, *J. Appl. Polym. Sci.*, **29**, 99 (1984).

465. B. Ameduri and R. E. Prud'homme, *Polymer*, **29**, 1052 (1988).
466. C. Belaribi, G. Martin, and Ph. Monge, *Eur. Polym. J.*, **22**, 487 (1986).
467. V. S. Shah, J. D. Keitz, D. R. Paul, *J. Appl. Polym. Sci.*, **32**, 3863 (1986).
468. J. W. Barlow and D. R. Paul, *Polym. Eng. Sci.*, **27**, 1482 (1987).
469. J. I. Equiazabal, J. J. Iruin, M. Cortazar, and G. M. Guzman, *J. Appl. Polym. Sci.*, **32**, 5945 (1986).
470. S. H. Goh, K. S. Siow, and K. S. Yap, *Thermochim. Acta*, **102**, 281 (1986).
471. S. H. Goh and K. S. Siow, *Thermochim. Acta*, **105**, 191 (1986).
472. J. Ch. Huang, K. Min, and J. L. White, *Polym. Eng. Sci.*, **28**, 1085 (1988).
473. J. Ch. Huang, K. Min, and J. L. White, *Polym. Eng. Sci.*, **28**, 1590 (1988).
474. D. Rigby, J. L. Lin, and R.- J. Roe, *Macromolecules*, **18**, 2269 (1985).
475. S. Klotz and H.- J. Cantow, *Polymer*, **31**, 315 (1990).
476. G. R. Brannock and D. R. Paul, *Macromolecules*, **23**, 5240 (1990).
477. T. Nishi, T. Suzuki, N. Matsuzawa, S. Mishizono, and H. Tanaka, *Am. Chem. Soc. Polym. Prepr.*, **28**, 110 (1987).
478. A. C. Su and J. R. Fried, *Polym. Eng. Sci.*, **27**, 1657 (1987).
479. W. H. Christiansen, D. R. Paul, and J. W. Barlow, *J. Appl. Polym. Sci.*, **34**, 537 (1987).
480. Q. Guo, *Eur. Polym. J.*, **26**, 1329 (1990).
481. Q. Guo, *Eur. Polym. J.*, **26**, 1333 (1990).
482. T. C. Ahn, S. Lee, H. M. Jeong, and K. Cho, *Angew. Makromol. Chem.*, **192**, 133 (1991).
483. D. Yu, E. H. Hellmann, and G. P. Hellmann, *Makromol. Chem.*, **192**, 2749 (1991).
484. A. E. Nesterov, V. V. Horichko, V. V. Muzhev, and Y. S. Lipatov, *Vysokomolek. soed.*, **B31**, 296 (1989).
485. A. E. Nesterov, V. V. Horichko, V.V. Muzhev, and Y. S. Lipatov, *Reports Acad. Sci. USSR*, **308**, 641 (1989).
486. A. E. Nesterov, V. V. Muzhev, and V. V. Horichko, *Vysokomolek. soed.*, **B31**, 855 (1989).
487. Y. S. Lipatov and A. E. Feinerman, *Reports Acad. Sci. USSR*, **271**, 896 (1983).
488. Y. S. Lipatov and A. E. Feinerman, *J. Colloid and Interface Sci.*, **98**, 126 (1984).
489. A. I. Rusanov in *Phase Equilibria and Surface Phenomena,* L. Chimiya, 1967.
490. S. Wu, *J. Macromol. Sci.*, **C10**, 1 (1974).
491. M. D. Fried and M.-C. Chein, *J. Appl. Polym. Sci.* **30**, 2105 (1985).
492. J. J. Jutier, E. Lemieux, and R. E. Prud'homme, *J. Polym. Sci., Phys. Ed.*, **26**, 1313, (1988).
493. R. B. Kambour, W. L. Nachlis, and J. D. Carbeck, *Polymer*, **35**, 809 (1994).
494. T. B. Repina, A. V. Kotova, A. V. Shashkov, G. M. Zeitlin, and M. S. Mezhjkovsky, *Vysokomol. soed.*, **B35**, 1557 (1993).
495. A. M. Bonfatti, M. Canetti, P. Sadocco, A. Seves, and E. Martuscelli, *Polymer*, **34**, 990 (1993).
496. C. Silvestre, M. Canetti, C. Lalla, A. Bonfatti, and A. Seves, *Makromol. Chem.*, **190**, 2615 (1989).
497. T. S. Ellis, *Am. Chem. Soc. Polym. Prepr.*, **28**, 181 (1987).
498. X. Yu, Y. Zeng, Z. Wu, X. Tang, and B. Jiang, *Polymer Physics Laboratory,* Academia sinica, Annual Reports, p. 125, 1989.
499. F. S. Bates and P. Wiltzius, *J. Chem. Phys.*, **91**, 3258 (1989).
500. H. Jinnai, H. Hasegawa, and T. Hashimoto, *Macromolecules*, **25**, 6078 (1992).
501. C. A.. Tras and C. M. Roland, *Polymer Commun.*, **29**, 332 (1988).
502. M. Takenaka and T. Hashimoto, *J. Chem. Phys.*, **98**, 6177 (1992).
503. P. H. Huo and P. Gebe, *Macromolecules*, **26**, 3127 (1993).

504. R. J. Composto, J. W. Mayer, E. J. Kramer, and D. M. White, *Phys. Rev. Lett.*, **57**, 1312 (1988).
505. S. H. Goh and S. Y. Lee, *Eur. Polymer J.*, **25**, 997 (1989).
506. R. Vukovic, M. Zuanic, G. Bogdanovic, V. Kuresevic, F. E. Karasz, and M J. MacKnight, *Polymer*, **34**, 1449 (1993).
507. E. Pedemonte and G. Bargisi, *Polymer*, **35**, 3719 (1994).
508. G. Pompe, E. Meyer, and H. Hamann, *9 Tag. Polym. Phys.*, Postsdam, *24-28 Apr., 1989: Kurfass., Vortr. und Poster-Teltow-Seehot,* p. 150, 1989.
509. Q. Guo, J. Huang, and T. Chen, *Polymer Physics Laboratory*, Academia sinica, Annual Reports, p. 53, 1989.
510. T. Kyu and J. M. Saldanha, *J. Polym. Sci.- Lett.*, **26**, 33 (1988).
511. T. Kyu and D.-S. Lim, *J. Chem. Phys.*, **92**, 3951 (1990).
512. T. Kyu and D.-S. Lim, *Macromolecules*, **24**, 3045 (1991).
513. T. Kyu, C. C. Ko, D.-S. Lim, S. D. Smith, and I. Noda, *J. Polym. Sci, Phys. Ed.*, **31**, 1641 (1993).
514. M. M. Coleman, X. Yang, P.C. Painter, and J. F. Graf, *Macromolecules,* **25**, 4414 (1992).
515. A. C. Fernandes, J. W. Barlow, and D. R. Paul, *J. Polym. Sci.*, **32**, 6073 (1986).
516. Q. Guo, *Polymer Physics Laboratory,* Academia sinica, Annual Reports, p. 57, 1989.
517. H.-L. Chen and R. S. Porter, *J. Polym. Sci, Phys. Ed.*, **31**, 1845 (1993).
518. K. Liang, J. Grembowicz, E. Valles, F. E. Karasz, and W. J. MacKnight, *J. Polym. Sci., Phys. Ed.,* **30**, 465 (1992).
519. M. C. Lin, T. Yang, and K. Y. Huang, *Polymer*, **35**, 594 (1994).
520. X. Luo, S. Zheng, N. Zhang, and D. Ma, *Polymer*, **35**, 2619 (1994).
521. Y. Nishio, N. Hirose, and T. Takahashi, *Polym. J. (in Japan),* **21**, 347 (1989).
522. Q. Guo, H. Xu, D. Ma, and S. Wang, *Eur. Polymer J.*, **26**, 67 (1990).
523. S. H. Jeon and T. Ree, *J. Polym. Sci., Phys. Ed.*, **26**, 1419 (1988).
524. Z. Mo, J. Lin, Y. Qi, H. Zhang, and D. Chen, *Polymer Physics Laboratory,* Academia sinica Annual Reports, p. 95, 1989.
525. C. Marco, M. A. Gomez, J. G. Fatou, A. Etxeberria, M. Elorza, and J. J. Iruin, *Eur. Polym. J.*, **29**, 1477 (1993).
526. S. Cimmino, E. Martuscelli, and Silvestre, *Polymer*, **30**, 393 (1989).
527. S. Cimmino, E. Martuscelli, C. Silvestre, C. de Lalla, and A. Seves, *J. Polym. Sci., Phys Ed.*, **27**, 1781 (1989).
528. Q. Guo, J. Huang, T. Chen, and Z. Feng, *Polymer Physics Laboratory,* Academia sinica, Annual Reports, p. 51, 1989.
529. A. K. Bhowmick, C. C. Kuo, A. Manzur *et all.*, *Macromol. Sci*, **B25**, 283 (1986).
530. T. A. Callaghan and D. R. Paul, *Macromolecules*, **26**, 2939 (1993).
531. V. Janarthanan and G. Thyagarajan, *Polymer*, **33**, 3593 (1992).
532. S. H. Goh, S. Y. Lee, and S. M. Low, *Macromoleceles*, **27**, 6736 (1994).
533. N. Lotti, M. Pizzoli, G. Ceccorulli, and M. S. Candola, *Polymer*, **34**, 4935 (1993).
534. P. C. Painter, J. E. Graf, and M. M. Coleman, *Macromolecules,* **24**, 5630 (1991).
535. S. Das and F. Rodriguez, *Am. Chem. Soc., Polym. Prepr.*, **28**, 146 (1987).
536. M. K. Neo and S. H. Goh, *Eur. Polymer J.*, **27**, 927 (1991).
537. S. M. Low, S. Y. Lee, and S. H. Goh, *Eur. Polymer J.*, **29**, 909 (1993).
538. S. H. Goh, S. Y. Lee, K. S. Siow, and M. K. Neo, *Polymer*, **31**, 1085 (1990).
539. J. F. Chong, S. Y. Lee, and S. H. Goh, *Eur. Polymer J.,* **26**, 1207 (1990).
540. M. K. Neo and S. H. Goh, *Polymer*, **33**, 3203 (1992).
541. S. M. Low, S. H. Goh, and S. Y. Lee, *Polymer*, **35**, 3290 (1994).
542. J. Peng, S. H. Goh, S. Y. Lee, and K. S. Siow, *Polymer*, **35**, 1482 (1994).
543. J. Peng, S. H. Goh, S. Y. Lee, and K. S. Siow, *Polymer*, **34**, 4930 (1993).
544. M. K. Neo, S. Y. Lee, and S. H. Goh, *J. Appl. Polym. Sci*, **43**, 1031 (1991).

545. S. M. Low, S. H. Goh, and S. Y. Lee, *Macromolecules*, **27**, 3764 (1994).
546. D. Garcia, *Am. Chem. Soc., Polym Prepr.*, **27**, 859 (1986).
547. B. Endres and J. H. Wendorf, *Ber. Bunsenges. Phys. Chem., 1985, n. 11: Morphol. und Eigenschaften Polym. Syst. Ber. 84 Hauptversamml., Aachen, 16-18 Mai,* 1985, s. 1176.
548. R. M. Briber and F. Khoury, *Polymer*, **28**, 38 (1987).
549. G. Ceccorulli, M. Pizzolli, M. Scandola, G. C. Alfonso, and A. Turturro, *Polymer*, **30**, 1251 (1989).
550. S. H. Goh and K. S. Siow, *Polym. Bull.*, **17**, 453 (1987).
551. M. B. Huglin and J. M. Rego, *Polymer*, **31**, 1269 (1990).
552. J. F. Manson and R. St. J. Manley, *Macromolecules*, **24**, 5914 (1991).
553. J. F. Manson and R. St. J. Manley, *Macromolecules*, **24**, 6670 (1991).
554. J. R. Isai, L. Cesteros, and I. Katime, *Polymer*, **34**, 2374 (1993).
555. C. Bhattacharyya, S. N. Bhattacharyya, and B. M. Mandal, *J. Indian Chem. Soc.*, **63**, 157 (1986).
556. C. Landry, D. Massa, D. M. Teegarden, M. R. Landry, P. M. Hendricks, R. H. Colby, and T. E. Long, *Macromolecules*, **26**, 6299 (1993).
557. G. Guerra, D. J. Williams, F. E. Karasz, and J. MacKnight, *J. Polym. Sci., Phys. Ed.*, **26**, 301 (1988).
558. Z. G. Gardlung, *Polymer*, **34**, 1850 (1993).
559. R. J. Karcha and R. S. Porter, *Polymer*, **33**, 4866 (1992).
560. Q. Trang-Cong, R. Kawakubo, and S. Sakurai, *Polymer*, **35**, 1237 (1994).
561. S. S. Chalraborty, N. Maiti, B. M. Mandal, and S. N. Bhattacharyya, *Polymer*, **34**, 111 (1993).
562. A. Brunacci, E. Pedemonte, J. M. G. Cowie, and I. J. McEwen, *Polymer*, **35**, 2893 (1994).
563. M. Takahashi, S. Kinoshita, and T. Nose, *J. Polym. Sci., Phys. Ed.*, **27**, 2159 (1989).
564. R. Register and T. R. Bell, *J. Polym. Sci., Phys. Ed.*, **30**, 569 (1992).
565. R. A. Weiss and X. Lu, *Polymer*, **35**, 1963 (1994).
566. M. He, Y. Lee, Y. Feng, M. Jiang, and C. Han, *Macromolecules*, **24**, 464 (1991).
567. Y. Nishio and R. St. J. Manley, *Macromolecules*, **21**, 1270 (1988).
568. Y. Nishio, T. Haratani, T. Takahashi, and R. St. J. Manley, *Macromolecules*, **22**, 2547 (1989).
569. X. Zang, K. Takegoshi, and K. Nikichi, *Polymer*, **33**, 718 (1992).
570. J. M. Machado and R. N. French, *Polymer*, **33**, 760 (1992).
571. H. Feng, Z. Feng, and L. Shen, *Polymer*, **34**, 2516 (1993).
572. X. Zang, K. Takegoshi, and K. Nikichi, *Polymer*, **33**, 712 (1992).
573. G. Thyagarajan and V. Janarthanan, *Polymer*, **30**, 1797 (1989).
574. Y. Nishio, T. Haratani, and T. Takahashi, *J. Polym. Sci., Phys. Ed.*, **28**, 355 (1990).
575. Z. H. Ping, Q. T. Ngyyen, and J. Neel, *Makromol. Chem.*, **190**, 437 (1989).
576. M. K. Neo and S. H. Goh, *Polymer*, **33**, 2012 (1992).
577. E. G. Lescano, C. Salom-Coll, and M. G. Prolongo, *Macromolecules*, **25**, 6849 (1992).
578. A. K. Nandi, B. M. Mandal, S. N. Bhattacharyya, and S. K. Roy, *Polymer Commun.*, **27**, 151 (1986).
579. Q. Guo, *Polymer Physics Laboratory*, Academia sinica, Annual Reports, p. 49, 1989.
580. T. E. Sincock and D. J. David, *Polymer*, **33**, 4515 (1992).
581. A. K. Sen and G. S. Mukheryee, *Polymer*, **34**, 2386 (1993).
582. M. Bosma, G. ten Brinke, and T. S. Ellis, *Polymer*, **29**, 86 (1988).
583. E. J. Vorenkamp and G. Challa, *Polymer*, **29**, 86 (1988).
584. S. Shen and J. M. Torkelson, *Macromolecules*, **25**, 721 (1992).

585. A. G. Margaritis and N. K. Kalfoglou, *J. Polym. Sci., Phys. Ed.*, **26**, 1595 (1988).
586. I. A. Katime, M. S. Anagasti, M. C. Peleteiro, and R. Valensiano, *Eur. Polymer J.*, **23**, 907 (1987).
587. A. G. Margaritis and N. K. Kalfoglou, *J. Polym. Sci., Phys. Ed.*, **27**, 1767 (1989).
588. I. Hopkinson, F. T. Kiff, R. W. Richards, S. M. King, and H. Munro, *Polymer*, **35**, 1722 (1994).

Chapter 4

HOMOPOLYMER-COPOLYMER AND COPOLYMER-COPOLYMER BLENDS

4.1 MEAN-FIELD BINARY INTERACTION MODELS

In Chapter 3 (3.2.2), the binary polymer mixtures, whose miscibility is determined mainly by the specific interactions, were discussed. Investigation of the homopolymer blends with copolymers shows[1-3] that their miscibility is not necessary connected with specific interactions. For example, poly-2,6-dimethyl-1,4-phenylene oxide is neither miscible with poly-*o*-chlorostyrene nor with poly-*p*-chlorostyrene.[1,2] At the same time, it is miscible with copolymer of *o*-chlorostyrene and *p*-chlorostyrene at the content of the latter comonomer in copolymer from 23 to 65%. It is also miscible with copolymers of *o*-fluorostyrene at the content of the latter 15-36% and *o*-chlorostyrene or *o*-chlorostyrene and *p*-chlorostyrene at 5-80% of the latter.[2] However, poly-2,6-dimethyl-1,4-phenylene oxide is not miscible with copolymers of *p*-chlorostyrene-*p*-fluorostyrene at any composition.[2,3] It was also discovered that in homopolymer mixtures with copolymers, the transition from miscibility to immiscibility occurs very rapidly even at small change of the copolymer composition. The general feature of such mixtures is the existence of the so-called "miscibility windows" (Fig. 4.1) which are observed in the case when the corresponding homopolymers are not miscible. For example, neither PS nor PAN are miscible with polymethylacrylates and polyacrylates. But styrene-*co*-acrylonitrile polymers are miscible with these homopolymers.[8,9] Ethylene-*co*-vinylacetate polymers and butadiene-*co*-acrylonitrile polymers are miscible with PVC but immiscible with the corresponding homopolymers.[10]

Kambour *et al.*[11] have explained the miscibility of such systems by the effects of repulsion between comonomeric units in copolymer chain. This repulsion was considered to be the driving force of miscibility. To describe the thermodynamic behavior of such systems, the so-called binary interaction model[11-13] was proposed that involves both intra- and intermolecular interactions. In this model, the mixture of homopolymer and copolymer is considered as the ternary system including comonomers of the copolymer as the independent compo-

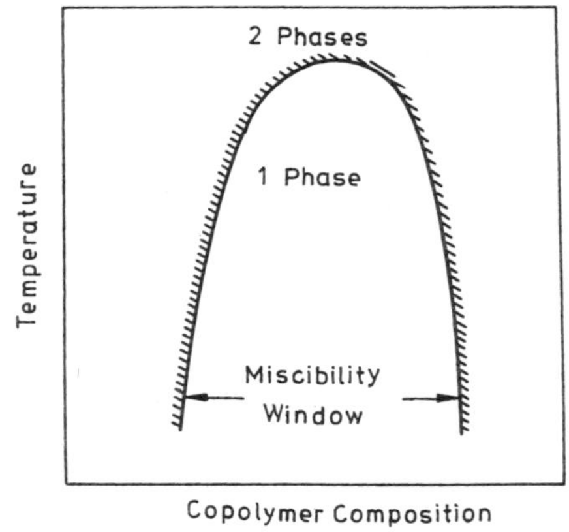

Figure 4.1. Diagram of phase behavior for blends of a copolymer with another polymer showing LCST behavior to illustrate the concept of "miscibility window". The amount of the copolymer in the blend is held fixed while the composition of the copolymer, i.e., the relative proportions of its two monomers, are varied. Adapted, by permission, from D. R. Paul and J. W. Barlow, *Polymer*, **25**, 487 (1984).

nents 1 and 2, and homopolymer 3 differing by composition from copolymer. Three types of interactions are accounted for: 1-2, 1-3, and 2-3. The expression for the parameter of thermodynamic interaction in the mixtures of homopolymer with copolymers at the phase separation point is presented[11] as

$$\chi_{AB} = (1 - \varphi_c)\chi_{ij} + \varphi_c\chi_{ik} - \varphi_c(1 - \varphi_c)\chi_{kj} \qquad [4.1]$$

where index i denominates segments of homopolymer A and indices j and k are the same for copolymer B, φ_C is the volume fraction of one of the monomers in copolymer chain. First two terms reflect the interaction between homopolymer and segments of copolymer whereas the third term characterizes the interaction between copolymer monomeric units. It was shown that the "inner repulsion" between monomeric units in copolymer may prevail over intermolecular attraction between monomeric units of homopolymer and copolymer and determine the miscibility.

The theory was used[12] to explain the so-called "miscibility windows", the existence of LCST, and to describe the thermodynamic behavior of copolymer mixtures. The expression was derived for the Gibbs free energy of mixing of two copolymers of the type $(A_{x-1}B_x)$ and (C_yD_{1-y}) in the form:[12]

$$\frac{\Delta G}{kT} = (\varphi_1/N_1)\ln\varphi_1 + (\varphi_2/N_2)\ln\varphi_2 + \varphi_1\varphi_2[xy\chi_{AC} + (1 - x)y\chi_{BC} + x(1-y)\chi_{AD} + (1-x)(1-y)\chi_{BD} - x(1-x)\chi_{AB} - y(1-y)\chi_{CD}] \quad [4.2]$$

where x and y are correspondingly the copolymer compositions, φ_i is volume fraction of copolymers in the mixture, N_i are polymerization degrees of copolymers, and χ_{ij} is the interaction parameter. The expression in the square brackets is analogous to the Flory-Huggins equation for the interaction parameter in the mixture of homopolymers:

$$\frac{\Delta G}{kT} = \frac{\varphi_1}{N_1}\ln\varphi_1 + \frac{\varphi_2}{N_2}\ln\varphi_2 + \varphi_1\varphi_2\chi_{mixture} \quad [4.3]$$

If to accept that $\chi_{mixture}$ in Eq 4.2 is expressed as

$$\chi_{mixture} = xy\chi_{AC} + (1-x)y\chi_{BC} + x(1-y)\chi_{AD} + (1 - x)(1-y)\chi_{BD} - x(1-x)\chi_{AB} - y(1-y)\chi_{CD}] \quad [4.4]$$

From Eq 4.4, it follows that intra- and intermolecular interactions may suppress each other and lead to the total negative value of parameter $\chi_{mixture}$. Temperature dependence of $\chi_{mixture}$ may be described[12] if to consider this parameter as consisting of two contributions: exchange interaction and excluded volume:

$$\chi_{mixture} = \frac{c}{\tilde{v}(T)T}[xX_{AB} + (1-x)X_{AC} - x(1-x)X_{BC}] \quad [4.5]$$

where c is a constant, $\tilde{v}$ is the reduced mixture volume, and X_{ij} is exchange interaction parameter. Having introduced the apparent interaction parameter:

$$X_{app} = xX_{AB} + (1-x)X_{AC} + x(1-x)X_{BC} \quad [4.6]$$

we have the same situation as for homopolymer mixtures. For $X_{app}>0$, $\chi_{mixture}$ decreases with temperature whereas for $X_{app} < 0$, it increases. Any of these cases can be obtained by the corresponding choice of copolymer composition x. On the other hand, the contribution of the free volume always increases with temperature growth as in the case of homopolymer mixtures. The typical situations are presented in Fig. 4.1.

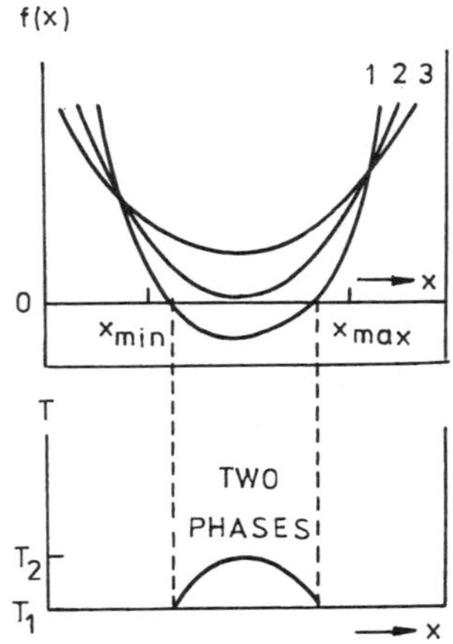

Figure 4.2. Schematic illustration of the connection between the behavior of f(x) as a function of copolymer composition, x, and temperature, T, and miscibility window in the temperature-copolymer composition plane. Curves 1, 2, and 3 are for the temperatures $T_1<T_2<T_3$; x_{max} are the zeros of the effective blend interaction. Adapted, by permission, from G. ten Brinke, F. E. Karasz, and W. J. MacKnight, *Macromolecules,* **16**,1827 (1983).

If $X_{app} > 0$, $\chi_{mixture}$ is positive, the system is immiscible. If $X_{app}<0$, $\chi_{mixture}$ increases with temperature leading to the appearance of LCST.

4.1.1 MISCIBILITY WINDOWS

The miscibility of systems, such as $A/(B_xC_{1-x})$ is determined by the quadratic function

$$f(x) = x^2\chi_{BC} + x(\chi_{AB} - \chi_{AC} - \chi_{BC}) + \chi_{AC} - \chi^{crit}_{mixture} \quad [4.7]$$

where

$$\chi^{crit}_{mixture} = \frac{1}{2}(N_1^{-1/2} + N_2^{-1/2}) \quad [4.8]$$

Neglecting value $\chi^{crit}_{mixture}$ because of its small magnitude for polymer mixtures, three different cases may be considered:[12]

- χ_{AB}, χ_{AC}, and χ_{BC} are positive. In this case, function f(x) is a concave function that may have two or one zero or non-zero values (Fig. 4.2). Any one-zero value f(x) may occur only in the composition range where $X_{app} < 0$. For values of x

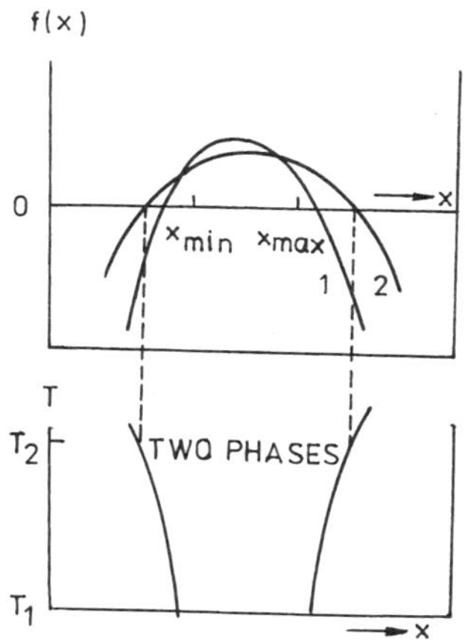

Figure 4.3. Schematic illustration of the connection between the behavior of f(x) as a function of copolymer composition, x, and temperature, T, and miscibility window in the temperature-composition plane. Curves 1 and 2 are for temperature $T_1<T_2$. x_{min} and x_{max} are the zeros of the effective blend interaction. Adapted, by permission, from G. ten Brinke, F. E. Karasz, and W. J. MacKnight, *Macromolecules*, **16**, 1827 (1983).

between possible zero values of f(x), χ_{min}, and χ_{max}, function f(x) decreases with temperature whereas for x outside this region it increases

- χ_{AB}, χ_{AC}, and χ_{BC} are negative. Because $\chi_{BC} < 0$, function f(x) is convex one (Fig. 4.3). Again, f(x) has two zero values, for example, at x_1 and x_2 at some temperature. Values x between x_1 and x_2 represent immiscible systems. Accounting for the temperature dependence of f(x), one has the immiscibility window. In this case, it is determined by the attraction effect contrary to repulsion effect. The mixture of poly(vinyl chloride-*co*-vinyl acetate) with poly(n-butyl methacrylate) may serve as an example of such system.[14] For a given mixture at the component ratio 1:1 and depending on copolymer composition, there was observed the immiscibility region, in spite of the fact that each pair: poly(vinyl chloride-co-vinyl acetate)-n-PBMA, poly(vinyl chloride-co-vinyl acetate)-iso-PBMA and n-PBMA-iso-PBMA are miscible. Another similar system is the mixture of PMMA with copolymer of epichlorohydrine- ethylene oxide[15]

- $\chi_{AB} < 0$, but χ_{AC} and $\chi_{BC} > 0$. Because $\chi_{AB} < 0$, the system is immiscible at all x close to 1. Non-zero values of f(x) may occur for $0 < x < 1$ and $x > 1$. The latter situation has no physical meaning.

There can be distinguished various types of binary mixtures of homopolymers with copolymers and copolymers with copolymers.[16]

1. Mixtures of homopolymer A with copolymer of the A_xB_{1-x} type. Here monomer A is common to both homopolymer and copolymer. Eq 4.4 can be reduced to

$$\chi_{mixture} = \chi_{AB}(1 - x)^2 \quad [4.9]$$

where x is the molar fraction of comonomer A in copolymer. There are only two different segments and only one interaction parameter is necessary.

2. Mixtures of two copolymers of the type A_xB_{1-x}/A_yB_{1-y} of different composition.

$$\chi_{mixture} = \chi_{AB}(x - y)^2 \quad [4.10]$$

where x and y are volume fractions of monomer A in mixed copolymers. There are only two segments and one interaction parameter.

3. Mixtures of homopolymer A with copolymer B_xC_{1-x}. Equation 4.4 becomes

$$\chi_{mixture} = x\chi_{AB} + (1 - x)\chi_{AC} - x(1 - x)\chi_{BC} \quad [4.11]$$

where x is the volume fraction of monomer B in copolymer.

4. Mixtures of copolymers A_xB_{1-x} and C_yB_{1-y}. Because of one common monomer, Eq 4.4 has the form:

$$\chi_{mixture} = x^2\chi_{AB} + (\chi_{AC} - \chi_{BC} - \chi_{AB})xy + \chi_{BC}y^2 \quad [4.12]$$

5. Mixtures of copolymers A_xB_{1-x} and C_yD_{1-y}. This case is the most general and contains four different comonomers. Interaction parameter is described by Eq 4.4.

In the first and the second case, the conditions of miscibility are slightly more favorable due to multiplication by factor $(1-x)^2$ or $(x-y)^2$ as compared with the homopolymer mixture A and B. In the third case the mixture consists of three monomers and is described by three pairs of interactions. Even if all three parameters, χ_{ij}, were positive (it

Table 4.1: The shape of the miscibility borders at various conditions

$c^2 - 4ab$	Shape of the border
< 0	ellipse
$= 0$	parabola
> 0	hyperbola

means the combinations of corresponding homopolymers are immiscible), the mixtures of homopolymer with copolymers may be miscible (have miscibility window) at some concentrations of copolymer due to the segmental repulsion inside copolymer. In the fifth case, six parameters of pair interactions are necessary.

For this general case of the mixture of two copolymers, the function f(x,y) has the following form:[16]

$$f(x,y) = \chi_{AB}x^2 + \chi_{CD}y^2 + (\chi_{AC} - \chi_{BC} - \chi_{AD} + \chi_{BD})xy +$$
$$+ (\chi_{AD} - \chi_{BD} - \chi_{AB})x + (\chi_{BC} - \chi_{BD} - \chi_{CD})y +$$
$$+ (\chi_{BD} - \chi^{crit}_{mixture}) = 0 \qquad [4.13]$$

or in the generalized quadratic form:

$$ax^2 + by^2 + cxy + dx + ey + f = 0 \qquad [4.14]$$

Possible types of miscibility borders[16] are determined by the solution of the generalized equation where the variables are compositions x and y of copolymers. If χ_{ij} are considered as independent variables of composition and concentration, various miscibility borders may be obtained according to Eq 4.14 with corresponding conditions (Table 4.1):

$$(c^2 - 4ab) \equiv [(\chi_{AB} - \chi_{BC}) - (\chi_{AD} - \chi_{BD})]^2 - 4\chi_A\chi_{CD} \qquad [4.15]$$

Schematically the border types are presented in Fig. 4.4. Coefficients d and e in Eq 4.14 determine the position of the focal points. The parabolic shape of the border is a very rare case. The analysis of conditions for the miscibility borders allows a conclusion to be drawn that the first term (c^2) characterizes the contribution of segment non-similarity of two copolymers. The greater non-similarity of two monomeric units in copolymer favors the hyperbolic shape of the miscibility border. If both χ_{AB} and χ_{CD} have the same signs, the elliptic border is observed for high values of these parameters. If they are

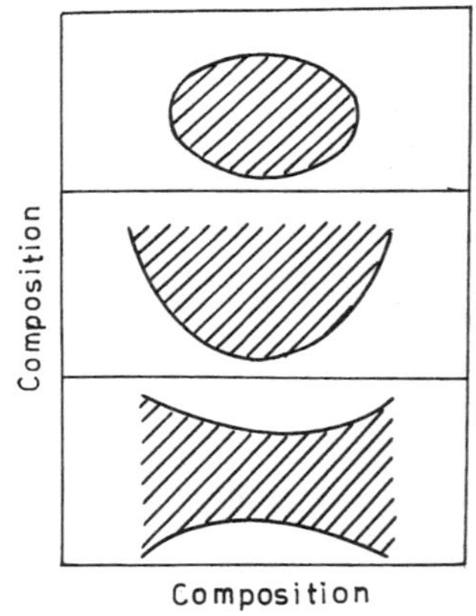

Figure 4.4. Schematic illustration of theoretical miscibility boundaries for (A-B)/(C-D) blends. The shaded and open areas differentiate the miscibility and immiscibility regions. Adapted, by permission, from Ch. Zhikuan, F. E. Karasz, *Macromolecules*, **25**, 4716 (1992).

equal to zero and at least one of the copolymers contains two similar segments, the border of miscibility has parabolic shape. At $\chi_{CD} < 0$ (i.e., when one copolymer contains repulsive segments and another attractive segments), the hyperbolic shape of border should be the case. The miscibility region may lay either inside or outside the ellipse and its positions are determined by the sign of the function f(x,y). It is typical that, although all four pairs of homopolymers may be immiscible, there exists an elliptic shape of miscibility borders and the regime at which the corresponding mixtures of copolymers are miscible. From the other side, the situation is probable when all four pairs of homopolymers are miscible but the mixture may demix at some compositions. Such a mixture may have the elliptic regions of immiscibility. The first case arises due to repulsion in copolymers, whereas the second case is determined by the segregation in copolymers which need high negative value of parameters of segmental interactions. The same arguments may be presented for parabolic and hyperbolic shapes of the miscibility border.

From Eq 4.4, it is seen that all six parameters of interaction fully determine the miscibility border for the mixtures of A_xB_{1-x}/C_yD_{1-y} type but not *vice versa*, i.e., the miscibility borders do not give the sole solution for all parameters χ_{ij} because only five of six coefficients (a,b,..., f) in Eq 4.14 are independent. The same is valid for the mixtures of the type A_xB_{1-x}/B_xC_{1-x} where only two of three parameters are independent. The only solution may be found[17] from the miscibility borders for the mixtures of the type A_xB_{1-x}/C_yD_{1-y}.

Using expression 4.13, for $\chi_{mixture}$ for the mixtures of two random copolymers, such as A_xB_{1-x}/C_yD_{1-y} and varying the relations between parameters including segmental repulsion, authors[17] have obtained various shapes and sizes of the miscibility regions. In particular, when the absolute values of parameters of interaction, χ_{ij}, between various monomeric units are very large (at least, for one copolymer), the borders between the regions of miscibility and immiscibility are restricted by two straight lines for infinite molecular mass of copolymers. In the systems that are immiscible at infinite molecular masses, at finite molecular masses, there can be observed the region of miscibility confined by the coordinate axis and by one of the following curves: ellipse, hyperbola, or straight line. In the same work, the experimental pictures of miscibility have been constructed for the mixtures of poly(vinyl chloride-*co*-vinyl acetate)-poly(ethylene-*co*-vinylacetate) and poly(vinyl chloride-*co*-vinyl acetate)-chlorinated PVC. The latter polymer was considered as random copolymer formed from vinyl chloride and 1,2-dichloroethylene units. The miscibility picture was constructed also for the mixture of poly(styrene-*co*-maleic anhydride)-poly(styrene-*co*-acrylonitrile).[18] For the first system, the miscibility border was found to be elliptic, whereas for the last one it is restricted by parabola.

4.2 EFFECT OF THE SEQUENCE DISTRIBUTION IN COPOLYMER ON THE PHASE BEHAVIOR OF HOMOPOLYMER-COPOLYMER MIXTURES

The relations given above describe the phase behavior of the mixtures of homopolymers and copolymers and do not allow for the effect of surrounding of monomers in copolymer on the pair interactions, i.e., do not consider the distribution of the sequences of monomeric units in copolymer. The distribution of monomeric units may very strongly affect the miscibility with homopolymers. It was shown[19] that copolymer of styrene and methylmethacrylate of the regular structure, having only 10% of SSS and MMM triad units, is miscible with PMMA at any ratios of components, whereas the non-regular copolymer has very restricted miscibility. To take account of the distribution of the units, the so-called order parameter, Θ, was introduced[20-23] that describes the distribution of the binary sequences of monomeric units in copolymer. Varying parameter Θ, one has the possibility to describe random, alternating and block-copolymers. The approximation used by the authors[20-23] is based on the following premises. The energy of interaction between monomeric units of copolymer and homopolymer is mediated by the surrounding of chemically bound structure units of copolymer. For the mixture of homopolymer C with copolymer AB, authors[20] introduce the concept of the probability of interaction of homopolymer C with the central atom of triplets of different structures, such as, AAB, BAA, BAB, AAA, etc. For this purpose the triplet probabilities are introduced expressed in the fractions of molecules f_A and f_B of a simple copolymer chain (f_A and f_B correspond to x and (1–x) in Eq 4.11). The pair probability for one and the same chain are denominated as f_{AA}, f_{BB}, and f_{AB}. The volume fractions, φ, of A, B, AB, and C are introduced as well. All these variables are connected by the series of equations:

$$f_A = f_{AA} + f_{AB};\ f_B = f_{BB} + f_{BA};\ f_{AB} = f_{BA};\ f_A + f_B = 1;$$

$$\varphi_A + \varphi_B = \varphi_{AB};\ f_A\varphi_{AB} = \varphi_A;\ f_B\varphi_{AB} = \varphi_B;\ \varphi_{AB} + \varphi_C = 1 \qquad [4.16]$$

The probabilities of interaction are expressed by these variables in the following manner. Interactions A-C, involving triplet BAA are determined by the probability of triplet BAA equal to the product of f_{AB} and f_{AA}/f_A. This means that A unit follows the next A unit. There are four types of triplets involving the central A copolymer unit, the sum of these four triplet probabilities is f_A. This implies that the normalized probability of BAA triplet is $f_{AB}f_{AA}/f_A^2$. The mean-field pair probability of interaction A-C is proportional to $\varphi_{AB}\varphi_C$ and in such a way the probability of interaction A-C involving triplet BAA is $(f_{AB}f_{AA}/fA)\varphi_{AB}\varphi_C$. Thus, there is a great number of parameters χ involving various triplets. To diminish their number, the following simplification have been introduced:[20] it is accepted that all A-A and B-B interactions are equivalent and equal to zero. All possible A-B interactions are supposed to be equivalent to the mean interaction parameter $\overline{\chi}_{AB}$ and $\overline{\chi}_{AC} = \overline{\chi}_{BAB,C} = \overline{\chi}_{AAB,C} = \overline{\chi}_{BAA,C} \neq \overline{\chi}_{AAA,C}$ and $\overline{\chi}_{BC} \equiv \overline{\chi}_{ABA,C} = \overline{\chi}_{ABB,C} = \overline{\chi}_{BBA,C} \neq \overline{\chi}_{BBB}$. Applying these approximations, the free energy of interaction C/AB in equation of the type of 4.3 contains the term $\chi_{AB}\varphi_C\chi_{mixture}$ where

$$\chi_{mixture} = \chi_{comp} + \chi_{dist} \qquad [4.17]$$

$$\chi_{comp} = f_A\overline{\chi}_{AC} + f_B\overline{\chi}_{BC} - f_Af_B\overline{\chi}_{AB} \qquad [4.18]$$

$$\chi_{dist} = (f_{AA}^2/f_A)\Delta\chi_a + (f_{BB}^2/f_B)\Delta\chi_b \qquad [4.19]$$

$$\Delta\chi_A = \chi_{AAA,C} - \overline{\chi}_{AC} \qquad [4.20]$$

$$\Delta\chi_B = \chi_{BBB,C} - \overline{\chi}_{BC} \qquad [4.21]$$

The dependence of χ_{comp} on the composition is determined by equation such as 4.11 whereas χ_{dist} is a correction of zero order in Eq

4.11. Negative value of $\Delta\chi_a$ implies that interactions AAA-C are more favorable energetically as compared to any other A-C interactions. Positive term $\Delta\chi_a$ implies that AAA-C interactions are less favorable as compared to all other A-C interactions. The same consideration is valid in relation to $\Delta\chi_b$.

The order parameter, Θ, and parameter of composition, δ, allow to express all other probabilities.[20] These parameters are determined by the relationships:

$$f_{AB} = 2\Theta f_A f_B \qquad [4.22]$$

$$f_A = \frac{1}{2}(1+\delta);\ \ f_B = \frac{1}{2}(1-\delta);\ \ -1 \le \delta \le 1 \qquad [4.23]$$

Block-copolymer corresponds to Θ approaching zero, random copolymer meets condition $\Theta = 1/2$. If $f_B \le f_A$, the maximum value of f_{AB} is equated to f_B. This situation corresponds to the case where each B-unit is bound to two A-units. Thus, maximum value of Θ is

$$\Theta_{max} = (1 + |\delta|)^{-1} \qquad [4.24]$$

For the symmetric block-copolymer (δ=0), Θ=1. These values describe the alternating copolymer. The general value $0 \le \Theta \le 1/2$ describes block-copolymers, $\Theta = 1/2$-random copolymer and $1/2 \le \Theta \le \Theta_{max}$ - copolymer with alternating sequences. Applying Eq 4.16, one can determine the χ_{dist} through parameters Θ and δ:

$$\chi_{dist} = \frac{1}{2}(1+\delta)[1 - \Theta(1-\delta)]^2\Delta\chi_a + \frac{1}{2}(1-\delta)[1 - \Theta(1 +$$

$$+ \delta]^2\Delta\chi_b \qquad [4.25]$$

As χ_{dist} is a quadratic function of Θ, it has an extremum at Θ_C found from the condition

$$\frac{\delta\chi_{dist}}{\delta\Theta} = 0 \qquad [4.26]$$

which comes to

$$\Theta = \frac{\Delta\chi_a + \Delta\chi_b}{(1+\delta)\Delta\chi_b + (1-\delta)\Delta\chi_a} = \frac{1}{1 + \delta(1-R)/(1+R)} \qquad [4.27]$$

$$R = \frac{\Delta\chi_a}{\Delta\chi_b} \qquad [4.28]$$

For symmetric copolymer, $\Theta_C = 1$ at all R. Either purely alternating or block-copolymer is the most stable. For asymmetric copolymer ($\delta \neq 0$), the analysis is done with assumption that component B is always chosen at condition $f_B \le f_A$ and δ is always positive. At $R > 0$ and $\delta \neq 0$, Θ_C is not in the allowed interval $[0, \Theta_{max}]$. If $R < 0$, Θ_C is written as

$$\Theta_C = \frac{1}{1 + \delta(1 + |R|)/(1 - |R|)} \qquad [4.29]$$

The extremum for χ_{dist} appears always in the region $[0, \Theta_{max}]$. This case is interesting because it predicts the existence of the miscibility windows. For this case, χ_{dist} is estimated at Θ_C:

$$\chi_{dist}(\Theta_C) = \frac{2\Theta_C \Delta\chi_a \Delta\chi_b}{\Delta\chi_a + \Delta\chi_b} \qquad [4.30]$$

If $R < 0$ and $|R| < 1$, we have the miscibility or immiscibility windows when

$$\chi_{dist}(\Theta_C) < 0; \quad \Delta\chi_a + \Delta\chi_b > 0 \qquad [4.31]$$

$$\chi_{dist}(\Theta_C) > 0; \quad \Delta\chi_a + \Delta\chi_b < 0 \qquad [4.32]$$

In the case when there is no extremum of χ_{dist} in the interval $[0,\Theta_{max}]$ (i.e., when $R > 0$) but when $|R| < 1$), χ_{dist} is a monotonous function of Θ in the interval $[0,\Theta_{max}]$. Any block-copolymer ($\Theta = 0$) or alternating copolymer ($\Theta = \Theta_{max}$) has at least more stable value of χ_{dist}. The sign of $\delta\chi_{dist}/\delta\Theta$ at $\Theta = 1/2$ may serve as a criterion of the stable distribution. The alternating copolymer is more stable at condition 4.33 and block-copolymer is more stable if Eq 4.34 is valid:

$$f_A \Delta\chi_a + f_B \Delta\chi_b > 0 \qquad [4.33]$$

$$f_A \Delta\chi_a + f_B \Delta\chi_b < 0 \qquad [4.34]$$

This criterion is valid for $R < 1$ and $|R| < 1$ because χ_{dist} has an extremum in the region $[0,\Theta_{max}]$.

In the case, where C is homopolymer of A or B (mixture of the type A/AB or B/AB), for all the possible interactions, the equal energy is given only to the triplet pairs AAA-BBB and ABA-BAB, i.e., $\chi_{AAA,BBB}$ for the first and $\chi_{BAB,ABA}$ for the second pair. Then, the term $\varphi'_A \varphi_{AB} \chi_{mixture}$ is a part of the equation for the free energy of mixing, where φ'_A is the volume fraction of homopolymer A in the mixture, and $\chi_{mixture}$ is expressed by Eq 4.17 and where

$$\chi_{dist} = \Delta\chi_b(f_{BB}^2/f_A)(1 - f_{AA}^2/f_A) + \Delta\chi_a f_{AB}^4/f_A f_B \quad [4.35]$$

$$\Delta\chi_b = \chi_{AAA,BBB} - \overline{\chi}_{AB} \quad [4.36]$$

$$\Delta\chi_a = \overline{\chi}_{AB} - \chi_{BAB,ABA} \quad [4.37]$$

Here indices a and b relate to the alternating and block-copolymer parts of the chain. Similar expressions may be obtained for the mixture AB/B by changing indices of A and B.

Because the interactions AAA-BBB and ABA-BAB are favorable (interactions of the type block-block or alternating-alternating triplet), one can expect that for the majority of the systems values $\Delta\chi_a$ and $\Delta\chi_b$ will have the same sign. The positive coefficient of $\Delta\chi_a$ changes monotonously with Θ^4 whereas the positive coefficients at $\Delta\chi_b$ complicate quadratic function of Θ and may show an extremum (or extrema) in the region [0, Θ]. Therefore, the windows of miscibility or immiscibility may occur for the mixtures AB/A (or AB /B) and depend on the sequence distribution.

The probability of diades (or, equivalently, Θ) can be determined by ^{1}H NMR or ^{13}C NMR and from estimation of reaction ability of comonomers.

4.3 THE DEPENDENCE OF THE PHASE BEHAVIOR OF COPOLYMER MIXTURES OF VARIOUS COMPOSITIONS ON THE SEQUENCE DISTRIBUTION

The model taking into account the sequence distribution was used to describe the mixtures of two copolymers differing only by composition[21] and to analyze the effect of sequence distribution and the critical compositions at which the mixtures become miscible.[22] The model was also applied to the ternary mixtures (two homopolymers-copolymer).[23] The mixture of two copolymers differing only by composition (one contains f_A of units A and f_B of units B and other - g_A of units A and g_B of units B). This is the mixture of the type A_xB_{1-x}/A_yB_{1-y} according to nomenclature.[12] Authors[21] denominate all A-B interactions by the mean value $\overline{\chi}_{AB}$ and consider only the interactions of triplet pairs AAA-BBB and BAB-ABA, i.e., $\chi_{AAA;BBB}$ and $\chi_{BAB;ABA}$. For this case, the term of interaction in the expression for the free energy of mixing is expressed as

$$\chi_{mix} = (f_A - g_A)^2\chi_{AB} \quad [4.38]$$

where χ_{mix} is given by Eq 4.17 and values of χ_{comp} and χ_{dist} correspondingly by the following expressions:

$$\chi_{comp} = \overline{\chi}_{AB}(f_A - g_A)^2 \quad [4.39]$$

$$\chi_{dist} = [(g_{AB}^2/g_B - f_{AB}^2/f_B)(g_{AB}^2/g_A - f_{AB}^2/f_A)]\Delta\chi_a +$$

$$+[(g_{BB}^2/g_B) - f_{BB}^2/f_B)(f_{AA}^2/f_A - g_{AA}^2/g_A)]\Delta\chi_b \quad [4.40]$$

where $\Delta\chi_a$ and $\Delta\chi_b$ are given by Eq 4.36 and 4.37, correspondingly. Indices a and b relate to the alternating and block sequences. It is expected that $\chi_{AAA;BBB} > \overline{\chi}_{AB} > \chi_{BAB;ABA}$ or $\chi_{AAA,BBB} < \overline{\chi}_{AB} < \chi_{BAB;ABA}$ and therefore $\Delta\chi_a$ and $\Delta\chi_b$ always have the same sign.

Using Eqs 4.16, 4.22 and 4. 23, all pair probabilities (f_{AA}, f_{BB}, etc.) can be expressed as functions of Θ and δ. Block-copolymer corresponds, as before, to $\Theta = 0$, random copolymer to $\Theta = 1/2$, and alternating copolymer to $\Theta = 1$. Authors[21] considered the case of mixture of 1:1 of alternating copolymer (δ=0, Θ=1) with 1:1 random copolymer ($\delta' = 0$, Θ'=1/2). For this system, $\chi_{comp} = 0$ and

$$\chi_{dist} = (9\Delta\chi_a - \Delta\chi_b)/64 \quad [4.41]$$

There are no possible interactions AAA-BBB between alternating and random copolymers in this case. This preference is reflected in Eq 4.41 by the coefficient at $\Delta\chi_a$ which is 9 times greater than coefficient at $\Delta\chi_b$. It becomes clear that χ_{dist} and, correspondingly, χ_{mix} may be positive showing the possible immiscibility of alternating and random copolymers of the same composition ($f_A = g_A$). Two specific cases follow from Eq 4.41. First, when $\chi_a = 0$, $\Delta\chi_b$ is positive and $\chi_{mix} < 0$ (miscible mixture). Here $\chi_{AAA;BBB} > \chi_{AB}$ and the presence of the alternating copolymer "dilutes" energetically unfavorable "block" interactions. The second case of immiscible mixture is characterized by $\Delta\chi_b = 0$ and $\Delta\chi_a > 0$.

Here $\overline{\chi}_{AB} < \chi_{BAB;AAA}$ and the presence of the random copolymer dilutes favorable interactions in the alternating triplets. For the mixture of two random copolymers (Θ=Θ'=1/2) varying only in composition ($\delta \neq \delta$), we have that $f_{AA} = f_A^2$; $f_{BB} = f_B^2$; $f_{AB} = f_A f_B$, etc. In this case

$$\chi_{dist} = (g_B g_A^2 - f_B f_A^2)(g_A g_B^2 - f_A f_B^2)\Delta\chi_a + (f_A^3 - g_A^3)(g_B^3 - f_B^3)\Delta\chi_b \quad [4.42]$$

The proposed model was used [21] to obtain the values of $\Delta\chi_a$ and $\Delta\chi_b$ for the mixtures of chlorinated PE of different composition. The critical compositions for these mixtures[24] at 423K have been used for the calculations. From the critical values of f_A, f_B, g_A and g_B there were calculated the critical f_{AA}, f_{AB}, f_{BB}, g_{AB}, g_{AA}, and g_{AB}. Unknown values of χ_{AB}, $\chi_{AAA;BBB}$, and $\chi_{BAA;ABA}$ have been found by the linear adjustment (using the least square method) of the apparent experimental values of $\chi_{crit,\ CHCl}$ using the following expressions:

$$\chi_{mixt,exp} = (f_A - g_A)^2 \chi_{CH_2,CHCl,app} = (f_A - g_A)^2 \overline{\chi}_{CH_2,CHCl} + \chi_{dist} \quad [4.43a]$$

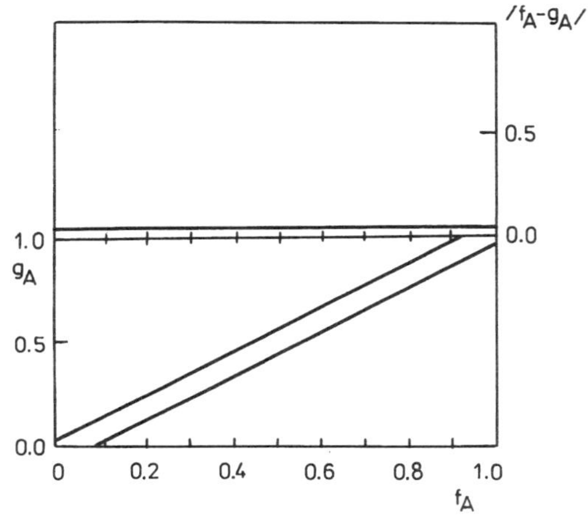

Figure 4.5. Critical composition difference f_A-g_A is plotted against f_A in the upper region of the figure. The lower region displays g_A versus f_A where the values of g_A are obtained for fixed values of the relevant χ, s, φ_1, φ_2, N, Θ_1 and Θ_2 and f_A for the 1:1 mixture (φ_1=φ_2=0.5) of random copolymers (Θ_1=Θ_2=0.5) with equal degrees of polymerization (N_1=N_2=1000) and $\bar{\chi}_{AB}$=0,3, $\Delta\chi_a$=1,1 and $\Delta\chi_b$=0. Adapted, by permission, from J. W. Hunsel, A. C. Balazs, R. Koningsveld, and W. J. MacKnight, *Macromolecules*, **21**, 1528 (1980).

$$\chi_{CH_2,CHCl,app} = \bar{\chi}_{CH_2,CHCl} + \chi_{dist}/(f_A - g_A)^2 \qquad [4.43b]$$

It was found that χ_{AB} (A = CH_2, B = CHCl) only slightly differs from zero; $\chi_{ABA,BBB} = 0.37$, $\chi_{BAA,ABA} = -4.34$. High negative values of $\chi_{BAA;ABA}$ imply that the alternating structure CH_2, CHCl is energetically less favorable as compared with statistical distribution of these sequences. Using the values of the interaction parameters, calculated values of $\chi_{CH2;\ CCl}$ were in agreement with experiment. The values of χ were used in Eq 4.41 to obtain $\chi_{mixture}$ for the mixtures of PVC with CPE of the equivalent composition (f_A=g_A=0.5). The value of $\chi_{mixture}$=0.60 shows that the mixture is immiscible as established experimentally[24] even when PVC and CPE have the same amount of Cl. From this point of view CPE may be considered as random copolymer formed by CH_2 and CHCl sequences, whereas PVC is an alternating copolymer, formed by the same entities.

The following equation was derived [22] for the critical difference of composition for two copolymers of the type $A_xB_{1-x}/A_y\ B_{1-y}$:

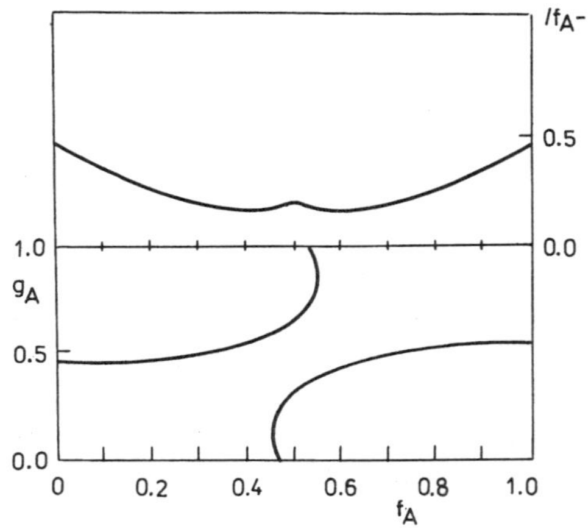

Figure 4.6. The same as in Figure 4.5, but with $\bar{\chi}_{AB}$=-0,3, $\Delta\chi_a$=0,5 and $\Delta\chi_b$=0,7. Adapted, by permission, from J. W. Hunsel, A. C. Balazs, R. Koningsveld, W. J. MacKnight, *Macromolecules*, **21**, 1528 (1980).

$$\Delta = (f_A - g_A)^2 = \frac{1}{2\bar{\chi}_{AB}}[\frac{1}{N_1\varphi_1} + \frac{1}{N_2\varphi_2} - 32\Delta\chi_a(\Theta_2^2 g_A^2 g_B^2 - \Theta_1^2 f_A^2 f_B^2) -$$

$$- 32\Delta\chi_b(\Theta_2^2 g_B^3 - \Theta_1^2 f_B^3)(\Theta_1^2 f_A^3 - \Theta_2^2 g_A^3)] \qquad [4.44]$$

The solution of this equation for the corresponding values of φ_1, N_1, φ_2, N_2, Θ_1, Θ_2, $\bar{\chi}_{AB}$, $\Delta\chi_a$, and $\Delta\chi_b$ allows to find one or more critical values of g_A for any value of f_A. At these g_A, phase separation occurs. The plot $|f_A - g_A| = \Delta$ vs. f_A or g_A may be constructed.

For the mixture of random copolymers ($\Theta_1 = \Theta_2 = 1/2$), Eq 4.44 was solved for various χ. The results for the 1:1 mixtures of copolymers with $N_1 = N_2 = 1{,}000$ are presented in Figs. 4.5 and 4.6. Both pictures are symmetric because χ_{dist} is invariant to changing f_A to g_A. Fig. 4.5 shows the normal behavior predicted by the Scott theory:

$$\Delta = \frac{1}{2\bar{\chi}_{AB}}\left(\frac{1}{N_1\varphi_1} + \frac{1}{N_2\varphi_2}\right)^{1/2} \qquad [4.45]$$

Fig. 4.6 shows very different behavior. Here Δ is neither constant, nor linear function of composition of one of the copolymers. Δ passes through the minimal values. These data show a very important role of the sequence distribution in the miscibility of two copolymers.

4.4 THE EFFECT OF THE SEQUENCE DISTRIBUTION IN COPOLYMER AND ITS COMPATIBILIZING EFFECT IN MIXTURE OF TWO IMMISCIBLE POLYMERS

The effect of the sequence distribution on the miscibility was studied for the system A/B/AB.[23] Copolymer AB was considered as compatibilizer. In its presence, there is a possibility the dilution of unfavorable interactions between two homopolymers and formation of one homogeneous phase. The enthalpy parameter of interaction in this case is expressed as

$$\chi_{mix} = \chi_{AB}(\varphi_1\varphi_2 + f^2\varphi_2\varphi_3 + (1-f)^2\varphi_1\varphi_3) \qquad [4.46]$$

where f and (1 – f) denote the copolymer composition $(A_fB_{1-f})_n$ with the volume fraction φ_3. For this system the model gives the following expressions for the energy of interaction without combinatorial terms:

$$\frac{\Delta G}{RT} = \varphi_1\varphi_2\chi_{AB} + \varphi_3\overline{\chi}_{AB}(\varphi_1 f_B^2 + \varphi_2 f_A^2) + \varphi_1\varphi_3[\Delta\chi_b \frac{f_{BB}^2}{f_B}\left(1 - \frac{f_{AA}^2}{f_A}\right) +$$

$$+ \Delta\chi_a \frac{f_{AB}^4}{f_A f_B}] + \varphi_2\varphi_3[\Delta\chi_b \frac{f_{AA}^2}{f_A}\left(1 - \frac{f_{BB}^2}{f_B}\right) + \Delta\chi_a \frac{f_{AB}^4}{f_A f_B}] \qquad [4.47]$$

The first two items contain terms depending only on composition whereas the both terms in the third item depend on the sequence distribution. Introducing the order parameter, Θ, (Eqs 4.22 and 4.23) and using Eq 4.16, two last terms of Eq 4.47, it may be expressed as the dependence on the sequence distribution in terms of Θ and δ:

$$\varphi_1\varphi_3\Delta\chi_b(1/2)(1-\delta)[1 - \Theta(1-\delta)]^2[1 - (1/2)(1+\delta)[1 - \Theta(1-\delta)]^2 +$$

$$+ \Delta\chi_a \frac{\Theta^4}{4}(1+\delta)^3(1-\delta)^3 + \varphi_2\varphi_3\Delta\chi_b(1/2)(1+\delta)[1 - \Theta(1-\delta)]^2[1 -$$

$$- (1/2)(1-\delta)[1 - \Theta(1+\delta)]^2] + \Delta\chi_a \frac{\Theta^4}{4}(1+\delta)^3(1-\delta)^3] \qquad [4.48]$$

Authors[23] have considered the case when $f_A = f_B = 0.5$ and $\delta = 0$. The total interaction energy is given as

$$\frac{\Delta G}{RT} = \varphi_1\varphi_2\chi_{AB} + \varphi_3(1-\varphi_3)\chi_{mixture} \qquad [4.49]$$

where $\chi_{mixture}$ is determined by Eq 4.17 and

$$\chi_{comp} = 0.25\overline{\chi}_{AB} \qquad [4.\ 50]$$

$$\chi_{dist} = [\Delta\chi_b(1/2)(1-\Theta)^2[1-(1/2)(1-\Theta)^2] + \Delta\chi_a\Theta^4/4] \qquad [4.\ 51]$$

The first term in Eq 4.49 relates to the homopolymer-homopolymer energy of interaction whereas $\chi_{mixture}$ describes the homopolymer-copolymer interaction. From Eq 4.51, it is seen that if $\Delta\chi_a$ and $\Delta\chi_b$ are positive, χ_{dist} is positive as well. If $\Delta\chi_a$ and $\Delta\chi_b$ are negative, χ_{dist} is negative. The extremum of χ_{dist} occurs when $\delta\chi_{dist}/\delta\Theta = 0$:

$$\frac{\delta\chi_{dist}}{\delta\Theta} = \Delta\chi_b\Theta(1-\Theta)(\Theta-2) + \Delta\chi_a\Theta^3 = 0 \qquad [4.52]$$

For $f_A = f_B = 0.5$, $\Theta = 0$ (block-copolymer), there is always an extremum. Another extremum is equal to

$$\Theta_C = \frac{-3 \pm (1+8R)^{1/2}}{2(R-1)} \qquad [4.53]$$

where R is given by Eq 4.28. If $R \geq 0$, the extremum of χ_{dist} appears in the interval [0,1], corresponding to $\Theta_c = [-3 + (1+8R)^{1/2}]/[2(r-1)]$. The estimation of the sign of $\delta\chi_{dist}/\delta\Theta$ at $\Theta = 0$ gives the condition of the second extremum:

$$\delta^2\chi_{dist}/\delta\Theta^2 = -2\Delta\chi_b; \quad \Theta_C = 0 \qquad [4.54]$$

Thus, if $\Delta\chi_b$ is positive, $\Theta = 0$ is the local maximum and other Θ_c is the local minimum. If $\Delta\chi_b$, $\Theta_c = 0$ corresponds to the local minimum and another Θ_c is the local maximum. Two cases may be considered. If $R = 0$, Eq 4.59 leads to $\Theta = 1$. Thus, one extremum corresponds to the block-copolymer and other to the alternating copolymer. If $R = 1$ ($\Delta\chi_a = \Delta\chi_b$), Eq 4.52 can be used to calculate the second Θ_c. If $-(1/8)<R<0$, there is no extremum in the interval [0,1] and then χ_{dist} is the monotonous function of Θ in this interval. Therefore, if $\Delta\chi_b$ is positive, value χ_{dist} diminishes from the maximum value at $\Theta_c = 0$ and increases from the minimum at $\Theta_c = 1$ at negative $\Delta\chi_b$ (see Fig. 4.7). For such a system, the microphase separation has been analyzed.[23] For the ternary system with $f_A = f_B = 0.5$, the equation for free energy is written as

$$\Delta G/RT = (\varphi_1/N_1)\ln\varphi_1 + (\varphi_2/N_2)\ln\varphi_2 + (\varphi_3/N_3)\ln\varphi_3 + \varphi_1\varphi_2\chi_{AB} +$$

$$+ \varphi_3(1-\varphi_3)\chi_{mixture} \qquad [4.55]$$

The stability condition is determined by Eqs 3.176 and 3.177. Inserting 4.55 into 3.176 and 3.177, one can find the value of χ corresponding to the formation of a stable, miscible system. Generally three conditions are obtained:

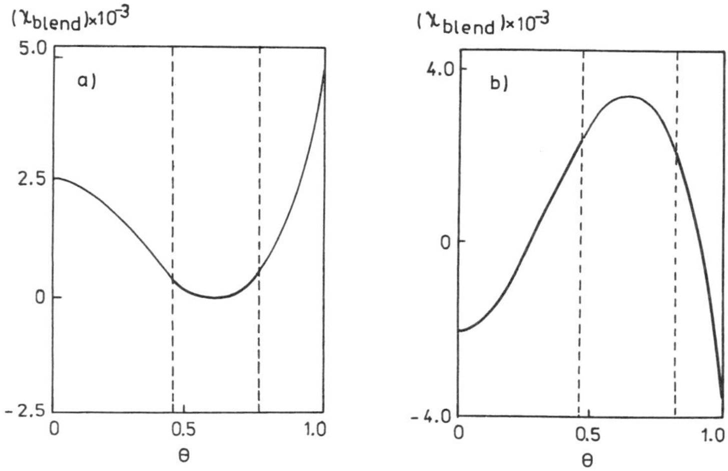

Figure 4.7. χ_{blend} vs. Θ for R>0, $\Delta\chi_a$=0.03, $\Delta\chi_b$=0.00004 and $\bar{\chi}_{AB}$=-0.01, $\varphi_1=\varphi_2$=0.2 and φ_3=0.6, N=5000 (a) and for R>0, $\Delta\chi_a$=-0.045, $\Delta\chi_b$=-0.039, $\bar{\chi}_{AB}$=0.0001, $\bar{\chi}_{AB}$=0.0302, $\varphi_1=\varphi_2$ =0.2, φ_3 =0.6, N=1000, (b) $\chi_{AAA,BBB} > \bar{\chi}_{AB}$. The section of curve drawn in the thick line indicates where polymers are miscible and form the single, homogeneous phase. Adapted, by permission, from A. C. Balazs, and M. T. De Meuse, *Macromolecules,* **22**, 4260 (1989).

$$\chi_{mixture} < \frac{1}{4}(Q_1 + Q_2 + 2Q_3) \qquad [4.56]$$

$$\chi_{AB} < \frac{1}{2}(Q_1 + Q_2) \qquad [4.57]$$

$$\chi_{AB}^2 + 2\chi_{AB}(Q_3 - 2\chi_{mixture}) + 2\chi_{mixture}(Q_1 + Q_2) - Q < 0 \qquad [4.58]$$

where $Q_i = 1/N_i\varphi_i$, $Q = Q_1Q_2 + Q_2Q_3 + Q_1Q_3$ and $\chi_{mixture}$ is given by Eq 4.17, 4.50, and 4.51. Eq 4.58 can be rewritten as

$$(2\chi_{mixture} - Q_3) - [(Q_1 + Q_3 - 2\chi_{mixture})(Q_2 + Q_3 - 2\chi_{mixture})]^{1/2} <$$

$$< \chi_{AB} < (2\chi_{mixture} - Q_3 + [(Q_1 + Q_3 - 2\chi_{mixture})\times$$

$$\times (Q_2 + Q_3 - 2\chi_{mixture})]^{1/2} \qquad [4.59]$$

At $N_1 = N_2 = N_3 = N$, these three conditions come to

$$N\chi_{mixture} < \frac{1}{4}\left(\frac{1}{\varphi_1} + \frac{1}{\varphi_2} + \frac{2}{\varphi_3}\right) \qquad [4.60]$$

$$N\chi_{AB} < \frac{1}{2}\left(\frac{1}{\varphi_1} + \frac{1}{\varphi_2}\right) \qquad [4.61]$$

$$\left(2N\chi_{mixture} - \frac{1}{\varphi_3}\right) - \left[\left(\frac{1}{\varphi_1} + \frac{1}{\varphi_3} - 2N\chi_{mixture}\right)\left(\frac{1}{\varphi_2} + \frac{1}{\varphi_3} - 2N\chi_{mixture}\right)\right]^{1/2} <$$

$$< N\chi_{AB} < \left(2N\chi_{mixture} - \frac{1}{\varphi_3}\right) +$$

$$+ \left[\left(\frac{1}{\varphi_1} + \frac{1}{\varphi_2} - 2N\chi_{mixture}\right)\left(\frac{1}{\varphi_1} + \frac{1}{\varphi_3} - 2N\chi_{mixture}\right)\right]^{1/2} \qquad [4.62]$$

One-phase mixture is formed when all three conditions are kept. Finally, in the case $N_1 = N_2 = N_3 = N$ and $\varphi_1 = \varphi_2 \neq \varphi_3$, the conditions come to:

$$4N\chi_{mixture} - \frac{2}{\varphi_3(1 - \varphi_3)} < N\chi_{AB} < \frac{2}{1 - \varphi_3} \qquad [4.63]$$

$$N\chi_{mixture} < \frac{\varphi_3 + 1}{2\varphi_3(1 - \varphi_3)} \qquad [4.64]$$

In the simplest case, $N_1 = N_2 = N_3 = N$ and $\varphi_1 = \varphi_2 = \varphi_3 = 1/3$, we have

$$(4N\chi_{mixture} - 9) < N\chi_{AB} < 3 \qquad [4.65]$$

$$4N\chi_{mixture} - 9 < 3; \quad N\chi_{mixture} < 3 \qquad [4.66]$$

Using equations 4.56-4.64, the ternary phase diagrams have been calculated[23] for the case when homopolymers are immiscible ($\chi_{AB} > 0$), i.e., when the copolymer plays a role of compatibilizer. The results show that mixtures A/B/AB may be miscible if parameter Θ varies. The part of the curve in Fig. 4.7 (thick line) shows the region where at least one condition is satisfied. Thin lines show the region where at least one condition is violated and the system is immiscible. For example, if $R > 0$, $\Delta\chi_a$ and $\Delta\chi_b$ have the same signs. If both are positive, then $\chi_{BAA;ABA} < \chi_{AB} < \chi_{ABA,BBB}$. The relatively high value of $\chi_{ABA,BBB}$ implies the repulsion between blocks of copolymer and homopolymer. Thus block copolymer does not promote miscibility in the system copolymer-homopolymer. Relatively low values of the same parameter imply the attraction between segments of copolymer. Adding the alternating copolymer promotes copolymer-copolymer interactions and favors copolymer-homopolymer interactions. Thus, random copolymer with $\Theta = 1/2$ is a better compatibilizer. This is seen from Fig. 4.7a. Here the concave shape of dependence of $\chi_{mixture}$ shows the

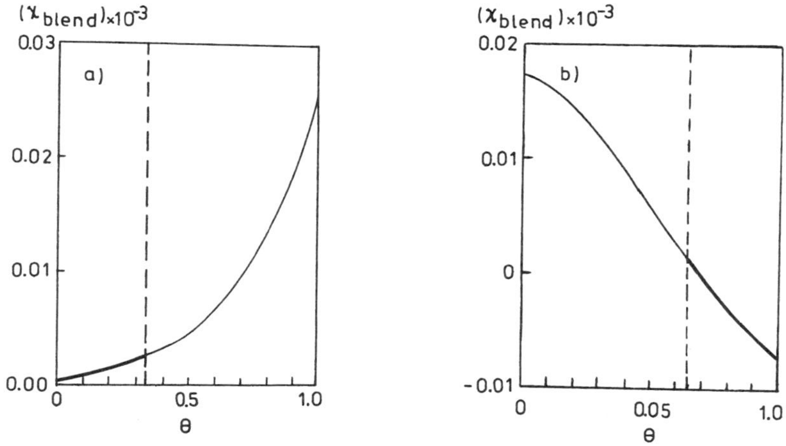

Figure 4.8. χ_{blend} vs. Θ for -1/8<R<0, $\Delta\chi_a$=0.08, $\Delta\chi_b$=0.001, $\overline{\chi}_{AB}$=0.022, φ_1=φ_2=0.25,φ_3 = 0.5, N=1000 (a), and for 1/8<R<0 $\Delta\chi_a$=-0.02, $\Delta\chi_b$=0.0001, $\overline{\chi}_{AB}$=-0.01, φ_1 =φ_2 =0.25, N=1000 (b); $\chi_{AAA,BBB}$ < $\overline{\chi}_{AB}$. The section of curve drawn in the thick line indicates where the polymers are miscible and form single, homogeneous phase. Adapted, by permission, from C. Balazs, and M. T. De Meuse, *Macromolecules*, **22**, 4260 (1989).

miscibility window: for Θ values close to 1/2, all the necessary conditions are fulfilled and the ternary mixture forms homogeneous system.

If both $\Delta\chi_a$ and $\Delta\chi_b$ are negative, $\chi_{ABA;BBB} < \chi_{AB} < \chi_{BAA;ABA}$. Relatively low value of $\chi_{ABA;BAA}$ implies the attraction between block-copolymer and homopolymer. In such a way, block-copolymer acts as a good compatibilizer. Simultaneously, high value of $\chi_{BAA;ABA}$ implies the repulsion between copolymers and promotes the miscibility between copolymer and homopolymer. Therefore, the alternating copolymer is a good compatibilizer as seen from Fig. 4.7b. The convex shape of $\chi_{mixture}$ shows immiscibility windows: for the region of intermediate values of Θ at least one of the specific conditions is not valid and polymers demix.

Fig. 4.8 shows the situation when $-1/8 \leq R \leq 0$ and $\chi_{mixture}$ is a monotonous function of Θ in the interval [0,1]. Fig. 4.8a presents the case $\Delta\chi_b < 0$, therefore $\Theta_C = 0$ is a local minimum and $\chi_{ABA,BBB} < \chi_{AB}$. Thus, copolymer for which1/2 $\leq \Theta \leq$ 1 is a better compatibilizer. With increasing the degree of polymerization, N, the entropy of mixing diminishes and lesser values of interaction parameters are necessary to obtain miscibility (see Eqs 4.60-4.66). In the limit $N \to \infty$, these equations show that both $\Delta\chi_{mixture}$ and χ_{AB} should be less than 0. Thus for the system A/B/AB, with high molecular masses, the condition of miscibility is the miscibility of both A and B homopolymers and homopolymer-copolymer.

The analysis of Eqs 4.63 and 4.64 allows one to estimate the effect of φ_3 on the miscibility in the A/B/AB system. The right side of Eq 4.63, $2/(1 - \varphi_3)$, monotonously increases with increasing φ_3. However, for the left side function $\{4N\chi_{mixture} - 2/[\varphi_3(1 - \varphi_3)]\}$ is a concave parabola with maximum at $\varphi_3 = 0.5$. Thus for $\varphi_3 < 0.5$, the increase in φ_3 always meets the first inequality but does not meet the left inequality. In other words, for fixed values $N\chi_{mixture}$ and $N\chi_{AB}$ and for $\varphi_3<0.5$, the increase of φ_3 (decrease of φ_1 and φ_2) does not lead inevitably to miscibility. However, if the system is miscible at $\varphi_3 = 0.5$, the mixtures stay miscible at increasing values of φ_3 (and diminishing φ_1 and φ_2). This equation shows also that because $N\chi_{AB}$ increases (due to increasing either N or immiscibility of A and B homopolymers), the higher amount of copolymer is necessary to obtain the miscible mixture.

The phase diagrams of the ternary mixtures A/B/AB show[23] that miscibility is extremely sensitive to the value Θ, i.e., to the distribution of the sequences in copolymer. From the calculated data, it follows that block-copolymer not always acts as a good compatibilizer for such systems. For the mixtures AB/C and AB/AB, the effect of the sequence distribution is determined by the additional terms $\Delta\chi_a$ and $\Delta\chi_b$ introduced in parameter, $\chi_{mixture}$ (Eqs 4.20 and 4.21).[20,21] Cantow and Schulz[25] have made an alternative approximation

$$\chi_{AAA,C} - \chi_{AAB,C} = \chi_{AAB,C} - \chi_{BAB,C} \qquad [4.67a]$$

$$\chi_{BBB,C} - \chi_{BBA,C} = \chi_{BBA,C} - \chi_{ABA,C} \qquad [4.67b]$$

to determine $\Delta\chi_a$ and $\Delta\chi_b$ as equal to these differences. Using Eq 4.67 for the mixtures AB/C, the expression for $\chi_{mixture}$ was found[26] in the form:

$$\chi_{mixture} = f_A\chi_{AAA,C} + f_A\chi_{BBA,C} - f_A f_B \overline{\chi}_{AB} - 4(\Delta\chi_a + \Delta\chi_b)\Theta f_A f_B \qquad [4.68]$$

At a fixed copolymer composition, the parameter $\chi_{mixture}$ is a linear function of Θ and then

$$\frac{\delta\chi_{mixture}}{\delta\Theta} = -4(\Delta\chi_a + \Delta\chi_b) f_A f_B \qquad [4.69]$$

being dependent on the sign of $\Delta\chi_a + \Delta\chi_b$. Any more blocks or more alternating structure favors mixing. From Eq 4.67 is seen, for example, that $\Delta\chi_a + \Delta\chi_b < 0$ shows that $\chi_{AAA,C} + \chi_{BBA,C} < \chi_{AAB,C} + \chi_{BBA,C} <$

Table 4.2: Triads in copolymers

Index no.	Triad	Index no.	Triad
1	BAB	4	ABA
2	ABB, BAA	5	ABB, BAA
3	AAA	6	BBB

$\chi_{BAB,C} + \chi_{ABA,C}$, i.e., the interaction between block triades and C units is lower, compared with alternating triads. In this case, parameter $\chi_{mixture}$ increases with Θ and block sequence ($\Theta = 0$) favors mixing. Because $\chi_{mixture}$ is not a quadratic function of Θ, it has no extremum and the miscibility window cannot be predicted.

For the mixtures AB/AB differing only in composition there are 6 triads and it is necessary to have 15 interaction parameters (see Table 4.2). If to accept that a and b correspond to the index number of triad sequences, given in Table 4.2, then 6 of 15 parameters may be taken equal zero, because the interacting pairs have identical central numbers: $\chi_{12} = \chi_{13} = \chi_{23} = \chi_{45} = \chi_{46} = \chi_{56} = 0$.

Using the approximation similar to Eq 4.67, one can obtain six equations, five of which are independent. From nine non-zero interaction parameters only four stay unknown:

$$\Delta_1 = \chi_{36} - \chi_{26} = \chi_{26} - \chi_{16};\ \Delta_2 = \chi_{35} - \chi_{25} = \chi_{25} - \chi_{15};$$

$$\Delta_3 = \chi_{34} - \chi_{24} = \chi_{23} - \chi_{14};\ \Delta_4 = \chi_{36} - \chi_{35} = \chi_{35} - \chi_{34};$$

$$\Delta_5 = \chi_{26} - \chi_{25} = \chi_{25} - \chi_{24};\ \Delta_6 = \chi_{16} - \chi_{15} = \chi_{15} - \chi_{14} \qquad [4.70]$$

It is seen that

$$\Delta_6 - \Delta_5 = \Delta_5 - \Delta_4 = \Delta_3 - \Delta_2 = \Delta_2 - \Delta_1 \qquad [4.71]$$

These relations have been used[26] for the mixtures AB/AB. The values φ'_{AB}, φ'_i, g'_i, and g'_{ij} (i,j = A,B) have been used, as earlier,[21] to describe the second copolymer where φ' and g' are the same values as φ and f for the first copolymer. The relations between these variables are the same as in Eq 4.16 but $\varphi_{AB} + \varphi_C = 1$ is changed to $\varphi_{AB} + \varphi'_{AB} = 1$. For the parameter $\chi_{mixture}$, the following equation has been obtained:

$$\chi_{mixture} = \chi_{25}(f_A - g_A)^2 + [\Delta_2(F_A - G_A) - \Delta_5(F_B - G_B)](f_A - g_A) +$$

$$+ (\Delta_6 - \Delta_5)(F_A - G_A)(F_B - G_B) \qquad [4.72]$$

where $F_A = f_{AA} - f_{AB}$, $F_B = f_{BB} - f_{AB}$, $G_A = g_{AA} - g_{AB}$, $G_B = g_{BB} - g_{AB}$. If both copolymers in the mixture have the same composition ($f_A = g_A$), Eq 4.72 comes to

$$\chi_{mixture} = 4(\Delta_6 - \Delta_5)(f_{AB} - g_{AB})^2 \qquad [4.73]$$

The miscibility is determined only by the sign of $\Delta_6 - \Delta_5$. From Eq 4.70 one can see that $\Delta_6 - \Delta_5 > 0$ leads to

$$\chi_{16} + \chi_{34} > \chi_{14} + \chi_{36}; \quad \chi_{15} + \chi_{24} > \chi_{14} + \chi_{25}; \quad \chi_{26} + \chi_{35} > \chi_{25} + \chi_{36}$$

In this case, the mixture can be immiscible. Parameters χ_{14}, χ_{25}, and χ_{36} represent the interactions between BAB and ABA, AAB and BBA, and AAA and BBB triads, correspondingly. Other parameters are cross-interactions between various types of triads. If both copolymers have the same Θ (i.e., the same microstructure), from condition $\delta\chi_{mixture}/\delta\Theta = 0$ at fixed f_A and g_A, one can get:

$$\Theta_C = \frac{(f_A - g_A)(\Delta_2 - \Delta_1)}{8(\Delta_6 - \Delta_5)(f_A f_B - g_A g_B)} \qquad [4.74]$$

where $\chi_{mixture}$ has an extremum at Θ_C. From condition

$$\frac{\delta^2\chi_{mixture}}{\delta\Theta^2} = 32(\Delta_6 - \Delta_5)\,(f_A f_B - g_A g_B)^2 \qquad [4.75]$$

it is seen that when $\Delta_6 > \Delta_5$, $\chi_{mixture}$ has a minimum. The system may have a miscibility window for some sequence distribution at the fixed compositions of two copolymers. In the opposite case, $\Delta_6 < \Delta_5$, any block or alternating copolymer mixtures should be more miscible. If both copolymers are random, $\Theta = 0.5$ and Eq 4.72 comes to

$$\chi_{mixture} = (f_A - g_A)^2[(\chi_{25} - \Delta_2 + 3\Delta_6) + (2\Delta_2 + 6\Delta_5 - 8\Delta_6)(f_A + g_A) +$$

$$+ 4(\Delta_6 - \Delta_5(f_A + g_A)^2] \qquad [4.76]$$

If $f_A = 1$, $g_A = 0$ or $f_A = 0$, $g_A = 1$ (i.e., for A/B homopolymer mixture), the $\chi_{mixture}$ (Eq 4.76) comes to χ_{36}, i.e., to χ_{AB}.

There has been conducted a comparative analysis[26] of the applicability of Eq 4.72 according to the Balazs[20,21] theory and Eq 4.76. The phase diagrams of the mixtures of chlorinated PVC, poly(methyl methacrylate-*co*-ethyl methacrylate) and mixtures of chlorinated PE show that the miscibility boundaries are better described by Eq 4.76.

Presently, there have been investigated many binary mixtures of homopolymers with copolymers and copolymer mixtures and the

Table 4.3: The shape of compatibility boundaries for some mixtures of copolymers

Component 1	Component 2	Shape of boundaries	Refs.
SAN	SMA	Two straights	18, 28-31
	SMMA		23
	SPhMI		18, 32
SMMA	SMA		100
CPE	ENRU		33
	TMGTPh-co-BDTPh		41
SPPO	S-co-pFS		34
SAN	SMA	Two parabolaes	36
SMMA	SAAc		35
	S-co-oFS		34
SAN	pMSAN	Ellipse	38
	MMA-co-tert-BMA		40
	MMA-co-CHMA		40
	MMA-co-PhMA		40
	BAN		16, 47
CPE	BAN		16
CPVC (1)	CPVC (2)		26
	EVA		39
a-MSAN	MAN		48

miscibility windows of different types have been discovered (Table 4.3). The regions of copolymer compositions have been established for copolymers which are miscible with homopolymers (Table 4.4); segmental interactions were also calculated (Table 4.5).

4.4.1 MIXTURES OF HOMOPOLYMERS WITH HOMOLOGUE SERIES OF POLYMERS OF ANOTHER CHEMICAL NATURE

Besides mixtures of homopolymer with copolymers and copolymer mixtures, the miscibility windows have been observed also for the mixtures of homopolymers with homologue series of polymers of various chemical nature.[72,93,94] In particular, miscibility windows

Table 4.4: Compatibility range in copolymer mixtures

First component	Second component	Compatibility range, wt%	Refs.
PMMA	SAN	10-38% AN	50, 51
	pMSAN	11-32% AN	52
	SMAN	8-63% MAN	49
PCMMA	SAN	12-37% AN	53
	pMSAN	13-42% AN	54
PTHP2MMA	SAN	0-30% AN	55
	pMSAN	5-29% AN	55
PEMA	pMSAN	12-32% AN	52
P2CEMA	SAN	12.5-43% AN	56
	pMSAN	18.3-46% AN	56
P2BrEMA	SAN	10-58% AN	57
	pMSAN	14-61% AN	57
PPMA	SAN	2-25% AN	58
	pMSAN	8-29% AN	52
PMBA	SVPh	0-97% Ph	59
PHMA	SVPh	32-94% Ph	59
PCHMA	SAN	0-20% AN	60
	pMSAN	0-22% AN	60
PACMA	SAN	33-58% AN	61
	pMSAN	43-61% AN	61
PPhMA	SAN	11.5-32% AN	27
PIP	cis-BD-co-1,2-vinyl BD	>32.3% vinyl	79
PCL	SAN	8-28% AN	66, 67
PPO	S-co-oCIS	-	12
	S-co-pCIS	>34.7% pCIS	74
	S-co-oFS	-	12
	S-co-pFS	-	12
	S-co-oBrS	-	12
	S-co-pBrS	-	12
	S4CIS	0-67% (mol) 4CIS	71
	oCIS-co-pCIS	23-64% pCIS	72
PPO	oFS-co-pCIS	15-17% pCIS	2, 73
	oFS-co-pFS	10-38% pFS	73
	oFS-co-pBrS	11-73% pBrS	86
	pMSAN	0-5% AN	75
PS	SDMS	60-90% S	68
P4HS	n-BA-co-tert-BA	>64% n-BA; 28-66% HS	69
PVME	SAAc	0-3% AAc	62

First component	Second component	Compatibility range, wt%	Refs.
	SAN	0-11% AN	62
	SMA	0-15% MA	62
PVP	MMA-co-2HEMA	>1.9% HEMA	70
	BMA-co-2HEMA	>17.4% HEMA	70
PVC	BDAN	23-45% AN	64
	SAN	11.5-26% AN	63
CPVC	EVA	35-37% E	39
PVF	SMMA	0-30% (vol) MA	65
TMPC	SAN	0-13% AN 0-20% AN	76 89
	SMA	0-8% MA	76
	SMMA	0-40% MMA	89
	OSAA	0-19.1% AA	76
	STHFMA	<50% THFMA	84
	pMSTHFMA	<53% THFMA	84
SAN	MMA-co-EMA	in SAN 22% AN; in MMA-co-EMA 60 and 30% MMA	78
	MMA-co-BMA	in SNA 22% AN; in MMA-co-BMA 70% BMA	78
	PTHFMA	9-48% AN in SAN	83
	MGI-co-MMA (0-6% acid and 0-1% anhydride groups)	17-30% AN	87
	SMMA (18% MMA)	8.5-30% AN	100
pMSAN	MMA-co-EMA	in pMSAN 22% AN; in MMA-co-EMA 60 and 30% MMA	78
	MMA-co-BMA	in pMSAN 22% AN; in MMA-co-BMA 70% BMA	78
	PTHFMA	12-48% AN in SAN	83

were found for the mixtures of hydroxy ethyl ether of bisphenol A with aliphatic polyethers,[93] chlorinated PVC with polyesters.[94] The miscibility windows in such systems are determined by the ratio CH_2/COO groups in homologue series of the second polymer. To explain such behavior Paul and Barlow[13] have used the same model of binary interaction, considering the homologue series as copolymers. In particular, polyesters may be presented as copolymer series containing CH_x (2) and COO (3) units. The volume fractions of each of these units can be computed from the increments which each adds to the molar volume.[95]

Table 4.5: The segment-segment interaction parameters

Component 1	Component 2	Segment pair	χ_{ij}	Refs.
PMMA	SAN	S/AN	0.1	50
		MMA/S	0.03	52
	pMSAN	MMA/pMS	0.037	50
		MMA/AN	0.56	50
PEMA	SAN	EMA/S	0.016	52
	pMSAN	EMA/pMS	0.036	52
		EMA/AN	0.56	52
PCMMA	SAN	CMMA/S	0.037	56
		CMMA/AN	0.48	56
	pMSAN	CMMA/pMS	0.049	56
P2CEMA	SAN	CEMA/S	0.045	56
		CEMA/AN	0.43	56
	pMSAN	CEMA/pMS	0.071	56
P2BrEMA	SAN	BrEMA/S	0.052	57
		BrEMA/AN	0.34	57
	pMSAN	BrEMA/pMS	0.078	57
PPMA	pMSAN	PMA/pMS	0.018	52
		PMA/AN	0.64	52
	SAN	PMA/S	0.013	52
PBMA	pMSAN	BMA/pMS	0.019	52
		BMA/AN	0.70	52
PCHMA	SAN	CHMA/AN	0.8	60, 69
		CHMA/S	-0.02	60, 69
	pMSAN	CHMA/pMS	-0.007	60
PACMA	SAN	ACMA/AN	0.25	61
		ACMA/S	0.15	61
	pMSAN	ASMA/pMS	0.22	61
PTHP2MA	SAN	THPMA/AN	0.64	55
		THPMA/S	-0.011	55
	pMSAN	THPMA/pMS	0.019	55
PTHFMA	SAN	THFMA/S	0.036	83
		THFMA/AN	0.42	83
PCL	SAN	CL/AN	0.049	77
		S/AN	0.12	77
PPO	S-co-oCIS	PO/S	-0.1	11, 12
		S/oCIS	0.005	11, 12
PPO	S-co-pCIS	S/pCIS	0.09	11, 12
PPO	S-co-pFS	S/pFS	0.08	11, 12
	S-co-oFS	S/oFS	0.09	11, 12
	S-co-pBrS	PO/pBrS	0.06	11, 12
	S-co-oBrS	PO/oBrS	0.13	11, 12

Component 1	Component 2	Segment pair	χ_{ij}	Refs.
	oCIS-co-pCIS	PO/oCIS	0.04	11, 12
		PO/pCIS	0.045	11, 12
		oCIS/pCIS	0.16	11, 12
	oFS-co-pFS	PO/pFS	0.09	11, 12
		PO/oFS	0.012	11, 12
		oFS/pFS	0.17	11, 12
	a-MSAN	PO/S	-0.1	81
		S/a-MS	0.032	81
		PO/a-MS	0.116	81
	pMSAN	PO/AN	0.96	75
PS, PPO and their derivatives		S/PO	-0.956	11
		BrS/BrPO	-0.0563	11
		PO/BrPO	0.0156	11
		S/BrPO	0.0395	11
		S/BrS	0.20	11
		PO/BrS	0.2215	11
PPhA	SAN	PhA/S	0.581	88
		PhA/AN	0.041	88
PVC	EVA	VA/E	16.27*	80
		VA/VC	4.3*	80
		E/VC	2.4*	80
CPVC	BDAN	CHCICHCI/CH2CHCI	0.042	16
		CHCICHCI/AN	0.144	16
		CHCICHCI/BD	0.027	16
		CH2CHCI/AN	0.085	16
		CH2CHCI/BD	0.019	16
		AN/BD	0.186	16
	EVA	CHCICHCI/CH2CHOCOCH3	0.173	39
		CH2CHCI/CH2CHOCOCH3	0.267	39
		CH2CH2/CH2CHOCOCH3	0.01	39
		CHCICHCI/CH2CHCI	0.042	39
		CH2CHCI/CH2CH2	0.155	39
		CHCICHCI/CH2CH2	0.1	39
		(CHCICCI2,CH2CCI2)/ CH2CHOCOCH3	0.65	39
CPE	BDAN	CHCI/CH2	0.092	16
CPE		CHCI/AN	0.113	16
		CHCI/BD	0.03	16
		CH2/AN	0.123	16
		CH2/BD	0.040	16
		AN/BD	0.186	16
TMPC	STHFMA	TMPC/THFMA	0.024	84
		TMPC/S	0.001	84

Component 1	Component 2	Segment pair	χij	Refs.
		TMPC/AN	0.78	84
	pMSTHFMA	TMPC/pMS	0.01	84
	SAN	S/AN	7.25	89
	pMSAN	S/MMA	0.19	89
ENRu	CPP	NRu/C(CH2)CI	0.035	33
		ENRu/C(CH2)CI	0.009	33
		CH2/C(CH2)CI	0.008	33
	CPE and PVC	NRu/ENRu	0.078	33
		NRu/CH2	0.006	33
		NRu/CHCI	0.071	33
		ENRu/CH2	0.126	33
		ENRu/CHCI	0.000	33
		CH2/CHCI	0.118	33
	PVDCI	NRu/CCI2	0.047	33
		ENRu/CCI2	0.004	33
		CH2/CCI2	0.086	33
SAN	BAN	S/B	2.6*	47
		B/AN	34.7*	47
	MMA-co-MA	MMA/S	0.01	36
		MMA/AN	0.05	36
		MA/AN	0.03	36
		MMA/MA	0.20	36
		S/AN	0.12	36
		MA/S	0.34	36
	MMAN and SMMA	S/AN	0.83	82
		MMA/AN	0.46	82
		S/MMA	0.03	82
	MMA-co-CHMA	S/AN	6.7*	40
		S/MMA	0.18*	40
		AN/MMA	4.1*	40
		CHMA/AN	5.9*	40
		CHNA/S	-0.13*	40
SAN	MMA-co-CHMA	CHMA/MMA	0.8*	40
	MMA-co-PhMA	PhMA/S	0.2*	40
		PhMA/AN	4.0*	40
		PhMA/MMA	0.3*	40
	aMSAN	aMS/AN	22.3*	38
		S/AN	22.8*	38
	SMA	S/MA	0.18	29
		AN/MA	0.0004	29
		S/AN	0.98	32
		AN/MA	0.04	32
		S/MA	1.85	32

Component 1	Component 2	Segment pair	χ_{ij}	Refs.
	SMMA	S/AN	0.12	100
		S/MMA	0.01	100
		MMA/AN	0.05	100
	SPhMI	AN/SPhMI	0.07	32
		S/SPhMI	0.29	32
	SPPO	S/AN	0.7	34
		AN/PO	0.745	34
		AN/SPO	0.097	34
SMMA	SMA	S/MA	0.34	100
		MMA/MA	0.20	100
		AN/MA	0.03	100
S4HS	n-BA-co-tert-BA	S/HS	0.68	69
		S/tert-BA	0.25	69
		HS/tert-BA	0.54	69
SPPO	S-co-oFS	S/oFS	0.040	38
		oFS/PO	0.004	38
		oFS/SPO	0.113	38
	S-co-pFS	S/pFS	0.08	38
		S/PO	-0.043	38
		S/SPO	0.20	38
		pFS/PO	0.053	38
		pFS/SPO	0.022	38
		PO/SPO	0.15	38
TMPC-co-PC	SAN, SMMA	TMPC/MMA	0.26	85
		TMPC/AN	6.1	85
		TMPC/S	-0.15- -0.21	85
		TMPC/PC	-0.32- -0.31	85
		PC/AN	3.8	85
TMPC-co-PC		PC/S	0.49- -0.61	85
TMPC-co-PC	SAN, SMMA	PC/MMA	0.05	85

*B_{ij} (cal/cm^3)

For the mixtures of aliphatic polyethers with poly(hydroxy ether) of bisphenol A and their low molecular weight analogues (considering polyether as copolymer), by variation of the pair interaction parameters in Eq similar to 4.11, the dependence of the total interaction parameter $B = \chi RT/V$ on the CH_X group amount was calculated. It was found that the miscibility window corresponds to the experimentally found values[13] (Fig. 4.12). In the same work,[13] the

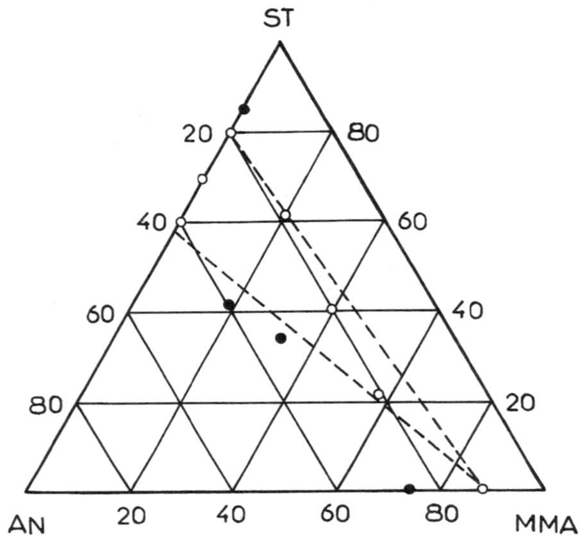

Figure 4.9. Phase behavior of SMAN terpolymers with a MAN copolymer containing 12.3% AN. • - immiscible, o - miscible. Adapted, by permission, from M. Nishimoto, H. Keskkula, and D. R. Paul, *Polymer*, **30**, 1279 (1989).

case of the mixture of two copolymers varying in composition was considered. The interaction parameter in this case is expressed by Eq 4.10. If $\chi_{AB} > 0$, $\chi_{mixture}$ is always > 0. Within the limit of the mixture of high molecular weight samples, the miscibility is not reached for any composition. Parameter $\chi_{mixture}=0$ when copolymers have identical composition. However, for the finite molecular masses (where there is the combinatorial entropy contribution to the three interaction energies) some composition differences between two copolymers may lead to miscibility. For the copolymers of the same structure and differing only by composition, the miscibility may be reached when the composition difference is

$$(x - y) = (2\rho RT/B_{ij}M)^{1/2} \qquad [4.77]$$

where ρ is the copolymer density, M is the molecular mass, and B_{ij} is the interaction parameter between copolymer components. Using value B_{ij} for comonomers styrene:acrylonitrile (1) = 6.74, methylmethacrylate:acrylonitrile (2) = 4.11 and styrene:methyl methacrylate (3) = 0. 181 cal/cm^3, the difference in copolymer compositions was

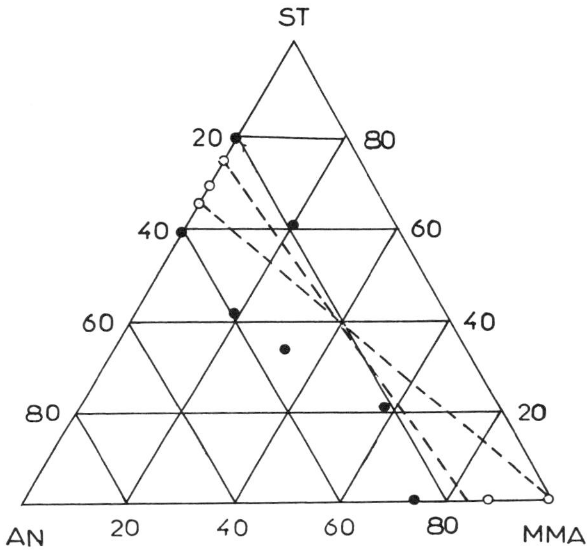

Figure 4.10. Phase behavior of SMAN terpolymers with another SMAN terpolymer containing 40% S, 39.1% M, and 20.9% AN, Adapted by permission, from M. Nishimoto, H. Keskkula, and D. R. Paul, *Polymer*, **30**, 1279 (1989).

calculated using Eq 4.77.[90] For copolymer 1, the difference was found to be 5.4%, for 2 - 7.0%, and for 3 - 33.3 %, respectively.

4.5 TERNARY MIXTURES

The miscibility windows were observed[90] for copolymers of styrene and acrylonitrile, methyl methacrylate and acrylonitrile with terpolymers styrene-methylmethacrylate-acrylonitrile and for terpolymer mixtures. The miscibility regions are restricted by the straight lines on the triangle diagrams presented in Figs. 4.9 and 4.10. For ternary mixtures of copolymers it was established,[91] that the miscibility region is situated in the region of crossing of two miscibility regions for one copolymer with other copolymer. For example, for the ternary mixture SAN/MMAN/SMMA, the miscibility region gets to the triangle cut off by the miscibility regions of SAN-MMAN and SAN-SMMA (straight lines, Fig. 4.11).

The latter system has been studied in detail.[92] It was shown that two binary pairs SAN(1) and SMMA(2), and SAN(1) and MAN(3) are miscible at some compositions, whereas the third mixture SMMA(2)-MAN(3) is immiscible. Using the Flory-Huggins theory, authors have

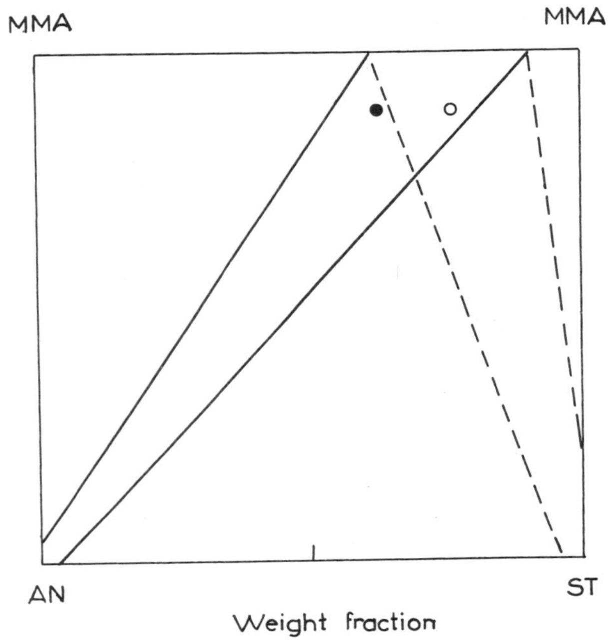

Figure 4.11. Experimental points for (50/25/25) blends of SAN/MAN/SMMA with fixed compositions MAN (13 wt% AN), SMMA (10 wt% styrene) mixed with SAN (23 wt% AN) (o) a miscible blend, and SAN (37 wt% AN)(•) an immiscible blend. Adapted, by permission, from J. M. G. Cowie, D. Lath, *Polym. Commun.*, **28**, 300 (1987).

constructed binodals and spinodals for the ternary system and have shown that more correct boundary between miscibility and immiscibility is given by the spinodal. It was demonstrated that in some ternary systems with fixed copolymer composition, the miscibility may be changed by the changing ratio of three components in the mixture. It was found that conditions for miscibility in this ternary system, and possibly a general feature of all such systems, are (a) that the binary interaction parameters χ_{ij} are smaller than the critical value, χ_{crit}, while the third should not be too much larger, that is, one of the copolymers may act as a compatibilizer for the other two components (b) that the difference $\Delta\chi = |\chi_{12} - \chi_{13}|$ is small. At χ_{12} and $\chi_{13} < \chi_{crit}$ and at χ_{23} = const, the region of immiscibility in the ternary phase diagram increases with increasing $\Delta\chi$. At $\chi_{12} = \chi_{13}$, the spinodal does not depend on χ_{12} and χ_{13} but is controlled by χ_{23}. If χ_{12} and $\chi_{13} > \chi_{crit}$ (outside the predicted miscibility window) at $\chi_{23} > \chi_{crit}$, the

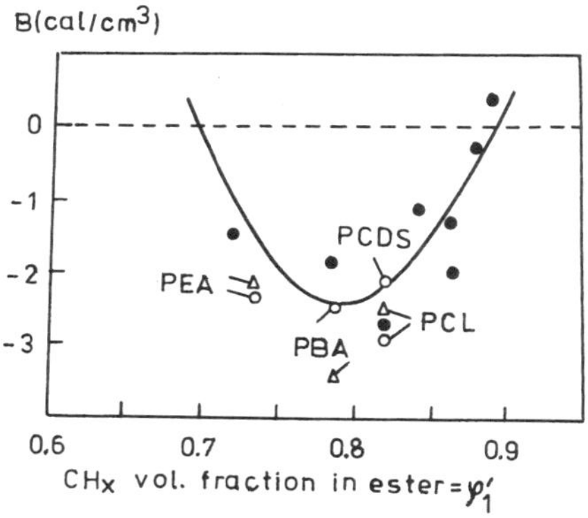

Figure 4.12. Interaction parameters for blends of aliphatic polyesters with the polyhydroxy ether of bisphenol-A and their low molecular weight analogues treating the polyesters as "copolymers". • - calorimetry, o – sorption, Δ - melting point depression. Adapted, by permission, from D. R. Paul and J. W. Barlow, *Polymer*, **25**, 487 (1984).

spinodal covers almost all phase diagrams. At $\chi_{23} = 0$, we have a special case of binary mixture SMMA/SAN and the miscibility depends on the values of χ_{12} and χ_{13}.

The effect of addition of copolymer on the position of spinodals was studied for the mixtures of poly(butadiene) with polystyrene[96] (the system with UCST). It was found that addition of a random copolymer butadiene-styrene lowers UCST. The analysis of phase behavior of the system was performed using equation for the free energy of mixing (4.46) where the interaction parameter was expressed in the form $\Lambda_{ij} = RT\chi/V_{ref}$ and instead of N_i values of V_i were used. For the case in which the composition of the initial mixture obeys the condition

$$\frac{\varphi_1^0 V_1}{f_A} = \frac{\varphi_2^0 V_2}{f_B} \qquad [4.78]$$

where φ_i^0 is the volume fraction of i-th component in the initial binary mixture of homopolymers, the equation of spinodals gives the following simple solution:

$$2\Lambda/RT = 1/f_B V_2(1 - \varphi_3)\varphi_1^0 \qquad [4.79]$$

$$\frac{T}{\Lambda} = \frac{T_0}{\Lambda_0}(1 - \varphi_3) \qquad [4.80]$$

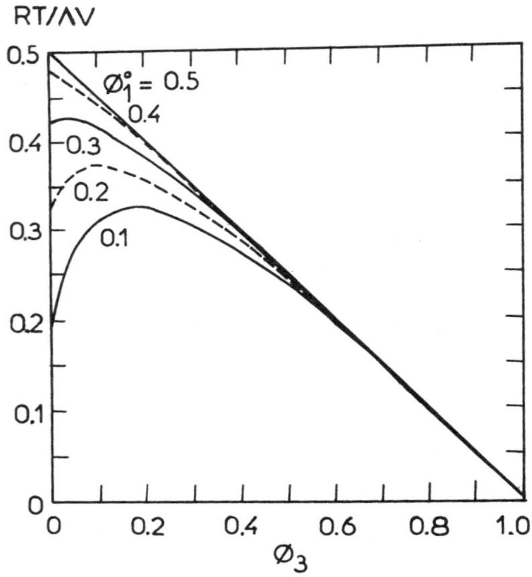

Figure 4.13. Spinodal temperature T_S against volume fraction of added copolymer, calculated for the system homopolymer A-homopolymer B-copolymer A-B with $V_1=V_2$, $V_3=10V_1$, $f_1=f_2$ and the initial volume fraction of homopolymer A as indicated. Adapted, by permission, from D. Rigby, J. L. Lin, and R. -J. Roe, *Macromolecules*, **18**, 2269 (1985).

where T_0 is the spinodal temperature of the initial binary mixture (φ_1^0, φ_2^0) and Λ_0 is the value at T_0. Eq 4.80 shows that, if the initial choice of the homopolymer mixture meets the condition 4.78, the addition of copolymer linearly decreases the spinodal temperature in accordance with the experimental data. If the relative amount of homopolymers does not meet the condition 4.78, the spinodal temperature diminishes slower and passes through the maximum (Fig. 4.13). Fig. 4.14 shows the isothermal spinodal curves for various values ΛV/RT. Here point A determines the composition above which the addition of copolymer increases the spinodal temperature. The point B is determined by the expression:

$$\frac{1}{V_3} = \frac{f_A^2}{\varphi_1^0 V_1} + \frac{f_B^2}{\varphi_2^0 V_2} - \frac{1}{\varphi_1^0 V_1 + \varphi_2^0 V_2} \qquad [4.81]$$

The straight line tying the point on the 1-2 axis with the apex 3 in Fig. 4.14 is the composition line. This composition line changes by the addition of copolymer. If this line crosses the isothermal spinodal

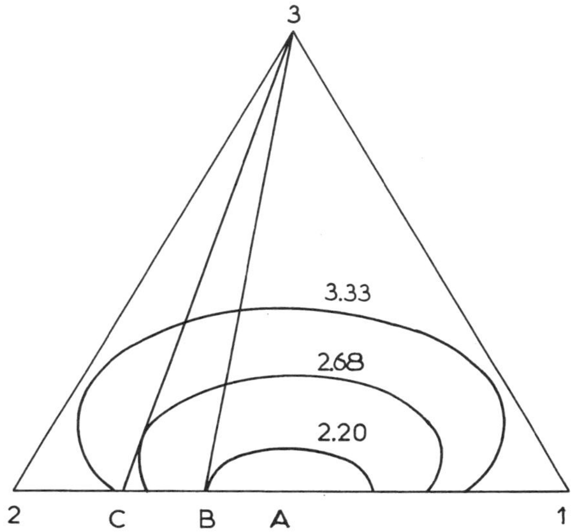

Figure 4.14. Spinodal isotherms calculated for the ternary mixture $V_1=V_2$, $V_3=10V_1$ and $f_1= f_2$. Indicated with the curves are the values of $\Lambda V/RT$. The straight line connecting a point on the 1-2 axis with the apex 3 is composition line which is traversed as we add copolymer. When the initial composition of the homopolymers is to the right of point B, the addition of the copolymer lowers the spinodal temperature, whereas when it is to the left of B, the spinodal temperature goes through a maximum before decreasing. Adapted, by permission, from D. Rigbi, J. L. Lin, and R. -J. Roe, *Macromolecules*, **18**, 2269 (1985).

curve twice, it means that the temperature of spinodal first increases and passes through the maximum. Eq 4.81 shows that with increasing volume V_3 of copolymer, the point B is pushed away from the pure homopolymer. In other words, when the size of copolymer is made larger, the range of composition (φ_1^0, φ_2^0) in which the added copolymer leads to monotonic decrease in the spinodal temperature which becomes smaller.

In the same work, it was found that the addition of block-copolymer leads first to lowering spinodal temperature and then to its increase, i.e., there is a minimum in the dependence of the spinodal temperature on the amount of additive. With concentration of block-copolymer increasing, the mixture becomes optically transparent at all temperatures above the room temperature. The block copolymer has a regular structure consisting of segregated polystyrene and polybutadiene domains and is capable of dissolution of large amounts of low molecular PS and PB in the corresponding domains. The mixture becomes optically transparent because domains dimensions are lower than the light wavelength. With further increase of the block-copolymer amount, the solubility limit is reached and the macroscopic phase separation occurs.

The data[97] for the mixtures of deuterated and protonated polybutadienes provide interesting example. The inversion of phase diagram from UCST to LCST has been observed. This inversion occurs by increasing the amount of 1,2-vinyl groups in protonated specimens holding their constant content in deuterated specimens. The explanation of the effect is based on the consideration of the PB mixtures as copolymers of AB/CD type, where A and C represent 1,2 units, whereas B and D are 1,4 units in protonated and deuterated specimens.[98] The effect of intermolecular repulsion between 1,2 and 1,4-units may lead to the negative values of interaction parameter (isotopic effect leads to positive values of χ) and therefore to LCST.

All the data considered in this section have been obtained using the approximation of the mean field theory and do not take into account changes in the free volume by mixing. Some attempts have been made to use the equation-of-state theories to describe the thermodynamic behavior of mixtures of homopolymers and copolymers.

4.6 MODIFIED THEORIES OF THE EQUATION-OF-STATE

The simplicity of the above considered models, describing thermodynamic behavior of the polymer blends, is reached due to many approximations that may not be correct and lead to some mistakes in predicting the phase behavior. The most important approximation includes the independence of χ_{ij} of composition and neglects effects of free volume. These shortcomings were the reason for development of other approaches to the phase behavior of the polymer blends.[99-102] The mixtures of random copolymers of the type A_xB_{1-x}/C_yD_{1-y} consisting of N_1r_1 mers and N_2r_2 mers may be described using the theory of equation-of-the-state with some modifications.

One of the modifications[99] uses the Flory theory with application of other set of rules of combination for mixtures of copolymers of the type mentioned above. According to this approach,[99] the characteristic and molecular parameters may be expressed through the same parameters for copolymer components:

$$P_1^* = [x\Theta_{1A}P_A^*v_A^{*2} + (1-x)\Theta_{1B}P_B^*v_B^{*2} + 2x\Theta_{1B}P_{AB}^*v_{AB}^{*2}]/v^*_1 \quad [4.82]$$

$$P_2^* = [y\Theta_{2C}P_C^*v_C^{*2} + (1-y)\Theta_{2D}P_D^*v_D^{*2} + 2y\Theta_{2D}P_{CD}^*v_{CD}^{*2}]/v^*_2 \quad [4.83]$$

$$P_{12}^* = [x\Theta_{2C}P_{AC}^*v_{AC}^{*2} + (1-x)\Theta_{2C}P_{BC}^*v_{BC}^{*2} + x\Theta_{2D}P_{AD}^*v_{AD}^{*2} +$$

$$+ (1-x)\Theta_{2D}P^*_{BD}v*_{BD} / v*_{12}] \quad [4.84]$$

where $v*_1$, $v*_2$, and $v*_{12}$ are determined from the relations:

$$v*_1 = x^2v*_A + 2x(1 - x)v*_{AB} + (1 - x)^2v*_B \quad [4.85]$$

$$v*_2 = y^2v*_C + 2y(1 - y)v*_{CD} + (1 - y)^2v*_D \quad [4.86]$$

$$v*_{12} = [(v*_1^{1/3} + v*_2^{1/3})/2]^3 \quad [4.87]$$

and corresponding freedom degrees are described by the equations:

$$c_1 = xC_A + (1 - x)C_B - x\Theta_{1B}C_{AB} \quad [4.88]$$

$$c_2 = yC_C + (1 - y)C_D - y\Theta_{2D}C_{CD} \quad [4.89]$$

$$c_{12} = x\Theta_{2C}C_{AC} + x\Theta_{2D}C_{AD} + (1-x)\Theta_{2C}C_{BC} + (1-x)\Theta_{2D}C_{BD} \quad [4.90]$$

The parameter of contact interaction is equal to

$$X_{12} = P_1^* + (v*_2/v*_1)(S_1/S_2)P_2^* - 2(v*_{12}/v*_1)P_{12}^* \quad [4.91]$$

The parameters of the reduction of the copolymer mixture are equal to

$$P^* = (\sigma_1^2\Theta_1P_1^*v_1^* + \sigma_2\Theta_2 + P_2^*v_2^*\sigma_1\Theta_2P_{12}^*v_{12}^*)/v^* \quad [4.92]$$

$$V^* = \sigma_1^2v_1^* + \sigma_2v_2^* + 2\sigma_1\sigma_2v_{12}^* \quad [4.93]$$

$$T^* = P^*v^*/cR \quad [4.94]$$

where

$$\sigma_1 = 1 - \sigma_2 - 2 = r_1N_1/(r_1N_1 + r_2N_2) \quad [4.95]$$

$$C = \sigma_1C_1 + \sigma_2C_2 - \sigma_1\Theta_2C_{12} \quad [4.96]$$

$$\Theta_1 = 1 - \Theta_2 = (r_1N_1xS_A + r_1N_1(1 - x)S_B)/\bar{r}NS \quad [4.97]$$

$$\Theta_{1A} = 1 - \Theta_{1B} = \Theta_A/\Theta_1 = xS_A + (1 - x)S_B] \quad [4.98]$$

$$\Theta_{2C} = 1 - \Theta_{2D} = \Theta_C/\Theta_2 = yS_C/[yS_C + (1 - y)S_D] \quad [4.99]$$

Value $1/T^*$ is equal to

$$1/T^* = (\sigma_1P_1^*v_1^*/T_1^* + \sigma_2P_2^*v_2^*T_2^* - \sigma_1\Theta_2RC_{12})/P^*v^* \quad [4.100]$$

Using these parameters, the equations may be derived[99] for the energy of mixing, chemical potential, heat of mixing, and heat of mixing at the infinite dilution:

$$(\mu_1 - \mu_0)^R = r_1RT[3\Theta_2^2C_{12} \ln [(2\pi mk_BT)^{1/2}/h(gv*)^{1/3}(\tilde{v}^{1/3} - 1)] +$$

$$+ 3C_1 \ln (m_1/m)^{1/2} + (3/2)C(m - m_1)/m] + P_1^*r_1v_1^*\tilde{T}_1[\ln (v*_1/v*) +$$

$$+ 3 \ln [(\tilde{v}_1^{1/3} - 1)/(\tilde{v}^{1/3} - 1)]] + 2P^*r_1v*\tilde{T}(v* - \sigma_1v*_1 - \sigma_2v*_{12}) /$$

$$/ (v* + P_1^*r_1v*_1/\tilde{v}_1 + (P_1^*r_1v*/\tilde{v})[(\Theta_1/\sigma_1) - 2(v* - \sigma_1v*_1 -$$

$$- \sigma_2v*_{12})/v* - 2(\Theta_1P_1^*v*_1^2 + \Theta_2P_{12}^*v*_{12}^2)/P^*v*^2 \qquad [4.101]$$

where g is the geometric factor[109] equal to $(4\pi/3)r^{1/2}$

$$\Delta H^M = (r_1N_1 + r_2N_2)[-(3/2)\sigma_1\Theta_2r_1C_{12}RT + \sigma_1P_1^*v_1^*/\tilde{v}_1 +$$

$$+ \sigma_2P_2^*v*_2/\tilde{v}_2 - P^*v^*/\tilde{v}] \qquad [4.102]$$

$$\Delta H^{M(\infty)} = (P_1^*v*_{sp,2}/\tilde{v}_1)(v*_1/v*_2)(1 + \alpha_1T)[(S_2/S_1) +$$

$$+ 2(v*_{12}/v*_1) - 2] + P_2^*v*_{sp,2}[\alpha_1T(T_1^*/T_2^*)/\tilde{v} + 1/\tilde{v}_2] -$$

$$- 2(P_{12}^*v*_{sp,2}/\tilde{v}_1)(v*_1/v*_2)(v*_{12}/v*_1]^2(S_2/S_1)(1 +$$

$$+ \alpha_1T) - RTC_{12}(v*_{sp,2}/v*_2)(S_2/S_1)[(3/2) + \alpha_1T/\tilde{v}_1\tilde{T}_1)] \qquad [4.103]$$

where $v^*_{sp,2}$ is the characteristic specific volume per one gram of component 2, α_1 is the volume thermal expansion coefficient of component 1. The excess free volume of the mixture is

$$\frac{V^E}{V_0} = \frac{\tilde{v}v^*}{\sigma_1\tilde{v}_1v_1^* + \sigma_2\tilde{v}_2v_2^*} - 1 \qquad [4.104]$$

The application of this theory to the mixtures of random copolymers[99] shows that parameter of interaction χ_1 depends on composition, x, in two ways: a) U-shape curve and b) monotonously increasing function (Fig. 4.15a,b). In both cases at the definite compositions and temperature, the sign of α_1 changes from positive to negative and again to positive. Therefore, there is observed the transition from immiscibility to miscibility and again to immiscibility. The miscibility windows are determined in the case a) by the parameter χ_1.

The modified Prigogine-Flory equation, describing the phase behavior of the mixtures of homopolymers with copolymers and co-

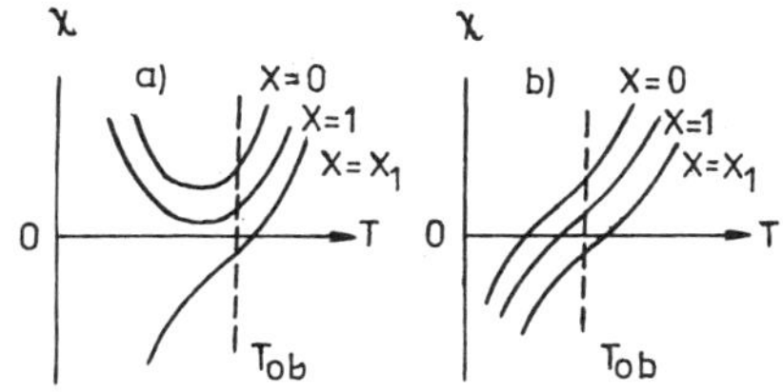

Figure 4.15. Schematic illustration of the temperature dependence of χ for the copolymer blends, whose miscibility changes from immiscible to miscible with copolymer composition, x, at the temperature, Tobs. Adapted, by permission, from T. Shiomi, H. Ishimatsu, T. Eguchi, and K. Imai, *Macromolecules*, **23**, 4978 (1990).

polymer with copolymer, was obtained by Kammer.[101,102] The Helmholz free energy of mixing is expressed in the following way:

$$\frac{\Delta F^M}{RT} = \frac{\varphi_A}{r_A}\ln\varphi_A + \frac{\varphi_B}{r_B}\ln\varphi_B - 2\frac{U_A}{RT}\varphi_A X_{AB} -$$

$$-\frac{C_{vA}}{2R}\left[\frac{\varphi_B^2\Gamma^2}{1+\varphi_B\Gamma} - \varphi_B\frac{\Gamma^2}{1+\Gamma}\right] \qquad [4.105]$$

where

$$\Gamma \equiv \frac{(\eta - \varepsilon)}{(1+\varepsilon)}; \quad \varepsilon = \frac{\langle\varepsilon^*_{AA}\rangle}{\langle\varepsilon^*_{11}\rangle} - 1 \qquad [4.106]$$

$$\eta = \frac{\langle\varepsilon^*_{BB}\rangle}{\langle\varepsilon^*_{11}\rangle} - 1 \qquad [4.107]$$

$$-U_A = \frac{P_A^* V_A^*}{\tilde{V}_A}; \quad C_{vA} = \frac{P_A^* V_A^*}{T_A^*}\frac{\tilde{v}_A^{1/3}}{4/3 - \tilde{v}_A^{1/3}} \qquad [4.108]$$

By using Flory's equation for $\tilde{T}$, the difference between Helmholz and Gibbs energies disappears and the conditions for spinodal and critical points are expressed in the following way:

$$\frac{\delta^2(\Delta G^M/RT)}{\delta\varphi^2} = 0 = \frac{1}{r_A\varphi_A} + \frac{1}{r_B\varphi_B} + \frac{U_A}{RT}4X_{AB} -$$

$$-\frac{C_{vA}}{2R}\frac{2\Gamma^2}{(1+\varphi_B\Gamma)^2} \qquad [4.109]$$

$$\frac{\delta^3(\Delta C^M/RT)}{\delta\varphi^3} = 0 = -\frac{1}{r_A\varphi_A^2} + \frac{1}{r_B\varphi_B^2} - \frac{C_{vA}}{2R}\frac{6\Gamma^3}{(1+\varphi_B\Gamma)^4} \quad [4.110]$$

where the interaction parameter X_{AB} is equal to:

$$X_{AB} = \frac{RT}{4U_A}\left[\frac{C_{vA}}{2R}\frac{2\Gamma^2}{(1+\varphi_B\Gamma)^3} - \frac{r_A\varphi_B + r_B\varphi_A}{r_A r_B \varphi_A \varphi_B}\right] \quad [4.111]$$

Parameters Γ and X_{AB} may be calculated from Eqs 4.110 and 4.111. In the case of the infinite molecular masses of components from Eq 4.110 and 4.111, it follows that $\Gamma = 0$ and $X_{AB} = 0$.

If to designate monomer of the homopolymer A as 1 and monomers of the copolymer as 2 and 3, according to Eq 4.106, $\varepsilon = 0$ and $\Gamma=\eta$, where η is given as

$$\eta = y\delta_2 + (1-y)\delta_3 - 2y(1-y)\chi_{23} \quad [4.112]$$

$$\delta_i = \frac{<\varepsilon^*_{ii}>}{<\varepsilon^*_{11}>} - 1 \quad [4.113]$$

(for monomer 1, $\delta_i = 0$). From equations 4.106-4.112 one can calculate X_{AB} and η whereas χ_{ij} may be found from the relation:

$$X_{AB} = y\chi_{12} + (1-y)\chi_{13} - y(1-y)\chi_{23} \quad [4.114]$$

For the mixtures of two copolymers with a common comonomer (the mixture of the type A_xB_{1-x}/B_yC_{1-y} values of ε, η, and X_{AB} are determined by the relations:

$$\varepsilon = x\delta_2 - 2x(1-x)\chi_{12} \quad [4.115]$$

$$\eta = y\delta_3 - 2y(1-y)\chi_{23} \quad [4.116]$$

$$X_{AB} = \frac{\Lambda}{1+\varepsilon} \quad [4.117a]$$

$$\Lambda = y^2\chi_{23} + xy(\chi_{13} - \chi_{23} - \chi_{12}) + \chi_{12}x^2 \quad [4.117b]$$

where x and y corresponds to copolymers A_xB_{1-x} and B_yC_{1-y} and are molar fractions of monomer 1 in each of copolymer. For the copolymer mixtures of the type A_xB_{1-x}/C_yD_{1-y} the values ε, η, and Λ are equal to:

$$\varepsilon = (1-x)\delta_2 - 2x(1-x)\chi_{12} \quad [4.118]$$

$$\eta = y\delta_3 + (1-y)\delta_4 - 2y(1-y)\chi_{34} \quad [4.119]$$

$$\Lambda = xy\chi_{13} + x(1-y)\chi_{14} + y(1-x)\chi_{23} + (1-x)(1-y)\chi_{24} - x(1-$$

$$- x)\chi_{12} - y(1 - y)\chi_{34} \qquad [4.120]$$

Kammer[103] has added a new parameter ρ to the generalized interaction parameter X_{AB}. This parameter ρ reflects the contribution of the difference in the segment dimension in the free energy of mixing. For this case, the equation has been derived which predicts both UCST and LCST for the mixtures of two polymers and mixtures of random copolymers. The final equation for the Gibbs free energy of mixing is written in the following form:

$$\frac{\Delta G^M}{RT} = 2C_A\varphi_A\varphi_B\left[\frac{\tilde{v}_A^{1/3}}{\tilde{v}_A^{1/3}-1}\left(X_{AB}^T+\frac{4/3-\tilde{v}_A^{1/3}}{\tilde{v}_A^{1/3}-1}\frac{9}{16}\rho^2\right)+\frac{\tilde{v}_A^{1/3}}{4/3-\tilde{v}_A^{1/3}}\frac{7}{16}\Gamma^2\right]+$$

$$+\frac{\varphi_A}{r_A}\ln\varphi_A+\frac{\varphi_B}{r_B}\ln\varphi_B \qquad [4.121]$$

where the following definitions are taken:

$$X_{AB}^T = X_{AB} - \frac{9}{2}\rho^2 - \frac{3}{4}\Gamma\rho \qquad [4.122]$$

$$\rho = \rho_A - \rho_B \qquad [4.123]$$

$$\rho_A = (1 - x)\delta_2^r - 2x(1 - x)R_{12} \qquad [4.124]$$

$$\rho_B = y\delta_3^r + (1 - y)\delta_4^r - 2y(1 - y)R_{34} \qquad [4.125]$$

$$\delta_i^r = \frac{\langle r^*_{ii}\rangle}{\langle r^*_{11}\rangle} - 1 \qquad [4.126]$$

$$\Gamma = (\eta - \varepsilon)/(1 + \varepsilon) \qquad [4.127]$$

$$\varepsilon = (1 - x)\delta_2 - 2x(1 - x)\chi_{12} - 18x(1 - x)(\delta_2^r)^2 \qquad [4.128]$$

$$\eta = y\delta_3 + (1 - y)\delta_4 - 2y(1 - y)\chi_{34} - 18y(1 - y)(\delta_3^r + \delta_4^r)^2 \qquad [4.129]$$

$$\delta_i = \frac{\langle\varepsilon^*_{ii}\rangle}{\langle\varepsilon^*_{11}\rangle} - 1;\ i = 2, 3, 4 \qquad [4.130]$$

Here x and y are the molar fractions of monomer 1 in A and 3 in B, if A and B are random copolymers consisting of monomers 1, 2 and 3, 4. To calculate the phase behavior of such systems, a simple relation for ρ was used.[103] Contribution of R_{ij} was neglected. Varying the parameters $X_{AB}T$, Γ, and ρ, it is possible to obtain both UCST and LCST and diagrams of the "sand-glass" type. Both CSTs were found experimentally.[42-45]

The model based on the theory of equation-of-state by Prigogine-Flory-Patterson was used[104] to establish in a quantitative way the effect of comonomer on the LCST behavior of copolymer mixtures. In the framework of the binary interaction model, it was shown that the intramolecular repulsions contribute both to the term of the free volume and Flory-Huggins term of interaction. Considering the mixture of homopolymer A and copolymer B, the expressions for the characteristic values of P_B^* and T_B^* have been obtained:

$$P_B^* = \varphi_B^{(2)} P_2^* + \varphi_B^{(3)} P_3^* - \varphi_B^{(2)} \Theta_B^{(3)} X_{23} \quad [4.131]$$

$$T_B^* = \frac{P_B^*}{\varphi_B^{(2)} P_2^*/T_2^* + \varphi_B^{(3)} P_3^*/T_3^*} \quad [4.132]$$

where φ_B is the volume fraction of comonomer i in copolymer, Θ_B is the fraction of sites of segments in copolymer P_B^* and T_B^* are characteristic pressure and temperature of comonomer i, X_{23} is the contact interaction parameter for comonomers. By analogy with the model of binary interactions,[13] the effective parameter of contact interactions is expressed in the following way:

$$X_{AB} = \Theta_B^{(2)} X_{12} + \Theta_B^{(3)} X_{13} - (S_A/S_B)\varphi_B^{(2)} \Theta_B^{(3)} X_{23} \quad [4.133]$$

For the case when $X_{ij} = X_{ji}$, one has:

$$X_{AB} = \varphi_B^{(2)} X_{12} + \varphi_B^{(3)} X_{13} - \varphi_B^{(2)} \varphi_B^{(3)} X_{23} \quad [4.134]$$

For the mixture of copolymer A formed by comonomers 1 and 2 with copolymer B (comonomers 3 and 4), the following relation is valid:

$$X_{AB} = \varphi_A^{(1)} \Theta_B^{(3)} X_{13} + \varphi_A^{(1)} \Theta_B^{(4)} X_{14} + \varphi_A^{(2)} \Theta_B^{(3)} X_{23} + \varphi_A^{(2)} \Theta_B^{(4)} X_{24} -$$

$$- \varphi_A^{(1)} \Theta_A^{(2)} X_{12} - (S_A/S_B) \Theta_B^{(4)} \varphi_B^{(3)} X_{34} \quad [4.135]$$

where

$$S_B = S_2 \varphi_B^{(2)} + S_3 \varphi_B^{(3)} \quad [4.136]$$

$$\varphi_B^{(i)} = \frac{r_i N_i}{r_B N_B} \quad [4.137]$$

$$\Theta_B^{(i)} = \frac{r_i N_i S_i}{r_B N_B S_B} \quad [4.138]$$

The temperature of the phase separation corresponds to the condition $\chi_{AB}(T) = \chi_{AB\,(crit)}$. Here these values are determined by the equations:

$$\frac{\chi_{AB}(T)}{C_A} = \frac{\tilde{v}_A^{1/3}}{\tilde{v}_A^{1/3} - 1}\left[\frac{X_{AB}}{P_A^*}\right] + \frac{\tilde{v}_A^{1/3}}{2(4/3 - \tilde{v}^{1/3})}\left(1 - \frac{T_A^*}{T_B^*}\right)^2 \quad [4.139]$$

$$\frac{\chi_{AB,crit}}{C_A} = \frac{RT_A^*}{2P_A^*}\left[\frac{1}{V_A^{*\,1/2}} + \frac{1}{V_B^{*\,1/2}}\right]^2 \quad [4.140]$$

$$C_A = \frac{P_A^* V_A^*}{RT_A^*} \quad [4.141]$$

The results of calculations according to the model under consideration were compared with experimental data for the mixtures PVME-styrene copolymers. A good correlation between theory and experiment was found.

REFERENCES

1. R. Vukovic, V. Kuresevic, and F. E. Karasz, *Int. Pure and Appl. Chem. 28th Macromol. Symp.*, Amherst, 1982, p.707.
2. R. Vukovic, V. Kuresevic, F. E. Segudovic, F. E. Karasz, and W. J. MacKnight, *J. Appl. Polym. Sci.*, **28**, 1379; 3079 (1983).
3. G. ten Brinke, F. E. Karasz, and W. J. MacKnight, *Macromolecules*, **16**, 1827 (1983).
4. V. R. Landi, *Rubber Chem. Thechnol.*, **45**, 222 (1975).
5. C. Trembley and R. E. Prud'homme, *J. Polym. Sci. Phys. Ed.*, **22**, 1857 (1984).
6. N. F. Bakeev, Z. F. Jarikova, Y. N. Malinsky, and A. L. Izyumnikov, *Vysokomolek.soed.*, **B19**, 823 (1977).
7. F. Kollinsky and G. Market, *Adv. Chem. Ser.*, **99**, 175 (1971).
8. D. J. Stein, R. H. Yung, K. H. Illers, and H. Hendus, *Angew. Makromol. Chem.*, **36**, 83 (1974).
9. J. S. Chiou, D. R. Paul, and J. W. Barlow, *Polymer*, **23**, 1543 (1982).
10. C. F. Hammer, *Macromolecules*, **4**, 69 (1971).
11. R. P. Kambour, J. T. Bendler, and R. C. Bopp, *Macromolecules*,**16**, 753 (1983).
12. G. ten Brinke, F. E. Karasz, and W. J. MacKnight, *Macromolecules*, **16**, 1827 (1983).
13. D. R. Paul and J. W. Barlow, *Polymer*, **25**, 487 (1984).
14. T. Shiomi, M. Suzuki, M. Tohyama, and K. Imai, *Macromolecules*, **22**, 3578 (1989).
15. A. S. C. Fernandes, W. J. Barlow, and D. R. Paul, *J. Appl. Polym. Sci.*, **32**, 5, 481 (1986).
16. C. Zhikuan and F. E. Karasz, *Macromolecules*, **25**, 4716 (1992).
17. T. Shiomi, F. E. Karasz, and W. J. MacKnight, *Macromolecules*, **19**, 2274 (1986).
18. W. J. Hall, R. L. Kruse, R. A. Mendelson, and Q. H. Trementozzi, *Am. Chem. Soc. Div. Org. Chem. Plast. Progr.*, **47**, 298 (1982).
19. M. E. Galvin, *Macromolecules*, **24**, 6354 (1991).
20. A. C. Balazs, I. C. Sanchez, I. B. Epstein, F. E. Karasz, and W. J. MacKnight, *Macromolecules*, **18**, 2188 (1985).

21. A. C. Balazs, F. E. Karasz, W. J. MacKnight, H. Ueda, and I. C. Sanchez, *Macromolecules*, **18**, 2784 (1985).
22. J. V. Hunsel, A. C. Balazs, R. Koningsveld, and W. J. MacKnight, *Macromolecules,* **21**, 1528 (1988).
23. A. C. Balazs and M. T. De Meuse, *Macromolecules*, **22**, 4260 (1989).
24. H. Ueda and F. E. Karasz, *Macromolecules*, **18**, 2719 (1985).
25. H.-J. Cantow and O. Schulz, *Polym. Bull.*, **15**, 449 (1986).
26. C. Zhikuan, S. Ruona, and F. E. Karasz, *Macromolecules*, **25**, 6113 (1992).
27. D. Rana, B. M. Mandal, and S. N. Bhattacharyya, *Polymer*, **34**, 1454 (1993).
28. R. A. Mendelson, *J. Polym. Sci. Phys. Ed.*, **23**, 1985 (1985).
29. C. Zhikuan and S. Ruona, *Polymer*, **24**, 1279 (1983).
30. J. H. Kim, J. W. Barlow, and D. R. Paul, *J. Polym. Sci. Phys. Ed.*, **27**, 223 (1989).
31. Y. Takeda and D. R. Paul, *J. Polym. Sci. Phys.Ed.*, **30**, 1273 (1992).
32. Y. Aoki, *Macromolecules,* **21**, 1277 (1988).
33. A. G. Margaritis, J. K. Kallitsis, and N. K. Kalfoglou, *Polymer,* **28**, 2123 (1987).
34. H. S. Kang, W. J. MacKnight, and F. E. Karasz, *Am. Chem. Soc. Polym. Prepr.*, **28**, 134 (1987).
35. G. R. Brannock, W. J. Barlow, and D. R. Paul, *J. Polym. Sci. Phys. Ed.*, **28**, 871 (1990).
36. B. Litauszki, G. Schmidt-Naake, J. Kressler, and H.-W. Kammer, *Polym. Commun.*, **30**, 359 (1989).
37. P. R. Kohl, A .M. Seifert, and G. P. Hellmann, *J. Polym. Sci. Phys. Ed.*, **28**, 1309 (1989).
38. J. M. G. Cowie, E. M. Elexpuru, and I. J. McEwen, *Polymer*, **33**, 1993 (1992).
39. T. Shiomi, F. E. Karasz, and M. J. MacKnight, *Macromolecules,* **19**, 2644 (1986).
40. M. Nishimoto, H. Keskkula, and D. R. Paul, *Macromolecules*, **23**, 3633 (1990).
41. J. K. Kallitsis, D. R. Sotiropoulou, and N. K. Kalfoglou, *Polymer,* **29**, 1233 (1988).
42. D. J. Walsh and S. Rostami, *Macromolecules,* **18**, 216 (1985).
43. D. J. Walsh, S. Laingle, and C. Zhikuan, *Polymer*, **28**, 1005 (1981).
44. H. Saito, Y. Fujita, and T. Inoue, *Polym. J. (Japan)*, **19**, 405 (1987).
45. G. Cong, Y. Huang, W. J. MacKnight, and F. E. Karasz, *Macromolecules,* **19**, 2765 (1986).
46. S. L. Zaharias, G. ten Brinke, W. J. MacKnight, and F. E. Karasz, *Macromolecules*, **16**, 381 (1983).
47. J. M. G. Cowie, J. H. Harris, and I. J. McEwen, *Macromolecules*, **25**, 5287 (1992).
48. J. M. G. Cowie and E. M. Elexpuru, *Eur. Polym. J.*, **28**, 693 (1992).
49. S. H. Goh, S. Y. Lee, K. S. Siow, and J. Chen, *Polymer*, **34**, 2898 (1993).
50. M. Suess, J. Kressler, and H.-W. Kammer, *Polymer*, **28**, 957 (1987).
51. J. Stein, P. H. Jung, K. H. Illers, and H. Hendus, *Angew. Makromol. Chem.*, **36**, 89 (1974).
52. S. H. Goh, K. S. Siow, and S. Y. Lee, *Eur. Polym. J.*, **27**, 921 (1991).
53. S. H. Goh, and S. Y. Lee, *J. Appl. Polym. Sci.*, **41**, 1391 (1990).
54. S. H. Goh and S. Y. Lee, *Eur. Polym. J.*, **26**, 715 (1990).
55. Y. F. Chong and S. H. Goh, *Polymer,* **33**, 127 (1992).
56. M. K. Neo, S. Y. Lee, and S. H. Goh, *Eur. Polym. J.*, **27**, 831 (1991).
57. M. K. Neo and S. H. Goh, *Eur. Polym. J.*, **27**, 915 (1991).
58. M. E. Fowler, W. J. Barlow, and D. R. Paul, *Polymer*, **28**, 1177 (1987).
59. Y. Xu, J. Graf, P. C. Painter, and M. M. Coleman, *Polymer*, **32**, 3103 (1991).
60. Y. F. Chong and S. H. Goh, *Polymer*, **33**, 132 (1992).
61. Y. F. Chong and S. H. Goh, *Polymer*, **33**, 1287 (1992).
62. K. E. Min and D. R. Paul, *J. Polym. Sci., Phys. Ed.*, **26**, 2257 (1988).

63. J. H. Kim, W. J. Barlow, and D. R. Paul, *J. Polym. Sci., Phys. Ed.*, **27**, 2211 (1989).
64. G. A. Zakrzewski, *Polymer*, **14**, 347 (1973).
65. W. H. Jo, J. T. Yoon, and S. C. Lee, *Polym. J. (Japan)*, **23**, 1243 (1991).
66. S.-H. Chiu and T. G. Smith, *J. Appl. Polym. Sci.*, **29**, 1797 (1984).
67. J. Kressler and H.-W. Kammer, *Polym. Bull.*, **19**, 283 (1988.)
68. B. Wang and S. Krause, *J. Polym. Sci. Phys. Ed.*, **26**, 2237 (1988).
69. K. J. Zhu, S. F. Chen, T. Ho, E. M. Pearce, and T. K. Kwei, *Macromolecules*, **23**,150 (1990).
70. S. Y. Lee, M. Y. Low, and S. H. Goh, *Eur. Polym. J.*, **27**, 1378 (1991).
71. J. R. Fried, F. E. Karasz, and W. J. MacKnight, *Macromolecules*, **11**, 150, (1978).
72. P. R. Alexandrovich, F. E. Karasz, and W. J. MacKnight, *Polymer*, **18**, 1022 (1977).
73. R. Vukovic, F. E. Karasz, and W. J. MacKnight, *Polymer*, **24**, 529 (1983).
74. A. R. Shultz and B. M. Beach, *Macromolecules*, **7**, 902 (1974).
75. S. H. Goh and L. S. Lim, *Eur. Polym. J.*, **26**, 711 (1990).
76. A. C. Fernandes, W. J. Barlow, and D. R. Paul, *Polymer*, **27**, 1799 (1986).
77. K. Schulze, J. Kressler, and H.-W. Kammer, *Polymer*, **34**, 3704 (1993).
78. S. H. Goh and K. S. Siow, *J. Appl. Polym. Sci.*, **29**, 99 (1984).
79. S. Kuwahara, S. Akijama, and A. Ueda, *Polym. J.(Japan)*, **21**, 221 (1989).
80. C. A. Cruz-Ramos and D. R. Paul, *Macromolecules*, **22**, 1289 (1989).
81. J. Kressler, H.-W. Kammer, K. Herzog, and H. Heyde, *Acta Polym.*, **41** (1990).
82. T. Fukuda, M. Nagata, and H. Inagaki, *Macromolecules*, **19**, 1411 (1986).
83. S. H. Goh, K. S. Siow, and S. Y. Lee, *Eur. Polym. J.*, **28**, 657 (1992).
84. S. H. Goh, S. Y. Lee, and S. M. Low, *Eur. Polym. J.*, **28**, 66 (1992).
85. C. K. Kim and D. R. Paul, *Polymer*, **33**, 4941 (1992).
86. R. Vukovic, G. Bogdanic, V. Kuresevic, *et al.*, *Eur. Polym. J.*, **24**, 123 (1988).
87. M. E. Fowler, D. R. Paul, L. A. Cohen, and W. T. Freed, *J. Appl. Polym. Sci.*, **37**, 513 (1989).
88. M. Sankarapandian and K. Kishore, *Macromolecules*, **24**, 3090 (1991).
89. C. K. Kim and D. R. Paul, *Polymer*, **33**, 2089 (1992).
90. M. Nishimoto, H. Keskkula, and D. R. Paul, *Polymer*, **30**, 1279 (1989).
91. J. M. G. Cowie and D. Lath, *Polym. Commun.*, **28**, 300 (1987).
92. J. M. G. Cowie, G.-G. Li, R. Ferguson, and I. J. McEwen, *J. Polym. Sci. Phys. Ed.*, **30**, 1351 (1992).
93. J. E. Harris, S. H. Goh, D. R. Paul, and W. J. Barlow, *J. Appl. Polym. Sci.*, **27**, 839 (1982).
94. J. S. Ziska, J. W. Barlow, and D. R. Paul, *Polymer*, **22**, 918 (1981).
95. J. E. Harris, *Ph. D. Dissertation*, University of Texas at Austin, 1981.
96. D. Rigby, J. L. Lin, and R.-J. Roe, *Macromolecules*, **18**, 2269 (1985).
97. H. Jinnai, H. Hasegawa, and T. Hashimoto, *Macromolecules*, **25**, 6078 (1992).
98. S. Sakurai, H. Hasegawa, T. Hashimoto, I. G. Hargis, S. L. Aggarwall, and C. C. Han, *Macromolecules*, **23**, 451 (1990).
99. T. Shiomi, H. Ishimatsu, T. Eguchi, and K. Imai, *Macromolecules*, **23**, 4970 (1990).
100. F. Hamada, T. Shiomi, K. Fujisawa, and K. Nakajima, *Macromolecules*, **13**, 729 (1980).
101. H.-W. Kammer, *Acta Polym.*, **37**, 1 (1986).
102. H.-W. Kammer, J. Kressler, B. Kressler, D. Scheller, D. Kroschwitz, and G. Schmidt-Naake, *Acta Polym.*, **40**, 75 (1989).
103. H.-W. Kammer, T. Inoue, and T. Ougizawa, *Polymer*, **30**, 888 (1989).
104. W. H. Jo and M. S. Lee, *Macromolecules*, **25**, 842 (1992).

Chapter 5

MECHANISM AND KINETICS OF PHASE SEPARATION IN POLYMER SOLUTIONS AND BLENDS

One of the fundamental problems of the polymer science is to establish the mechanism of self-organization in polymer systems under the non-equilibrium conditions. From practical point of view, this phenomenon conveys the possibility to manipulate and control the phase morphology in polymer blends and alloys.

The general approach to the description of the phase separation in non-homogeneous systems was developed by van-der-Waals, Hart, Cahn and Hilliard, and others.[1-22] Two mechanisms of phase separation are usually considered: nucleation and growth and spinodal decomposition. The first type is described in detail in classical Wunderlich's monograph *Physics of Macromolecules*. Cahn and Hilliard[5-11] described the second type of the phase separation, namely, spinodal decomposition of the non-homogeneous solution leading to the formation of typical network structures in isotropic medium, for example in inorganic glasses.[7,8] Later it was established that this type of the phase separation is typical of both polymer solutions and blends.

5.1 NUCLEATION AND GROWTH

Presently it is well known that the mechanisms of the phase separation of the liquid-liquid type are qualitatively different in the regions of metastability and instability of the phase diagram. Inside each region, depending on the closeness to binodal, spinodal, or critical point, there are observed qualitative and quantitative distinctions in the kinetics of the phase separation.

In the metastable region, the phase separation proceeds according to the mechanism of nucleation and growth.[7,9] The transition of the system from one-phase to two-phase state is connected with amplification of fluctuations in the composition and with development of the microregions of a new phase. This process proceeds when the system is transferred into the metastable region of the phase diagram (Ch. 1). The diffusion flows of components in the metastable region are directed towards diminishing fluctuations. The evolution of a new phase begins in this region. The energetically unfavorable effect of the appearance of a new interface determines the system instability only in relation to fluctuations whose sizes exceed some critical values. The composition of fluctuations is close to composition for a new phase. Such fluctuations are considered as critical nuclei. The process of phase separation in this case is similar to the appearance, growth, and coalescence of microregions of evolved new phase by evaporation, condensation, or crystallization. These processes were described by Frenkel[3] and Lifshitz and Pitaevskii.[4] Only main features of the nuclei growth and their distribution at the latest stages of separation are considered here, when fluctuational appearance of new nuclei is practically impossible. As an example we may remember the deposition of the solute from supersaturated solution, where the criterion of the metastability is the degree of supersaturation of the solution.[4]

When the formation of new nuclei is excluded, the critical sizes of the existing nuclei are increased due to the "eating" of small grains by the big ones (so-called Ostwald ripening). In this stage, the average size of the grain is increased asymptotically with time:

$$\bar{a} = (4\sigma Dt/9)^{1/3} \qquad [5.1]$$

where D is the diffusion coefficient of solute, $\sigma = 2\gamma V'C_\infty/T$, γ is the coefficient of surface tension at the interface, V' is molecular volume of solute, C_∞ is the concentration of saturated solution over the plane surface of the solute, t is time.

The mechanism of "eating" of small grains of a new phase by the big grains was described by Lifshitz and Slyozov.[23] The distribution of the grains by their dimensions at any time is determined by the function:

$$P(u) = \frac{3^4u^2\exp[-1/(1-2u/3)]}{2^{5/3}(u+3)^{7/3}(3/2-u)^{11/3}} \qquad [5.2]$$

where $u = a/a_k(t)$, $a_k(t)$ is critical grain radius. At $u > 3/2$, we have $P(u)=0$, i.e., the function P(u) differs from zero only in the region $u<3/2$. As long as $a = a_k(t)$ and the mean dimensions coincide with critical dimensions, maximum P(u) corresponds with values u close to unity. The total number of grains per volume unit diminishes in time since $N(t) = 0.5Q/D\sigma t$, where Q is total initial supersaturation.

The analogous effects can be observed in polymer solutions and at concentrations of the order of a few percents due to transition into the region between binodal and spinodal. Active nucleation may proceed under such conditions. This explains the existence of the so-called cloud points for highly concentrated solutions below binodal.

The concentrated solution of a polymer may be considered as ensemble of polymer segments and solvent molecules with practically homogeneous distribution of segments in solution.[24] Concentration fluctuations may be described as density fluctuations of segments or concentration fluctuations of solvent molecules. The effect of individual properties of macromolecules (for example, molecular-mass distribution) on the nucleation and on the cloud point curve becomes non-essential at high concentrations, as distinct from dilute solutions.

Nucleation mechanism of phase separation in polymer solutions at the initial stages was thoroughly investigated[24] for the system: poly-2,6-dimethyl-1,4-phenylene oxide-toluene, using critical light scattering induced by thermal impulses and analyzing the experimental data on light scattering within the framework of the concept of heterogeneous crystallization.[25]

5.2 SPINODAL MECHANISM OF PHASE SEPARATION

The theory of spinodal decomposition of homogeneous system at the initial stages of the process was developed in the fundamental works done by Cahn and Hilliard.[5-11] According to this theory at the initial stages of phase separation, the concentration fluctuations appear. They can be considered as a set of sinusoidal waves with fixed length. Wavelengths represent the dimensions of the structures formed during the course of spinodal decomposition. Following Cahn,[6-8] consider the basic concepts of the theory.

Suppose we have a solution with a small compositional gradient. For such solution, the free energy can be presented in the following form:

$$F = \int [g(\varphi) + K(\nabla\varphi)^2]dV \qquad [5.3]$$

where $g(\varphi)$ is free energy of mixing for volume unit of homogeneous solution of composition φ, $K(\nabla\varphi)^2$ is additional energy determined by the compositional gradient. Factorizing $g(\varphi)$ in series by small concentration fluctuation, we have:

$$g(\varphi) = g(\varphi_0) + (\varphi - \varphi_0)\frac{\delta g}{\delta\varphi} + \frac{1}{2}(\varphi - \varphi_0)^2\frac{\delta^2 g}{\delta\varphi^2} + \dots \qquad [5.4]$$

Taking into consideration that $\int(\varphi - \varphi_0)dV = 0$, we can obtain the expression for the difference in free energy between homogeneous and inhomogeneous solution as

$$\Delta F = \int\left[\frac{1}{2}\frac{\delta^2 g}{\delta\varphi^2}(\varphi - \varphi_0)^2 + K(\nabla\varphi)^2\right]dV \qquad [5.5]$$

Constant K is supposed to be close to 0. It is seen that solution is stable in relation to all extremely small fluctuations, if $\delta^2 g/\delta\varphi^2 > 0$. On the other hand, if $\delta^2 g/\delta\varphi^2 < 0$, the solution is unstable to such fluctuations, for which the first term in Eq 5.5 is prevailing. Using this equation, the Fourier analysis of local fluctuations is performed. From the physical point of view this analysis is a description of the heat movement in liquid as superposition of supersonic waves of different directions and wavelengths. To determine the contribution of fluctuations to ΔF, the contributions of each Fourier component are summarized. For example, contribution from one component Acosqx can be represented by

$$\frac{1}{4}VA^2\left[\frac{\delta^2 g}{\delta\varphi^2} + 2Kq^2\right]; \quad q = \frac{4\pi}{\lambda}\sin\frac{\Theta}{2} \qquad [5.6]$$

The existence of Fourier component with sufficiently high wavelength (or small wavenumber q) leads to decrease in free energy, if the solution is in metastable region. The maximum wavenumber is expressed as $q_{crit} = [(-\delta^2 g/\delta\varphi^2)/2K]^{1/2}$.

The kinetics of the initial stage of decomposition can be easily estimated by a simple solution of diffusion equation:

$$J_B = -J_A = M\nabla(\mu_A - \mu_B) \qquad [5.7]$$

where J_i is diffusion flow, μ_i is chemical potential, M is mobility. The latter should be positive, if diffusion, with a gradient of chemical potential present, leads to diminishing free energy. Value $\mu_A - \mu_B$ can be found from a partial derivative of F in Eq 5.3:

$$\mu_A - \mu_B = \frac{\delta F}{\delta\varphi_A} = \frac{\delta g}{\delta\varphi_A} - 2K\nabla^2\varphi_A + \ldots \qquad [5.8]$$

Substituting 5.8 into 5.7, we find the diffusion equation in the form:

$$\frac{\delta\varphi}{\delta t} = M\frac{\delta^2 g}{\delta\varphi^2}\nabla^2\varphi - 2MK\nabla^4\varphi \qquad [5.9]$$

The coefficient $\nabla^2\varphi$ can be identified with diffusion coefficient, D. It is seen that diffusion coefficient on the spinodal changes its sign from positive to negative (for spinodal $\delta^2 g/\delta\varphi^2=0$). The solution of Eq 5.9 is found as

$$\varphi - \varphi_0 = \exp[R(q)t]\cos(\vec{q}, \vec{r}) \qquad [5.10]$$

where the so-called amplification factor R(q) is equal to

$$R(q) = -M\frac{\delta^2 g}{\delta\varphi^2}q^2 - 2MKq^4 \quad [5.11]$$

This equation is uniform and linear with φ. The sum of solutions gives:

$$\varphi - \varphi_0 = \sum_{\text{all } q} \exp R(q)t[A(\vec{q})\cos(\overrightarrow{q,r_i}) + B(\vec{q})\sin(\overrightarrow{q,r_i})] \quad [5.12]$$

where q_i and r_i are vectors of the wavenumber and volume unit i. Cahn has established[8] that in the course of spinodal decomposition, the fluctuations with wavenumber, q, below critical value, q_{crit}, are increased in amplitude with time, whereas fluctuations with $q<q_{crit}$ diminish their amplitudes and disappear. Amplitudes A and B are evaluated at t = 0, i.e., at the moment of transition of the system through spinodal, which can be found from condition $\delta^2 g/\delta\varphi^2 = 0$. Values A and B do not change with time or temperature. Amplification factor R(q) acquires maximum value at

$$q_m = \frac{q_c}{\sqrt{2}} = \frac{1}{2}\left[-\frac{\delta^2 g}{\delta\varphi^2}/K\right]^{1/2} \quad [5.13]$$

As the dependence of the amplitude of the sound wave on time is exponential (exponent is equal to R(q)), all Fourier components can be ignored but component which has maximum wavenumber, q_m. The following transformation can be done:

$$\varphi - \varphi_0 = \sum_{\text{all } q} \exp[R_m(q)t][A(\vec{q})\cos(\overrightarrow{q,r_i}) + B(\vec{q})\sin(\overrightarrow{q,r_i})] \quad [5.14]$$

where R_m can be given in various forms:

$$R_m = M\left(\frac{\delta^2 g}{\delta\varphi^2}\right)^2/2K = 2MKq^4 = -\frac{1}{2}M\frac{\delta^2 g}{\delta\varphi^2}q_m^2 \quad [5.15]$$

From the above, it follows that at the initial stages of spinodal decomposition, one has to observe the exponential growth in the intensity of sinusoidal fluctuations of composition. Their wavelength is determined by the relation 5.13. Optimum and critical wavelength of spinodal decomposition are interconnected as[8] $\lambda_m = 2^{1/2}\lambda_c$.

For later stages of decomposition, i.e., at large values of difference $(\varphi - \varphi_0)$, expression 5.3 can be considered only as a rough approximation, as it predicts continuous increase of this difference with time. The concentration difference cannot exceed the limiting values, corresponding to evolution of microregions of pure components. It was shown[26] that for the late stages of spinodal decomposition, the diffusion equation acquires the following form:

$$\frac{\delta\varphi}{\delta t} = D_1\nabla^2\varphi - D_2\nabla^4\varphi + D_1(\nabla\varphi)^2 \qquad [5.16]$$

where $D_1 = (M/N_v)(-\delta^2 g/\delta\varphi^2)$, $D_2 = 2MK/N_v$, $D'_1 = \delta D_1/\delta\varphi$, where N_v is the number of atoms in volume unit. When the volume of the system depends on its composition, the sequence of infinitely small fluctuations of composition in solid should lead to the appearance of elastic deformations. Cahn[10] has shown that, as a result, an additional contribution into the free energy of the system appears. Its value is equal to $\eta^2 E(\varphi - \varphi_0)^2/(1 - \nu)$, where E is Young modulus, ν is Poisson ratio, η is linear expansion coefficient, corresponding to a single composition change. Taking this relation into account one can express the contribution to ΔF of Fourier component with the wavelength $2\pi/q_m$ as follows:

$$\frac{1}{2}VA^2\left(\frac{\delta^2 g}{\delta\varphi^2} + \frac{2\eta^2 E}{1-\nu} + 2Kq^2\right) \qquad [5.17]$$

Therefore, for the system with elastic deformations, arising as a result of the local composition change, the stability limit is determined by the following expression:

$$\frac{\delta^2 g}{\delta\varphi^2} + Y = 0 \qquad [5.18]$$

where $Y = 2\eta^2 E/(1-\nu)$. The elastic stress stabilizes the system. The relation 5.18 determines the position of the so-called coherent spinodal as distinct from chemical spinodal, determined only by the chemical nature of components and by condition $\delta^2 g/\delta\varphi^2 = 0$.

Cahn theory forms the fundamental basis for studying kinetics of phase separation by spinodal mechanism in aging of metal alloys,[27] inorganic glasses,[28] etc. Corresponding experiments are based on the application of small angle X-ray scattering, SAXS, and angular dependence of light scattering. The application of the Cahn-Hilliard theory to data interpretation allows to calculate such important characteristics, as the optimum wavelength of spinodal decomposition, second derivative of free energy by composition, value $\eta^2E/(1-\nu)$ and R(q). Calculations are based on the analysis of the time dependence of the intensity of scattered light or X-rays. As shown,[11] such dependence can be found from the ratio:

$$I(q,t) = I(q,0)\exp[2R(q)t] \qquad [5.19]$$

where R(q) is found as

$$R(q) = -\left(D \Big/ \frac{\delta^2 g}{\delta\varphi^2}\right)\left(\frac{\delta^2 g}{\delta\varphi^2} + \frac{2\eta^2 E}{1-\nu} + 2Kq^2\right)q^2 \qquad [5.20]$$

D is the coefficient of interdiffusion of components. From the theory, it follows that the main features of spinodal decomposition are the following: rectlinearity of dependencies $R(q)/q^2$ on q^2 and lnI(q,t) on t, the presence of interference maximum on the angular dependence of scattering intensity. Revealing these effects allows one to find the presence or absence of spinodal decomposition. The analysis of corresponding dependencies allows to evaluate some parameters of decomposition (see below).

When establishing the most characteristic feature of spinodal decomposition, the mathematic modelling of the phase separation in the region of unstable states is very important. In spite of relatively small amount of such investigations, their results have given the possibility to retrace in time the processes of the appearance, enrichment, and, in some cases, coalescence of microregions of heterogeneous structure.[18,19] In particular, it was established[19] that at a definite composition of the system, there can be found temperature corresponding to the maximum rate of formation of interconnected microregions of phase separation. The development of modulated structures by spinodal decomposition corresponds to a short period of development of interacting nuclei. During later stages of decomposition, the important role belongs to hydrodynamic initiated coalescence. Numerical solution of equations of type 5.16 gives the possibility of modelling phase separation both in metastable and unstable regions of phase diagram and evaluation of the microregion coalescence during late stages of decomposition.

Recently, some theoretical works were aimed at more general description of the phenomenon of spinodal decomposition than possible with Cahn-Hilliard theory. Such approach is necessary because the main equation of Cahn theory (Eq 5.3) cannot be applied generally. Based on symmetry consideration, it takes no account of contributions proportional to $\nabla\varphi$. As an alternative to describe the phase separation in the region of unstable states, it was proposed[20] to use the theory based on the set of phenomenological coefficients. The latter can be determined from the condition of correlation with experimental data. Another approach is based on dissemination of theory of disturbance in liquid state of non-uniform liquids.[21] The numerical solution of the corresponding diffusion equation for one-dimensional case shows essential deviations for non-linear regime from Cahn-Hilliard theory.

Modern theory of phase separation was developed[17,22] which includes description of processes of nucleation and spinodal decomposition. These processes are considered as relaxation processes obeying Markov equation. For thermodynamic system subjected to such relaxation process (for example, liquid in a definite state), the rate of changes in local density $\rho(r,t)$ is determined by the evolution equation

$$\frac{\delta\rho(r,t)}{\delta t} = -K[1 - \exp(\lambda^2\nabla^2/r)\frac{\delta\Omega}{\delta\rho(r)}|_{\rho(r,t)}] \qquad [5.21]$$

where K is a constant, connected with temperature and characteristic transition time, Ω is the great thermodynamic potential, and λ is the distance which liquid molecule passes in characteristic time. Eq 5.21 describes the rate of the most probable evolution of density fluctuation. In approximation of small variation of $\rho(r,t)$ in comparison with λ, the next Cahn equation is valid:

$$\frac{\delta\rho}{\delta t} = K'\nabla^2\frac{\delta\Omega}{\delta\rho(r)}|_{\rho(r,t)} \qquad [5.22]$$

From the analysis of Eqs 5.21 and 5.22, it is seen that Ω may be identified with Lyapunov function. The investigation of the number and properties of corresponding stationary solutions is of a special interest. The direction of the most probable evolution of the system is determined by the condition:

$$\frac{d\Omega}{dt}[\rho(r,t)] \le 0 \qquad [5.23]$$

In particular, if we have two stationary states: $\rho_1(r)$ and $\rho_2(r)$, then on the way from the first to second state fluctuation, $\Delta\rho(r)$ is described as

$$\Omega[\rho_1(r) + \Delta\rho(r)] > \Omega\rho_2(r) \qquad [5.24]$$

The stationary state is conditioned as

$$\frac{\delta\Omega}{\delta\rho(r)} = \frac{\delta F}{\delta\rho(r)} - \mu = 0 \qquad [5.25]$$

The uniform states correspond to the trivial solution of Eq 5.25: $\rho(r)=\rho$. At the point where such solution becomes unsteady, bifurcation takes place.

The correlation 5.25 gives the condition of unsteadiness for such uniform solution in spinodal region. This means that an infinite family of unstable stationary states is branched off from each point below spinodal. The unsteadiness of uniform solutions is determined either within the framework of mechanical analogy (a great response to outer action) or thermodynamic analogy (the lack of a simple maximum of $\Omega[\rho(r)]$). It was established that periodical stationary states are branched off with a definite magnitude of initial wavelength, λ_c. Some spontaneous fluctuations are capable to transfer the system from becoming unstable, one-phase state, to metastable through periodical states mentioned above. Later due to nucleation, the transition from metastable to equilibrium uniform state becomes possible. The property of $\Omega[\rho(r)]$ known as Lyapunov function (Eq 5.21) determines the direction in the sequence of stationary states, passed through the most probable path after transferring the system into spinodal region of phase diagram. In such a way, the phenomenon of spinodal decomposition is associated with a great number of pathways of the system relaxation to equilibrium. After quenching, the system develops periodical non-uniformity with a wave vector, which corresponds to the initial state. With time this non-uniformity increases by amplitude and wavelength up to the moment, where nucleation becomes possible after reaching metastable state. In metastable state, the fluctuations in various parts of the space are not dependent on each other. These fluctuations lead to the development of nucleation process, which naturally cannot be described within the framework of the local approximation by Cahn-Hilliard.

5.3 APPLICATION OF THE THEORY OF SPINODAL DECOMPOSITION TO POLYMERIC SYSTEMS

To describe the kinetics of spinodal decomposition in polymer solutions, van Aarsten[29] has used the model of concentrated solution, i.e., a totality of segments uniformly distributed in solvent medium. In this case, equation of Flory-Huggins for free energy of mixing can be applied. Then, the expression for free energy proposed by Debye[30] was used:

$$F_e = kT \int_V \Delta V[\frac{\varphi_0}{v_1}\ln\varphi_0 + \frac{(1-\varphi_0)}{v_2}\ln(1-\varphi_0) + \frac{\Omega}{2kT}\varphi_0(1-\varphi_0) +$$

$$+\left(\frac{1}{v_1\varphi_0} + \frac{1}{v_2(1-\varphi_0)} - \frac{\Omega}{kT}\right)\frac{(\varphi-\varphi_0)^2}{2} + \frac{H}{2kT}\nabla^2\varphi] \qquad [5.26]$$

where $(\varphi\text{-}\varphi_0)$ is the difference between local concentration, φ, and average concentration, φ_0, of solution, ΔV is a volume element, v_1 and v_2 are molecular volumes of solvent and polymer, Ω and H are interaction parameters. The contribution of fluctuations to free energy ΔF can be presented as

$$\Delta F = \int_V \Delta V \left[\frac{1}{2} \left(\frac{\delta^2 g}{\delta \varphi^2} \right)_{\varphi_0} (\varphi - \varphi_0)^2 + \frac{1}{H} \nabla^2 \varphi \right] \qquad [5.27]$$

where g is free energy of mixing homogeneous solution (the first term in Eq 5.26). The comparison of expressions obtained by Cahn (Eq 5.5) and Debye shows they are identical, if K = (1/2)H.

From Eqs 5.27 and 5.5, ΔF is negative if the first term in brackets is negative and greater by absolute magnitude than the second term. According to Cahn, concentration fluctuations with $q < q_{crit}$ should spontaneously increase with time. The rate of increase depends on q. Most often the fluctuations with a wavenumber q_m are realized (Eq 5.12). To evaluate q_m, it is necessary to find $\delta^2 g/\delta\varphi^2$. Using Flory-Huggins Eq 5.16, we have

$$\left(\frac{\delta^2 g}{\delta \varphi^2} \right)_{\varphi_0} = RT \left[\frac{1}{V_1 \varphi_0} + \frac{1}{V_2 (1 - \varphi_0)} - \frac{2\chi_1}{V_1} \right] \qquad [5.28]$$

Assuming that $\chi_T = A + (B/RT)$ and considering that at some temperature T_{sp} (temperature on spinodal), the expression 5.28 should be equal to zero, we have

$$\left(\frac{\delta^2 g}{\delta \varphi^2} \right)_{\varphi_0} = \frac{2RT}{V_1} (\chi_s - \chi_T) = 2B \left(\frac{T}{T_s} - 1 \right) \qquad [5.29]$$

If value K is known, one can calculate q_m from Eqs 5.13 and 5.29. To find K, the Debye concept on concentration fluctuations as periodic functions of the type $\eta(q) = \eta_0(q)\sin(qx)$ was used. By measuring the intensity of the scattered light, its intensity should be proportional to $\eta_0^2(q)$, where $q = (2\pi/\lambda)s = 4(\pi/\lambda)\sin\theta/2$ and $\eta = \varphi - \varphi_0$. Value $\eta_0^2(q)$ may be found as

$$\eta_0^2(q) = \int_0^\infty \eta_0^2 \exp(-\Delta F/VkT) d\eta_0 \Big/ \int_0^\infty \exp[-\Delta F/VkT] d\eta_0 =$$

$$= 2RT \left[\left(\frac{\delta^2 g}{\delta \varphi^2} \right)_{\varphi_0} + \frac{K}{B} \frac{4\pi^2}{\lambda^2} s^2 \right]^{-1} \qquad [5.30]$$

where $s = 2\sin\Theta/2$. Substituting 5.29 in 5.30, we obtain

$$\eta_0^2(q) = RT/B[(T/T_s - 1) + (4K\pi^2 s^2)/B\lambda^2]^{-1} \quad [5.31]$$

Comparing 5. 31 with corresponding Debye equations,[30] we find that

$$K/B = l^2/6 \quad [5.32]$$

where l is Debye radius of action of intermolecular forces. Combining 5.13, 5.30, and 5.32, we can find the magnitude of q as

$$q_m = l^{-1}[3\,|\,1 - T/T_s\,|\,]^{1/2} \quad [5.33]$$

The wavelength of the most often realized fluctuation is

$$\lambda_m = 2\pi q_m^{-1} = 2\pi l[3\,|\,1 - T/T_s\,|\,]^{-1/2} \quad [5.34]$$

It is worth noting that value λ_m depends not simply on l, but is a function of thermodynamic quality of the solvent.

Using Cahn equation for concentration fluctuations of non-homogeneous solution 5.12, the expression for the intensity of scattered light was derived[31] as a function of time:

$$I_{v_V} = K_1 \exp[2R(q)t] + K_2 \quad [5.35]$$

where

$$K_1 = \frac{1}{4}\left(\frac{4\pi^2}{\lambda_0^2}\right)^2 I_0 V \left(\frac{\delta\alpha}{\delta\varphi_2}\right)_{\varphi_2} [A^2(q) + B^2(q)] \quad [5.36]$$

λ_0 is the wavelength of incident light, $\delta\alpha/\delta\varphi_2$ is fluctuation of solution polarizability due to concentration fluctuation. Value K_2 represents the scattering intensity by glass-made walls of the cell in which the specimen is placed. The angular dependence of the scattered light can be written as follows:

$$P = \frac{1}{q^2}\frac{\delta}{\delta t}\left[\ln\left(\frac{\delta I_{v_V}}{\delta t}\right)_{s_2}\right] = 4MK(q_{crit}^2 - s^2) \quad [5.37]$$

where λ is the wavelength in a medium.

From equation 5.35, dependence ln (dI_{v_V}/dt) on t should be linear and the curve should have a slope equal to $2R_2(q)$. As follows from Eq 5.11, dependence of $R(q)/q^2$ on q^2 by spinodal decomposition should be linear, as well. Having determined value $MK = M(\delta^2 g/\delta\varphi^2) = D$ from the slope of dependence $R(q)/q^2 = f(q^2)$, one can calculate from Eq 5.30 the value q_{crit} and from Eq 5.34 the value λ_m.

De Gennes[32] and Pincus[33] have developed the theory of spinodal decomposition for a mixture of two high molecular components. To describe the dynamics of fluctuations in such a systems, de Gennes used the concept of chain reptation in a tube formed by labile knots of entanglements.[34-37] Free energy of the system was taken as expansion of Flory-Huggins equation for non-uniform systems

$$\frac{F}{kT} = \frac{\varphi_A}{r_A}\ln\varphi_A + \frac{\varphi_B}{r_B}\ln\varphi_B + \frac{a^2}{36\varphi_A\varphi_B}(\nabla\varphi_A)^2 + \chi_1\varphi_A\varphi_B \qquad [5.38]$$

On this basis, the equation describing the relaxation of concentration fluctuations with wavenumber, q, was derived:

$$\frac{d}{dt}\delta\varphi_q = -q^2\Lambda(q)\left(\frac{1}{N\varphi_A\varphi_B} + \frac{a^2q^2}{36\varphi_A\varphi_B} - 2\chi\right)\delta\varphi_q \qquad [5.39]$$

The time dependent changes of the local composition of the system are connected with component A flow through the continuity equation

$$\frac{\delta\varphi_A}{\delta t} + \text{div } J_A = 0 \qquad [5.40]$$

The Fourier component of the flow for wave vector $J_A(q)$ can be determined on the basis of corresponding Fourier component of the chemical potential gradient of the component $\nabla\mu(q)$ using equation

$$J_A(q) = -[\Lambda(q)/kT]\nabla\mu(q) \qquad [5.41]$$

where $\Lambda(q)$ is Onsager coefficient. Having found $\nabla\mu$ and leaving in corresponding equation only linear terms by $\delta\varphi(r)$, we can write

$$\frac{\nabla\mu}{kT} = \nabla(\delta\varphi_A)\left(\frac{1}{N\varphi_A\varphi_B} - 2\chi\right) - \frac{a^2}{36\varphi_A\varphi_B}\nabla^3(\delta\varphi_A) \qquad [5.42]$$

From Eqs 5.40-5.42, the so-called central relaxation formula can be derived:

$$\frac{1}{\tau_q} = \frac{1}{\delta\varphi_A}\frac{\delta(\delta\varphi_A)}{\delta t} = q^2\Lambda(q)\left(\frac{1}{N\varphi_A\varphi_B} - 2\chi + \frac{a^2q^2}{36\varphi_A\varphi_B}\right) \qquad [5.43]$$

Relaxation spectrum (according to Eq 5.43) determines the initially unstable mode in binary system and the rate of its growth. If to assume the existence of only single relaxation time for the wave vector mode, q, then the Onsager coefficient in melt of two polymers is derived by de Gennes as

$$\Lambda(q) \approx \varphi_A\varphi_B N_e D_1 (qa)^2 \qquad [5.44]$$

where D_1 is microscopical diffusion coefficient and N_e is the amount of polymer links between entanglements. From this relation follows that Onsager coefficient is proportional to the squared wave vector, i.e., sharply increases with diminishing wavelength. When the system is transferred to two-phase region, where $N^{-1} << \chi_1 << 1$, we have

$$-\frac{1}{\tau_q} = \text{cost } D_1 N_e \chi q^4 a^2 (2 - q^2 l^2) \qquad [5.45]$$

where $l = (1/3)a\chi_1^{-1/2}$ is the thickness of the interfacial layer in the theory of Helfand-Tagami.[38,39] From Eq 5.45, it is seen that at $ql<2^{1/2}$, the fluctuations are enhanced. Because $(l/R_g)^2 = 1/\chi_1 N << 1$ (R_g is unperturbed gyration radius of a coil), value l is much smaller than coil dimension. Expression 5.45 differs from Cahn equation for spinodal decomposition by introducing factor q^4 instead q^2. The stable fluctuation growth proceeds at

$$q = 2\frac{\sqrt{3}}{9}a\chi^{-1/2} \qquad [5.46]$$

This means that the optimum wavelength is much smaller than gyration radius of a polymer coil. Taking Eq 5.46 into account, the optimal rate of the fluctuation growth can be presented by the relation

$$-\frac{1}{\tau_{opt}} \approx D_1 N_e \chi a^2 l^{-4} \approx (N_e/\tau_1)\chi^3 \qquad [5.47]$$

where τ_1 is the time of microscopical diffusion of the chain element.

For the latest stages of growth (Ostwald ripening), the average dimension of droplets increases according to equation

$$R^2\frac{dR}{dt} = \frac{2\Lambda(q)\gamma a^2}{kT} \qquad [5.48]$$

where γ is interfacial energy equal to

$$\gamma = kTa^{-3}\chi l = \frac{1}{3}kT\chi^{1/2}a^{-2} \qquad [5.49]$$

whereas $\Lambda(q)$ is determined at $\varphi = \varphi_{crit}$ by expressions

$$\Lambda(0) = D_1 N_0 N^{-1}\varphi_A\varphi_B \quad (qR_g < 1) \qquad [5.50]$$

$$\Lambda(q) \approx \varphi_A\varphi_B N_e D_1 (qa)^2 \quad (qR_g >> 1) \qquad [5.51]$$

For $qR_0 > 1$, the following expression is valid:

$$R^4\frac{dR}{dt} = \mathrm{const}\frac{D_1 N_e a^5 \gamma \varphi_c(1-\varphi_c)}{kT} \quad [5.52]$$

In this regime, R(t) increases proportional to $t^{1/5}$. When R(t) becomes greater than R_0 (i.e., $\Lambda(q) = \Lambda(0)$), we have

$$R^2\frac{dR}{dt} = \frac{2ND_{rept}\gamma a^3}{kT}\varphi_c(1-\varphi_c) \quad [5.53]$$

and dimensions increase proportionally to $t^{1/3}$. Value $\delta\varphi_q$ entering Eq 5.39 can be measured by neutron, X-ray, or light scattering experiments. The scattering intensity measured at a given q is determined by structure factor

$$S(q) = \langle|\delta\varphi_q|^2\rangle \quad [5.54]$$

If the solution of Eq 5.39

$$\delta\varphi_q(t) \sim \exp(-\tau_q^{-1}t) \quad [5.55]$$

together with Eq 5.43 is inserted to Eq 5.39, one can obtain the expression for time dependence of structure factor S(q,t):

$$S(q,t) = S(q,0)\exp(-2\tau_q^{-1}t) \quad [5.56]$$

It was shown[40] that calculation of the total structure factor, S(q,T), accounts for effects of the random thermal forces, responsible for concentration fluctuations. These effects were later considered[13,33,42] to explain the spinodal decomposition. For example, Binder[13] has fundamentally revised the dynamics of concentration fluctuations in polymer blends. The equations derived for the time dependence of structure factor include terms connected with thermal fluctuations. Following the procedure,[40] Binder[13] derived the following equation for movement of structure factor S(q,t):

$$\frac{dS(q,t)}{dt} = -2q^2\Lambda(q)\left[\left(\frac{1}{N\varphi(1-\varphi)} - 2\chi + \frac{a^2q^2}{18\varphi(1-\varphi)}\right)S(q,t) - 1\right] \quad [5.57]$$

This equation has a non-vanishing stationary solution:[42]

$$(dS/dt) \to 0 \quad S_T(q) = 1/\left(\frac{1}{N\varphi(1-\varphi)} - 2\chi + \frac{a^2q^2}{18\varphi(1-\varphi)}\right) \quad [5.58]$$

In one-phase region, S_T is finite and positive for all q. Stationary solution gives structural factor at thermal equilibrium. It shows the well-known Ornstein-Zernike form.

Eq 5.57 describes correctly the structural relaxation after temperature jump in one-phase region. The equation, such as 5.57 should be used for initial stage of spinodal decomposition. For values χ in spinodal region of phase diagram, function $S_T(q)$ (Eq 5.58) shows singularity at $q=q_C$. This value is negative for small wavenumber $q<q_C$ and positive for high wavenumbers $q>q_C$. In the latter case, function $S_T(q)$ describes metastable structural factor approached at early stages of spinodal decomposition.[48]

Equation of movement 5.57 is restricted to wavenumbers $q<<R_g^{-1}$. The generalization and expansion to greater region of values $q<<a^{-1}$ can be obtained by substitution of Eq 5.58 by the complete equilibrium structure factor as given by the dynamical random phase approximation, RPA:[37]

$$S_T^{-1}(q) = \frac{1}{N\varphi(1-\varphi)} f_D^{-1}(x) - 2\chi \quad [5.59]$$

where $f_D(x)$ is Debye structure factor for non-interacting ideal chains:

$$f_D = \frac{2}{x}[1 - (1 - \exp(-x))/x] \quad [5.60]$$

with $x = q^2R_g^2$. The general equation of movement has the form:

$$dS(q,t)/dt = -2q^2\Lambda(q)S_T^{-1}(q)[S(q,t) - S_T(q)] \quad [5.61]$$

with $S_T^{-1}(q)$ given by Eq 5.59. Binder has shown that equation of movement 5.57, valid for low wavenumber, follows from the expansion in series $S_T^{-1}(q)$ in Eq 5.59 by q up to the second order terms in q^2. Eq 5.61 has a typical form of irreversible equation of movement in linearized approximation and may be solved immediately

$$S(q,t) = S_T(q) + [S(q,0) - S_T(q)]\exp(-2\tau_q^{-1}t) \quad [5.62]$$

$$\tau_q^{-1} = q^2\Lambda(q)S_T^{-1}(q) \quad [5.63]$$

In one-phase region, $S_T(q)$ constitutes the equilibrium structure factor, whereas, in spinodal region, it corresponds to a "virtual" structure factor which sets the direction of change at the early stages of unmixing.

For $q << R_g^{-1}$, Cahn-Hilliard model leads to reasonable rates of relaxation, τ_q^{-1}. However, as seen from Eq 5.62, these rates should not be associated with an exponential growth or decay of the initial structure factor, $S(q,0)$, (Eqs 5.43-5.48), but with the transition to the virtual final structure factor, $S_T(q)$.

To neglect value $S_T(q)$, it is possible only when $S(q,t=0) >> |S_T(q)|$, which is a very rare case. Cahn-Hilliard model predicts $S(q,t)$=constant for $\tau^{-1} = 0$, which occurs for $q=q_c$. The condition introduced by Cook[40] treatment according to Eq 5.62 is different, namely $S(q,0) = S_T(q)$.

In some works,[13,32,33] Onzager's coefficient has been used, $\Lambda(q)$, that of ideal non-interacting chains which perform a Rouse or reptation motion. The expressions given by Binder are:

$$\Lambda(q) \approx wa^2\varphi(1-\varphi)f_D \qquad \text{Rouse motion} \qquad [5.64]$$

$$\Lambda(q) \approx (wd^2/N)\varphi(1-\varphi)f_D \qquad \text{reptation} \qquad [5.65]$$

Here w denotes the elementary jump rate of the segments, d gives the diameter of the tube, which restricts the reptative motion. The relaxation rates τ_q^{-1} follow from Eq 5.63 together with Eq 5.59:

$$t_q^{-1} \approx (6/\tau_R)[1-(\chi/\chi_s)f_D(x)] \qquad [5.66]$$

where $\chi_S = 1/2N\varphi(1-\varphi)$.The fundamental time τ_R is $\tau_R = N^2/w$ for Rouse motion and $\tau_R = N^2a^2/wd^2$ for reptation.

In both cases, Rouse motion and reptation, τ_R, describes the time required by a polymer molecule to diffuse over a length comparable to its own size, $N^{1/2}a$. The two modes show a different molecular weight dependence.

Eq 5.66 is valid after all interactional modes in the molecules have relaxed, i.e., for $t > \tau_R$. The limiting behavior of τ_q^{-1} for small and large x becomes $\tau_q^{-1} \approx (6/\tau_R)(1-\chi/\chi_S)x$ for $x<<1$ and $\tau_q^{-1} \approx (6/\tau_R)x$ for $1<<x<<(a^2)$. For small x, a critical slowing down is expected on approaching the spinodal. For large values of x, the times lie in the order τ_R and became independent on the interaction. From the analysis of the early stages of spinodal decomposition, such important parameters can be found, as D_{app}, D_c, $q_m(0)$, and ε_T which characterize collective diffusivity, self-diffusivity, maximum wavenumber, and thermodynamic motive power $\varepsilon_T = (\chi - \chi_S)/\chi_S$. From the dependence $R(q)/q^2$ on q^2 values D_{app} and $q_m(0)$ can be found:

$$R(q)/q^2 = D_{app}[1-(q^2/2q_m(0)^2)] \qquad [5.67]$$

For symmetric polymer blends, i.e., for $N_A = N_B$, values D_c and ε are connected with D_{app} and q_m as[32]

$$D_{app} = D_C\varepsilon_T; \quad q_m^2 = 3\varepsilon_T/2R_g^2 \qquad [5.68]$$

The ratio of two parameters is:

$$q_m^2/D_{app} \sim D_C^{-1}R_g^{-2} \qquad [5.69a]$$

and does not depend on ε but depends on D_c, i.e., on dynamics of polymer chains due to phase transition. If dynamics obeys reptation mode,[42,43] then

$$q_m^2/D_{app} \sim N \qquad [5.69b]$$

because $D_c \sim N^{-2}$ and $R_g^2 \sim N$. If the dynamics obeys Rouse mode, then

$$q_m^2/D_{app} \sim N^0 \qquad [5.69c]$$

Molecular weight dependence of q_m^2/D_{app} may serve as characteristics of molecular dynamics and transport mechanism. The characteristic time

$$t_C = \zeta^2/D_{app} \qquad [5.70]$$

where ζ^2 and D_{app} are thermal correlations length and mutual diffusivity of the system at a given temperature, T. Because t_C depends on the type of the system, the time of evolution of fluctuations during phase separation is usually characterized by reduced scale time:

$$\tau = t/t_C \qquad [5.71]$$

where t is a real time. Critical reduced time, t_C, for which spinodal decomposition is described by linearized theory with good accuracy, is a universal constant[44] equal approximately to 2. Increasing t_C allows to broaden the real time scale, where the early stages of spinodal decomposition are observed. This case is usually realized for polymer blends.[45,46]

Pincus,[33] using the central relaxation formula by de Gennes, has corrected his conclusions as for the value of the optimal wavelength of spinodal decomposition. Pincus considers that, for the wave vector of non-interacting chains in a melt, there is not only one, but at least two independent relaxation modes.[41] For Onsager coefficient, he has found, at $qR_0 >> 1$:

$$\Lambda(q) = 2T\mu_1(ql)^{-2}[1 - \exp(-q^2R_g^2/6)]\varphi_A\varphi_B \qquad [5.72]$$

where μ_1 is a mobility of monomeric unit. The Onsager coefficient is proportional to the reciprocal value of squared wave vector. Combining Eqs 5.42 and 5.72, we can find the expression describing the rate of relaxation to the equilibrium state:

$$\frac{1}{\tau_q} = \frac{1}{2}(T\mu_1/L^2)[1 - \exp(-q^2R_g^2/6)][N^{-1} +$$

$$+ 2\chi\varphi_A\varphi_B + (qa)^2/36] \qquad [5.73]$$

where L is a contour length of a chain. At $\tau_q < 0$, there is observed exponential growth of concentration fluctuations. By sharp transition from $\chi < 0$ to $\chi > \chi_S$, the initially unstable mode is characterized by extremum, i.e., by fast transition of the system into the region below spinodal, we have:

$$q_m^{-1} \approx (R_g/6)[\ln(6\chi/\chi_s)]^{-1}; \quad \chi > \chi_s \qquad [5.74]$$

In the unstable region of the phase diagram near spinodal:

$$q_m^{-1} \approx (R_g/6\sqrt{3})(\chi/\chi_s - 1)^{-1/2}; \quad (\chi - \chi_s)/\chi_s << 1 \qquad [5.75]$$

For the first case, we have the following expression for the optimum rate of fluctuation growth:

$$\frac{1}{\tau_{q_m}} \approx T_{rept}^{-1}(\chi/\chi_s); \quad \chi >> \chi_s \qquad [5.76]$$

In the second case:

$$\frac{1}{\tau_q} \approx T_{rept}^{-1}(\chi/\chi_s - 1); \quad (\chi - \chi_s)/\chi_s << 1 \qquad [5.77]$$

where T_{rept} is the time necessary for the chain to go out of the entanglement tube. Pincus has shown that unstable mode has a wavelength comparable with radius of molecular coil. It changes proportionally to $N^{1/2}$. The rate of fluctuation growth, corresponding to this mode, is proportional to the constant of rotation diffusion in a melt, changing as N^{-2}. Its concentration dependence reflects the shape of a spinodal.

The analysis of phenomenon of spinodal decomposition in homopolymer blends, made by Pincus, shows that microphase separation on a scale of the order N is possible only for the case of very deep cooling of one-phase blend of polymers. The expressions given above are restricted by linear regime of spinodal decomposition, i.e., by its early stages. In this case, value q_m does not depend on time and maximum of intensity of structure factor (or light scattering intensity) exponentially increases with time. However, it is known that for low-molecular systems,[15] the duration of early stages is very short, followed by the regime, characterized by development of periodicity of modulated structure according to power law with exponent close to 1/5. It is necessary to consider what are the time intervals restricting the applicability of linear approximation for the blends of high molecular compounds. According to reference,[13] it may be expressed as

$$t_m << (\tau_{q_m}/2) \ln \sqrt{N} \frac{1 - (\chi_{crit}/\chi)}{(1 - \chi/\chi_s)^{1/2}} = \tau_{q_m} \ln N \qquad [5.78]$$

The calculation shows that at N = 3,000, this time interval approximately coincides with with τ_{qm}.

Linearized theories by Cahn-Hilliard,[5-10] Hillert,[12] and others[11,17-22, 41,42] describe the early stages of spinodal decomposition and predict the exponential growth of intensity of light scattering, the appearance and growth of maximum of light scattering at a constant angle with increasing time of decomposition. New theories and experimental investigations reveal[47-57] that the amplitude of composition fluctuations (maximum of light scattering intensity) is shifted with time to smaller angles (intermediate stage) and then amplitude reaches its equilibrium value. Interfacial profile and the heterogeneity length are continuously developed (transition stage) and the last stage occurs. During this stage only heterogeneity length is changed[58,59] whereas the structure change includes its "self-similarity".[60] To describe this nonlinear evolution, it was proposed[53] to introduce additional combination terms into diffusion equation. An additional term due to hydrodynamic flow was introduced[54] as well. The mechanism of coalescence, in the framework of constant-volume evaporation, was considered by Lifshitz and Slyozov.[23]

Izing model[55] may be considered as the next approximation. Model includes calculation of a set of interacting spins using Izing hamiltonian. The calculations[56] include cluster expansion of droplets with a given number of molecules. The rate of cluster growth is calculated by accounting for simple molecular exchange, binary dividing of droplets and their melting. To explain the acceleration of the domain growth at the early stages, two factors were taken into account:[57] diffusion and hydrodynamic flow. It was shown[57] that the domain growth proceeds through three stages:

- Initial stage, $\lambda \sim AT^{1/3}$, $\lambda_0 \le \lambda \le 5\lambda_0$
- Flow stage, $\lambda \sim Bt$, $(k_BT/\gamma)^{1/2} \le \lambda \le (\gamma/g\Delta\rho)^{1/2}$
- Gravitation stage, $\lambda \ge \gamma/g\Delta\rho$

Here k_B is Boltzman constant, $\Delta\rho$ is the difference in density between two phases, γ is surface tension, g gravitation constant,

$A = [(6/5)k_BT/\pi\eta]^{1/3}$, η is viscosity, $B = [p\gamma/\eta]$, p is a dimensionless value.

The general feature of all the theories is the application of the universal scaling concept. The time-dependent heterogeneity length (or characteristic length scale) is predicted to evolve a single function of time, t, for all systems that belong to particular dynamic class. For example, binary liquid mixtures are characterized by a conserved order parameter with a long time ($t \to \infty$) limiting growth law[57] given by $\lambda_m \sim t^1$. Universal scaling should collapse all spinodal decomposition results into a unique master curve by rescaling the heterogeneity length $\lambda_m(t)$ with its value at t = 0 and plotting it against a scaled time $\tau = tD_{app}q_m^2(t{=}0)$.

The wavenumber of characteristic length scale $q_m(t{=}0)$ can be calculated using Eq such as 5.43, or, according to the de Gennes,[32] Pincus,[33] or Binder[13] theories using equation

$$q_m^2(t = 0) = \frac{3}{2R_g}[(\chi/\chi_s) - 1] \qquad [5.79]$$

where χ is the parameter of segment-segment interaction of Flory-Huggins. In the framework of the mean field theory, the critical point is given[32] as $\chi_C = 2/N$. From scaling analysis of values $q_m(t)$, $I_m(t)$, and the scale structure function F(x,t), it follows that, if spinodal decomposition develops to the stage where F(x,t) for various t are superimposable (i.e., universal for t) over the entire range of x, then

$$\beta = 3\alpha \qquad [5.80]$$

where α and β are exponents determined by

$$q_m(t) \sim t^{-\alpha} \qquad [5.81]$$

$$I_m(t) \sim t^{-\beta} \qquad [5.82]$$

and F(x,t) is determined as

$$F(x,t) = I(q,t)q_m^3(t) \qquad [5.83]$$

with $x = q/q_m(t)$. Eqs 5.80-5.82 follow from the demand of invariability of composition of evolved microphases with time at the latest stages of spinodal decomposition. According to the phenomenological theory by Furukawa[61] values α or β depend on the volume fraction of phase, changing from $\alpha = 1$ for critical composition of a mixture to $\alpha = 1/3$ for non-critical composition of liquid mixtures. Various approximations predict different values of the constant α depending on the effects

Table 5.1: Theoretical values of α

Theory	α
Linearized, Cahn,[5] Cook[40]	no growth
Nonlinear, Langer and others[52,53]	0.21
Nonlinear, hydrodynamic effects[54]	no scale
Monte Carlo Izing model[55,56]	0.21-0.35
Coalescence through constant volume evaporation (Lifshitz & Slyozov)[23]	0.333
Coalescence due to the hydrodynamic effect[57]	1.0

considered in theories (Table 5.1). The scattering intensity I(q,t) may be expressed as[47]

$$I(q,t) \sim \langle\eta^2\rangle \lambda_m(t)^3 S(x,t) \qquad [5.84]$$

where $\langle\eta^2\rangle$ is mean square fluctuation of the refraction index and S(x,t) is a scaling function. Its value changes between 0 and 1 when x increases from zero to infinity. This parameter characterizes q-dependence of the scattered intensity and therefore the pattern self-assembled during spinodal decomposition. Value $\langle\eta^2\rangle$ is proportional to concentration fluctuation $\Delta\varphi(t)^2$ and may be experimentally found from integrated scattering intensity

$$\langle\eta^2\rangle = \int_{q_1}^{q_2} I(q,t) q^2 dq \qquad [5.85]$$

where q_1 and q_2 are wavenumbers at which the intensity of scattering disappears. The scaling function depends mainly on time and temperature, as self-assembled pattern may change with time. However, at the latest stages of decomposition, where the pattern increases with dynamic self-similarity, this function becomes independent of time.

If to compare Eq 5.84 and 5.83, and note the definition of $q_m(t)$, function S(x,t) can be connected with F(x,t) by the relation

$$F(x,t) = \langle\eta^2\rangle S(x,t) \qquad [5.86]$$

According to Hashimoto[47] the process of spinodal decomposition may be considered as consisting of three stages: early, intermediate, and late (Fig. 5.1). At the early stage, composition distributions are described by linearized theory. At the intermediate stage function F(x,t) becomes more expressed with a peak at x = 1, which increases

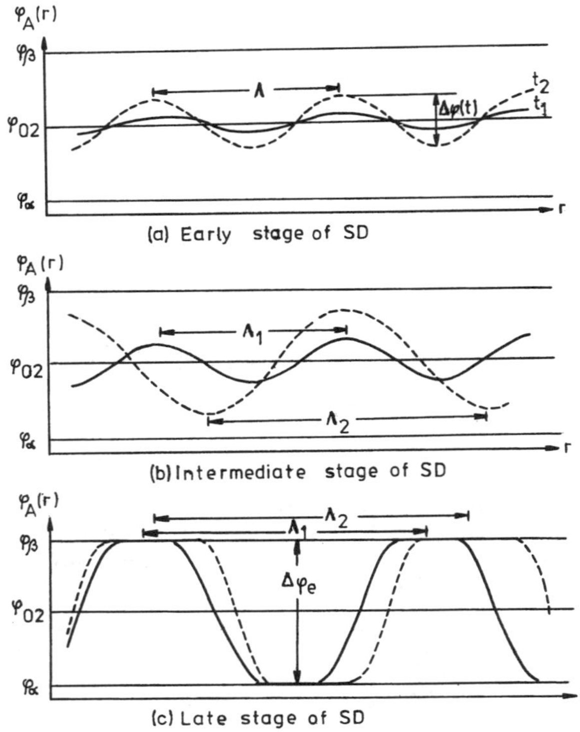

Figure 5.1. Schematic diagram showing the spatial segmental density profile of component A in the early (a), intermediate (b), and late (c) stages of spinodal decomposition. Adapted, by permission, from T. Hashimoto, M. Takenaka, and H. Jinnai, *J. Appl. Crystal.*, **24**, 457 (1991).

with t. However with following development of the process, F(x,t) turned out to be universal for t, first in the range of x smaller than about 2 and then over the entire range of x. The time interval where the former picture occurs, is defined as a late stage I and the one at which the latter was realized is called late stage II. For PB-PI system, it was found[47] that if the experimental data for F(x) at $x > 2$ at the stage I are fitted to relation of the kind

$$F(x,t) \sim x^{-n} \quad [5.87]$$

then the exponent n diminishes with time from 7.2 to 4.0 (dynamic crossover). Similar change in exponent can occur for F(x,t) at a given time, when x increases. This may be referred to as a special crossover, which can be observed at the late stage II. The time of evolution of interphase structure can also be derived.[47] The scattering intensity I(q,t) at large q for self-assembled structure with diffuse interface may be presented as

$$I(q,t) \sim \langle\eta(t)^2\rangle \sum(t) q^{-4} \exp[-\sigma(t)^2 q^2] \qquad [5.88]$$

where σ(t) is a time-dependent profile which characterizes the spread of segment density profile ρ(z) across the interface:

$$\rho(z) = \rho_0(z) * h(z) \qquad [5.89]$$

In this equation

$$h(z) = (2\pi\sigma^2)^{-1/2} \exp(-z^2/2\sigma^2) \qquad [5.90]$$

where z is coordinate normal to interface, $\rho_0(z)$ is a profile of segment density for a sharp interface and (*) denotes a convocation product. Value σ relates to a characteristic thickness of an interface t_I, which can be determined as a width of gaussian curve h(z):

$$t_I = \int_{-\infty}^{\infty} h(z)dz / h(z=0) = (2\pi)^{1/2}\sigma \qquad [5.91]$$

Together with equation 5.88 we obtain

$$\bar{I}(q,t) = I(q,t) / \int_{q_1}^{q_2} I(q,t) q^2 dq \sim \sum(t) q^{-4} \exp[-\sigma(t)^2 q^2] \qquad [5.92]$$

which shows than $\ln[\bar{I}(q,t)q^4]$ at fixed t, plotted against q^2, gives a straight line, whereas parameters Σ(t) and σ(t) can be found from the intercept at q = 0 and slope of line. Value Σ(t) can be expressed by an exponential law

$$\sum(t) \sim t^{-\gamma} \qquad [5.93]$$

Combining Eqs 5.93 and 5.73, we obtain

$$\sum(t)/q_m(t) \sim t^{-(\gamma - \alpha)} \qquad [5.94]$$

$$\sum(t) t_I(t) \sim t^{-\delta} \qquad [5.95]$$

It was shown[47] that non-universality of F(x,t) at high x (> 2) at the late stage I can be explained by the fact that $q_m(t)$ and $\Sigma(t)$ follow different laws of dynamic scaling, whereas $t_I(t)$ still decreases at the late stage I. From Eqs 5.73 and 5.8, it follows that F(x,t) at high x at late stages (where $<\eta^2>$ becomes independent of time), may be presented as

$$F(x,t) \sim \sum(t) q_m(t)^{-1} x^{-4} \exp[-\sigma(t)^2 q_m(t)^2 x^2] \quad [5.96]$$

$$F(x,t) \sim t^{-(\gamma - \alpha)} x^{-4} \exp[-\sigma(t)^2 q_m(t)^2 x^2] \quad [5.97]$$

As the term $t^{-(\gamma-\alpha)}$ diminishes with time, whereas exponential term increases, the growth in F(x,t) at x > 2 at the late stage is to outweigh the latter over the former. Crossover from non-universality to universality of F(x,t) occurs at t, at which $\gamma = \alpha$ and $\sigma_e q_m(t) << 1$, where σ_e is an equilibrium interphase thickness. At the late stages, where the interface contribution in $<\eta^2(t)>$ is small and where the interface thickness reaches its equilibrium magnitude, an equilibrium value of $<\eta^2(t)> = <\eta^2>_e$ is reached as well, which is determined by the phase diagram and temperature. In this case $\tilde{F}(x,t)$ at a given t becomes universal with t. The reduced value of $\tilde{F}(x,t)$ can be written as

$$\overline{F}(x,t) = F(x,t) / \int_{q_1}^{q_2} q^2 I(q,t) dq \sim F(x,t)/<\eta^2>_e \quad [5.98]$$

It becomes universal with T also:

$$\overline{F}(x,t) \sim S(x) \quad [5.99]$$

Reduced intensity $\overline{I}(q,t)$ (Eq 5.92) in Porod regime is expressed as:[47,62,63]

$$\overline{I}(q,t) = [\pi\varphi(1-\varphi)]^{-1} \sum(t) q^{-4} \exp[-\sigma(t)^2 q^2] \quad [5.100]$$

where $\sigma(t)$ is a parameter, connected with $t_I(t)$ by relation:[64]

$$t_I(t) = (2\pi)^{1/2} \sigma(t) \quad [5.101]$$

From the dependence of $\ln[q^4 \overline{I}(q,t)]$ on q^2 both $\sigma(t)$ and $\Sigma(t)$ at q =0 may be found. $\Sigma(t)$ has a dimension of a wavenumber and may be denoted as "local wavenumber" as opposite to $q_m(t)$, characterizing "global wavenumber". An attempt[47] has been made to scale $\Sigma(t)$ and characteristic wavenumber, $q_m(0)$, and characteristic time, t_C. By analogy with analysis of $q_m(t)$, one can determine a dimensionless value $\bar{\Sigma}(t)$ and $\bar{t}_I(t)$:

$$\bar{\Sigma}(t) = \Sigma(t)/q_m(0) \qquad [5.102]$$

$$\bar{t}_I(t) = t_I(t)q_m(0) \qquad [5.103]$$

Another scaling law was proposed by Furukawa.[61] The dynamic scaling form of structure function S(q,t) or intensity I(q,t) was presented in terms of cluster dimension R(t) as:

$$S(q,t) \sim R(t)^d \tilde{S}(x) \qquad [5.104]$$

with universal function $\tilde{S}(x) \sim x^2/(y/2 + x^{2+\gamma})$. Here $x = qR(t)$ and d is a space dimension. For the last stage of spinodal decomposition, where concentration fluctuations reach an equilibrium composition, the universal scaling function $\tilde{S}(x)$ for small and large x is expressed as

$$\tilde{S}(x) \sim x^2 \quad \text{for large x} \qquad [5.105]$$

$$\tilde{S}(x) \sim x^{-y} \quad \text{for small x} \qquad [5.106]$$

For critical composition of the mixture y = 2d, whereas for non-critical compositions y=d+1. For critical compositions, we have

$$\tilde{S}(x) \sim x^2/(3 + x^8) \qquad [5.107]$$

and for non-critical compositions:

$$\tilde{S}(x) \sim x^2/(3 + x^6) \qquad [5.108]$$

For three-dimensional space (d = 3), the structure function should change with exponent 2 at $q < q_m$ and exponent –4 and –6 at $q > q_m$ for non-critical and critical mixtures, correspondingly. If at the late stages of decomposition, the function $\tilde{S}(x)$, calculated for various times, fits the master curve, it means that the process of decomposition is accompanied by the development of self-similarity structure. Values of the exponent –4 assumes the existence of isolated phase-separated clusters. They may be expected for non-critical quenches. Value of the exponent –6 is connected with a percolation structure and may be expected for critical quenches.

5.4 SOME EXPERIMENTAL RESULTS ON KINETICS OF PHASE SEPARATION IN SOLUTIONS AND MIXTURES OF POLYMERS

The theories by Cahn-Hilliard[5-10] and many others[11,12,17-22] describe mainly the initial stage of spinodal decomposition. If to fix the decomposition pattern at later stages by sharp increase in the system viscosity (rapid quenching), one can see a uniform structure with the distances between polymer-enriched regions equal to λ_m. For polymer solutions, the morphology of a system is formed by the interconnected regions of concentrated phase dispersed in a dilute polymer phase or pure solvent. Cahn has assumed that interconnected structure may be maintained if the volume fraction of the second phase is no less than 15%. This means that for polymer solutions such interconnected structures should be observed at mass fraction of polymer of an order of some percents (for example, at polymer concentration in solution of 4.0%, the interconnected structures reach 10% of polymer phase). If polymer-enriched phase is labile at all concentrations, the initially formed regions may aggregate and at last the separation into two liquid phases proceeds.

Experimental studies on mechanism of spinodal decomposition in solutions and mixtures of polymers were conducted by many authors.[67-121] One of the first works was done by van Emmerik and Smolders.[68] For solutions mixtures of poly(2,6-dimethyl-1,4-phenylene oxide) (PPO) with molecular mass $M = 2.1\times10^5$ and caprolactam, it was found, by using light scattering, that at a definite conditions, the phase separation proceeded according to the mechanism of spinodal decomposition. Values λ_m decrease with increase in the difference $\Delta T = |T - T_S|$ in accordance with predictions of Eq 5.33. At $T_S = 400K$ and $l = 20$ nm, values λ_m were calculated for various temperatures. In the range of 399-360K, λ_m changes from 0.23 to 1.43 mcm in a good agreement with the electron microscopy data.[69] In the system under consideration, the phase separation proceeds according to spinodal mechanism. In the beginning, the interconnected structure is developed with a non-equilibrium composition. Later, concentration continues to change. One of the phases is enriched with one of the components. This process is accompanied by the volume changes. At relatively high temperature (increased molecular mobility), some spheres are formed from the polymer-enriched phase.

The mechanism of phase separation was studied for the system PPO-toluene using electron microscopy.[70] When the solution temperature was decreased to the temperature below binodal but above spinodal (metastable region), the nuclei of a new phase appeared, i.e., the separation proceeded according to nucleation mechanism. However, if the same solution was rapidly transferred into the unstable region below spinodal, the interconnected structures were formed together with a small amount of nuclei. This process was caused by a finite time of passing through the metastable zone by cooling. The higher the rate of this transition, the smaller the number of nuclei. Therefore, the structure at the initial stage of phase separation depends on the rate of transition through metastable zone. Some induction time is needed for nuclei formation by cooling to temperature of metastable state. This time increases with temperature increasing. At the temperature, T_S, the induction time approaches zero. For l = 20 nm, T_S = 311 K, the dimensions of spinodal structures at 258 K for 30% solution of PPO in toluene were calculated from Eq 5.33.[70] Values λ_m were in accordance with the data from electron microscopy. The nuclei rate dr/dt was 0.15 mcm/ min.

The spinodal mechanism of phase separation was discovered also for solutions of ethylene-vinyl acetate copolymer, EVA, (40% VA) in caprolactam.[69] For PS-decaline system, the criterion of spinodal decomposition proposed by Cahn-Hilliard is valid only for solutions at critical concentration.[71] At $c > c_{crit}$, the criterion is valid only for the initial stages. The lifetime of spinodal structure increases with temperature difference decreasing (i.e., $|T - T_S|$). The dimensions of spinodal structures and diffusion coefficient met theoretical prediction.

The more detailed study of spinodal decomposition in polymer solutions was conducted for PS solutions in cyclohexane.[72] The time evolution of structure factor, S(q,t), was studied vs. time after overcooling the system to temperatures below T_S. At the initial stages, the intensity of light scattering increases exponentially. The analysis in terms of linear Cahn-Hilliard theory was used to determine the extrapolated rates of growth at zero time and to determine the initial diffusion coefficient. Comparing initial rates for two PS specimens with different molecular weights ($M_1/M_2 = 12$), the evidence of Zimm dynamics was discovered[37] ($D \sim M_w^{0.5}$ and $\tau_0 = 1/(Dq_m^2(0) \sim M_w^{1.5})$. Zimm dynamics was preferred over Rouse or reptation dynamics.

The analysis of the exponential dependence of the domain dimensions vs. time has shown that the exponent for an intermediate stage of decomposition is larger than 1/3 for high molecular weight specimen — untypical of the data for polymer mixtures, where exponents are smaller than 1/3. It was argued that $\alpha > 1/3$ is a sign of hydrodynamic interactions in the system. The diminishing α with time, approaching asymptotically the value 0.4-0.5, means that the hydrodynamic effects are more important for the initial stages (hydrodynamic effects of Siggia[57] give $\alpha \rightarrow 1$). The initial solution concentration is situated in a dilute regime, where hydrodynamic interactions do not screen and polymer dynamics is described by Zimm model. However, immediately after the beginning of phase separation, the majority of polymer molecules is in concentrated or semi-diluted regions, where hydrodynamic screening becomes effective. Thus, hydrodynamic effects become less important with process progressing as compared with binary liquids for which these effects are important in later stages. If it is correct, then α should be equal to 1/3 for initial times at spinodal decomposition of concentrated solutions. Another explanation can be based on the concept considering that mobility of polymer in solution is a strong function of polymer concentration and changes during phase separation. Thus, characteristic time scale changes in the course of phase separation, becoming longer. There is no theoretical model describing the mobility and the characteristic time scale as time-dependent phenomena.

The analysis of asymptotic scale factor by Furukawa[73] $F(q/q_m)$ has shown that for a specimen of a low molecular weight, it is proportional to x^{-2} for a deep quench and shallow quench. For $q/q_m>1$, the exponent is equal to 4 for a deep quench and 3.5 for a shallow quench. According to Furukawa,[73] the exponent equal to 4 assumes the existence of isolated phase-separated clusters and is expected for non-critical superquenching. Value of exponent 6 predetermines the percolation structure and should be expected for critical superquenching. The value of exponent equal to 4, found in[72] for deep quench supposes the cluster isolation and formation of smooth interface. This case is a deep quench when the late stage of decomposition is already reached. Because the mixtures are nearly critical, one would expect the exponent equal to 6. The discrepancy with Furukawa theory was found in many works,[58,59] including the results of computer simulations, giving exponent 6 instead of 4 for critical superquenching. For shallow quench, where the experiments have been done for an intermediate stages of spinodal decomposition exponent 3.5 was found, close to 4. Exponents smaller than 4 have been observed for polymer mixtures as well.[74] The exponent equal to 3 may indicates morphology change during decomposition: in intermediate stage in polymer solutions the morphology is diffuse with a rough interface. With increasing time, the interface becomes less diffuse and exponent reaches 4.

The comparison of theoretical functions $F(x) = 3x^2/(2 + x^6)$ and $F(x) = 4x^2(3 + x^8)$ for critical and non-critical supercooling with experimental findings shows a good agreement between $q/q_m < 1$ and divergence for high values of $x = q/q_m$. The determination of evolution time of structure factor[74] by integration of spinodal decomposition equation for polymer blends gives a result incompatible with Furukawa function. It is evident that the theoretical interpretation is necessary which takes into account the asymmetric shape of phase diagram[72] and includes concentration-temperature dependence of mobility and non-linear effects in polymer solutions.

In some works,[75-80] the spinodal decomposition has been studied for polymer gels. The appearance of a diffraction maximum of light scattering intensity in hydrogels of agarosa was attributed to the formation of regularly arranged regions of phase separation.[75] The phase separation was also studied during transition sol-gel.[76] The mechanism of phase separation in solutions capable to gel, composed of isotactic and atactic PS, PPO, and their mixtures was investigated[77] in different solvents. The gel formation was determined by formation of a thin dispersion of enriched regions when the system is transferred into the region of metastable states rather than by crystallization. Glass transition of these regions is sufficiently high to fix the ends of various polymer chains in one and the same vitrificated region. This concept was verified by measurements of the microregion dimensions and the rate of their formation. When the microregions pass to the elastic state, the segments bound become free and gel melts. It is a phase separation of liquid-liquid type.

Tanaka *et al.*[78-80] studied the spinodal decomposition in gels. The light scattering intensity increases as a function of excluded volume $I \sim (v - v^*)^{-1}$ on approaching spinodal (see Chapter 2). Data on photon correlation spectroscopy[79,80] show also that, on approaching spinodal, the coefficients of cooperative diffusion become negligibly small. This result is in accordance with the mean field increase of correlation length according to $D(T)/(6\pi\eta_s\xi_s)$.

First data on the spinodal decomposition in polymer mixtures were obtained by McMaster[81] and Nishi.[82] McMaster discovered spinodal structures by transition from one-phase to two-phase region for the copolymer of acrylonitrile (28%) with styrene (SAN)-PMMA system at critical concentration. The domain dimensions formed at the initial stages are equal to 50-100 nm according to the electron microscopy data. The destruction of interconnected structures occurs when domains increase their dimensions. Fully destroyed structure consists of droplets having size about 500 mcm. The rate of particle growth is constant until the structure is preserved and then it increases exponentially.

At concentrations far from critical concentrations, the phase separation proceeds according to the mechanism of nucleation and growth. After two-hour exposure near T_S, the droplet dimensions increase only twofold. There is no essential coalescence due to the high viscosity. A highly dispersed structure is formed. The interdiffusion coefficients, depending on the mechanism of separation, change within the limits of 10^{-12} cm^2/s (538K, nucleation) and 10^{-14} cm^2/s (453K, spinodal decomposition). The scale level of phase separation is higher than for low-molecular substances.

Nishi[82] studied the mechanism of phase separation, using NMR and optical microscopy. The PS-PVME system was used which has LCST. The morphological data show that for this system two types of phase separation are possible. By elevating temperature, the microphase structure becomes finer but size of microregions increases at the later stages of the process. After this stage, the droplets begin to join and form macrophase. For slow phase separation, the dimensions of microregions are small as compared with a period of interconnected structure. NMR data confirm the continuous change of microregion composition with time in accordance with the spinodal mechanism of decomposition.

The value of $R(q_m)$, characterizing the kinetics of changes of microregion composition is another evidence of spinodal decomposition:

$$\varphi - \varphi_0 = A(q_m)\exp R(q_m)t \cos xq_m \qquad [5.109]$$

where $R(q_m) = (1/2)q_m^2D$. In this equation, φ is a concentration of PVME in PS-enriched phase. From Eq 5.109, it follows that the total decrease Q of amount of PVME in a given phase during phase separation can be calculated as:

$$Q = \left[\left|\int_{\pi/2q_m}^{3\pi/q_m}(\varphi - \varphi_0)dx\right|\right]^3 = \left[\frac{2}{q_m}A(q_m)\right]^3 \exp[3R(q_m)t] \qquad [5.110]$$

Therefore:

$$\ln Q = 3\ln\left[\frac{2}{q_m}A(q_m)\right] + 3R(q_m)t \qquad [5.111]$$

From the slope of the dependence of lnQ on t, value $3R(q_m)$ is found. Based on Q from NMR data, it was found that in the system with 50% of PVME value $R(q_m)$ is positive. This means the increase in concentration fluctuations with time is in accordance with the theory of spinodal decomposition. From estimation of q_m from morphological data, value of diffusion coefficient is found, $D=-2R(q_m)/q_m^2=-2.8\times10^{-17}$ m^2/sec at $R(q_m)=3.82\times10^{-4}$ sec^{-1} and $q_m = 5.24\times10^{-17}$ m^{-1}.

The kinetics of decomposition of PS-PVME mixtures have been studied in many works.[83-85,87,98] The experimental data are described by the theory in its linear approximation.[13] At the early stages, the exponential growth of light scattering intensity was observed at constant position of the maximum. The characteristic relaxation time was of the order of some seconds. The transition to non-linear regime of decomposition occurred in the time interval a few times exceeding τ_q. The dimensions of microregions at the early stages are in order of a few nanometers and diminish with moving of T_s. Diffusion coefficients are of the order of $2\text{-}9\times10^{-13}$ cm^2/s and by approaching the spinodal (in the region of 2^0) they decrease more than three times.[85] At large distance from spinodal to the region of unstable states, the diffusion coefficient is equal to $10^{-11}cm^2/sec$. At the latest stages of decomposition, the exponents α and β are close to 0.4 and 0.95, correspondingly.[85] The formal consideration shows that these data are in a good agreement with cluster theory by Binder,[16] but do not correspond to the values following from the Langer theory.[15]

Value $\alpha = 1$ for polymer mixtures was obtained by McMaster[86] and Stein[87] for the first time. Nishi and Wang[82] have found $\alpha = 0.3$. In the precritical region, the processes of phase separation were also studied.[88-91] For critical composition of the system PDMS-diethylmalonate, it was found[88] that cooling the system by 0.002K below T_S (the system has UCST), the typical rings of light scattering appeared which show the origin of correlated compositional fluctuations. The rings from the beginning had a constant diameter at simultaneous increase in light scattering intensity and then the stepwise compression of a discrete halo occurred in the primary beam.[88] The dependence of initial diameter of ring on the degree of overcooling was described as $d \sim (\Delta T)^{1/3}$. The invariability of the ring diameter was observed in the interval of reduced time $\tau = tD/\xi^2 = 1.5\text{-}3.2$. The following change of reduced values of wave vector, corresponding to the optimal wavelength of spinodal decomposition, was in agreement with Kawasaki-Ohta theory,[54] accounting for hydrodynamic interaction. The diffusion coefficient $D = 10^{-9} - 10^{-8}$ cm^2/s meets universal law of critical diffusion, $v^* = 0.62$. Value D found from linear approximation exceeds three times this value at the same $(T - T_S)/T$ for one-phase regime, exponent $\alpha = 1$. These data are consistent with experimental $f = 0.90$. For the mixtures of PS with poly-2-chlorostyrene the changes of I_m and q_m are described by an exponential law. For the late stages of decomposition, $\alpha \rightarrow 1$ and almost does not depend on the composition and temperature.[51] However, the ratio β/α was found to be less than 3. Value $\alpha=1$ assumes the Siggia mechanism[57] for the last stage of cluster growth in critical liquid mixtures, where the dominant factors are hydrodynamic interaction and interfacial tension. However, for the system studied, the decomposition does not necessary follow Siggia mechanism, as the same α was obtained for non-critical conditions as well and because the condition $\beta/\alpha = 3$ is not fulfilled. The observed departure of the system behavior from the theory is because LCST is close to the T_g of both components, and only slightly depends on the composition, as different, for example, from PS-PVME system, for which a good agreement was obtained.[87.98] At relatively small differences between the temperature of experiment and T_g (40-50°), the blends have high viscosity and elasticity, which prevents establishing the local equilibrium and has influence on the kinetics of cluster growth (in analogy with the effect at the initial stages discussed by Binder[99]). It was found that the smaller is the difference between the experimental temperature and T_{sp}, the higher is the deviation of the value β/α from the theoretical.

Kyu and Saldanha[48-50] have established for the PC-PMMA blend that the early stage of spinodal decomposition at low degrees of supercooling $\Delta T = T - T_s$ obeys the linearized theory by Cahn-Hilliard, whereas at high ΔT there are departures from this theory. At the latest stage, $\alpha = 1$ and $\beta/\alpha = 3$ are in accordance with Siggia mechanism of the ripening. In this regime, there is a good superposition of curves S(x), pointing to dynamic self-similarity of the structure change. The latest stage of spinodal decomposition for this system was studied by the same authors.[50] The main attention was paid to the analysis of the scale law by Furukawa[73] at different T-jumps, i.e., at various distances from LCST. As before,[48] a good superposition of curves S(x) indicates dynamic self-similarity. This stage corresponds to the time scale at which structure factor becomes universal with time. Values 2 and –4 have been obtained correspondingly for small and large q, in a good agreement with experiment. Meanwhile, Hashimoto[44] obtained the slopes 2 and –6 for critical mixtures PVME-PS. Furukawa[100] has mentioned that the slope –6 at $q > q_m$ can be valid only for intermediate wavenumbers. At higher q, the exponent reverts to Porod value –4. Oono and Puri[98] predicted that the exponent can be steeper than –4 at intermediate wavenumbers in percolation regime. These data allow to conclude that for PC-PMMA system both the theory and experiment are in a good agreement for asymptotic behavior of scale function.

For PC-PMMA blend, it was also established[102] that at an abrupt increase in temperature from one-phase state (423K) to temperatures above the phase separation (478K), the dynamics of the decomposition is characterized by three periods. First (<20 min), a sharp increase in intensity of light scattering is observed with the formation of a maximum at a definite value of wavenumber, q. During the second period (< 110 min), the maximum shifts to a shorter wavelengths. At the last period (> 110 min), the intensity of scattering diminishes. It was found that during the first and the second period the demixing of the system proceeds, whereas during the last period the reverse process of mixing two phases takes place. Early stages of spinodal decomposition are described by Cahn-Hilliard theory and the latest agrees with the prediction by Binder-Stauffer.[46]

The PB-PI blends were studied[47,103] and the structure self-assembling during spinodal decomposition was explored for non-critical mixture by a time-resolved light-scattering technique with a time evolution of the interface structure. The mixture was investigated at a constant width of supercooling $\Delta T = T - T_{sp}$. The analysis of a scaled structure factor $F(x,t) = I(q,t)q^3(t)$ in an intermediate region of a reduced scattering vector $x = q/q_m(t)$ and time was conducted. The late stage of spinodal decomposition can be divided into two stages I and II. It was established that at the intermediate stage preceding the late stage, F(x,t) becomes sharper with its peak at x = 1. However, as time elapsed, F(x,t) depended on t. In the beginning, the range of x was smaller than about 2 and then over the entire range of x of experimental conditions. The time interval for these two regimes was defined as late stage I and late stage II.

The exponent n in Eq 5.87 at x > 2 changes from 7.2 to 4 with time increasing (a type of dynamic crossover). The same change was observed for n at a given t with increasing x (space crossover). Both crossover types were confirmed for the stage II. For the exponents in Eqs 5.94 and 5.95, the following values were found $\gamma - \alpha = 0.23$ and 0 and $\delta = 1.3$ and 1.0. These data allowed one to conclude that at the late stage II the change of $\Sigma(t)$ and $q_m(t)$ with time follows different laws and $t_I(t)$ keeps diminishing with time. At the late stage II, $\Sigma(t)$ and $q_m(t)$ follow the same law and $t_I(t)$ reaches its equilibrium value. Thus, for the late stage II, the evolution time of self-assembled structure is scaled with a single length parameter over the entire length scale from the local to the global. For the late stage of spinodal decomposition, the maxima of the second order of the function F(x,t) at x = 3 have been found by these authors and in other works.[47,74,104-106]

The data on F(x,t) for the last stage II were compared with the Ohta and Nozaki theory[10] and simulation results.[74] It was found that F(x,t) is not universal for x > 2, and diminishes with time. The experiment and calculations[107] reveal the existence of a shoulder at the function F(x,t) at x = 3. The experimental data meet also the numerical calculations.[74] The shoulder at F(x,t) appears, according to calculations, later. This time agreement between the experimental curve obtained for a hydrodynamic regime ($\alpha = 1/3$) indicates that the behavior of F(x,t) is not sensitive to the detailed mechanism of coarsening. On the contrary, the change in $\lambda_m(t)$ or $1/q_m(t)$ remains sensitive to the coarsening mechanism. Similar comparison was done also[58] for the mixture of deuterated and protonated PB. Although, the presence of such shoulder was observed at x = 2 instead x = 3, the data also met theoretical curve.[107]

An attempt to confirm the data and conclusions[47] was done,[103] using results for PB-SBR mixtures.[110] Spinodal decomposition has been studied at various temperatures from $\Delta T = 5.5$ to 34.5^o (or ε_T from 4.5×10^{-2} to 2.80×10^{-1}). Spinodal decomposition was divided into four stages - initial, intermediate, late I, and latest II. The initial stage was still divided into two more substages - very early and early. Very early stage is the stage where thermal and concentration fluctuations differ from these described by linearized theory. The crossover from the very early to early stages has not been discovered in this work. However, it was found and analyzed by Strobl[42] and implied experimentally to some extent.[111] For intermediate to late stage of spinodal decomposition, the postulate by Chou-Goldburg was confirmed[65] by plotting reduced wavenumber $Q_m(\tau)$ and reduced maximum of intensity $\tilde{I}_m(\tau)$ against reduced time, τ, at various temperatures and thermodynamic driving force ε_T and by constructing master curves. F(x,t) happened to be non-universal in relation to t for intermediate stage, but universal at the late stage I, where F(x,t) determined by Eq 5.98 does so with t and T at x > 2. At higher x > 2, F(x,t) becomes non-universal and increases with t. In this regime, the local wavenumber $\Sigma(t,T)$ (Eq 5.94) relaxes faster than global wavenumber ($\gamma > \alpha$). Interfacial thickness $t_I(t,T)$ (Eq 5.95) still makes non-essential contribution to F(x,t) or $\overline{F}(x,t)$ (Eqs 5.96 and 5.97). Therefore, the development of a pattern includes temporary change of three parameters of length, namely $1/q_m(t,T)$, $1/\Sigma(t,T) \sim R_m(t,T)$ and $t_I(t,T)$. For the late stage II, $\gamma = \alpha$ and $t_I(t,T) = t_{Ie}(T) \ll 1/q_m(0,T)$. The interphase boundary does not contribute to F(x,t) or $\overline{F}(x,t)$ and, consequently, pattern growth is scaled with the time and temperature-independent, single length parameter $1/q_m(t,T)$, assuring the dynamic scaling hypothesis,[16,112] a time- and temperature-independent universal scale function $\overline{F}(x,T) = S(x)$, but its small temperature dependence F(x,t) at high x, associated with $I_{Ie}(T)$. In all cases for the late stage, it was found that $\Sigma(t,T) \sim t^{-1}$ whereas time dependence of $\Sigma(t,T)$ at all temperatures is scaled with temperature-independent characteristics $q_m(0,T)$ and $t_C(T)$. It was found that $\Sigma(t,T)/q_m(t,T)$ is also independent of temperature (Eq 5.102), i.e., Chon-Goldberg scaling postulate was found applicable to the local structure.

The following pattern growth was observed:[103] the composition difference $\Delta\varphi(r,t)$ first reaches an equilibrium value in the beginning of the late stage, which is followed by equilibrium of $t_I(t,T)$. The local and global structures are still changing with time towards equilibrium. At the late stage I, the local structure relaxes faster towards equilibrium than the global structure ($\gamma > \alpha$) and at the late stage II both structures eventually relax at the same rate ($\gamma = \alpha$).

As for the observed maximum of the second order for $\overline{F}(x,t)$ or S(x) at x = 3, authors suppose that the local structure at the scale length $r < r_C = R_m$ comprises a stack of alternating lamellae rich in PB and PI with such a regularity as to give the third order maximum. The appearance of the second order maximum at x = 2 cannot be attributed[113] to the fact that the volume fraction of lamellae is very close to 0.5. At $r > r_C$, lamellae are curved, branched, or intersected, resulting in tangled, tortuous, discontinuous, and interconnected domain structures with a mean spacing $2\pi/q_m(t,T)$.

Some works[114-117] have been dedicated to the studying the processes of spinodal decomposition for the PB-poly(styrene-*co*-butadiene). To establish the influence of molecular mass on spinodal decomposition and transport mechanism, the initial stage was investigated.[115] The mixture was situated in the dissipative limit where $t \geq t_R \geq \varepsilon_T^2$ (ε_T is thermodynamic driving force) and, therefore, the fluctuation growth is the consequence of translational diffusion of the center of molecule mass. Value t_R is the time necessary for the center of molecule mass to translate over its end-to-end distance (so-called reptation time, when diffusion mobility obeys the reptation mechanism)[32,43] $t_R = R_g^2/D_C$, where D_C is self-diffusivity, $\varepsilon_T = (\chi - \chi_S)/\chi_S$. Parameters D_{app} and q_m were found from the analysis of the initial stage of spinodal decomposition, using Eq 5.67 for various molecular masses. It was discovered that the ratio q_m^2/D_{app} depends on M in power 1 at the limiting molecular mass. Analysis of the Eqs 5.69a-5.69c has allowed the conclusion that in a given case demixing includes mutual diffusion of the center of mass according to the reptation mechanism (the correspondence to Eq 5.69b). It was also found that in spite of a broad supercooling ($\Delta T = T - T_S$), value ε_T is very small due to limited temperature dependence of χ.

The late stages of spinodal decomposition were studied[110,116] at various temperatures. It was established[116] that variations of I_m and q_m with t at different temperatures fall onto master curves on the reduced plots, thus assuring the scaling postulate that the data obtained at discreet t and T for given mixtures are properly scaled with the temperature-dependent characteristic wavenumber $q_m(0,T)$ and characteristic time $t_C(T)$. It was also shown that dependencies of Q_m on $\tilde{I}_m$ nicely fall onto a master curve, suggesting that the coarsening processes, occurring at these different temperatures, are the same and that temperature affects only the time scale (through $t_C(0,T)$ of growing fluctuations). Thus, the scale postulate was confirmed.[44,49,85,112,118]

For the intermediate stage ($2 \leq \tau \leq 10$) values β=1.9-2.1 and α=0-0.7, whereas for the late stage β=2.1-2.6 and α=0.7-0.87, correspondingly. New method for testing the Chou-Goldberg scaling result was proposed.[65] Two experimentally determined values were introduced X(t,T) and Y(t,T) defined by

$$X(t,T) \equiv q_m(t,T)[D(T)t]^{1/2} \qquad [5.112]$$

$$Y(t,T) \equiv \bar{I}_m(t,T)[D(T)t]^{-3/2} \qquad [5.113]$$

where

$$\bar{I}_m = I_m(t,T) / \int_{q'}^{q''} I(q,t,T)q^2 dq \qquad [5.114]$$

and q′ and q″ are values of q at which I(q,t,T) virtually becomes zero. For this purpose the postulate by Chou and Goldberg was used according to which $q_m(t,T)$ and $\bar{I}_m$ (t,T) at T near critical temperature, T_{cr}, are scaled by T-dependence correlation length ξ(T) and a time constant $t_C(T)$ with the latter defined by Eq 5.115

$$t_C(T) \equiv [\zeta_c(T)]^2 / D(T) \qquad [5.115]$$

where D(T) is the interdiffusion coefficient of the system at T. This postulate means that the dimensionless peak position and height, defined as

$$\tilde{q}_m(t,T) = q_m(t,T)\zeta_c(T) \qquad [5.116]$$

$$\tilde{I}_m(t,T) = \bar{I}_m(t,T)[\zeta_c(T)]^{-3} \qquad [5.117]$$

depends only on the reduced time $t/t_C(T)$. This implies that

$$\tilde{q}_m(t,T) = f(t/t_c) \qquad [5.118]$$

$$\tilde{I}_m(t,T) = g(t/t_c) \qquad [5.119]$$

In this case, together with Eqs 5.115, 5.118, and 5.119, the values of X(t,T) and Y(t,T) can be written as

$$X(t,T) = [t/t_c(T)]^{1/2} f(t/t_c(T)) \qquad [5.120]$$

$$Y(t,T) = [t/t_c(T)]^{-3/2} g(t/t_c(T)) \qquad [5.121]$$

Eq 5.120 shows that X(t,T) depends on a simple variable $t/t_C(T)$. Therefore, the dependence of X(t,T) on ln t for various T should be plotted on a master curve by horizontal shifting. The master curve allows one to find the shape of the following value $[(t/t_C)(T')]^{1/2} f(t/t_C)(T')$ at a reference temperature T′ in a broad region of t. If the master curve is multiplied by $t^{-1/2}$, the resulting curve yields the values of $(t'_C)^{-1/2} f(t/t'_C)$ where t'_C is the value of $t_C(T)$ at the reference temperature T′, i.e., $t'_C = [\xi_C(T')]^2/D(T')$.

From Eq 5.121, it follows that the dependence of Y(t,T) on ln t for various T can be reduced to a master curve by horizontal shifting and multiplication of the master curve Y(t,T′) by $t^{3/2}$ gives the value of $(t'_C)^{3/2} g(t/t'_C)$. If one finds that the data for X(t,T) and Y(t,T) can be reduced to master curves, the conclusion can be drawn that the system obeys the Chou-Goldberg scaling laws.[65]

The reexamination of scaling postulate has been done[116] in terms of variables X(t,T) and Y(t,T) (Eqs 5.112 and 5.113) introduced in ref. 66. These values have been defined in the following way:

$$X(t,T) = [t/t_c(T)]^{1/2} Q_m(t,T) \qquad [5.122]$$

$$Y(t,T) = [t/t_c(T)]^{-3/2} \tilde{I}_m(t,T) \qquad [5.123]$$

If the scaling postulate is valid, Q_m and $\tilde{I}_m$ become universal functions $\tau = t/t_C(T)$. Therefore, the values X(t,T) and Y(t,T) should be universal in relation to τ. In such a way, the values X(t,T) and Y(t,T), estimated at different temperatures, may be superimposed and reduced to the respective master curves by horizontal shifting when they are plotted against log tX(t,T) and Y(t,T) dependent on logt at a given T. These values can be reduced to those at a reference temperature T_r by a horizontal shifting by log $a_T(T)$, where $a_T(T)$ is temperature dependent shift factor defined as

$$a_T(T) = t_c(T)/t_c(T_r) \qquad [5.124]$$

The master curves for $X(t/a_T,T_r)$ and $Y(t/a_T,T_r)$ reduced at T_r have been constructed. If to multiply X and Y by $[D_{app}(T_r)t]^{-1/2}$ and $[D_{app}(T_r)t]^{3/2}$ correspondingly, one can obtain resulting master curves for $q_m(t/a_T,T_r)$ and $\tilde{I}_m(t/a_T,T_r)$ with reducing at $T = T_r$, since the following conditions are valid:

$$[D_{app}(T_r)t]^{-1/2} X(t/a_T,T_r) = q_m(t/a_T,T_r) \qquad [5.125]$$

$$[D_{app}(T_r)t]^{3/2} Y(t/a_T,T_r) = \tilde{I}_m(t/a_T,T_r) \qquad [5.126]$$

The absolute values of $Q_m(\tau)$ and $\tilde{I}_m(\tau)$ cannot be determined, since $q_m(0,T)$ and $t_C(T_r)$ are unknown. In spite of this, the master curves $q_m(t/a_T,T_r)$ and $\tilde{I}_m(t/a_T,T_r)$ can be superimposed on $Q_m(\tau)$ and $\tilde{I}_m(\tau)$ by horizontal and vertical shifts when they are plotted in double logarithmic scales. In this sense, the master curves, given by Eqs 5.125 and 5.126, are "relative" reduced curves corresponding to Q_m and $\tilde{I}_m$.

Parameters X(t,T) and Y(t,T) for five temperatures have been obtained[66] as a function of logt for mixtures of PI and SBR. For this mixture, the critical temperature is 673K and reference temperature 313K. By horizontal shift, the master curves were derived, from which functions f and g can be obtained (Eqs 5.120 and 5.121). From these data, value α was estimated to be equal 0.2-0.8 and value $\beta = 4$. These results meet the relation $\beta = 3\alpha$ for the time scale > 340 min.

For the same mixtures of PB/SBR, it was found[110] that the function F(x) is not universal at the intermediate stage of spinodal decomposition, i.e., at reduced time, τ, shorter than 10. F(x = 1) increases and F(x) becomes sharper with x when τ increases. The growth of F(x = 1) with τ is connected with increasing compositional fluctuations between matrix and domains. However, at $\tau \geq 10$, F(x) at x = 1 becomes universal, implying that the global structure grows with a dynamic self-similarity and scales with a time- and temperature-dependent single length parameter $\lambda_m = (t,T) = 2\pi/q_m(t,T)$. A spectacular feature of F(x) was discovered at the late stage of spinodal decomposition. F(x) has the second-order maximum and, in some cases, even the third-order shoulder at x = 2 and x = 3, respectively, implying existence of the domains with a sufficiently fine periodicity. Although, $F(x \sim 1)$ at the late stage is universal with t, it is not universal at higher x, indicating that the local structure, i.e., the interfacial structure of the phase-segregated domains, is still changing with t or cannot be scaled with $1/q_m(t,T)$.

To describe non-universal nature of F(x,T) an additional characteristic length scale $1/S_{sp}(t,T)$ was introduced.[110] Taking into account that at high x the Porod law is valid ($I(q) \sim q^{-4}$ for $q \geq q_C(t,T)$), the function I(q) can be expressed as

$$I(q) = C < \eta(t)^2 >_e S_{sp}(t,T) q^{-4} \quad \text{for } q > q_c(T) \qquad [5.127]$$

where C is a constant, independent of q and t, and S_{sp} is the interface area per volume unit, i.e., specific interface area. From Eqs 5. 83 and 5.127, it follows that

$$F(x) = C < \eta(t)^2 >_e S_{sp}(t,T) q_m(t,T)^{-1} x^{-4} \qquad [5.128]$$

Value S_{sp} should generally decrease with t and it scales as

$$S_{sp}(t,T) \sim t^{-\gamma} \qquad [5.129]$$

$$F(x) \sim t^{\alpha-\gamma} x^{-4} \qquad [5.130]$$

Therefore, if $\alpha \neq \gamma$, F(x) becomes non-universal with t and self-assembling structure has two characteristic wavenumbers, $q_m(t,T)$ and $S_{sp}(t,T)$. If $\alpha > \gamma$, F(x) increases with t. This situation corresponds to the case that growth of the length relevant to the global structure $\lambda_m(t,T) \sim 1/q_m(t,T)$ occurs faster than the growth of the length relevant to the local structure $1/S_{sp}(t,T)$. Only in the case $\alpha = \gamma$, the function F(x) (Eq 5.130) is universal with t and is scaled with the single length parameter $\lambda_m(t,T) \sim 1/S_{sp}(t,T)$. From Eq 5.127, $S_{sp}(t,T)$ can be determined by

$$S_{sp}(t,T) = [\lim_{q\to\infty} q^4 I(q)] / [\int_{q_1}^{q_2} q^2 I(q) dq] \quad [5.131]$$

where

$$\int_{q_1}^{q_2} I(q) q^2 dq \approx C <\eta(T)^2>_e \quad [5.132]$$

The non-universality of F(x) implies that the evolution time of parameter $S_{sp}(t,T)$ is slower than that $q_m(t,T)$. Asymptotic behavior of F(x) with q implies dynamic cross-over for rather late stage of spinodal decomposition. Crossover of q-dependence in the region from q^{-6} to q^{-4} at $q > q_m$ was discovered by analyzing the theoretical scaling function proposed by Furukawa in some works[73,120,121] (Eq 5.107) and derived later for a deep quench as[122]

$$F(x) \sim \frac{x^3}{2 + x^9} \quad [5.133]$$

Scaling analysis of the structure function by Furukawa was done for the mixture PMMA/SAN.[123] It was shown that at the late stages of decomposition, the scaled structure factors $F(x) = \beta^3{}_m I(\beta/\beta_m)$, obtained at different time, may be presented by the master curve, which implies the development of self-similarity of structure. For mixtures PMMA/PVA, at critical and non-critical compositions, function $\tilde{S}(x)$ found experimentally[125] meets theoretical prediction for critical composition (Eq 5.107) and almost corresponds to the predicted function for non-critical composition (Eq 5.108). Asymptotic behavior of experimental and theoretical functions $\tilde{S}(x)$ at small x shows that at $q > q_m$ function $\tilde{S}(x)$ is approximately proportional x^{-4} and x^{-6} for critical and non-critical compositions, respectively. Scattering function for this system was analyzed in accordance with Porod law [62,126] according to which

$$I(q) \sim q^{-4}[1 - q^{-2}/R_m^2 + ...] \quad [5.134]$$

where R_m is an average radius of interfacial wave. For ideal two-phase structure, I(q) is proportional to q^{-4} but not to q^{-6}. Therefore to determine where the formation of domain droplets proceeds in immiscible system, light scattering invariant was considered as a function of time for non-critical concentration. This invariant is determined as[44]

$$Q_{LS} = \int_{q_{min}}^{q_{max}} q^2 I(q) dq \qquad [5.135]$$

and can be expressed as

$$Q_{LS} = (32\pi^6/\lambda_0)\varphi_1\varphi_2(a_1 - a_2)^2 \qquad [5.136]$$

where φ_i and a_i are volume fraction and polarizability of i-th phase. If only domain droplets are formed, then $\varphi_1\varphi_2$ and $(a_1 - a_2)^2$ are constant values and thus Q_{LS} is also constant and does not depend on time. This case occurs[125] if scaling function law holds.

To compare data with theoretical predictions, results on phase separation in the mixtures of polyethersulfone and poly(ethyl oxazoline)[124] and PS/PVME[127] were plotted in coordinates of reduced variables $\tau = (2\pi)^2 Dt/\xi^2$ and $Q_m = q_m(t)\xi/2\pi$, where $\xi = 2\pi/q_m$. It was found that all results are described by one curve, showing cross-over regions for various processes, such as Cahn linear regime, cluster regime by Binder-Stauffer, and regime of interfacial instability by Siggia (Fig. 5.2). The late stage of spinodal decomposition may be well described by cluster regime[127] ($\alpha = 1/3$). At this stage, the self-similar structure was maintained during the ripening process.

The initial stage of spinodal decomposition was investigated also for the mixtures of poly(methylmethacrylate-co-butylmethacrylate) with SAN,[128] PS and PB,[129] PMMA and SAN,[130] and water solutions of PVME.[131] The initial stage of decomposition for all these systems is rather well described by the Cahn-Hilliard theory. For the PVME-water system, the unusual great size of initial fluctuations (10 mcm) was observed, explained by a jump of temperature close to T_S. The morphology of the system at the early stages was characterized by the existence of droplets, probably, due to asymmetry of phase diagram. For poly(MMA-co-BMA), the early stage of spinodal decomposition can be described by Cahn-Hilliard theory in the region of small values of the wavenumber ($\sim 10^5$ cm^{-1}).

Phase separation in a ternary system of two polymers and solvent were studied by Monte-Carlo simulation of dynamics in the lattice.[132] The mixture of partially incompatible polymers with equal number of monomeric units, $N_A = N_B = 32$, and volume fractions, $\varphi_A = \varphi_B = 0.2$, was analyzed. This mixture was cooled from initially homogeneous state at the "infinite" temperature to temperature $\varepsilon/kT = 0.6$, corresponding to the two-phase region. It was shown that at the initial stage of decomposition, substantial compression of coils should occur which is not described by a linear Cahn theory.

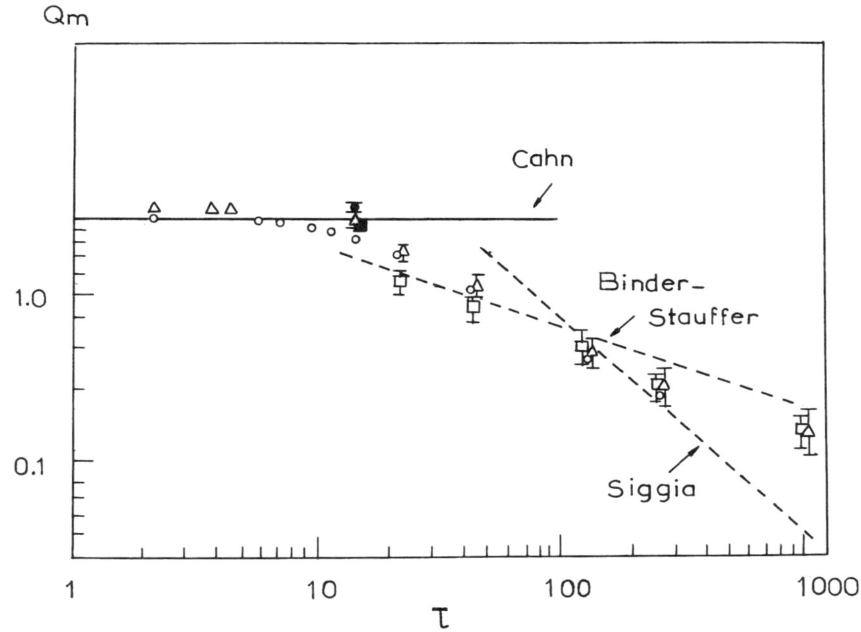

Figure 5.2. The development of wavenumber, Q_m, by differential techniques (o - LC, Δ - SEM, □ - OM, • STEM, ■ - SM). Adapted, by permission, from H. Nakamura, J. Maruta, T. Ohnaga, and T. Inoue, *Polymer*, **31**, 307 (1990).

For the system with strong hydrogen bonds (mixture of PBMA with PS containing 1.5 mol% of n-(1,1,1,3,3,3-hexafluoro-2-hydroxy-isopropyl)d-methylstyrene, it was found[135] that the early stages may be described by Cahn-Hilliard-Cook theory. For the late stages, especially at temperatures close to T_g, the curves of the light-scattering intensity have two peaks which implies coexistence of two alternative mechanisms of spinodal decomposition.

Critical and non-critical regimes of spinodal decomposition were compared by studying phase separation of oligostyrene-oligomethylphenylsiloxanes.[89,90] In both cases, there was no linear regime of phase separation: the dependencies of the intensity of light-scattering were found to depart from the exponential relationship predicted by Cahn theory. Different values were obtained for exponents β and α for non-critical and critical regimes: in the first case, $\beta = 0.69$ and $\alpha = -0.29$ and, in the second case, $\beta = 2.66$ and $\alpha = -0.9$. Thus for the dependence $q \sim t^{\alpha}$ in the non-critical region, the exponent is close to 1/3, whereas in critical region it is close to +1, in accordance with the data for low-molecular systems.[65,92-97] The experimental data in the non-critical region are well described by Binder theory.[16] The interdiffusion coefficients, measured by photon correlation spectroscopy and calculated from the light-scattering within the framework of linear approximation by Cahn-Hilliard, are -9.6×10^{-11} and -1.7×10^{-11} cm^2/sec for critical and non-critical compositions, correspondingly. For the late stages of phase separation in the system atactic PS-diisodecyl phthalate, an increase in time of the microdomain size was discovered to be $\lambda_m \sim t^{1/3}$ which corresponds to the mechanism of coalescence and Ostwald ripening.

As distinct from flexible macromolecules, for the mixtures of rigid rod-like molecules (blends of poly-*p*-phenyl benzo-*bis*-tiazol and polyamide-66), the intensity of light scattering and the position of a maximum obey the power laws with exponents 0.33 and 0.91-0.95, correspondingly[134] at temperatures above melting point of polyamide. The stage of rapid slowing down in the changes mentioned above is not typical of the mixtures of flexible macromolecules. The aggregation of different domains does not lead to coalescence because simultaneous ordering of rigid molecules in aggregates. Scaling analysis of Furukawa structure function gives an exponent –2.4, i.e., intermediate between predicted values for the systems with one-dimensional and two-dimensional diffusion as a result of rod anisotropy.

The development of modulated structures proceeds also by formation of binary polymer mixtures from the common solvent during drying.[135-137] The detailed study of such process was performed[136] using small-angle light-scattering. At relatively small concentrations of polymers in common solvent, the regular structure is developed which is later preserved in the whole range of concentrations of dried film. More than 10 polymer pairs were studied[136] for which the modulated structure is formed during drying. The higher the rate of solvent evaporation, the smaller are periodic distances in modulated structure and *vice versa*. The rate of phase separation decreases and periodic distances are smaller when the component ratio in the mixture deviates from 1:1. With diminishing drying rate, the peaks of light-scattering are shifted to smaller angles, i.e., larger size of periodic structure corresponds to a slower drying. It is typical that during drying, the position of scattering peaks practically does not change, whereas the intensity of scattering increases. Besides, the dimensions of periodic structure depend on the so-called critical concentration of polymers, i.e., on concentration corresponding to the beginning of detection of scattering peak. This critical concentration grows with evaporation rate. With increasing molecular weights of polymers, the phase separation is hampered, especially at rapid drying. The latter

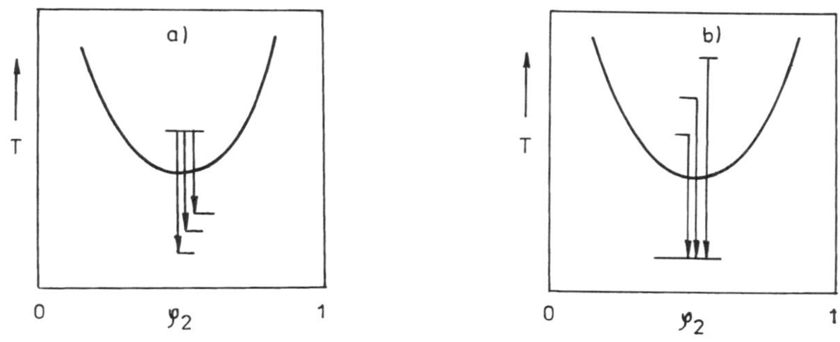

Figure 5.3. Courses of experimental procedure in phase-dissolution experiments. (a) the thermodynamic driving force is kept constant; (b) the mobility remains constant. Adapted, by permission, from H. -W. Kammer, *J. Macromol. Sci-. Chem.*, **A27**, 1713 (1990).

may lead to the formation of pseudo-homogeneous structure without appreciable concentration fluctuations.

Linearized theory by Cahn was used to analyze the processes of dissolution of phases at fast transfer of the system from the region above T_{sp} to the region of single-phase solutions. It was discovered[138,139] that at the latest stages of phase dissolution, the light-scattering intensity exponentially diminishes with time. The rate depends on the effective diffusion coefficient for center-of-mass motion of chain molecules in the melt. After the temperature jump, the phase-separated structures are dissoluted by a continuous presence of the thermodynamic driving force responsible for phase separation. Thus, phase dissolution may be considered as a process which is reverse to phase separation. It allows the phase separation theory to be used in this case. According to Cahn,[8] the apparent diffusion coefficient is given as

$$D_{app} = M\frac{\delta^2 \Delta G}{\delta \varphi^2} \qquad [5.137]$$

where M is the mobility constant having a positive value. D_{app} is directed by the mobility as well as by thermodynamic driving force, $\delta^2\Delta G/\delta\varphi^2$. It was assumed that M relates to the final state of temperature jump, whereas $\delta^2\Delta G/\delta\varphi^2$ characterizes the initial state. Then authors have used various experimental approaches to study the value D_{app} (Fig. 5.3a,b). The jump from one temperature in two-phase region to any temperatures in a single-phase region (Fig. 5.3a) corresponds to the change in mobility, M, as a function of temperature, whereby thermodynamic driving force is kept constant. The route indicated in Fig. 5.3b focuses attention on the influence of the thermodynamic driving force at the constant mobility. It was shown[138] that the diffusion coefficient increases with temperature, i.e., with diminishing the difference $\Delta T = T - T_S$, the mobility is described by Arrhenius-type equation. In other case, the behavior of D_{app} differs from the first case:[139] At first, D_{app} increases with ΔT in accordance with the mean-field theory. However, with further increase in ΔT, the value of D_{app} decays. This effect was attributted[139] to the formation of interconnected structure in some temperature range just above LCST. However, this structure rapidly decays with increasing temperature. Irregular two-phase morphology results. The decrease in diffusivity presumably begins when the temperature approaches the stability limit of the regular two-phase morphology. This result supposes that the thermodynamic driving force depends on the dispersal of the phases. Therefore, it is necessary to account for the dependence of the driving force on interfacial energy.[139] In both figures, the dissolution is accompanied by decrease of light-scattering intensity with time of annealing, the intensity decay is described by the exponential equation:[140]

$$\ln \frac{I}{I_0} = -2D_{app}q^2t \qquad [5.138]$$

For some mixtures, such as PMMA-SAN, PVC-PMMA, PVC-PBMA, PVC-MA-*co*-BMA, the intensity change follows Eq 5.138, at least during the latest stage of annealing. The coefficient D_{app} depends on the quench depth $\Delta T = |T - T_0|$ and increases with ΔT. However, the following increase in ΔT leads to diminishing D_{app}, which may proceed when the temperature approaches the stability limit of two-phase structure (T_0 is the same). Such change in D_{app} with T reflects the influence of the thermodynamic driving force on the kinetics of phase dissolution. The difference in the free energies of dispersed phase and single-phase melt may be expressed as

$$\Delta G(d) - \Delta G(\infty) = \varphi(1 - \varphi)A\gamma k \qquad [5.139]$$

where d is the length, characterizing the dimension of dispersed phase, A is the interfacial surface, γ is the interfacial tension, K is the equilibrium constant. The latter value characterizes the dispersity degree as

$$k = \exp\left(-\frac{A\gamma}{RT}\right) \qquad [5.140]$$

When applying the Flory-Huggins expression for $\Delta G(\infty)$, the expression for the thermodynamic driving force may be derived in the following form:

$$\frac{\delta^2 \Delta G(d)}{\delta \varphi^2} = -2RT(\chi - \chi_S) - 2A\gamma \exp\left(-\frac{A\gamma}{RT}\right) \qquad [5.141]$$

where χ_S is the parameter of thermodynamic interaction at T_{sp}. For compatible mixtures with LCST, the interaction parameter has been taken as

$$\chi = a - \frac{b}{T} \qquad [5.142]$$

and then

$$\chi = \chi_S = b\frac{\Delta T}{T_S} \qquad [5.143]$$

According to linearized Cahn theory, the partial Fourier component of concentration fluctuation with the wavelength λ_m has the maximum growth rate. Therefore, it was supposed that the interfacial surface A $\sim q_m^{-2}$. Then, within the framework of mean-field approximation for the spinodal decomposition, we have:

$$q_m^2 \sim \frac{\chi - \chi_S}{\chi_S} \qquad [5.144]$$

and the interfacial tension between two polymers is expressed as

$$\gamma \sim RT(\chi - \chi_S)^{1/2} \qquad [5.145]$$

therefore, we obtain the equation, showng the change of $\delta^2\Delta G(d)/\delta\varphi^2$ as a function of $\Delta T = |T - T_S|$:

$$\frac{1}{RT}\left|\frac{\delta^2 \Delta G(d)}{\delta \varphi^2}\right| = \frac{\alpha}{Y^2} + \beta^{-\beta Y} \qquad [5.146]$$

where $Y = [1/(\Delta T/T_S) + 1]^{1/2}$, α and β are connecting parameters characterizing the magnitude of contribution in thermodynamic driving force of bulk and interfacial structure, correspondingly. At $\Delta T/T_S$, the left part of Eq 5.138 linearly changes with $\Delta T/T_S$:

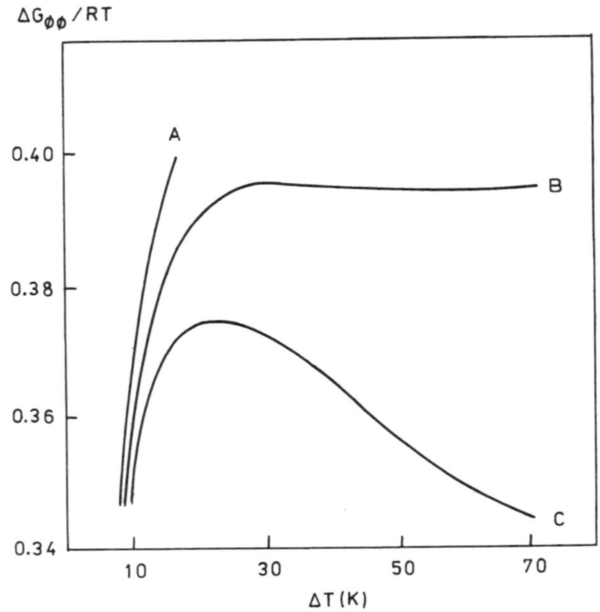

Figure 5.4. Thermodynamic driving force according to Eq 5.146 for β=0.2 and α=1(A), 0.6 (B), and 0.2 (C). T_s = 473K. Adapted, by permission, from J. Piglowski, H. -W. Kammer, and J. Kressler, *Polymer*, **30**, 1708 (1989).

$$\frac{1}{RT}\frac{\delta^2\Delta G}{\delta\varphi^2} \sim \frac{\Delta T}{T_S} \qquad [5.147]$$

According to Eq 5.146, an essential change in D_{app} is observed only at $\alpha/\beta = 1$. With increase in α/β, the effect becomes less pronounced and then disappears. The calculations have shown that the maximum on the dependence of ΔG/RT on ΔT is observed at $\alpha/\beta = 1$ and levels off with increase of this ratio at a definite ΔT (Fig. 5.4). Therefore, at a given condition, the diffusion coefficient D_{app} passes through a maximum depending on ΔT. Such dependence of D_{app} on ΔT has been observed also[139] for the following mixtures: PVC/P(MMA-*co*-BMA), PMMA/SAN, PVC/PBMA, and PVC/PMMA.

5.5 FACTORS INFLUENCING THE KINETICS OF PHASE SEPARATION

The kinetics of phase separation in polymer mixtures depends also on some other factors, connected with the structure of macromolecules.

5.5.1 TACTICITY

The role of tacticity of one of the components on the rate of phase separation has been studied for mixtures of PS and isotactic and atactic PVME [141] and PC with isotactic and heterotactic PMMA.[142] For the first mixture, the thermodynamic driving force has been calculated as

$$\varepsilon_T = \frac{g - g_S}{g_S} \quad [5.148]$$

where g is interaction parameter in Flory-Huggins-Staverman equation for the free energy of mixing,[144-148] which depends on composition:

$$\frac{\Delta G}{NRT} = \frac{\varphi_1}{m_1} \ln \varphi_1 + \sum \frac{\varphi_2}{m_2} \ln \varphi_i + g\varphi_1\varphi_2 \quad [5.149]$$

where g is

$$g = a + \frac{b_0 + b_1/T}{1 - c\varphi_2} \quad [5.150]$$

Here a, b_0, and b_1 are correcting parameters accounting for possible incorrectness in the first terms of combinatorial entropy of mixing in Flory-Huggins-Staverman equation, c is Staverman parameter. Values a, b_0, b_1, and c are calculated from the conditions of spinodal in the critical point:

$$\frac{1}{m_{w_1}\varphi_1} + \frac{1}{m_{w_2}\varphi_2} = 2a + \frac{2D(1 - 1)}{(1 - c\varphi)^3} \quad [5.151]$$

$$\frac{\alpha_1}{m_{w_1}\varphi_1^2} + \frac{\alpha_2}{m_{w_2}\varphi_2^2} = \frac{6bc(1 - c)}{(1 - c\varphi_2)^4} \quad [5.152]$$

To analyze the light-scattering data, the linearized theory by Cahn-Hilliard was applied for the initial stages.[5,8,41] Thermodynamic driving force is estimated from the ratio:[32,45]

$$\varepsilon_T = \frac{g - g_S}{g_S} = \frac{D_{app}}{D_c(T)} \quad [5.153]$$

where $D_C(T)$ is determined from Eq 5.154

$$R(q) = D_c q^2 \left(\frac{g - g_S}{g_S} - \frac{R_g^2 q^2}{36} \right) \quad [5.154]$$

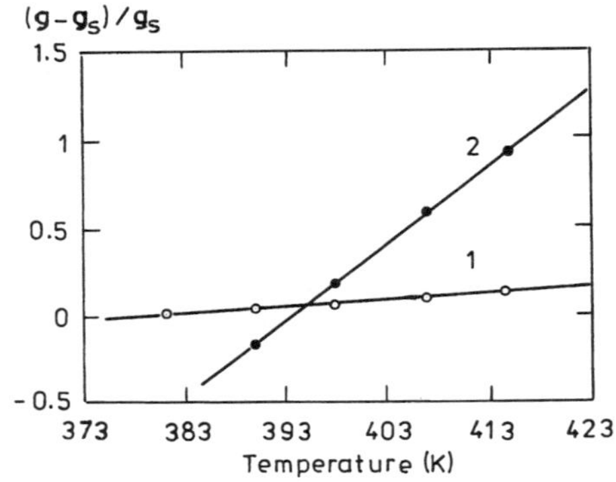

Figure 5.5. Predicted behavior for (g-gs)/gs from F-H-S analysis: 1 - isotactic PVME, 2 - atactic PVME. Adapted, by permission, from G. Beaucage and R. S. Stein, *Macromolecules*, **26**, 1610 (1993).

whereas D_{app} may be found from the dependence of $R(q)/q^2$ on q^2 using Eq 5.67. Both values are related:

$$D_{app} = D_c(T)\frac{g - g_S}{g_S} \qquad [5.155]$$

It was shown[141] that the dependence of ε_T on T is different for various compositions of PS and isotactic and atactic PVME (Fig. 5.5). In particular, for atactic PVME, this value is smaller than for isotactic PVME. The dependence of the rate of concentration fluctuation growth on the extent of quenching is less pronounced for isotactic PVME (the slope of ε_T on T is lower) and the limit of compatibility is smaller. Value ε_T for atactic PVME drops very rapidly with temperature to very low level for atactic PVME (Fig. 5.5). At similar quench depth, isotactic PVME shows lower rate of phase separation than atactic PVME. The diffusion coefficient, D_C, is practically the same for both polymers. Therefore, the main reason for differences in the behavior of these systems is related to thermodynamic driving force, ε_T.

The functional shape of Flory-Huggins-Staverman parameter, g, was used for analysis of the intermediate stages of spinodal decomposition based on modified approximation by Cahn-Hilliard (Eq 5.154). Functions of g, obtained from cloud point curves, well describe the intermediate stages. For the mixtures of PC with isotactic and heterotactic PMMA no tacticity effect was discovered, when studying the kinetics of growth in the latest stages of phase separation.[142]

5.5.2 EFFECTS OF ADDITION OF DIBLOCK-COPOLYMERS

It is known (Ch. 4) that the miscibility of two polymers can be changed by addition of diblock-copolymers, consisting of blocks, corresponding to homopolymers. The effect of diblock-copolymer on the kinetics of the latest stages of spinodal decomposition has been studied[143,149] for the mixtures of low molecular PS and PB. It was found that in presence of styrene-butadiene diblock copolymer, the rate of growth of phase-separated particles becomes smaller. The addition of diblock copolymer affects the kinetics of separation more extensively than equilibrium properties. The diminished dimensions of particles in copolymer presence are explained by interfacial effect of block-copolymer. The exponent α in expression for evolution time is equal approximately to 1/3 until the end of experiment, whereas exponent β is equal to 1. The ratio $\beta/\alpha = 3$ should be expected for the late stage of phase separation when the compositions of separated phases and their relative ratio do not change markedly with time. Another evidence of the effect of block-polymer may be obtained by estimation of the so-called invariant Q_{LS} (Eq 5.135). There is an agreement between calculated and experimental structure functions,[61] although at $x=q/q_{max} > 1$, the experimental curve reaches its asymptotic slope earlier compared with calculated value. This asymptote of structure function corresponds to

$$I_{max} \sim \left(\frac{q}{q_{max}}\right)^{-4} \qquad [5.156]$$

known as Porod law[62,126] in analysis of small-angle scattering of X-rays. A good correspondence of experimental data with Porod law shows that in a given case the mixtures consist of two phases with sharply defined interface.

The kinetics of phase separation was studied for non-critical composition of mixture of low molecular weight components in the presence of diblock-copolymer.[143] Hashimoto[149] studied the mixtures with compositions close to the critical for the same components but with higher molecular masses. The kinetics of phase separation was measured from initial to late stages of the process. The influence of diblock-copolymer on the kinetics was determined by two main factors. First, by the effect of addition of block-copolymer on the change of phase diagram, and, therefore, on the quench depth at a given temperature of phase separation. Secondly, the change of interfacial tension at the interface between two separating domains due to localization of block-copolymer also contributed to phase separation kinetics. The first effect exerts its influence both at the initial and the late stages of phase separation, whereas the second effect is the most pronounced at the late stage. At this stage, the interface boundary develops in the system where molecules of diblock-copolymer are segregated. The thickness of the interfacial region diminishes with time until equilibrium. The number density of block-copolymer junctions at the interphase increases with time as well. As a result, the interfacial tension decreases with time, leading to the increasing effect of block-copolymer on the coarsening, which depends on interfacial tension.

The analysis of the initial stages of spinodal decomposition in the framework of linearized Cahn theory has shown that self-diffusion coefficient, D_C, diminishes with addition of block-copolymer. This effect may be connected with higher molecular weight of block-copolymer as compared with homopolymers.

The coarsening at the intermediate and late stages was studied on the basis of analysis of power dependencies of $q_m(t,T,\varphi)$ and $I_m(t,T,\varphi)$ (Eqs 5.81 and 5.82). It was discovered that for each mixture with a given volume fraction of copolymer, φ, and at a given temperature, T, the dynamic exponents α and β obey the condition $\beta > 3\alpha$ on the time scale lesser than cross-over time, t_{cross}, between intermediate and late stages of decomposition. At the late stage at $t > t_{cross}$, the correlation $\beta \approx 3\alpha$ is valid. This indicates that the mean square fluctuation of the refraction index $<\eta^2>$ between domains and matrix reaches the equilibrium value of $<\eta^2>_{eq}$, whereas the structure scale factor, F(x), becomes universal with time.

Scaled characteristic time, τ_{cross}, can be defined as equal to t_{cross}/t_C, where $t_C^{-1} - 1 = q_m^2(0)D_{app} = 2R_m$ and $R_m = R(q)$ at $q_m(0)$. At a given temperature, τ_{cross} tends to increase, whereas at the late stage tends to diminish with an increase in φ. This indicates that the coarsening process is retarded at the late stages on small additions of block-copolymer. To test scaling postulate for the domain growth at the late stage, the scaled wavenumber, Q_m, scaled intensity, I_m, and scaled time, τ, were determined:[53,65]

$$Q_m = q_m(t,T,\varphi)/q_m(0,t,\varphi) \qquad [5.157]$$

$$\tilde{I}_m = I_m(t,T,\varphi)q_m^3(0,T,\varphi)/<\eta^2(T,\varphi)>_{eq} \qquad [5.158]$$

$$\tau = t/t_c(T,\varphi) \qquad [5.159]$$

where $q_m(0,T,\varphi)$ and $t_C(T,\varphi)$ have been determined from the analysis of the initial stage of spinodal decomposition, whereas $<\eta^2(T,\varphi)>_{eq}$ mean square fluctuation of the refraction index at thermal equilib-

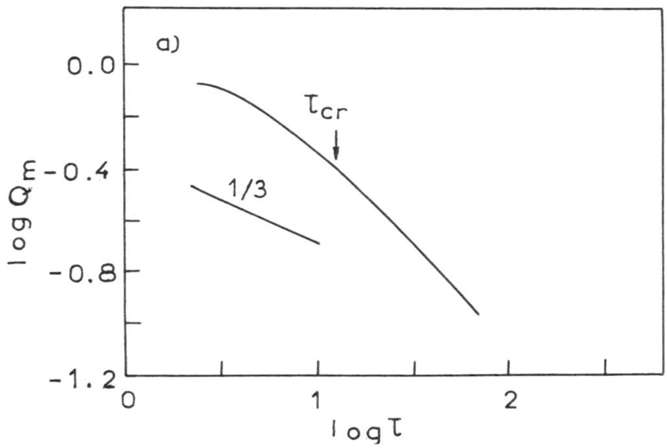

Figure 5.6a. Reduced characteristic wavenumber, Q_m, plotted against reduced time, τ, in double logarithmic scales for block 0 wt% (a). Solid line on parts *b* and *c* shows $Q_m(\tau)$ for block 0 wt% for composition. Temperature, K: 323 (---), 333 (-.-.-), 343 (. . .). Adapted, by permission, from T. Hashimoto and T. Izumitani, *Macromolecules*, **26**, 3631 (1993).

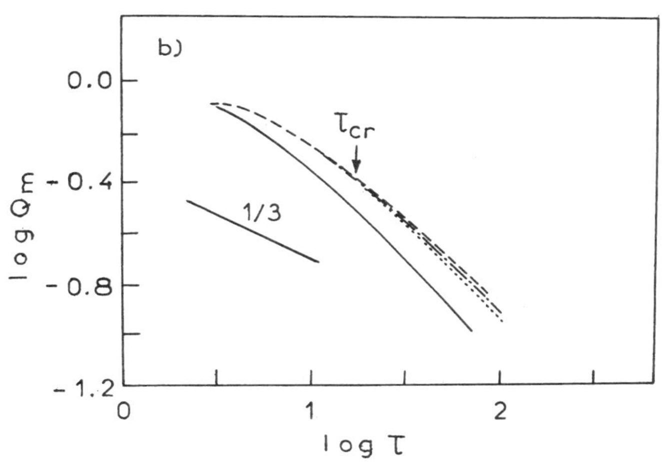

Figure 5.6b. Reduced characteristic wavenumber, Q_m, plotted against reduced time, τ, in double logarithmic scales block 3 wt% (b). Solid line on parts *b* and *c* shows $Q_m(\tau)$ for block 0 wt% for composition. Temperature, K: 323 (---), 333 (-.-.-), 343 (. . .). Adapted, by permission, from T. Hashimoto and T. Izumitani, *Macromolecules*, **26**, 3631 (1993).

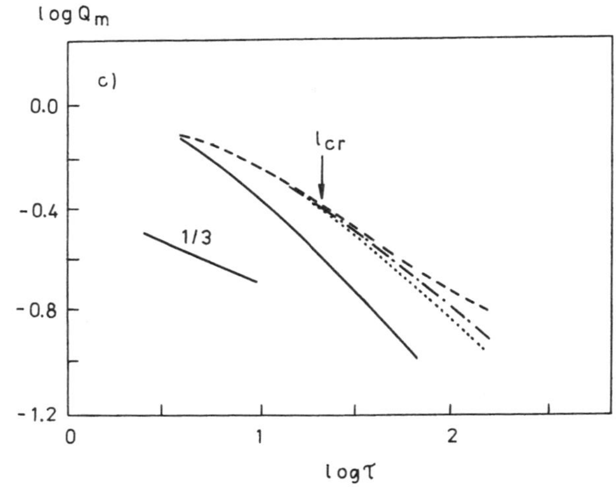

Figure 5.6c. Reduced characteristic wavenumber, Q_m, plotted against reduced time, τ, in double logarithmic scales for block 6 wt% (c). Solid line on parts *b* and *c* shows $Q_m(\tau)$ for block 0 wt% for composition. Temperature, K: 323 (---), 333 (-.-.-), 343 (. . .). Adapted, by permission, from T. Hashimoto and T. Izumitani, *Macromolecules*, **26**, 3631 (1993).

rium is supposed to be independent of temperature, because the temperature region of quenching is narrow and remote from the critical point. For the mixture without copolymer, values $Q_m(\tau)$ and $\tilde{I}_m(\tau)$ at three temperatures are described by a master curve, meeting the dynamic scaling postulate.[53,65] However, the postulate becomes invalid when block-copolymer is added: the curve $Q_m = f(\ln \tau)$ deviates upwards, whereas $\ln \tilde{I}_m = f(\ln \tau)$ deviates downwards as compared with the system without copolymer (Fig. 5.6). The additions of copolymer tend to slow down the coarsening even at scaled τ. Here T-branching is characterized by the greatest effect of copolymer occurring at lower temperatures. Deviations due to copolymer introduction, called B-branching are more pronounced than T-branching. Non-universal nature of coarsening change with φ shows that the block-copolymer has an effect not only on time and spacial scales of the ordering process but on the very mechanism of ordering, as elucidated from the copolymer influence on the initial stages. This effect is related to increased fluctuations and formation of an interface between adjacent domains where block-copolymer molecules are localized. This effect leads to the diminishing the interfacial tension and influences the interfacial dynamics and rate of coarsening.[54,57,151] The diminished coarsening rate cannot be scaled by the dependence of $q_m(0)$ and τ_C on φ. These effects are the key for interpretation of B-branch and T-branch because higher number density at the interface corresponds to a lower temperature at a given scaled time, τ.

In the process of coarsening, the interface area diminishes[47,103] and the number density of block-copolymer at the interface increases. When the interface becomes saturated, the following coarsening is impeded, leading to the pinning of domain growth. Non-universality of the reduced variables, observed during coarsening, i.e., the appearance of B- and T-branches corresponds to the cross-over stage.

The branching effect was discussed in terms of the equation for non-linear evolution time. The exponent α at the late stage is higher than 1/3 which indicates that the hydrodynamic effects should be taken into account by explaining the coarsening effects.[54,57,60] In critical binary liquid mixtures (true viscous liquids), the hydrodynamic effects were included in the time-dependence in Ginzburg-Landau equation:

$$\frac{\delta S(\vec{r},t)}{\delta t} = L\nabla^2\mu[S(\vec{r},t)] +$$

$$+ f_T - \nabla S(\vec{r},t) \int T(\vec{r} - \vec{r}')\nabla' S(\vec{r}',t)\mu S(\vec{r},t)d\bar{r}' \qquad [5.160]$$

where $S(\vec{r},t)$ is spatial concentration fluctuation at a given time t, L is Onsager kinetic coefficient, μ is chemical potential, which is a function of the order parameter $S(\vec{r},t)$, f_T is a contribution of the random thermal force in the evolution time of $S(\vec{r},t)$ and $T(\vec{r})$ is Oseen tensor:

$$T(\vec{r}) = \frac{1}{8\pi\eta}[1/\vec{r} + \vec{r}/r^3] \qquad [5.161]$$

where η is the viscosity of a system, 1 is a single tensor. First two terms of Eq 5.160 describe the evolution time of the order parameter S(r,t) accounting for molecular motions ruled by thermodynamic driving force $\varepsilon_T = (\chi(T) - \chi_S)/\chi_S$ (first term) and random thermal noise (second term), whereas the third term describes the evolution time of local velocity field (i.e., the hydrodynamic effect). Block-copolymer exerts an effect on the evolution time equation through interfacial tension, γ. The interfacial tension influences two terms of the equation: thermodynamic term (third term in Eq 5.160), which depends on γ/η, and term of individual molecular mobility $\nabla^2\mu$ (first term). In the latter term, the free energy of the system, F, includes square-gradient term $(\nabla S)^2$ which depends on the interfacial tension.[5,8,32] Variational derivative F by S(r,t) gives value $\mu = \delta F/\delta S$.

The experimental results in a qualitative form indicate that B-branching markedly increases during the late stage of spinodal decomposition, where hydrodynamic effects become very important.

The equation, describing the evolution time of the order parameter, expressed in terms of scaled wavelength, Q_m, intensity, I_m, and time, τ, becomes non-universal due to hydrodynamic effect, depending on both γ and η.[57] Based on computing results,[152] the non-universality may take place even if hydrodynamic interactions are absent in systems containing block-copolymers or surfactants. The non-universality is a consequence of diminishing interfacial tension due to block-copolymer localization at the interface. Localization affects both the square-gradient term $(\nabla S)^2$ in the expression for functional of free energy $F\{S(\vec{r,t})\}$ of the system[5,8,32] and value $\mu\{S(\vec{r,t})\}$ in the first term of Eq 5.160. However, the appearance of hydrodynamic terms gives bigger contribution to the behavior of the system. T and B-branches appear at the time scale, where hydrodynamic interactions prevent coarsening process.

5.5.3 SPINODAL DECOMPOSITION NEAR SOLID SURFACE

The spinodal decomposition is characterized by an isotropic structure factor, reflecting translational and rotational symmetry of bulk fluid. There may be cases that the symmetry is broken, for example, near the surface.[153-157] For the first time, Ball and Essery[153] suggested such effect based on theoretical grounds. Then, Jones *et al.*[156] experimentally confirmed this effect. They discovered a strong anisotropic structure factor by studying spinodal decomposition of thin films of isotopic mixtures of poly(ethylene propylene), PEP, with perdeuterated poly(ethylene propylene), d-PEP. The surface was enriched with d-PEP and oscillating concentration profile was observed normal to the surface. This means that, near the surface, there are observed composition waves with vector normal to, and propagating from, the surface, maintaining coherence for several wavelengths. These surface-directed spinodal decomposition waves are caused by the preferential attraction of d-PEP to the surface. The characteristic length of such lamellar structure grows much slower than the corresponding value for bulk.

Later for the same system, it was shown[157] that compositional profile can be modeled by two spinodal waves originating from two film surfaces - polymer-vacuum and polymer-solid substrate. The interference effects are observed by changing the film thickness. Two solid supports were used: the stripped Si, where SiO_2 layer was removed by etching with water solution of HF and SiO_2 surface treated by octadecyl-chlorosilane. The annealing of the mixture on Si-surface leads to the formation of bilayer structure with PEP-rich phase near the stripped Si surface. Silanized surface changes the system morphology. The d-PEP phase is situated near both surfaces (at the polymer-vacuum and polymer-surface interface). Most measurements of the film thickness during spinodal decomposition were made for Si-surface. The lamellar structure was stable over time longer than several years. Si-surface attracts PEP more than d-PEP. The study of the concentration profile $\varphi_{d\text{-}PEP}(z)$ on film thickness has shown that for thick films (150-1000 nm) only surface spinodal wave is observed, decaying into bulk of the films over a characteristic length of some 250 nm. No oscillations originating from Si-surface were observed. With diminishing film thickness, the compositional profile changes essentially. The third maximum appeared at around 500 nm. This maximum is due to a second spinodal wave, originating from the polymer-substrate interface. With diminishing the film thickness, the second maximum is destroyed because of interference of two spinodal waves. Below a critical film thickness, the coarsening of the phase structure is severally altered in comparison with a thick film behavior. Cell dynamical simulations were performed which closely resembled the experimental findings. The effect of solid support on the processes of phase separation in a thin film were observed in some works[158-160] and are discussed in Chapter 3.

5.5.4 SHEAR-INDUCED PHASE SEPARATION

The shear effect on the phase separation in flowing polymer solution was discussed many years ago.[161,162] For concentrated water solutions of poly(methacrylic acid), it was found that increasing shear rate leads to a rapid increase in viscosity.[162] The effect occurs in a definite temperature interval. The transition from conformation of coil to more extended conformations and the change in the ratio of intra- and intermolecular bonds between extended chains causes changes in viscosity. Such transition may be thermodynamically favorable because diminishing entropy is compensated by the gain in enthalpy. As a result of shear influence, the gel is formed and the system undergoes phase separation.[163-166]

Experimental observation concerning the influence of shear fields on demixing of polymer-containing fluid already date back more than 40 years.[167,168] With two immiscible polymers in a common solvent, it was found that homogeneous region of the system may extend by the order of 10K when the solutions flow. Hindawi *et al.*[169] studied flow-induced mixing of blends of poly(ethylene-co-vinyl acetate) and solution chlorinated polyethylene under simple shear flow. Light scattering experiments during flow of blends indicated that stress improves the miscibility of the two polymers. This was interpreted in terms of stress-dependent phase diagram in which the phase boundary of a polymer mixture that exhibits LCST is shifted upward in temperature due to the application of stress.

Barham and Keller[170] based on their own and many other experimental data reported that two processes may proceed in polymer solution under shear flow-induced nucleated liquid-liquid phase separation and spinodal decomposition induced by flow. Roland and Bohm[171] using small-angle neutron scattering measurements have studied shear induced coalescence in two-phase polymeric systems (PB/polychloroprene blends). Flow-induced coalescence occurs at a very rapid rate during the mixing of polymer fluids. It allows one to achieve optimal blending of immiscible polymers with respect to the morphology of the final material. The effect of shear on the phase behavior of polymer blends was discovered also for the mixture[172] poly(hydroxybutyrate-co-hydroxy valerate)-poly(hydroxybutyrate). The shear effect on phase separation behavior of a polymer blend (PS and PB in dioctyl phthalate) was studied by small angle neutron scattering.[173] Authors discovered differences in the correlation length parallel and perpendicular to the flow direction (by assuming mean field behavior of the solution). Dramatic decrease in the spinodal temperature occurred as a function of shear rate in the direction of flow. The results are consistent with the notion of shear stabilization and decoupling of the flow field in orthogonal direction.

The phase-separation behavior of dilute and semidilute polymer solutions under shear flow attracted considerable attention during last years.[174-177] The hydrodynamics of polymer solutions offers a quantitative description of the coupling between elastic stresses and concentration and conformation changes in flow.

Thermodynamic theory of flow of polymer solutions was developed by Wolf.[178] The phenomenological aspect of the theory is based on two premises: 1) it is possible to define a generalized Gibbs energy of the sheared system as the sum of its value at rest plus the energy stored during flow and 2) equilibrium thermodynamics can be applied to this function despite of the fact that streaming systems present stationary states only.

The thermodynamic description of polymer solution is performed by means of Flory-Huggins equation

$$\overline{G}_z/RT = x_1\ln\varphi_1 + x_2\ln\varphi_2 + gx_1\varphi_2 \qquad [5.162]$$

where $\overline{G}_z$ is the molar Gibbs energy at zero-shear conditions, x and φ are mole fractions and volume fractions, respectively, 1 stands for the solvent and 2 for polymer and g constitutes an interaction function. The reduced molar energy, E_S, which polymer solution is able to store during stationary flow, can be expressed as

$$\overline{E}_s/RT = \frac{\overline{V}_1(1 + (N - 1)x_2)}{RT} J_e^0 \eta^2 \dot{\gamma}^2 \quad [5.163]$$

where V_1 and NV_1 are the mole volumes of solvent and polymer, respectively, J_e^0 is the steady-state shear compliance, and η is the viscosity of the solution at the shear rate, γ. If a generalized reduced Gibbs energy is introduced according to

$$\overline{G}_\gamma = \overline{G}_z + \overline{E}_s \quad [5.164]$$

it is possible to calculate by a standard procedure the thermodynamic behavior of flowing solutions. The dependence of $J_e^0\eta^2$ on M, x_2 and γ can be obtained by combination of a number of known theoretical and empirical equations. The simple expression for the reduced stored energy of a flowing polymer solution may be derived from measured flow curves, if zero-shear viscosity of the solution, Rouse relaxation time, and some other values are determined. To estimate generalized reduced Gibbs energy, one has to know also the temperature and concentration dependence of the interaction parameter, g.

Wolf studied the dependence of demixing conditions of polymer solutions on flow for the system *trans*-decaline-PS and obtained the reduced Gibbs energy $\tilde{G}_\gamma$ of the flowing solution as a function of the mole fraction of the polymer. Together with the determination of the cloud point at a given shear rate, the conditions of demixing were found and composition of the coexisting phases was determined. At temperatures only 1-2K below the critical temperature of the system at rest and at shear rates of a few hundred to a few thousand s^{-1}, the miscibility gap observed for zero shear rate can be completely eliminated. The phenomenon of shear dissolution is explained by storage of the energy, $\overline{E}_S$, which can completely undo the “upturn” in the function $\overline{G}_z$ responsible for the phase separation in the absence of shear. This result agrees with theoretical calculations on the basis of a forced subdivision of the droplets of the coexisting phases down to molecular level. Phase diagrams of the system under various shear rates are presented in Fig. 5.7. The region of homogeneity is extended by shear. The demixing curves of the flowing solutions normally exhibit two maxima instead of one. The largest additional miscibility that can be achieved by a given shear rate is represented in the diagrams by solid circles. These points of maximum shear dissolution are called eulitic points by analogy with the term eutectic, which indicates the maximum extension of the liquid state with respect to the solid. At the eulitic temperature, three phases coexist as shown in Fig. 5.7 for $\gamma = 5{,}000\ s^{-1}$.

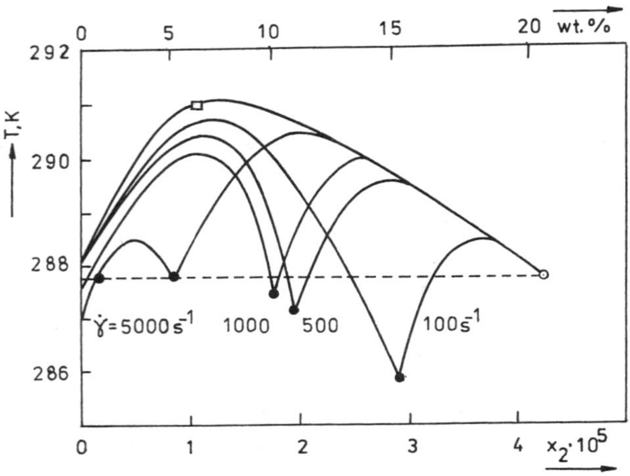

Figure 5.7. Demixing temperature of the system *trans*-decalin/polystyrene as a function of the mole fraction x_2 of the polymer at the indicated constant shear rates, $\dot{\gamma}$: 1 - 100, 2 - 500, 3 - 1,000, 4 - 5,000 sec^{-1} (at the upper abscissa, the composition is also given in terms of weight percent polymer). The solid circles represent the eulitic points, the dashed line shows, as an example, the coexistence of three phases at $\dot{\gamma} = 5{,}000$ s^{-1}. Adapted, by permission from, B. A. Wolf, *Macromolecules*, **17**, 615 (1984).

Another approach to the problem was proposed by Vrahopolou-Gilbert and McHugh.[179] The mechanism of phase separation under flow is intimately connected with high molecular weight chain entanglements. Under flow, frictional forces cause the macromolecules to uncoil from their random quiescent conformation to a more extended state of lower conformational entropy. The effect of these processes on the chemical potential of the system can be considered in one of two ways. Either, a term for so-called "stored free energy" can be added to the Flory-Huggins expression for the free energy of mixing polymer and solvent, as done by Wolf. Or, one can visualize the effect of chain stretching as equivalent to introducing a degree of rigidity to the polymer coils whose thermodynamic properties then become those of a system of semiflexible chains. Frenkel[180] suggested that stretching of macromolecular coil is equivalent to increasing its rigidity, thus shifting the binodal to higher concentrations and a higher critical temperature. To describe the situation, the effective flexibility parameter, f, proposed by Elyashevich[181] is introduced, which depends on the extending force, F. The free energy change per mole of polymer associated with the application of flow (stored free energy) may be shown to be

$$\Delta G_f = RT(x - 2) \ln [(1 - f)(z - 1)] \qquad [5.165]$$

where x is a number of segments in each polymer molecule and z is the lattice coordination number. In this work, it was theoretically shown that the change in flexibility has an effect on the binodal, if the chains in the more concentrated phase are less randomly coiled. Construction of thermodynamic diagrams requires a distinction between the cases of equal flexibility in the two phases of unequal flexibility. In both cases equations for binodal include flexibility parameter. The binodal equation is presented as

$$[1 - v_2(A + 1)/2] \ln [(1 - Av_2)/(1 - v_2)] + [v_2(A + 1)] \ln A/2x +$$

$$+ Bv_2^3(A - 1)^3/2 + v_2(A - 1)(1 - 1/x) =$$

$$= (G'_s - G_s)v_2(A + 1)/2RTx = 0 \qquad [5.166]$$

where A is ratio of polymer volume fractions in the concentrated and dilute phases, B is parameter describing the polymer volume fraction dependence on the interaction parameter, G'_s and G_s are stored free energies per polymer mole. If the free energy changes upon stretching, the macromolecules are the same in both phases after phase separation occurs, then no shifting of the binodal can be predicted (the case of equal flexibility in the two phases). In general, G'_s–G_s are given by

$$G'_s - G_s = RT(x - 2) \ln \frac{1 - f'}{1 - f} \qquad [5.167]$$

where f' is the flexibility in concentrated phase. Reduced flexibility in two phases contributes to stored free energy difference and position of binodal. Thus, thermodynamic phase diagram depends on the parameters connected with chain flexibility. A solution for shifted binodal exists only if $f' > f$. Authors concluded that thermodynamics of the reduced flexibility and stored free energy, used to describe flow-induced liquid-liquid phase separation, are fundamentally different.

There were published other approaches to the problem. Onuki Akira[182] has made theoretical analysis of the spinodal decomposition in semidiluted polymer solutions induced by shear deformation. Kammer[183] considered the effect of shear on the phase behavior of polymer-polymer mixtures with both UCST and LCST. Under shear the intermolecular interactions responsible for miscibility are changed due to stretching of chains and diminishing their flexibility. Depending on the ability of the system to store energy under shear, the miscibility may either increase or diminish.

5.6 PHASE SEPARATION ACCOMPANYING REACTIONS OF POLYMER BLEND FORMATION

A special case of polymer blends are semi- and full interpenetrating polymer networks,[184,185] and other systems that are formed from two polymers during the chemical reactions. In these systems, during their formation at a definite stage of conversion, due to arising thermodynamic incompatibility of constituent network fragments, there proceed the processes of phase separation, accompanying chemical reactions. It was shown that IPNs similar to individual network polymers possess clearly pronounced structural heterogeneity, caused by a thermodynamic incompatibility. A high degree of mechanical entanglements in IPNs hinders phase separation processes, which are always incomplete. This leads to the formation of thermodynamical non-equilibrium structures which depend on the prehistory of the system and essentially affect the IPN properties.[186,187]

5.6.1 SPINODAL DECOMPOSITION IN IPNs

In a number of articles, it was established[188-193] that the processes of microphase separation in IPNs proceed according to the spinodal mechanism. For a number of systems (network polymers on the basis of oligoisoprene dihydrazide and epoxy oligomers, styrene-divinyl benzene copolymer-PBMA, and others), from an analysis of their structure, it follows that, due to microphase separation, a periodic structure appears during curing. This periodicity is largely preserved until the end of microphase separation. Since the heterogeneity microregions is very different in their composition than for pure components, it is possible that phase separation in such systems initiates and proceeds in the regions of unstable states (inside a spinodal) and it is subject to spinodal decomposition peculiarities.[193]

A difficulty arising in the investigation of microphase separation processes by formation of IPN is that the composition of the system varies continuously during the conversion. This impedes the use of the available theories to describe phase separation. In order to avoid complications caused by simultaneous reactions, the reaction conditions should be selected so that the reaction rates are minimal compared with the rate of microphase separation. The experimental data show that the formation of semi- and full IPNs occurs according to a spinodal mechanism which is realized in spite of the proceeding reaction and, consequently, the non-equilibrium conditions of the process. It was also shown that under these conditions the reaction

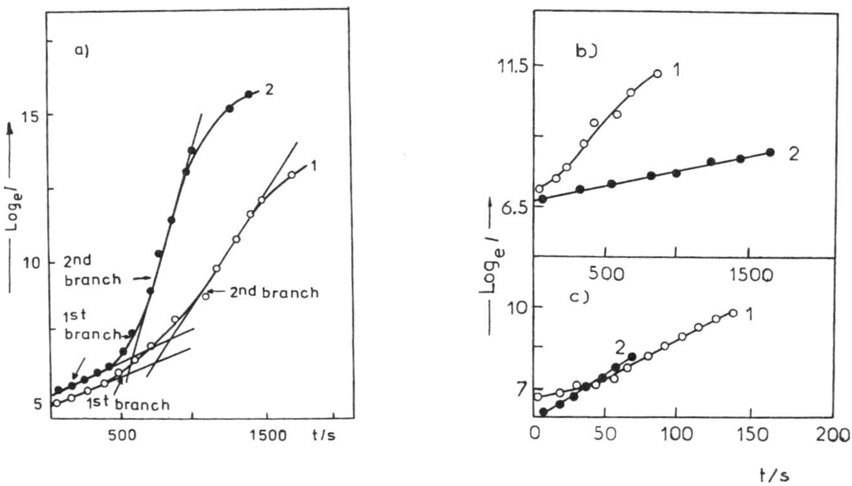

Figure 5.8. Dependence of the logarithm of the scattered light intensity (I, in arbitrary units) on time, t, for styrene-divinylbenzene-PBMA systems with 10 wt% PBMA undergoing phase separation, for incident angle 10° (curve 1, 0) and 20° (curve 2) at (a) 333K, (b) 343K, and (c) 363K. Adapted, by permission, from Y. S. Lipatov, O. P. Grigor'eva, G. P. Kovernik, V. V. Shilov, and L. M. Sergeeva, *Makromol. Chem.*, **186**, 1401 (1985).

kinetics determine the beginning of microphase separation, i.e., the thermodynamics of formation of the system and its thermodynamic state are determined by the reaction kinetics.[186,187]

The phase separation mechanism of IPNs depends on the position of a figurative point on the phase diagram. Unfortunately, to obtain the latter for multicomponent systems is a matter of a great difficulty. For simple systems, when semi-IPNs are obtained from a linear polymer and bifunctional polymer, they form a three-dimensional network. It is possible to plot a phase diagram for three-component system.[190] However, it is very difficult to obtain phase diagrams for four- and more component systems. Such diagrams should be obtained for different degrees of conversion of the constituent networks, i.e., multiplicity is needed for estimating the system state. If the interrelation between the reaction rate of constituent networks is taken into account, the solution of such a problem seems to be unrealizable.

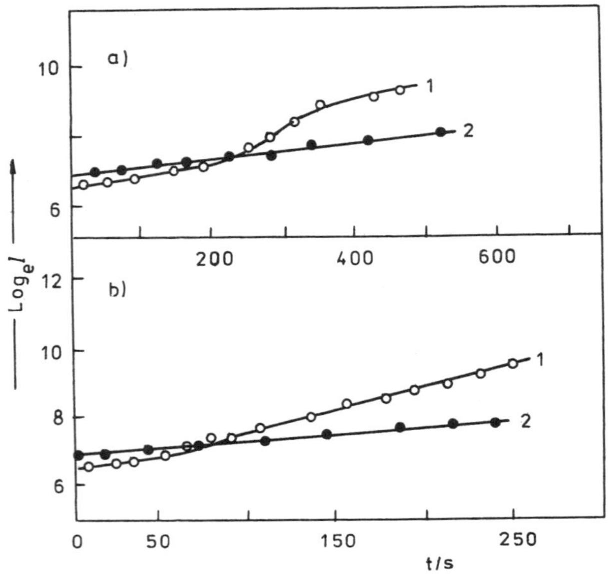

Figure 5.9. Dependence of the logarithm of the scattered light, I, on time, t, for styrene-divinylbenzene-PBMA system with 20 wt% PBMA undergoing phase separation for incident angle 10° (curve 1) and 20° (curve 2) at (a) 333K, (b) 343K. Adapted, by permission, from Y. S. Lipatov, O. P. Grigor'eva, G. P. Kovernik, V. V. Shilov, and L. M. Sergeeva, *Makromol. Chem.*, **186**, 1401 (1985).

Consider experimental data for the formation of semi-IPN based on crosslinked PS and linear PBMA, introduced into the reaction mixture.[194,195] In this system, during crosslinking reaction, the solidifying proceeds under conditions of continuous changes of its composition (with degree of conversion). As a result, thermodynamic compatibility changes continuously, as well. However, it was shown that during period of time when phase separation is observed (and especially at the very beginning), the composition of the system changes only slightly and can be considered constant. In this case, the Cahn theory may be applied. The time dependence of the light-scattering intensity I(q,t) has been studied. Figs. 5.8 and 5.9 show the time dependence of the intensity of scattered light ln I = f(t) for the systems with 10 and 20% PBMA, undergoing phase separation at various temperatures. These dependencies are similar for different scattering angles and reaction temperatures. For spinodal decomposition, the dependence of lnI on time should result in a straight line. From Figs. 5.8 and 5.9, it is seen that the time dependence is in fact linear at elevated temperatures and in the system with higher PBMA content, i.e., where the duration of phase separation does not last longer than 200-300 s. At lower temperatures and for lower PBMA fractions (i.e., for phase separation prolonged up to 200-2,000 s) the dependence becomes non-linear. However, two linear branches can be distinguished. The first branch corresponds to the initial stage of phase separation. After this time, there is an inflection point and the second more prolonged branch can be seen.

Cahn theory was developed for the initial stages of phase separation (the duration does not exceed several minutes) and may be applied to the case under consideration because phase separation takes no more than 200-300 s (the first linear branch in Fig. 5.8a). However, formally the theory can be applied also to the second linear branch. Fig. 5.10 shows the dependencies of $R(q)/q^2$ on q^2 for both branches for the system with 10 wt% PBMA (Fig. 5.10a). For the systems with 15 and 20% PBMA, these dependencies are similar. It can be seen that the linearity of this dependence is improved with increasing temperature. If the dependence is approximated by a straight line, some important parameters of Cahn theory can be calculated. From the slope of the plot lnI vs. t, the amplification factor R(q), and from the dependence of $R(q)/q^2$ on q^2 the interdiffusion coefficient and the optimum wavelength were found. The experimental data show that the interdiffusion coefficients are negative which corresponds to the spinodal mechanism. With growing temperature, D increases, i.e., the interdiffusion of the components in forming semi-IPN becomes easier due to a lower viscosity at elevated temperatures. The absolute values of D, as calculated for the second stage of the phase separation, are higher. This means that the effective diffusion of the components of the system is different for various stages of phase separation. In the second stage, the diffusion proceeds faster. It was also found that the optimum wavelength, λ_m, is almost the same for all temperatures. The dimensions of the microregions of phase separation are dependent on the composition and are at maximum in the region of small additions of PBMA.

It is important to note that phase separation in the reacting system does not affect the kinetic curves that is the total reaction rate does not change after the onset of phase separation. Such situation may be observed only when the microregions of phase separation only slightly differ in their composition which is typical of spinodal decomposition.

The data on the formation of a microphase structure for the system discussed above and for some other semi-IPNs based on crosslinked polyurethanes and linear PBMA[196,197] allows some important conclusions to be drawn. The process of microphase separation

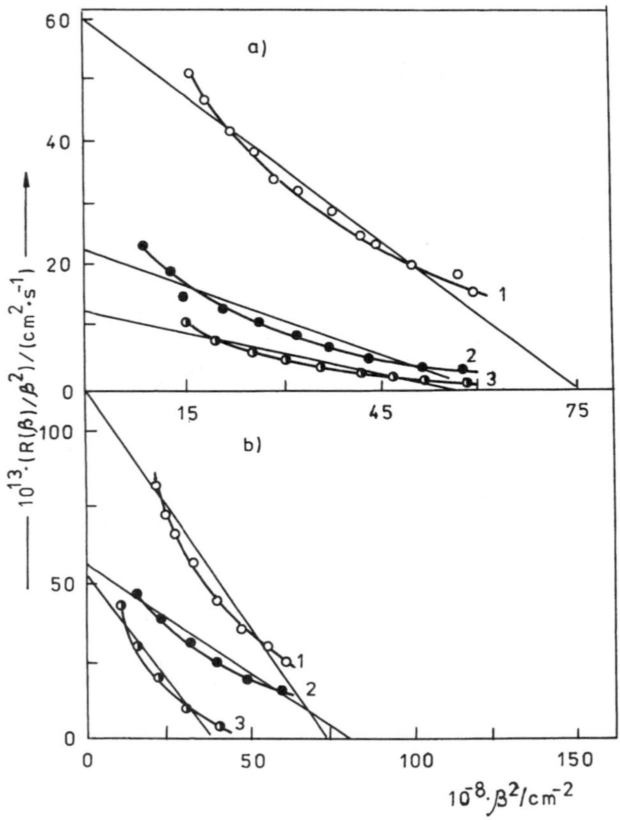

Figure 5.10. Dependence of $R(\beta)/\beta^2$ for semi-IPN with 10% PBMA as calculated for the first (a) and the second (b) linear branches in Figure 5.9: (a) at 363K (1), 343K (2), 353K (3). Adapted, by permission, from Y. S. Lipatov, O. P. Grigor'eva, G. P. Kovernik, V. V. Shilov, and L. M. Sergeeva, *Makromol. Chem.*, **186**, 1401 (1985).

in semi-IPNs proceeds according to the spinodal mechanism, in spite of the simultaneously proceeding chemical reactions of crosslinking and, in some cases, of polymerization of monomer, forming linear component of semi-IPN. This means that phase separation in the system proceeds under non-equilibrium conditions. The phase separation begins at a low degree of conversion before gel point and is enhanced by increasing molecular mass and volume fraction of polymers formed. The kinetics of chemical reactions determines the onset of phase separation and influences this process as a whole, whereas phase separation does not influence the reaction kinetics.

Thus, properties of IPNs are determined by the kinetics of their formation and by processes of microphase separation accompanying the reaction. In some works, the interconnection between the reaction kinetics and the onset of microphase separation was established.[198,199] The formation of IPNs is characterized by superposition of two processes: chemical reaction of crosslinking and physical process of microphase separation proceeding either according to nucleation and growth mechanism or due to the mechanism of spinodal decomposition. It is evident that these factors determining the reaction kinetics and phase separation determine the morphology and physical properties of IPNs.

From the point of view of microphase structure of IPNs, three cases may be distinguished. If one network is formed much faster than the other, the initial stage of the formation of IPN proceeds practically in liquid phase of components of the second network. In this case, the phase separation may proceed relatively fast and the evolution of the swollen first network in components of the second network is possible. The second network is formed later and as a result the phase structure appears which is characterized by a high degree of phase separation or segregation degree (ref. 199). This case is typical of IPNs obtained by sequential curing.

The second case is when both networks are formed very rapidly. Microphase separation, as the process governed by diffusion, has no time to proceed and the structure is frozen -- typical for one-phase state of initial mixture of reaction components. This state is not in thermodynamical equilibrium. The third case is the most frequent. It takes place when both networks are formed with comparable rates of reaction. Microphase separation begins during reaction before gel point and proceeds simultaneously with chemical reactions. The process is impeded in the course of reaction and it is stopped after reaching gel-point, i.e., rather high degree of crosslinking. The two-phase structure with incomplete phase separation appears as a result of these two overlapping processes which depend on the two reaction rates and on the kinetics of microphase separation. The last two cases are typical of simultaneous IPN's.

The peculiarity of phase separation in IPN consist of the existence of two-stage process. The initially "pure" spinodal mechanism becomes more complex due to the simultaneously proceeding processes of phase separation according to the nucleation and growth mechanisms (see below). These processes do not lead to a full separation into two independent phases because of chemical crosslinking and concomitant networks formation. As a result, systems with diffuse regions of microphase separation are formed. It is worth noting, that superposition of the chemical crosslinking reaction and phase separation makes it difficult to determine the mechanism of phase separation. In some cases one mechanism is replaced by the other. As a result, thermodynamically unstable diffuse microregions of incomplete phase separation arise, i.e., the segregation degree is not at its equilibrium value.

5.6.2 NUCLEATION AND GROWTH DURING IPN FORMATION

Phase separation during IPN formation may proceed by nucleation growth.[200,201] Rozenberg[202-204] studied the microphase separation in a number of oligomer-oligomer systems (epoxide oligomers and rubbers). Phase separation, due to constantly varying thermodynamic and kinetic parameters during curing, proceeds according to the nucleation mechanism. Constants, relating the quantity of evolved rubber particles to thermodynamic compatibility and kinetics of the process, were established. Sperling[205-207] investigated the theoretical conditions of formation of domains in IPNs. A degree of crosslinking of each network, the thermodynamics of mixing, and interfacial tension for successive IPNs were studied. Separation occurred by the nucleation mechanism.

The derivation of the basic equation for IPN domain diameter is based on a physical model of sequential IPN, according to which polymer II, formed in a swollen network I, constitutes a spherical core and exists in a contracted (deformed state), while polymer I surrounds the core and it is in an expanded (deformed state). Several assumptions were made:

- thermodynamic equilibrium process exists throughout the development of the domain formation
- the domains have identical diameters with a spherical shape
- the polymer network obeys Gaussian statistics
- a sharp interfacial boundary exists between the two phases.

The validity of these assumption is rather uncertain. The process of domain formation is presented in Fig. 5.11.

Initially, in the state 1, network I is completely separated from monomer II (including crosslinker). In the state 2, polymer network is swollen with the monomer mixture II. The pathway from the state 1 to the state 2 includes mixing between polymer I and monomer II and mutual concomitant expansion of polymer I caused by swelling with monomer II mixture. The free energy of polymerization from state 2 to state 3 is ignored. Also, the enthalpic energies of 1-2 contacts between monomer II and polymer I are assumed to be the same as the polymer II-polymer I enthalpic contact energies. State 3 is the hypothetical, mutually mixed state, where polymers I and II are mixed and

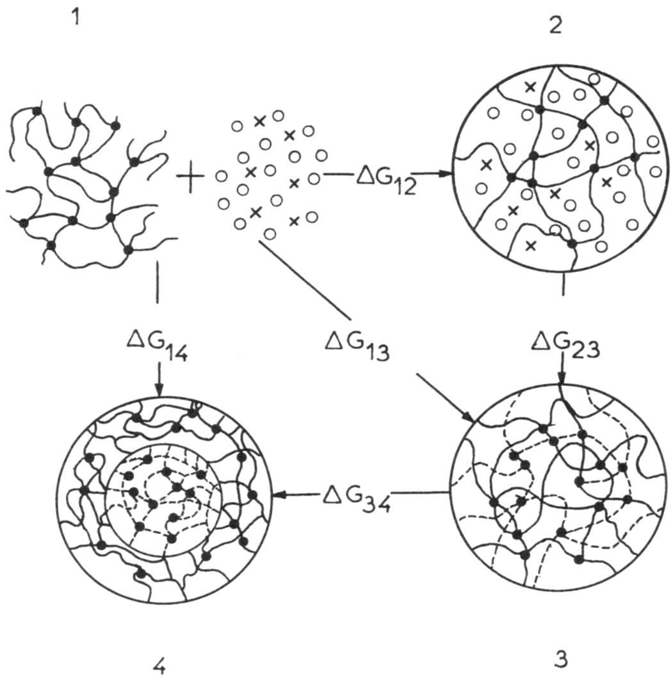

Figure 5.11. Simplified path of domain formation. Adapted, by permission, from J. K. Yeo, L. H. Sperling, and D. A. Thomas, *Polymer*, **24**, 307 (1983).

mutually diluted. Demixing (phase separation) between polymer I and polymer II with concomitant deformation of polymer II with further deformation of polymer I into a shell leads to the state 4. The state 4 is a phase-separated state with a spherical domain of polymer II surrounded by polymer I deformed into a spherical shell.

Referring to Fig. 5.13, the molecular rearrangements taking place on transforming from the state 3 to the state 4 requires amplification. No covalent bond is broken during the process, as would be required from a literal interpretation of the model. Instead, phase separation occurs at an early stage of polymerization of monomer II, when the free energy of mixing becomes positive and the second derivative of the free energy of mixing with respect to composition is negative. This is probably at or before the gel point. Thus, the molecular migration begins earlier that is illustrated in the model, and hence the stage 3 is hypothetical.

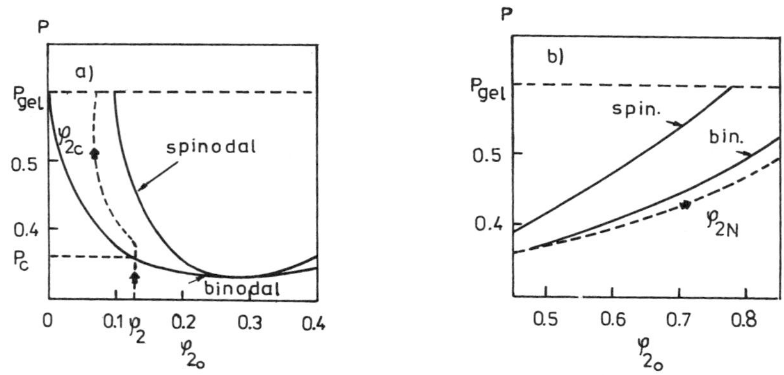

Figure 5.12. Binodal and spinodal curves for CTBN dissolved in a stoichiometric BADGE/DDS system plotted in a conversion vs. rubber volume fraction diagram at 393K. The dotted trajectory is referred to in the text. Adapted, by permission, from A. Vazquez, A. J. Rojas, H. E. Adabbo, J. Borrajo, and R. J. Williams, *Polymer,* **28**, 1159 (1987).

The thermodynamics of the process was described by Yeo, Sperling, and Thomas[205] as follows. For a closed system at constant pressure and temperature, the Gibbs free energy change, ΔG, is given by Eq 5.168:

$$\Delta G = \sum \Delta H_{i,i+1} - T \sum \Delta S_{i,i+1} \quad (i = 1,2,3) \qquad [5.168]$$

where $\Delta H_{i,i+1}$, $\Delta S_{i,i+1}$ represent the enthalpy and entropy changes involved in the process from state i to state i+1, respectively. Extending Eq 5.168 to the domain formation process, the free energy change for polymer II domain formation, ΔG_d, can be expressed as

$$\Delta G_d = \sum_{i=1}^{3} \Delta H_{i,i+1} - T \sum_{i=1}^{3} \Delta S_{i,i+1} + \Delta G_i \qquad [5.169]$$

where ΔG_i represents the interfacial free energy change for domain formation and relates to the processes taking place on transition from the state 3 to the state 4. From Fig. 5.13, it is seen that the pathway from the state 1 to the state 3 via the state 2 can be replaced by the direct path from the state 1 to 3. The following equation is valid:

$$\Delta G_d = \Delta H_{13} + \Delta H_{34} - T(\Delta S_{13} + \Delta S_{34}) + \Delta G_i \qquad [5.170]$$

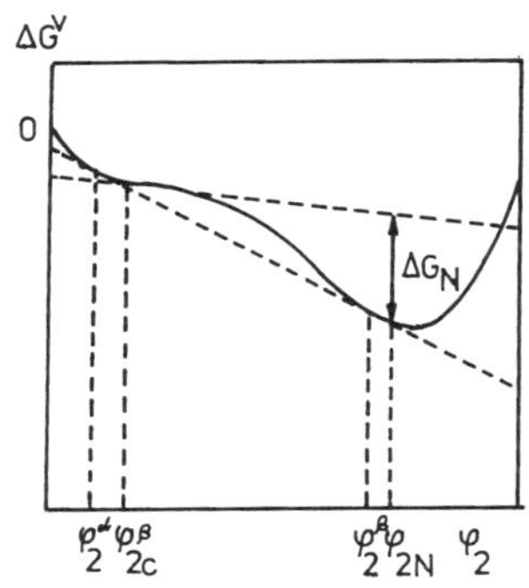

Figure 5.13. Free energy of mixing per unit volume as a function of rubber volume fraction, for a reaction extent where phase separation is taking place. φ_2^α and φ_2^β are the conjugated points of the binodal curve, φ_{2C} is the actual rubber concentration in the continuous phase, φ_{2N} represents the actual rubber concentration of the segregated phase, and ΔG_N is the free energy change per unit volume, associated with the phase separation process. Adapted, by permission, from A. Vazquez, A. J. Rojas, H. E. Adabbo, J. Borrajo, and R. J. Williams, *Polymer*, **28**, 1159 (1987).

In fact, ΔH_{13} and ΔH_{34} are the heat of mixing and the heat of demixing between polymer I and II, chosen that the sum of these two terms is zero. The quantity ΔS_{13} is equal to the sum of entropy of mixing, ΔS_m, and the entropy change for the elastic deformation of polymer I swollen with polymer II, ΔS^I_{sw}. The quantity ΔS_{34} is equal to the sum of entropy change of demixing, ΔS_{dm}, the rearrangement entropy change for elastic deformation of polymer II network upon deswelling, ΔS^{II}_{dsw}, and the entropy change for elastic deformation of polymer I network, ΔS^I_{df}. Here, ΔS_m and ΔS_{dm} cancel each other. The interfacial free energy change, ΔG_i, consists of the interfacial free energy change for domain formation, ΔG_i^0. A term for the entropy change on placing polymer I and II molecules in each domain, ΔS_p, must also be added. In summary, the free energy change for polymer II domain formation is expressed in the following way:

$$\Delta G_d = -T(\Delta S^I_{sw} + \Delta S^I_p + \Delta S^{II}_{dsw} + \Delta S^I_{df}) + \Delta G_i^0 \qquad [5.171]$$

On the basis of the model proposed, authors have derived the expressions for all terms in Eq 5.171. They are given by Eqs 5.172-5.175.

$$\Delta S^I_{sw} = -\frac{\pi}{12}\frac{\varphi_1}{\varphi_2}\nu_1 R(3\varphi_1^{-2/3} - 3 + \ln\varphi_1)D_2^3 \qquad [5.172]$$

$$\Delta S_p = \frac{\pi}{6} T\left(\frac{\varphi_1\rho_1}{\varphi_2 M_1} \ln \varphi_1 + \frac{\rho_2}{M_2} \ln \varphi_2\right) D_2^3 \qquad [5.173]$$

$$\Delta S_{dsw}^{II} = -\frac{\pi}{12} \nu_2 R(3\varphi_2^{2/3} - 3 - \ln \varphi_2) D_2^3 \qquad [5.174]$$

$$\Delta S_{df}^{I} = \frac{\pi}{4} \nu_1 R(1/\varphi_2)(2\varphi_1^{1/3} - \varphi_1^{4/3} - \varphi_1) D_2^3 \qquad [5.175]$$

The interfacial free energy change for domain formation is brought about from the intrinsic interfacial tension between the two polymers, γ^0, and is equal to $\Delta G_i^0 = \pi\gamma^0 D_2^2$ for spheres. Inserting all these quantities into Eq 5.171 and taking first partial derivative with respect to domain size D_2, $\delta(\Delta G_d)/\delta D_2$, the following expression for D_2 was derived

$$D_2 = 4\gamma^0[RT(A\nu_1 + B\nu_2 - C)]^{-1} \qquad [5.176]$$

where values A, B, and C are given by equations

$$A = (1/2)(1/\varphi_2)(3\varphi_1^{1/3} - 3\varphi_1^{4/3} - \varphi_1 \ln \varphi_1) \qquad [5.177]$$

$$B = (1/2)(\ln \varphi_2 - 3\varphi_2^{2/3} + 3) \qquad [5.178]$$

$$C = \frac{\varphi_1}{\varphi_2}\frac{\rho_1}{M_1} \ln \varphi_2 + \frac{\rho_2 \ln \varphi_2}{M_2} \qquad [5.179]$$

In these equations, φ is volume fraction of polymer, ρ is polymer density, γ is the number of moles of effective network chains, and M molecular mass which goes to infinity for full IPN. In the same work,[205] authors compared theoretical predictions with experimental data for IPN based on crosslinked poly(n-butyl acrylate) and PS. In spite of many arbitrary assumptions (including that spinodal decomposition process, rather than binodal phase separation, takes place), there was found rather good agreement between calculated and experimentally found domain diameters for various compositions of IPN and various crosslinking levels. It was noted that the domain size is very sensitive to the value of interfacial tension.

To describe the formation of binary systems where, in the course of reaction, phase separation occurs, another model was proposed[208] which allows to predict the fraction, composition, and average radius of disperse phase (domains) evolved during reaction.The model is based on the application of Flory-Huggins equation, in which the entropy term diminishes with conversion degree due to increase in polymerization degree. As a result of such diminishing, phase separation occurs at a definite stage. On the base of the model, the equations describing nucleation and growth rate (coalescence) have been derived. The model predicts the possibility of phase inversion, slowing down the separation at the gel point, increasing fraction of domains and their dimensions with me as a function of the curing temperature. Later [209] an expanded version of the model was proposed allowing to calculate the domain dimension distribution and establish factors, determining the distribution. The model was developed for the system epoxy resin-rubber (statistic copolymer of butadiene with acrylonitrile with end carboxylic groups and cycloaliphatic epoxy resin with amine hardener). Such system is characterized by the evolution of elastomeric phase with bimodal distribution of particles. Let rubber component 2 be dissolved in epoxy solvent 1. Then, the Flory-Huggins equation has the following form:

$$\Delta G^N = \frac{RT}{\overline{V}_{10}}[(\varphi_1/z_1)\ln\varphi_1 + (\varphi_2/z_2)\ln\varphi_2 + \chi\varphi_1\varphi_2] \qquad [5.180]$$

where φ_1 and φ_2 are volume fractions of components, V_{10} is the initial volume of the network defined as

$$\overline{V}_{10} = \overline{M}_{n,10}/\rho_1 \qquad [5.181]$$

where $\overline{M}_{n,10}$ is the initial mean number molecular weight and ρ_1 is density. Let ξ = (epoxy equivalent)/(amine equivalent) = $2B_2/4A_4$, where B_2 and A_4 are numbers of moles of each component. Then

$$\overline{M}_{n,10} = (M_{A_4} + 2\zeta M_{B_2})/(1 + 2\zeta) \qquad [5.182]$$

For the stoichiometric mixture of components, $\underline{\xi} = 1$. If z_1 and z_2 are ratios of molar volumes of both components to V_{10}, then, we have

$$z_1 = \overline{V}_1/\overline{V}_{10}; \quad z_2 = \overline{V}_2/\overline{V}_{10} \qquad [5.183]$$

It is supposed that z_2 does not change in the course of reaction, whereas z_1 increases. The conversion of amine and epoxy equivalents, determined as p_A and p_B are interconnected as $p_A = \xi p_B$. Then

$$z_1 = \overline{V}_1/\overline{V}_{10} = \overline{M}_{n,1}/\overline{M}_{n,10} = (A_4 + B_2)/(A_4 + B_2 - 4p_AA_4) \qquad [5.184]$$

and from Eq 5.172 and the definition of ξ, it follows that

$$z_1 = \frac{1 + 2\zeta}{1 + 2\zeta - 4p_A} \qquad [5.185]$$

For stoichiometric system $\zeta = 1$, $p_A = p_B = p$, and $z_1 = 1/(1 - 4p/3)$. The next step consists of determination of parameter χ. It may be found from the cloud points or from equation for excess chemical potentials. Rewrite Eq 5.168 as

$$\Delta G_v = \frac{RT}{\overline{V}_{10}} (\Delta\mu_1 + \Delta\mu_2) \tag{5.186}$$

where

$$\Delta\mu_1 = \frac{1}{z_1} \ln \varphi_1 + \left(\frac{1}{z_1} - \frac{1}{z_2}\right)\varphi_2 + \chi\varphi_2^2 \tag{5.187}$$

$$\Delta\mu_2 = \frac{1}{z_2} \ln \varphi_2 + \left(\frac{1}{z_2} + \frac{1}{z_1}\right)\varphi_1 + \chi\varphi_1^2 \tag{5.188}$$

At equilibrium, at the cloud point, the following relations are valid:

$$\Delta\mu_1^\alpha = \Delta\mu_1^\beta \tag{5.189}$$

$$\Delta\mu_2^\alpha = \Delta\mu_2^\beta \tag{5.190}$$

where superscript α relates to continuous phase (usually epoxy polymer) and β to disperse phase (rubber). At small additions of rubber, φ_{20}, the amount of phase β at the cloud point is small and $\varphi_2^\alpha = \varphi_{20}$ and $\varphi_1^\beta=1-\varphi_{20}$. Eqs 5.89 and 5.90 contain two unknown values: $\varphi_2^\beta(\varphi_1^\beta=1-\varphi_2^\beta)$ and χ. By solving them, one can find value of χ for any φ_{20}. Now, parameters found are put into Eq 5.168, from which binodal and spinodal can be found by the usual way. Authors[209] have calculated these curves in coordinates: conversion degree-composition (Fig. 5.12). For these binodals, the connecting lines are horizontal.

To describe phase separation, the dependence of free energy of the system per volume unit on the composition (at any conversion degree) is considered (Fig. 5.13). To derive driving potential of phase separation, it is necessary to come into metastable region of phase diagram. Let φ_{2C} be the true concentration of rubber in continuous phase at some conversion degree and φ_2^α and φ_2^β be the conjugated points of binodal (equilibrium compositions of phases). The sign in metastable region is that of $\varphi_{2C}>\varphi_2^\alpha$. Tangential line at φ_{2C} gives free energy per volume unit of a solution, segregated from a matrix of composition φ_{2C}. Vertical distance between tangential line at φ_{2C} and curve of free energy (Fig. 5.13) represent the change of free energy by evolution of a new phase of any composition. For particular composition, for example φ_{2N}, the maximum value of free energy, ΔG_N, is obtained. Thus, value φ_{2N} really represents the composition of a segregated phase, whereas ΔG_N the corresponding change of free energy. It is seen that φ_{2N} is always greater than φ_2^β. The experimental data indicate that the systems under consideration are characterized by separation according to the nucleation and growth mechanism. With the exception of incompatible systems, which may separate even at a very small conversion degree, the possibility of coalescence of dispersed domains is restricted by a strong increase in the system viscosity. The change in free energy due to the formation of a spherical particle with composition φ_{2N} is given by

$$\Delta G = (4/3)\pi r^3 \Delta G_N + 4\pi r^2 \gamma \quad [5.191]$$

where r is radius of a particle (domain), γ is the interfacial tension. Maximum value of ΔG_{crit} is reached at a critical domain radius, r_{crit}. From Eq 5.191, we have:

$$r_{crit} = 2\gamma \mid \Delta G_N \mid \quad [5.192]$$

$$\Delta G_{crit} = \frac{16\pi\gamma^3}{3 \mid \Delta G_N \mid^2} \quad [5.193]$$

The growth of dispersed domains with $r = r_{crit}$ diminishes free energy, as is shown in Fig. 5.14. The rate of homogeneous nucleation from condensed phases is written in the form:

$$\frac{dP(r_{crit})}{dt} = N_0 D \exp\left(-\frac{\Delta G_{crit}}{k_B T}\right) \quad [5.194]$$

where $P(r_{crit})$ is the volume concentration of particles with critical radius, r_{crit}, D is diffusion coefficient for rubber in epoxy network, N_0 is the fore-exponent factor and k_B is Boltzman constant.

To calculate the nucleation rate, one has to evaluate γ and D (N_0 is found from experimental data on particle concentration). Value γ was taken[209] as the difference between surface tension of epoxy resin and rubber, i.e., modelling epoxy-enriched phase as a pure epoxy resin and rubber-enriched phase as a pure rubber. Such simplification has no effect on the final results. Diffusion coefficient is determined using Stocks-Einstein equation:

$$D = \frac{D_0 T}{\eta(p,T)} \quad [5.195]$$

where η is the viscosity of continuous phase and depends both on temperature and conversion degree. At p = p(gel) value $D \rightarrow 0$ and nucleation is restricted. Near the gel point, strong increase in viscosity leads to very small nucleation rates. When the system passes through the metastable region, the particle growth occurs which is determined

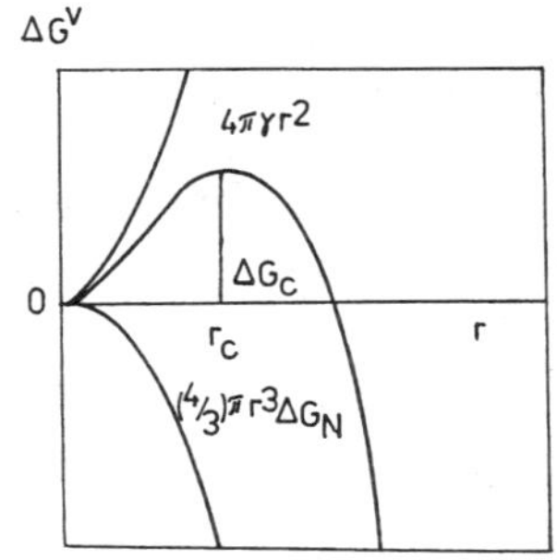

Figure 5.14. Different contributions to the free energy change associated with the formation of spherical domains. Adapted, by permission, from A. Vazquez, A. J. Rojas, H. E. Adabbo, J. Borrajo, and R. J. Williams, *Polymer*, **28**, 1159 (1987).

by the driving force ($\varphi_{2C} - \varphi_2^\alpha$). This force should transfer the system to the equilibrium state. Let dP(r′) correspond to the number of particles per unit volume when conversion degree changes from p to p+Δp. At $p = p'$, we have a set of particles with radius $r' = r_{crit}(p')$. For $p > p'$, radius grows to some value $r' > r_{crit}(p')$. Value dP(r′) is not a function of time. Growth rate is determined by the increase in volume fraction of particles in unit time and should be proportional to interfacial area per unit volume and to driving force:

$$(4\pi/3)dp(r')(dr')^3/dt = K_\varphi 4\pi(r')^2 dp(r')(\varphi_{2c} - \varphi_2^\alpha) \qquad [5.196]$$

where K_φ is mass transfer coefficient for the sphere in stationary medium $K_\varphi = D/r'$. When D→0 at p→p(gel), the rate of growth is restricted by gelation as well. Based on these premises, the structure of a system can be described in the following way. The volume fraction of evolved dispersed phase in the region of conversion degree from p to p + Δp includes new particles which appear at p^* plus additional fraction of particles which are formed at $p < p'$:

$$dV_D(p) = (4\pi/3)[r_{crit}(p)]^3 dP(r_{crit}) + 4\pi \int_{p'<p} (r')^2 dP(r')dr' \qquad [5.197]$$

All material segregated in the region of conversion degree from p to p+Δp has the composition φ_{2N}. The total concentration of particles of dispersed phase per volume unit at any conversion degree is given by Eq 5.198

$$P = \int_{p' \le p} dP(r') \quad [5.198]$$

The distribution function by particle dimensions in dispersed phase can be written as dP(r)/dr. Volume fraction of dispersed phase at any conversion P is

$$V_D = (4\pi/3) \int_{p' \le p} (r')^3 dP(r') \quad [5.199]$$

A volumetric average diameter of the distribution of dispersed phase particles may be calculated as

$$\overline{D} = (6V_D/\pi p)^{1/3} \quad [5.200]$$

The average rubber concentration in dispersed phase is

$$\overline{\varphi}_{2D} = (1/V_D) \int_{p' \le p} \varphi_{2N}(p') dV_D(p') \quad [5.201]$$

If φ_{20} is the initial volume fraction of rubber in the system. Its concentration in the continuous phase can be found from equation

$$\varphi_{2C} = \frac{\varphi_{20} - V_D \overline{\varphi}_{2D}}{1 - V_D} \quad [5.202]$$

The applicability of the model discussed above to the processes of phase separation was tested[209] for the thermocuring system of epoxy resin, based on bisphenol A-diglycidil ester of diamino-diphenylsulfone in which the statistic copolymer CTBN was dissolved. The binodal and spinodal were constructed (Fig. 5.15) and dependencies of rates of nucleation and growth were calculated. The model predicts the trajectories of φ_{2C} and φ_{2N} given in Fig. 5.12. The amount of rubber in continuous phase, φ_{2C}, does not change essentially up to conversion degree, p = 0.5, due to a rapid increase in viscosity in this conversion region. When the compatibility of the resin and rubber diminishes (using the same rubber with smaller amount of acrylonitrile), the phase separation proceeds in the medium with lower viscosity. The trajectory of φ_{2C} in the metastable region rapidly approaches binodal due to a high diffusion coefficient. Small driving force ($\varphi_{2C} - \varphi_s^{\alpha}$) determines marked diminishing rate of growth, whereas the nucleation rate is high. As a result, the bimodal distribution of particle dimensions takes place (Fig. 5.15). At the early stage of phase separation, very sharp distribution of particles is realized (curve A). In the course of the reaction, the fast growth of particles occurs, whereas small particles appear as a front tail of distribution (curve B). By

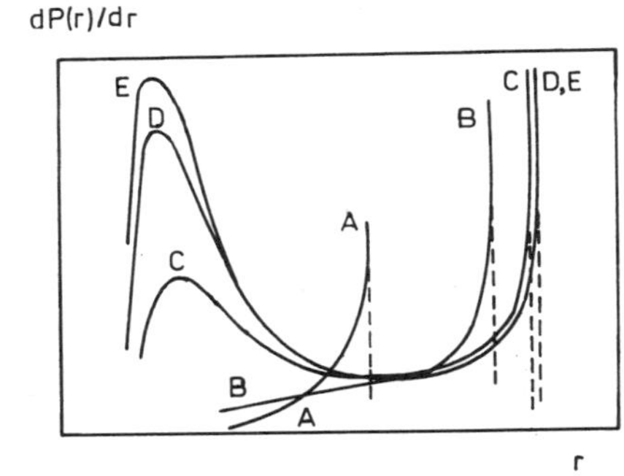

Figure 5.15. Partial size distribution of dispersed domains for different conversions of the thermosetting matrix: curve A, p=0.255, B, p=0.295, C, p=0.395, D, p=0.495, E, p=0.553. Adapted, by permission, from A. Vazquez, A. J. Rojas, H. E. Adabbo, J. Borrajo, and R. J. Williams, *Polymer*, **28**, 1159 (1987).

approaching the gel point, the growth of particles to larger sizes slows down and the concentration of small particles increases (curves C - E). This happens because the growth rate is inversely proportional to the dimension of dispersed domains. The decreasing miscibility, i.e., increasing parameter χ, leads to:

- increasing concentration of dispersed phase
- diminishing the rubber amount in continuous phase
- diminishing average concentration of small particles in total distribution.

All these effects are in operation because the process is far from the gel point and it occurs in low-viscosity medium.

The analysis of the interfacial tension effect on the distribution has shown that by changing γ from 2γ to $\gamma/2$, there is no essential change in final distribution. This occurs because, in metastable region, factor $\Delta G_{crit}/RT$ (Eq 5.194) is very small and has no influence on the nucleation rate. From the other side, although r_{crit} is directly proportional to γ (Eq 5.192), the mass transfer coefficient is inversely proportional to the particle dimension. As a result, the final dimension is almost independent of the critical radius, r_{crit}.

The phase separation in the same system was studied using SALS method.[210] The phase separation on curing proceeds due to increasing UCST. The angle dependence of scattering shows that the process proceeds by the mechanism of spinodal decomposition. The spherical domain structure of cured polymer is the result of fixation by crosslinking of some domains in epoxy-enriched phase. The structure formed is determined by the competition of two processes: phase separation and chemical reaction. Fast curing causes that the interconnected structure is fixed at the initial stages of separation. After some time, modulated structure is formed. If the hardener has a low reactivity,[210] small and large spherical domains are formed, i.e., bimodal structure. Small domains are distributed among large domains. Due to competition between phase separation and chemical reaction, two-phase morphology can be controlled, as well as physical and mechanical properties.[211,212]

The possibility of fixation of spinodal structures at different stages of spinodal decomposition was demonstrated.[213,214] For the mixtures of anthracene-labeled PS with poly(α-chlorostyrene), it was shown that because of transfer from single-phase into the spinodal region, the interconnecting structures were formed. UV light irradiation, applied after temperature jump (T-jump) due to crosslinking of anthracene, causes PS network formation and restricts diffusion of poly(chlorostyrene) chains during the evolution of spinodal decomposition. As a consequence, the interconnecting structure generated by the spinodal process was frozen. These structures were observed by the phase-contrast optical microscopy. At shallow jump, the morphology of the irradiated blends could not be observed by optical microscopy, however the light scattering profile shows peaks of higher order. This profile changes with time after irradiation. Taking into account non-equilibrium thermodynamic conditions caused by irradiation, especially by the gradient of radiation between the two surfaces of the sample, authors concluded that this experimental fact strongly suggests the existence of a new kind of instability originating from the photo-crosslinking reaction. For a constant duration time of irradiation, the sizes of ordered structures became larger if the blend was crosslinked with longer delay. The ordered structures revealed in these irradiated blends are different from these experiencing the same T-jump conditions without irradiation. Fixation of spinodal structures by crosslinking reactions was observed also for PB-PI blends.[215]

5.6.3 SOME FEATURES OF MICROPHASE STRUCTURE OF IPNs

Microphase separation is simultaneously the function of miscibility of constituent networks and reaction kinetics.[190,192,193,194,216-230] High degree of entanglement promotes inhibition of phase separation which usually remains incomplete. The periodicity arising at the initial stages of spinodal decomposition in IPN is preserved up to the end of separation.[192,222,226] The sequential IPN based on styrene with divinylbenzene in the matrix of swollen PU network[223] has the period of modulated structure of gel formation, characterized by the following dependencies on the reaction time, t, and conversion degree, P:

$$L \sim t^2; \quad L \sim P^{0.8} \tag{5. 203}$$

whereas IPN is characterized by relations:

$$L \sim t^1; \quad L \sim P^{0.4} \tag{5.204}$$

The calculations show that the segregation degree of components in IPN is about 70% at maximum conversion degree of monomer equal to 88%. Considering the existence of diffuse boundaries, one can conclude that such systems are fully separated into two microphases.[224] The evolution of pure microphases proceeds when 2-3 times increase in dimensions of microphase region proceeds during the process. However, the periodicity during the process does not change. This shows that for IPN the processes of spinodal decomposition proceed in accordance with classical regularities only at the early stages of reaction. Later (after transition of the separation process to non-linear regime, which is characterized by increasing dimensions of microregions), the three-dimensional structure of the system prevents phase separation, restricting the diffusion of components and slowing down the rate of growth of microregions. It was also shown[224] that phase separation in IPN strongly depends on composition. For IPN based on PU, it was found[227] that, at small amount of initial monomer in network matrix, more intense interpenetration takes place and therefore smaller segregation degrees are observed. An apparent compatibility may arise which may be considered as forced compatibility.[228] The non-equilibrium states are frozen.

If the component with high reaction rate, PUA, is taken in larger amount than the second component, PU, then segregation degree is very high (above 70%). This occurs because fragments of PU network are pushed out from PUA network when the conversion degree is not high and unreacted component, serving as a solvent, is present in the system. On the contrary, if the component with lower rate of network formation is taken in excess, microgel particles of PUA are in the medium formed by branched fragments of PU which restricts their motions necessary to form an isolated phase.

Thus, the reaction kinetics, composition, and diffusion are the most essential parameters which control IPN structure and phase separation. These processes are very complicated and interconnected. It is thus logical that in such systems pure mechanisms of nucleation and growth or spinodal decomposition can hardly occur in their pure form.

The effects of forced compatibility in IPN are of great importance for understanding the phase state. Forced compatibility should lead to an additional contribution to the free energy of the system. The contribution of elastic energy is small at the onset of phase separation and increases with increase in composition difference between phase separated regions.[195] As a result, the system is stabilized when the thermodynamic driving force for phase separation is balanced by elastic forces from entanglements of network fragments. The system stays in a state of incomplete phase separation. The stability limit of such polymer system is characterized by a curve below the chemical spinodal. In metal alloys, the elastic stresses stabilize the system and the decomposition of one-phase solution is determined by the coherent spinodal, also situated below the chemical spinodal. For the systems with UCST, the real spinodal is situated below the chemical spinodal. For some IPNs, the real spinodal is so remote from the chemical spinodal that the unstable state cannot be reached.

Analysis of the behavior of some IPNs shows[195] a state of forced compatibility and a state of forced microphase separation. If such systems have UCST, the end of the transition to one-phase system corresponds to the curves situated above the chemical spinodal (upper real spinodal). Its position is determined by the values of elastic forces needed for the system to transform from a state with microphase separation to one-phase system.[195] The transition of the system from one-phase state to two-phase state is hindered at a particular stage of IPN formation (lowering chemical spinodal, i.e., lower real spinodal) whereas transition from a two-phase state to one-phase state is hindered as well (upper real spinodal). The state of the system is determined by the chemical conditions of IPN formation.

The only effort to develop strictly statistical theory of microphase separation in IPN formation was made by Binder and Frisch.[229] Phase separation proceeding during reaction is hindered by simultaneous crosslinking, considered by authors as "chemical quench". Authors analyze the case of deep quench when the reaction rate essentially exceeds the rate of relaxation of the system into the equilibrium state. The expression for functional of free energy of mixing was derived, which strictly depends on the method of IPN formation (sequential or simultaneous). In both cases the expression contains the following contributions: elastic energy and entropy of crosslinking of two networks, the energy of interaction between monomeric units of both networks, and free energy of entanglements. The estimation of the limits of stability of one-phase state of various IPN (spinodals) shows that its stability is higher in the region of small amount of one of the networks. The theory of spinodal decomposition in the case of chemical quenching is valid, if t_{AB} (curing time) is smaller than τ_{AB} (system relaxation time). In long wave limits, the following conditions are valid:

$$a_A^2 q^2 << N_{AC}^{-1}; \quad a_B^2 q^2 << N_{BC}^{-1} \qquad [5.205]$$

where N_{AC} and N_{BC} are numbers of monomeric units between junction points in networks A and B, a_A and a_B are the dimensions of statistical segments in chains A and B, q is wavenumber, determining characteristic period of spatial inhomogeneity. Composition fluctuations begin to increase with time at wave vectors smaller than q_C:

$$q_C = a^{-1}\left(\frac{-(\delta^2 f/\delta\varphi^2)_{\varphi_0}\,\varphi_0(1-\varphi_0)}{K(\varphi_0)}\right)^{1/2} \qquad [5.206]$$

where φ_0 is the fraction of monomeric units of network A. The separation stops at the early stages of formation of modulated structures. Time period does not exceed the dimensions of the cell of network structure. In the case of a deep quench, value q_C is determined by expression

$$q_C = \left[\frac{6\chi^{1/2}}{a_A^2/\varphi_0 + a_B^2/(1-\varphi_0)}\right]^{1/2} \qquad [5.207]$$

The local separation degree can be very high, whereas the distance between microphase regions does not exceed the size of network cell. Further increase of microregions dimensions is hindered by forces distorting Gaussian conformations of chains forming entanglements. Thus, full separation of IPN, corresponding to equilibrium state, becomes impossible.

The analysis of glass transition temperatures in phase-separated IPN (the presence of two glass transition temperatures, corresponding to two phases enriched with one of the components), determination of segregation degree, and other properties allow the following conclusion to be drawn.[187] The microphase structure of IPNs can be described using the concept of the formation of two different clusters in the system: physical clusters due to phase separation of constituted networks under conditions of non-equilibrium phase transition and chemical clusters determined by crosslinking. The real phase structure of IPN should be considered as a multiphase structure that is determined by coexistence of at least three phases (none in a true thermodynamic sense). Two phases are formed from networks due to phase separation. Each phase can be considered as an independent IPN in which phase separation did not take place (the state of forced compatibility) and mixing near the molecular level is preserved. The composition of these two phases is determined by reaction conditions. Each phase has an average composition that does not correspond to network ratio in IPN. The third phase is a non-equilibrium transition zone from one phase to another. Its dimensions depend on the conditions of phase separation. This zone can also be considered as non-equilibrium IPN, because by spinodal decomposition there is no sharp border between coexisting regions of phase separation. This transition zone can be chosen that its composition corresponds to the average composition of IPN.

In general, the microphase structure of IPN is non-equilibrium. However, in such a non-equilibrium IPN, one can still distinguish various microregions that can be described as quasi-equilibrium that is microregions with a near molecular level of mixing. This concept meets experimental data showing comparatively low segregation degrees in IPNs. These microregions can be considered as quasi-equilibrium phases because they are the result of microphase separation and each phase preserves the composition that almost corresponds to one-phase state (one glass transition temperature), i.e., the equilibrium state of mixing. The level of mixing that is inherent to the state of mixing at earlier reaction stages is substantially preserved. The level of mixing corresponds to the onset of phase separation. Microphases are in quasi-equilibrium state. Thus, totally non-equilibrium microphase structure of IPNs may be presented as two-phase microheterogeneous system that lacks molecular mixing of two constituent networks throughout the bulk and with near molecular level of mixing in each phase and transition zone.

Introduction of fillers also has an influence on the phase separation in IPN.[230,231] It was shown that in presence of a filler, kinetic regularities of IPN formation are changed. By studying kinetics of phase separation in sequentially curing system of oligoisoprene dihydrazide-diglycidil ether of bisphenol A,[232] it was found that linear part of the time dependence of log of scattered light is observed only for unfilled system. Filling slows down the increase in light scattering intensity and the process cannot be described by the regularities of spinodal decomposition. At the same time, analysis of kinetics of formation of microphase structure has shown that microdomains are formed faster in filler presence. Filler presence in IPN increases thermodynamic stability of final system.[233-235] The main reason for increasing miscibility of IPN components in the presence of filler consists of changing conditions of thermodynamic interaction between components in surface layers at the interface with solid. Simultaneously, filler influences the kinetics of chemical reaction and changes also the conditions of phase separation. The restriction of molecular mobility of chains at the interface hinders phase separation as well. The filler plays a role of an agent regulating the microphase separation and microphase structure of filled IPN.[236,237]

REFERENCES

1. E. W. Hart, *Phys. Rev.*, **113**, 412 (1959).
2. S. P. Papkov in *Gel-State of Polymers* (in Russ.), Khimiya, Moscow, 1974.
3. S. Y. Frenkel in *Kinetic Theory of Liquids* (in Russ.), Nauka, Leningrad, 1975.
4. I. M. Lifshitz and L. P. Pitaevski in *Theoretical Physics* (in Russ.) v. 10, Nauka, Leningrad, 1979.
5. J. W. Cahn and J. E. Hilliard, *J. Chem. Phys.*, **28**, 258 (1958).
6. J. W. Cahn, *J. Chem. Phys.*, **30**, 1121 (1959).
7. J. W. Cahn, *J. Chem. Phys.*, **31**, 658 (1959).
8. J. W. Cahn, *J. Chem. Phys.*, **42**, 93 (1965).
9. J. W. Cahn, *Trans. Met. Soc. ASMC*, **242**, 166 (1968).
10. J. W. Cahn, *Acta Met.*, **9**, 795 (1962).
11. K. B. Rundman and J. E. Hilliard, *Acta Met.*, **15**, 1035 (1967).
12. M. Hillert, *Acta Met.*, **9**, 825 (1961).
13. K. Binder, *J. Chem. Phys.*, **79**, 638 (1983).
14. J. S. Langer and A. J. Schwartz, *Phys. Rev.*, **A21**, 948 (1980).
15. J. S. Langer and M. Bar-on, *Ann. Phys. (USA)*, **78**, 421 (1973).
16. K. Binder, *Phys. Rev.*, **B15**, 4425 (1977).
17. C. Varea and A. Rodledo, *J. Chem. Phys.*, **75**, 5080 (1981).
18. L. A. Swanger, P. K. Gupta, and A. R. Cooper, *Acta Met.*, **18**, 9 (1970).
19. L. H. Shendelman and J. T. O'Toole, *J. Colloid & Interface Sci.*, **27**, 145 (1968).
20. W. A. Tiller, G. M. Pound, and J. P. Hirth, *Acta Met.*, **18**, 225 (1970).
21. W. E. Langlois and F. F. Abraham, *Chem. Phys. Lett.*, **52**, 129 (1977).
22. M. Metiu, K. Kitahara, and J. Ross, *J. Chem. Phys.*, **65**, 393 (1976).
23. I. M. Lifshitz and V. V. Slyozov, *J. Phys. Chem. Solids*, **19**, 35 (1961).
24. D. M. Koenhen, C. A. Smolders, and M. J. Gordon, *J. Polymer Sci.*, **C61**, 93 (1977).
25. D. Turnbull in *Solid State Physics*, Heach Press, New York, 1956.
26. J. W. Cahn, *Acta Met.*, **14**, 1685 (1966).
27. K. V. Chuistov in *Modulated Structures in Aging Systems (in Russ.)*, Naukova Dumka, Kiev, 1975.
28. N. S. Andreev, O. V. Lazurin, and E. A. Poray-Koshitz in *Liquation Phenomena in Glasses (in Russ.)*, Nauka, Leningrad, 1974.
29. J. J. van Aartsen, *Eur. Polym. J.*, **6**, 919 (1970).
30. P. Debye, *J. Chem. Phys.*, **31**, 680 (1959).
31. J. J. van Aartsen and C. A. Smolders, *Eur. Polym. J.*, **6**, 1105 (1970).
32. P. G. de Gennes, *J. Chem. Phys.*, **55**, 572 (1971); **72**, 4756 (1980).
33. D. Pincus, *J. Chem. Phys.*, **75**, 1996 (1981).
34. P. G. de Gennes, *J. Phys. Lett (France)*, **38**, 441 (1977).
35. M. Daoud and P. G. de Gennes, *J. Polymer Sci. Phys.*, **17**, 1971 (1979).
36. J. Klein, *Macromolecules*, **11**, 852 (1978).
37. P. G. de Gennes in *Scaling Concept in Polymer Physics*, Cornell Univ. Press, New York, 1979.
38. E. Helfand and Y. Tagami, *J. Polymer Sci. Lett.*, **9**, 761 (1971).
39. E. Helfand and A. M. Sapse, *J. Polymer Sci.*, **C54**, 289 (1976).
40. H. E. Cook, *Acta Met.*, **18**, 297 (1970).
41. P. G. de Gennes, *J. Phys. (France)*, **42**, 735 (1981).
42. G. R. Strobl, *Macromolecules*, **18**, 555 (1985).
43. M. Doi and S. F. Edwards, *J. Chem. Soc. Faraday Trans.*, **2**, 74, 1789, 1802, 1818 (1978).
44. T. Hashimoto, M. Itakura, and N. Shimidzu, *J. Chem. Phys.*, **85**, 6773 (1986).
45. T. Hashimoto, J. Kumaki, and H. Kawai, *Macromolecules*, **16**, 641 (1983).
46. R. G. Hill, P. E. Tomlins, and J. S. Higgins, *Macromolecules*, **18**, 2555 (1985).
47. T. Hashimoto, M. Takenaka, and H. Jinnai, *J. Appl. Cryst.*, **24**, 457 (1991).

48. T. Kyu and M. Saldanha, *Amer. Chem. Soc. Polym. Prepr.*, **28**, 124 (1987).
49. T. Kyu and M. Saldanha, *Macromolecules*, **21**, 1021 (1988).
50. T. Kyu and M. Saldanha, *J. Polymer Sci. Phys.*, **28**, 97 (1990).
51. M. Takahashi, S. Kinoshita, and T. Nose, *J. Polymer Sci. Phys.*, **27**, 2159 (1989).
52. J. S. Langer, *Acta Met.*, **21**, 1649 (1973).
53. J. S. Langer, M. Bar-on, and H. D. Muller, *Phys. Rev.*, **A11**, 1417 (1975).
54. K. Kawasaki and T. Ohta, Progr. *Theor. Phys.*, **59**, 362 (1978).
55. J. Marro, A. B. Bortz, J. L. Lobowitz, and M. A. Zendejas, *Phys. Rev.*, **B10**, 535 (1974).
56. K. Binder and D. Stauffer, *Adv. Phys.*, **25**, 343 (1976).
57. E. D. Siggia, *Phys. Rev.*, **A20**, 595 (1979).
58. F. S. Bates and P. Wiltzius, *J. Chem. Phys.*, **91**, 3258 (1989).
59. P. Wiltzius, F. S. Bates, and W. R. Heffur, *Phys. Rev. Lett.*, **60**, 1538 (1988).
60. T. Hashimoto, M. Itakura, and M. Hasegawa, *J. Chem. Phys.*, **85**, 6118 (1986).
61. H. Furukawa, *Adv. Phys.*, **34**, 703 (1985).
62. G. Porod, *Kolloid-Z.*, **124**, 83 (1951); **125**, 51, 108 (1952).
63. W. Ruland, *J. Appl. Cryst.*, **4**, 70 (1971).
64. T. Hashimoto, M. Shibayama, and H. Kawai, *Macromolecules*, **13**, 123 7 (1980).
65. Y. C. Chou and W. Goldburg, *Phys. Rev.*, **A20**, 2105 (1979).
66. H. Fujita, T. Hashimoto, and M. Takenaka, *Macromolecules*, **22**, 4663 (1989).
67. I .G. Voight-Martin, K. H. Leister, R. Rosenau, and R. Koningsveld, *J. Polymer. Sci. Phys.*, **24**, 723 (1986).
68. P. T.. van Emmerik and C. A. Smolders, *J. Polymer Sci.*, **C38**, 73 (1972).
69. C. A. Smolders, J. J. van Aartsen, and A. Steenbergen, *Koll. Z. Z. Polymere*, **243**, 14 (1971).
70. P. T. van Emmerik, C. A. Smolders, and W. Geymeyer, *Eur. Polymer J.*, **9**, 309 (1973).
71. V. M. Andreeva, A. A. Tager, I. S. Tyukova, and L. F. Golenkova, *Vysokomolek. soed.*, **A19**, 2604 (1977).
72. J. Lal and R. Bansil, *Macromolecules*, **24**, 290 (1991).
73. H. Furukawa, *Physica*, **A123**, 497 (1984).
74. A. Chakrabarti, R. Torel, J. Gunton, and M. Muthukumar, *J. Chem. Phys.*, **92**, 6899 (1990).
75. E. Pines and W. Prins, *Macromolecules*, **6**, 888 (1973).
76. G. T. Feke and W. Prins, *Macromolecules*, **7**, 527 (1974).
77. S. Wellinghoff, J. Shaw, and E. Beer, *Macromolecules*, **12**, 932 (1979).
78. T. Tanaka, *Phys. Rev. Lett.*, **40**, 820 (1978).
79. I. Nishio, G. Switlow, S. T. Sun, and T. Tanaka, *Nature*, **300**, 243 (1982).
80. T. Tanaka, I. Nishio, S. T. Sun, and S. Ueno-Nishio, *Science*, **218**, 467 (1982)
81. J. P. McMaster, *Amer. Chem. Soc., Polym. Prep.*, **15**, 254 (1974).
82. T. Nishi, T. T. Wang, and T. K. Kwei, *Macromolecules*, **8**, 227 (1975).
83 R. Gelles and C. W. Frank, *Macromolecules*, **16**, 1448 (1983).
84. R. Gelles and C. W. Frank, *Macromolecules*, **15**, 1486 (1982).
85. H. L. Snyder and P. Meakin, *J. Chem. Phys.*, **79**, 5588 (1983).
86. J. P. McMaster in *Copolymers, Polyblends and Composites*, Washington, D.C., 1975, p. 43.
87. J. Gilmer, N. Coldstein, and R. S. Stein, *J. Polymer. Sci. Phys.*, **20**, 2219 (1982).
88. K. Hamano, N. Kuwahara, and M. Kaneko, *Phys. Rev.*, **A21**, 1322 (1980).
89. Sh. Noima, K. Tsutsumi, and T. Nose, *Polymer J.*, **14**, 907 (1982).
90. Sh. Noima, Y. Ohyama, and T. Yamaguchi, *Polymer J.*, **14**, 225 (1982).
91. Sh. Noima, K. Shizoshita, and T. Nose, *Polymer J.*, **14**, 289 (1982).
92. N. G. Wong and C. M. Knobler, *J. Chem. Phys.*, **66**, 4707 (1977).
93. N. G. Wong and C. M. Knobler, *J. Chem. Phys.*, **69**, 725 (1978).

94. K. M. Wong, A. J. Schwartz, and W. I. Goldburg, *Phys. Rev. Lett.*, **41**, 657 (1978).
95. J. S. Huang, W. I. Goldburg, and A. W. Bjerkaas, *Phys. Rev. Lett.*, **32**, 921 (1974).
96. W. I. Goldburg, C. H. Shaw, and J. S. Huang, *J. Chem. Phys.*, **68**, 484 (1978).
97. W. I. Goldburg, A. J. Schwartz, and K. M. Won, *Progr. Theor. Phys. Suppl.*, **64**, 477 (1978).
98. Y. Oono and S. Puri, *Phys. Rev. Lett.*, **58**, 836 (1987).
99. K. Binder, H. L. Frisch, and J. J. Jackle, *J. Chem. Phys.*, **85**, 1505 (1986).
100. H. Furukawa, *J. Appl. Cryst.*, **21**, 805 (1988).
101. J. Koberstein, T. P. Russell, and R. S. Stein, *J. Polymer Sci. Phys.*, **17**, 1719 (1979).
102. T. Kyu and D. S. Lim, *J. Chem. Phys.*, **92**, 3981 (1990).
103. M. Takenaka and T. Hashimoto, *J. Chem. Phys.*, **96**, 6177 (1992).
104. A. Shinozaki and Y. Oono, *Phys. Rev. Lett.*, **66**, 173 (1991).
105. T. Koga and K. Kawasaki, *Phys. Rev.*, **A44**, 817 (1991).
106. T. Ohta and N. Nazaki in *Space-Time Organization on Molecular Fluids*, Eds. F. Tanaka, M. Doi, and T. Ohta, Springer, Berlin, 1989.
107. T. Hashimoto, M. Takenaka, and T. Izumitani, *Polymer Commun.*, **30**, 45 (1959).
108. R. Toral, A. Chakrabarti, and J. D. Gunton, *Phys. Rev. Lett.*, **60**, 311 (1988).
109. S. Puri and Y. Oono, *Phys. Rev.*, **A38**, 1542 (1988).
110. M. Takenaka, T. Izumitani, and T. Hashimoto, *J. Chem. Phys.*, **92**, 4566 (1990).
111. D. Schwahn, K. Hahn, and T. Springer, *J. Chem. Phys.*, **93**, 8383 (1990).
112. K. Binder and D. Stauffer, *Phys. Rev. Lett.*, **33**, 1006 (1974).
113. T. Hashimoto, M. Shibayama, and H. Kawai, *Macromolecules*, **13**, 1237 (1980).
114. T. Izumitani and T. Hashimoto, *J. Chem. Phys.*, **83**, 3694 (1985).
115. M. Takenaka, T. Izumitani, and T. Hashimoto, *Macromolecules*, **22**, 2257 (1987).
116. T. Izumitani, M. Takenaka, and T. Hashimoto, *J. Chem. Phys.*, **92**, 3213 (1990).
117. T. Hashimoto, T. Izumitani, and M. Takenaka, *Macromolecules*, **22**, 2293 (1989).
118. S. Yang, M. Shibayama, R. S. Stein, N. Shimidzu, and T. Hashimoto, *Macromolecules*, **19**, 1667 (1986).
119. N. Inaba, K. Sato, S. Suzuki, and T. Hashimoto, *Macromolecules*, **19**, 1690 (1986).
120. H. Furukawa, *Progr. Theor. Phys.*, **59**, 1072 (1978).
121. H. Furukawa, *Phys. Rev. Lett.*, **43**, 136 (1979).
122. H. Furukawa, *Phys. Rev. Rapid Commun.*, **33**, 638 (1986).
123. T. Ohnaga, J. Maruta, and T. Inoue, *Polymer*, **30**, 1845 (1989).
124. H. Nakamura, J. Maruta, T. Ohnaga, and T. Inoue, *Polymer*, **31**, 303 (1990).
125. M. Song, Y. Huang, G. Cong, H. Liang, and B. Jiang, *Polymer*, **33**, 293 (1992).
126. H. Tomita, *Progr. Theor. Phys.*, **75**, 482 (1986).
127. H. L. Snyder and P. J. Meakin, *J. Polymer Sci. Symp.*, **73**, 217 (1985).
128. H.-W. Kammer and J. Piglowski, *Z. Phys. Chem. (DDR)*, **269**, 721 (1988).
129. F.-J. Tsai and J. M. Torkelson, *Macromolecules*, **21**, 1026 (1988).
130. I. Kiyoshi, U. Akio, H. Satoru, K. Toshiko, H. Hideo, C. Kuzuchiko, and A. Yoshiyuki, *Polym. J.*, **23**,1291 (1991).
131. H. Tanaka and T. Nishi, *Jap. J. Appl. Phys.*, **(2)27**, L 1753 (1988).
132. A. Sariban and K. Binder, *Polym. Commun.*, **30**, 205 (1989).
133. M. He,Y. Liu, Y. Feng, M. Jiang, and C. Han, *Macromolecules*, **24**, 7464 (1991).
134. H. H. Chuanh, T. Kyu, and T. E. Helminiak, *Polymer*, **30**, 1591 (1989).

135. J. B. Gardner, *Rubber Chem. Technol.*, **43**, 370 (1970).
136. T. Inoue, T. Ougizawa, O. Yasuda, and K. Miyasaka, *Macromolecules*, **18**, 57 (1985).
137. H.-W. Kammer, J. Kressler, C. Kummerloewe, and B. Morgenstern, *Polimery (Poland)*, **35**, 199 (1990).
138. H.-W. Kammer, *J. Macromol. Sci. Chem.*, **A27**, 1713 (1990).
139. J. Piglowski, H.-W. Kammer, and J. Kressler, *Polymer*, **30**, 1705 (1989). J. Piglowski, J. Kresser, H.-W. Kammer, *Polym. Bull.*, **16**, 493 (1986).
140. T. Hashimoto, T. Tsukahara, and H. Kawai, *Macromolecules*, **14**, 708 (1981).
141. G. Beaucage and R. S. Stein, *Macromolecules*, **26**, 1609 (1993).
142. T. Kyu and D. S. Lim, *Macromolecules*, **24**, 7345 (1991).
143. R.-J. Roe and C. M. Kuo, *Macromolecules*, **23**, 4635 (1990).
144. R. Koningsveld and A. Staverman, *J. Polymer Sci. Phys.*, **6**, 305, 325, 349 (1968).
145. R. Koningsveld and A. Staverman, *Kolloid Z. Z. Polymere*, **218**, 114 (1968).
146. R. Koningsveld, *Adv. Colloid Interface Sci.*, **2**, 151 (1968).
147. A. J. Staverman in *Integration of Fundamental Polymer Science and Technology*, Eds. L. A. Kleintjens and P. J. Lemstra, Elsevier Publ., London, 1986, p. 19.
148. R. Koningsveld, *Disc. Faraday Soc.*, **49**, 144 (1970).
149. T. Hashimoto and T. Izumitani, *Macromolecules*, **26**, 3631 (1993).
150. T. Kawakatzu and K. Kawasaki, *Physica*, **A167**, 690 (1990).
151. K. Kawasaki, *Ann. Phys. (N.Y.)*, **154**, 319 (1984).
152. T. Kawakatzu, K. Kawasaki, M. Furukawa, H. Okabayashi, and T. Kaneya, *Report*, 1993.
153. R. C. Ball and R. L. H. Essery, *J. Phys. Condens. Matter.*, **2**, (1990).
154. G. Brown and A. Chakrabarti, *Phys. Rev.*, **A46**, 4829 (1992).
155. S. Puri and K. Binder, *Phys. Rev.*, **A46**, 4487 (1992).
156. R. A. L. Jones, L. J. Norton, E. J. Kramer, F. S. Bates, and P. Wiltzius, *Phys. Rev. Lett.*, **66**, 1326 (1991).
157. G. Krausch, C. Dai, E. J. Kramer, J. E. Marko, and F. S. Bates, *Macromolecules*, **25**, 5566 (1993).
158. F. Bruder and R. Brenn, *Phys. Rev. Lett.*, **69**, 624 (1992).
159. A. E. Nesterov, V. V. Horichko, and Y. S. Lipatov, *Makromol. Chem. Rapid Commun.*, **12**, 571 (1991).
160. S. Reich and Y. Cohen, *J. Polymer Sci. Phys.*, **19**, 1255 (1981).
161. J. Eliassaf, A. Silberberg, and A. Katchalsky, *Nature*, **176**, 1119 (1958).
162. J. Eliassaf and A. Silberberg, *J. Polymer Sci.*, **41**, 33 (1959).
163. Y. S. Lipatov and P. I.. Zubov, *Vysokomolek. soed.*, **1**, 425 (1959).
164. Y. S. Lipatov, P. I. Zubov, and E. A. Andruschenko, *Koll. Zh.*, **21**,598 (1959).
165. Y. S. Lipatov, P. I. Zubov, and E. A. Kanevskaya, *Reports of Acad. Sci. USSR*, **141**, 387 (1961).
166. Y. S. Lipatov, *Thesis*, Moscow, 1963.
167. A. Silberberg and W. Kuhn, *Nature*, **170**, 450 (1952).
168. A. Silberberg and W. Kuhn, *J. Polymer Sci.*, **13**, 21 (1954).
169. I. Hindawi, J. S. Higgins, A. F. Galambos, and R. A. Weiss, *Macromolecules*, **23**, 670 (1990).
170. P. J. Barham and A. Keller, *Macromolecules*, **23**, 30 (1990).
171. C. M. Roland and G. A. Bohm, *J. Polym. Sci. Phys.*, **22**, 79 (1984).
172. S. J. Organ, *Polymer*, **35**, 86 (1994).
173. A. I. Nakatani, H. Kim, Y. Takahashi, and C. C. Han, *Polym. Commun.*, **30**, 143 (1989).
174. E. Helfand and G. H. Fredrickson, *Phys. Rev. Lett.*, **62**, 2468 (1989).
175. A. Onuki, *J. Phys. Soc. Japan*, **59**, 3423 (1990).
176. S. T. Milner, *Phys. Rev. Lett.*, **66**, 1477 (1991).

177. V. G. Mavrantzas and A. N. Beris, *Phys. Rev. Lett.*, **69**, 273 (1992).
178. B. A. Wolf, *Macromolecules,* **17**, 615 (1984).
179. E. Vrahopoulou-Gilbert and A. J. McHugh, *Macromolecules,* **17**, 2657 (1984).
180. S. Y. Frenkel, *Pure & Appl. Chem.*, **38**, 117 (1974).
181. G. Elyashevich, *Adv. Polymer Sci.,* **43**, 205 (1982).
182. A. Onuki, *J. Phys. Soc. Japan,* **59**, 3427 (1990).
183. H.-W. Kammer, *33rd IUPAC Int. Symp. Macromol.*, Montreal, Book Abstr., 1990, p. 14.
184. Y. S. Lipatov and L. M. Sergeeva in *Interpenetrating Polymer Networks,* Naukova Dumka, Kiev, 1979.
185. L. H. Sperling in *Interpenetrating Polymer Networks and Related Materials,* Plenum, New York, 1981.
186. Y. Lipatov, *J. Macromol. Sci. Rev. Chem. Phys.*, **C30**, 209 (1990).
187. Y. S. Lipatov in *Interpenetrating Polymer Networks,* Eds. D. Klempner, L. Sperling, and L. Utracki, ACS, Washington DC, 1994, p. 125.
188. Y. S. Lipatov, O. P. Grigor'eva, V. V. Shilov, and L. M. Sergeeva, *Reports of Acad. Sci. USSR,* **278**, 409 (1984).
189. Y. S. Lipatov, V. V. Shilov, Y. P. Gomza, G. P. Kovernik, O. P. Grigor'eva, and L. M. Sergeeva, *Makromol.. Chem.*, **185**, 347 (1984).
190. Y. S. Lipatov, O. P. Grigor'eva, G. P. Kovernik, and V. V. Shilov, *Makromol.. Chem.*, **186**, 1401 (1985).
191. Y. S. Lipatov and O. P. Grigor'eva, *Reports Acad. Sci. Ukraine,* **B11**, 43 (1983).
192. Y. S. Lipatov, V. V. Shilov, and Y. P. Gomza, *Polymer,* **25**, 459 (1984).
193. Y. S. Lipatov and V. V. Shilov, *Adv. Chem. (USSR),* **53**, 1197 (1984).
194. Y. S. Lipatov, O. P. Grigor'eva, L. M. Sergeeva, and V. V. Shilov, *Vysokomolek. soed.*, **A28**, 335 (1986).
195. Y. S. Lipatov in *Advances in Interpenetrating Polymer Networks,* Eds. D. Klempner and K. Frisch, Technomic Publ., Lancaster-Basel, 1989, v. 1, p. 261.
196. Y. S. Lipatov, T. T. Alekseeva, and V. V. Shilov, *Polymer Networks & Blends,* **1**, 129 (1991).
197. Y. S. Lipatov and T. T. Alekseeva, *Adv. Chem. (Russia),* **61**, 2187 (1992).
198. Y. S. Lipatov and T. T. Alekseeva, *Polym. Commun.*, **32**, 254 (1991).
199. Y. S. Lipatov, T. T. Alekseeva,V. F. Rosovitsky, N. V. Babkina, *Vysokomolek. soed.*, **A35**, 652 (1993).
200. S. Visconti and R. H. Marchessault, *Polym. Prepr. (ACS),* **15**, 66 (1974).
201. S. Visconti and R. Marchessault, *Macromolecules,* **7**, 913 (1974).
202. G. F. Roginskaya,V. P. Volkov, A. E. Chalykh,V. V. Matveev, B. A. Rozenberg, and N. S. Enikolopyan, *Reports of Acad. Sci. USSR,* **252**, 402 (1980).
203. G. F. Roginskaya, V. P. Volkov, A. E. Chalykh, and B. V. Rozenberg, *Vyssokomol. soed.*, **A25**, 1979 (1983).
204. G. F. Roginskaya,V. P. Volkov, A. I. Kuzaev, A. E. Chalykh, and B. A. Rozenberg, *ibid,* **A26**, 1020 (1984).
205. J. K. Yeo, L. H. Sperling, and D. A. Thomas, *Polymer,* **24**, 307 (1983).
206. J. Michel S. C. Hargest, and L. A. Sperling, *J. Appl. Polymer Sci.*, **26**, 743 (1981).
207. J. K. Yeo, L. H. Sperling, and D. A. Thomas in *Polymer Compatibility and Incompatibility,* Ed. K. Solc, Harwood, 1982.
208. R. J. Williams, J. Borrajo, H. E. Adabbo, and A. J. Rojas, *Adv. Chem. Ser.*, **128**, 195 (1984).
209. A. Vazquez, A. Rojas, H. E. Adabbo, J. Borrajo, and P. J. Williams, *Polymer,* **28**, 1156 (1987).
210. K. Yamakawa, Y. Takagi, and T. Inoue, *Polymer,* **30**, 1839 (1989).
211. K. Yamakawa and T. Inoue, *Polymer,* **30**, 662 (1989).
212. H.-S.-Y. Hsich, *Adv. Polymer Technol.*, **10**, 8 (1990).

213. Q. Tran-Cong, T. Nagaki, T. Nakagawa, O. Yano, and T. Soen, *Macromolecules*, **22**, 2720 (1989).
214. Q. Tran-Cong, T. Nagaki, O. Yano, and T. Soen, *Macromolecules*, **24**, 1505 (1991).
215. T. Hashimoto, M. Takenaka, and H. Jinnai, *Polym. Commun.*, **30**, 177 (1989).
216. L. M. Sergeeva, G. M. Semenovich, L. V. Karabanova, T. E. Lipatova, and Y. S. Lipatov, *Reports of Acad. Sci. USSR*, **288**, 158 (1986).
217. Y. S. Lipatov, V. V. Shilov, O. P. Grigor'eva,V. A. Bogdanovich, and V. N. Bliznyuk, *Reports Acad. Sci. USSR*, **285**, 394 (1985).
218. Y. S. Lipatov in *Mechanics of Composite Materials* (Riga), **5**, 771 (1983).
219. V. V. Shilov, Y. S. Lipatov, and V. K. Grischenko, *Bulletin of Acad. Sci. Ukraine*, No 4, 18 (1985)
220. Y. S. Lipatov, V. V. Shilov,V. A. Bogdanovich, D. P. Kochetov, *Reports of Acad. Sci. Ukraine*, **B7**, 50 (1981).
221. V. V. Shilov, Y. S. Lipatov, L. V. Karabanova, and L. M. Sergeeva, *J. Polym. Sci. Chem.*, **17**, 3083 (1979).
222. V. V. Shilov, L. V. Karabanova, Y. S. Lipatov, V. A. Bogdanovich, and L. M. Sergeeva, *Vysokomolek. soed.*, **A22**, 1354 (1980).
223. V. V. Shilov and Y. S. Lipatov in *Physical Chemistry of Multicomponent Polymer Systems*, Naukova Dumka, Kiev, 1986, v. 2, p. 25.
224. V. V. Shilov, *XII Mendele'ev Congress, Baku, Nauka*, Moscow, 1981, p.159.
225. Y. S. Lipatov, V. V. Shilov, B. A. Bogdanovich, and V. N. Bliznyuk, *Reports Acad. Sci. Ukraine*, **B2**, 34 (1983).
226. V. V. Shilov, Y. S. Lipatov, and V. A. Bogdanovich, *Vysokomolek. soed.*, **A25**, 921 (1983).
227. I. Sakurada, A. Nakajima, and H. Aoki, *J. Polymer Sci.*, **35**, 507 (1959).
228. Y. S. Lipatov, L. M. Sergeeva, A. E. Nesterov, and L. V. Karabanova, *Reports Acad. Sci. USSR*, **220**, 637 (1975).
229. K. Binder and H. L. Frisch, *J. Chem. Phys.*, **81**, 2126 (1984).
230. Y. S. Lipatov, L. M. Sergeeva, L. V. Mozzhukhina, and N. P. Apukhtina, *Vysokomolek. soed.*, **A16**, 2290 (1974).
231. Y. S. Lipatov, L. M. Sergeeva, and L. V.. Karabanova, *Vysokomolek. soed.*, **A18**, 1025 (1976).
232. V. V. Shilov, G. P. Kovernik, and Y. S. Lipatov, *Reports Acad. Sci. Ukraine*, **B7**, 52 (1986).
233. Y. S. Lipatov, L. V. Karabanova, and L. M. Sergeeva, *Composite Polymer Materials (Kiev)*, **18**, 26 (1983).
234. Y. S. Lipatov, L. M. Sergeeva, and L. V. Karabanova, *Reports Acad. Sci.*, **289**, 153 (1986).
235. Y. S. Lipatov, L. V. Karabanova, and L. M. Sergeeva, *Polymer Int.*, **34**, 7 (1994).
236. Y. S. Lipatov, T. T. Alekseeva, V. F. Rosovitsky, and N. V. Babkina, *Polymer Networks & Blends*, **4**, 9 (1994).
237. Y. S. Lipatov, T. T. Alekseeva, V. F. Rosovitsky, and N. V. Babkina, *Polymer International*, **35**, (1995).

Chapter 6

INTERFACE IN DEMIXING SOLUTIONS AND POLYMER MIXTURES

The formation of any heterogeneous system is always accompanied by the appearance of an interface or boundary layer. The latter determines many important properties of polymeric systems. There are many studies dedicated to the mechanism of formation and properties of the interface layers (the thickness, concentration profile, etc.) which have been reviewed in detail and generalized.[1-3] In this chapter we analyze only the problems connected with the interfacial layer in demixing solutions and immiscible blends.

The boundary or interfacial layers possess an effective thickness, beyond which the deviations of local properties from their bulk values become negligible.[1-3] The introduction of such a concept is possible because of the small radius of action of intermolecular forces, which causes a rapid decrease in the influence of one of the phases on any property of the neighboring phase. At the same time, in polymeric systems the experimentally found thickness of the interfacial layer (or region) may be rather high due to the chain structure of polymer molecules and consequently it is determined by the specific behavior of polymers as compared with low molecular mass substances.

It is worth noting that the interfacial layer has no definite border with bulk polymer and its thickness has no constant value. Recently, Sharpe[4,5] noted a fundamental difference between two definitions: interface and interphase. He puts the question: what plays the major role in determining the properties of binary system? Is it interface? Evidently, it is not so, as two-dimensional array of atoms or molecules, which is the interface, is impossible to measure, whereas interfacial layer has a large enough assembly of atoms or molecules to have a modulus, strength, etc. According to Sharpe[4] interphase is “a region intermediate for two (usually solid) phases in contact, the composition and/or structure and/or properties of which may be variable across the region and which also may differ from the composition and/or structure and/or properties of either of the two contacting phases”. In this chapter, we use both definitions: interfacial layer and interphase bearing in mind the equivalence of concepts.

The theory of interfacial tension and structure of an interfacial layer has been developed by many authors.[6-16] Vrij and Roebersen[6,7] derived equations for the interfacial tension between two demixing solutions based on the Cahn-Hillard theory[8] of interfacial tension in mixtures of low molecular weight substances. Helfand and Tagami[9,10] developed the theory of the interface boundary (interfacial region, interphase) for immiscible symmetric mixtures using the approximation of the mean field theory. It was predicted[10] that the thickness of the interfacial layer in the polymer-polymer system is about 1 nm and depends on the Flory-Huggins interaction parameter. The theory of the interfacial boundary between solutions and mixtures of polymers was developed by Roe.[11] It was based on the theory of non-uniform systems, developed on the basis of modified Flory-Huggins theory. According to Roe theory, the interfacial tension and the thickness of the interfacial layer (interphase) depends on the chain length and value of $(\chi - \chi_{crit},)$ where χ_{crit} is the interaction parameter at the critical solution temperature. Nose[12] has developed the theory of the interfacial tension and the thickness of the interfacial layer in demixing polymer systems, accounting for the change in the polymer coil dimensions at the interface. Sanchez[13,14] has used the gradient theory[15] and the model of compressible liquid lattice.[13,16,17] The scaling approach to describe the properties of an interface (or interphase) was also applied in some works.[18-20] The main problems solved were related to the establishment of the dependence of an interfacial tension on temperature and molecular mass of polymer. Some of the above mentioned theories are discussed below.

6.1 VRIJ-ROEBERSEN THEORY

Vrij and Roebersen[7] analyzed the concentration profile of an interphase and interfacial free energy (interfacial tension) for solutions of demixing polymers near the critical temperature of mixing, where the coil deformation is small. To characterize the homogeneous phases, they used Flory-Huggins equation for combinatorial entropy of mixing, S_k, per volume unit as:

$$-S_k/k = n_2 \ln (v_2 n_2) + n_1 \ln (v_1 n_1) \qquad [6.1]$$

where n_1 and n_2 are the numbers of molecules per unit volume of solvent and polymer, correspondingly, v_1 and v_2 are the volumes occupied by solvent and polymer molecules. It was assumed that the free energy of mixing depends also on the local structure of solution and is a function of the local density of segments:

$$g/kT = n_2 \ln (v_2 n_2) + n_1 \ln (v_1 n_1) + h(n_s, T) \qquad [6.2]$$

where n_S is the local number of segments. The densities of segments and solvent are connected by the following relationship:

$$v_2n_2 + v_1n_1 = mv_sn_2 + v_1n_1 = 1 \qquad [6.3]$$

where $m = v_2/v_S$. For non-homogeneous system $n_S \neq mn_2$. The mean spatial distribution of segments around the mass center of a coil is spherically symmetric and is determined by the dependence of $\xi(r)$ (r is the radial distance from the coil center). The local density of segments in volume element $d\tau_2$ at x_2 follows from the linear superposition of contributions of polymeric molecules with centers at $d\tau_1$ near $d\tau_2$:

$$n_s(x_2) = \int \xi \, | \vec{r_1} - \vec{r_2} | \, n_2(x_1) d\tau_1 \qquad [6.4]$$

where the integral is taken by the whole system volume. When the variation of $n_2(x)$ at the distance x_1 comparable with the coil dimension is small, one can write:

$$n_2(x_1) = n_2(x_2) + \left(\frac{\delta n_2}{\delta x}\right)(x_1 - x_2) + \frac{1}{2}\left(\frac{\delta^2 n_2}{\delta x^2}\right)^2 (x_1 - x_2)^2 + \ldots \qquad [6.5]$$

Integration of Eq 6.4 over angle coordinates leads to disappearance of a linear term of Eq 6.5 and the remaining terms are expressed as

$$n_s(x) = m\,n_2(x) + 1/6m\left(\frac{\delta^2 n_2}{\delta x^2}\right)R^2 + \ldots \qquad [6.6]$$

where the gyration radius is determined by the expressions:

$$R^2 = m^{-1}\int_0^\infty r^2\xi(r)4\pi r^2 dr; \quad m = \int_0^\infty \xi(r)4\pi r^2 dr \qquad [6.7]$$

In the interfacial region, $n_2(x)$, has S-shape character (Fig. 6.1). It implies that at some position of x_A, $d^2n_2/dx^2 = 0$ and $n_S(x) = mn_2(x)$. In more dilute region $n_C > mn_2$ and in more concentrated $n_s < m_2n_2$. The total number of segments in the interphase is m times larger than the total number of centers of macromolecules:

$$\int_{I}^{II} n_s dx = m \int_{I}^{III} n_2 dx \qquad [6.8]$$

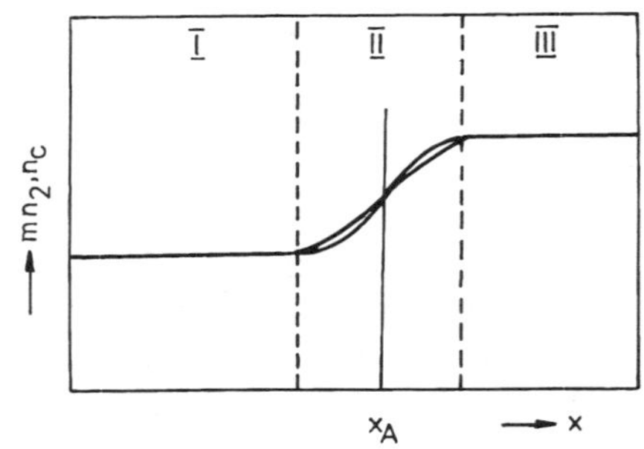

Figure 6.1. Schematic representation of the density change of polymer segments in demixing polymer solution: I - dilute solution, II - an interphase, III - concentrated solution. Adapted, by permission, from A. Vrij and G. J. Roebersen, *J. Polym. Sci., Phys.*, **15**, 109 (1977).

This result follows from the partial integration of d^2n_2/dx^2 and because in homogeneous phases $dn_2/dx = 0$. To calculate the free surface energy and concentration profile, the non-homogeneous interfacial layer is divided into layers of a thickness, dx, the layers are perpendicular to the coordinate x with $(v_1)^{1/3} << dx < R$. Gibbs free energy of mixing is found by integration of g (Eq 6.2) over entire interface:

$$\frac{G}{AkT} = \int_{I}^{III} \frac{g}{kT} dx \qquad [6.9]$$

where A is the surface of an interface. Then, it is supposed that d^2n_2/dx^2 is sufficiently small and therefore the second term in Eq 6.6 is small as well. Inserting Eq 6.6 into 6.2, we obtain:

$$g/kT = (g^0 + g')/kT \qquad [6.10]$$

where

$$g^0/kT = n_2 \ln v_2n_2 + v_1^{-1}(1 - v_2n_2) \ln (1 - v_2n_2) +$$

$$+ h[(1 - v_2n_2),T] \qquad [6.10a]$$

$$g'/kT = -\frac{R^2}{6} m \left(\frac{dn_2}{dx}\right)^2 \left[\ln (1 - v_2n_2) - v_1^2 \frac{dh}{d_cn_s}\right]_{n_s = mn_2} \qquad [6.10b]$$

Insertion of g′/kT into Eq 6.9 shows that this term, after partial integration, can be replaced (by integration) by expression:

$$g*/kT = -\frac{R^2}{6}v_2^2\left(\frac{d_2n_2}{dx}\right)\left[\frac{1}{v_1(1-v_2n_2)} + \frac{d^2h}{v_1^2d_sn^2}\right]_{n_s=mn_2} \quad [6.11]$$

Eq 6.9 can be transformed to expression:

$$G/AkT = \int_{I}^{III}(g^0/kT + g*/kT)dx \quad [6.12]$$

The term g*/kT can be considered as “gradient” contribution to the Gibbs free energy of mixing in the framework of van der Waals or Cahn concepts. Now, we have to subtract from G the Gibbs free energy of mixing for homogeneous system, G_0:

$$G_0 = N_1\mu_1(e) + N_2\mu_2(e) \quad [6.13]$$

where N_1 and N_2 are the number of molecules of components 1 and 2 in the system, $\mu_1(e) = \mu_1(I) = \mu_1(III)$ and $\mu_2(e) = \mu_2(I) = \mu_2(III)$ are the chemical potentials of mixing in two equilibrium phases. Eq 6.13 can be written in the following form:

$$G_0/A = \frac{\mu_1(e)}{v_1}\int_{I}^{III}(1-v_2n_2)dx + \mu_2(e)\int_{I}^{III}n_2dx \quad [6.14]$$

Values of $\mu_1(e)$ and $\mu_2(e)$ can be obtained by differentiation of Eq 6.2. From Eqs 6.12 and 6.14, it follows that

$$\frac{G-G_0}{AkT} = \int_{-\infty}^{+\infty}\left(\frac{\Delta g}{kT} + \frac{g*}{kT}\right)dx \quad [6.15]$$

where

$$\frac{\Delta g}{kT} = n_2\ln\frac{v_2n_2}{v_2n_2(e)} + \frac{1-v_2n_2}{v_1}\ln\frac{1-v_2n_2}{1-v_2n_2(e)} + h - h(e) -$$

$$-\left(\frac{\delta h}{v_2\delta n_2}\right)_{(e)} v_2[n_2 - n_2(e)] + (m-1)[n_2 - n_2(e)] \quad [6.16]$$

The limits of integration may be chosen between -∞ and +∞ because the expression under integral does not disappear, except for the

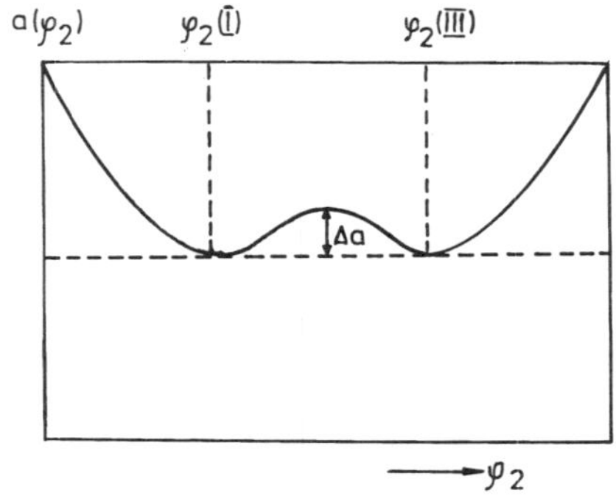

Figure 6.2. Change of excess free energy of mixing by transition from dilute (region I) to concentrated (region III) solution by phase separation (Δg is the change of free energy in the interphase). Adapted, by permission, from A. Vrij and G. J. Roebersen, *J. Polym. Sci.*, **15**, 109 (1977).

interfacial region. n_2(I) or n_2(III) can be used to determine the n_2(e) (Fig. 6.2). However, the value $(G - G_0)$ does depend on $n_2(x)$, i.e., on the choice of a profile. $(G - G_0)$ takes minimal value and it is equal to γA (γ is the interfacial tension or free interfacial energy). The variation calculations give Euler equation as a necessary condition for extremum:

$$\frac{\Delta g(n_2)}{kT} = -\frac{R^2}{6} v_2^2 \left(\frac{dn_2}{dx}\right)^2 \left[\frac{1}{v_1(1 - v_2 n_2)} + \frac{\delta^2 h}{v_2^2 \delta n_2^2}\right] = \frac{g*}{kT} \qquad [6.17]$$

Legendre criterion is an additional condition of true minimum; the second derivative of the under integral expression – dn_2/dx should not be negative between the integration limits. In a given case this conditions is expressed as

$$\frac{\delta^2}{\delta(\delta n_2/\delta x)^2}\left[\frac{\Delta g(n_2)}{kT} + \frac{g*(n_2, dn_2/dx)}{kT}\right] =$$

$$= -\frac{R^3}{3}\left[\frac{1}{v_1(1 - v_2 n_2)} + \frac{\delta^2 h}{v_1^2 \delta n_s^2}\right] \qquad [6.18]$$

This condition is valid only if

$$\frac{1}{1 - n_2 v_2} + \frac{\delta^2 h}{v_1 \delta n_s^2} << 0 \qquad [6.19]$$

for all $n_2(I) < n_2 < n_2(III)$. This condition is satisfied in the region of small deviations (1% and less) from the critical temperature. The applicability of the theory under consideration is very limited. Inserting 6.17 into 6.15, we obtain:

$$\gamma = 2\int_{-\infty}^{+\infty} \Delta g(n_2)dx = 2\int_{-\infty}^{+\infty} g*dx \qquad [6.20]$$

Using Eq 6.17 for changing variable x to n_2, we obtain:

$$\gamma = \frac{3Rv_2kT}{6^{1/2}} = \int_{n_2(I)}^{n_2(III)} \left[-\frac{\Delta g(n_2)}{v_1kT}\left(\frac{1}{1-v_2n_2} + \frac{v_1\delta^2 h}{v_2^2\delta n_2^2}\right)\right]^{1/2} dn_2 \qquad [6.21]$$

Eqs 6.17, 6.20, and 6.21 were used to calculate γ and $n_2(x)$ from two expressions of function h:

$$v_1h(\varphi_2,T) = (\Omega v_1/2kT)\varphi_2(1-\varphi_2) \qquad [6.22]$$

where Ω is a constant. In homogeneous phase, $\varphi_S = v_Sn_S = v_2n_2 \equiv \varphi_2$. The second expression for h is

$$v_1h(\varphi_2,T) = [g_{00} + g_{01}/T + g_1\varphi_2 + g_2\varphi_2^2]\varphi_2(1-\varphi_2) \qquad [6.23]$$

where g does not depend on φ_2. The empirical expression 6.23 was used in some previous works.[21,22] Based on the definition of g given in Eq 6.2, which is valid near the critical point, one can expand it into Taylor series. This implies that we are in the framework of classical theories of critical state.[8] In this case

$$g = g_{crit} + \varepsilon_1(T - T_{crit}) + \varepsilon_2(T - T_{crit})(\varphi_2 - \varphi_{crit})^2 +$$

$$+ \varepsilon_3(\varphi_2 - \varphi_{crit})^4 + \ldots \qquad [6.24]$$

where $\varepsilon_1 = (\delta g/T)_{crit}$; $\varepsilon_2 = (1/2)(\delta^3 g/\delta T\delta\varphi_2^2)_{crit}$; $\varepsilon_3 = (1/4!)(\delta^4 g/\delta\varphi_2^4)_{crit}$ (index "crit" relates to critical point). Eq 6.24 is symmetric in relation to $(\varphi - \varphi_{2\,crit})$ and consequently the binodal and spinodal can be obtained from conditions $\delta g/\delta\varphi^2 = 0$ and $\delta^2 g/\delta\varphi_2^2 = 0$, respectively. Designating $\Psi = \varphi_2 - \varphi_{2,crit}$, we have for binodal:

$$\Psi_b^2 = \frac{1}{2}(\varepsilon_1/\varepsilon_2)(T_{crit} - T) \qquad [6.25]$$

and for spinodal

$$\Psi_{sp}^2 = \frac{1}{6}(\varepsilon_2/\varepsilon_3)(T_{crit} - T) \qquad [6.26]$$

Then expression for Δg (Eqs 6.16 and 6.2) can be written as

$$\Delta g = \varepsilon_3(\Psi_b^2 - \Psi^2)^2 \qquad [6.27]$$

$$g* \to g*_{crit} = \frac{1}{6}(kT_{crit}\, R^2 v_2/\varphi_{2crit})\left(\frac{dn_2}{dx}\right)^2 \qquad [6.28]$$

From Eq 6.11, 6.17, and 6.21, it follows that

$$|\varphi_{2b} - \varphi_{2crit}| = \frac{1}{2}(\varphi_{2,III} - \varphi_{2,I}) = \frac{1}{2}\Delta\varphi_2 \qquad [6.29]$$

$$\gamma = A_1(\Delta\varphi_2)^3 \qquad [6.30]$$

$$A_1 = \left[\frac{kT_{crit}R^2\varepsilon_3}{54 v_2 \varphi_{2crit}}\right]^{1/2} \qquad [6.31]$$

or $\gamma = A_2(T_{crit} - T)^{3/2}$; $A_2 = (2\varepsilon_2/\varepsilon_3)^{3/2}A_1$. For the profile of the interfacial surface, we have

$$n_2 - \frac{1}{2}(n_{2,I} - n_{2,III}) = \frac{1}{2}\Delta n_2 \tan h[2(x - x_0)/L_p] \qquad [6.32]$$

In Eq 6.32, the value L_P is the thickness of an interfacial region and it is expressed as

$$L_p^2 = \frac{8kT_{crit}R^2}{3\varepsilon_3^2\varphi_{2,crit}\Delta\varphi_2^2} = \frac{4kT_{crit}R^2}{3\varepsilon_3 v_2\varphi_{2,crit}(T_{crit} - T)} \qquad [6.33]$$

According to Debue[23]

$$\varphi_{2,crit}^{-1} = 1 + m^{1/2}; \quad (\Omega v_1/kT_{crit}) = (1 + m^{-1/2})^2 = \varphi_{1,crit}^{-2} \qquad [6.34]$$

Then, using Eq 6. 22, one can write

$$\gamma = \frac{kT_{crit}R}{18\sqrt{2}\, v_1\varphi_{1,crit}^{5/2}}(\Delta\varphi_2)^3 = \frac{4RkT_{crit}\varphi_{1,crit}}{\sqrt{6}\, v_2\varphi_{2,crit}^{1/2}}\left(1 - \frac{T}{T_{crit}}\right)^{3/2} \qquad [6.35]$$

$$L_p = \frac{4\sqrt{2}\, R\varphi_{2,crit}^{3/2}}{m^{1/2}\Delta\varphi_2} = \frac{(8/3)^{1/2}R\varphi_{2,crit}^{1/2}}{(1 - T/T_{crit})^{1/2}} \qquad [6.36]$$

These equations predict weak dependence of γ on m, i.e., on molecular mass. In particular, $\gamma \sim m^{1/2}(\Delta\varphi_2)^3 \sim m^{-1/4}(1 - T/T_{crit})^{3/2}$ and $L_P \sim$ const $(\Delta\varphi_2)^{-1} \sim m^{1/4}(1 - T/T_{crit})^{-1/2}$ for m>>1, taking $R \approx m^{1/2}$. The value of exponent 1/4 was found also by Nose.[12] He also established the correlation between the thickness of the interphase and persistent length of fluctuation correlations. The possibility of such correlation follows from the fact that the free energy of concentration fluctuations is similar to the interfacial free energy. According to the theory,[24] the angle dependence of the intensity of light scattering is determined as

$$I_\Theta^{-1} \sim \varphi_2^2 \frac{\delta^2 g}{\delta \varphi_2^2}\left[1 + \frac{8\pi^2 L_s^2 \sin^2\frac{\theta}{2}}{3\lambda^2}\right] \quad [6.37]$$

where θ is the scattering angle, λ is the wavelength of light in vacuum, L_s is the persistent length of fluctuation correlations that can be found from the following equation:

$$L_s^2 = \frac{2R_k^2 kT}{v_2\varphi_2(\delta^2 g/\delta\varphi_2^2)} \quad [6.38]$$

Near, but above, T_{crit} and at $\varphi_2 = \varphi_{2,crit}$ from Eq 6.24, it follows that

$$L_s = \frac{kT_{crit}R^2}{\varepsilon_2 v_2 \varphi_2 (T - T_{crit})} \quad [6.39]$$

The comparison with Eq 6.33 allows to establish similarity of physical natures of L_s and L_p. By introducing the radius of intermolecular force action, l, according to Debue[23]

$$l^2 = L_s^2\left(\frac{T}{T}crit - 1\right) \quad [6.40]$$

one can find, for the regular solutions,[24] that $l^2 = 2R^2\varphi_{2,crit}$. From this it follows that both l and L_s are proportional to $m^{1/4}$. Such dependence of l on m was found experimentally by studying the critical opalescence in polymer solutions.[24]

6.2 KAMMER'S THEORY

The theory of an interface between two polymers was developed by Kammer[25-27] on the basis of fundamental principles developed by Gibbs. According to Gibbs, there exists an equilibrium between two phases separated by a boundary layer. The chemical potentials of both phases μ_i and an interfacial layer μ^s are equal $\mu_i^s = \mu_i^1 = \mu_i^2$. It is accepted that the composition of an interphase is a function of temperature and, correspondingly, the interfacial tension depends on temperature and composition. For the interphase, there are introduced the molar values of the interface surface of constituent components, A_{1s} and A_{2S}. The following expression, characterizing the dependence of the interphase composition on temperature and pressure, was derived for application to any binary system:

$$\left[S_{12} - S_{1S}\left(1 - \frac{A_{2S}}{A_{1S}}\right)\right] dT - \left[V_{12} - V_{1S}\left(1 - \frac{A_{2S}}{A_{1S}}\right)\right] dP +$$

$$+ \sum_{i,n=1}^{n-1} [(x_i^{(2)} - x_i^{(1)}) + (x_i^{(1)} - x_i^{(s)})(1 - A_{2S}/A_{1A})] G_{ik}^{S} dx_k^{S} = 0 \qquad [6.41]$$

where indices 1, 2, and s relate to both phases and an interphase, and i, k, and n are the system components, x is the molar fraction of a component. The molar free energy and entropy are expressed as

$$G_{ik}^{(s)} = (\delta^2 G/\delta x \delta x_k)^s; \quad S_{12} = S^2 - S^1 + (\delta x/\delta x_1)^s \qquad [6.42]$$

Here $V_{12} = V^2 - V^1 + (\delta V/\delta x_1)^s$, V is the molar volume. Values S_{1s} and V_{1s} can be found from the following relations:

$$Y_{1i} = Y^l - Y^1 + \sum_{i=1}^{n=1} x_i^l - x_i^l \left(\frac{dY}{dx_i}\right)^s \qquad [6.43]$$

where Y = S, V, and l = s, 2. Kammer assumes that both polymers are in a liquid state and are immiscible, whereas, in the interphase, only a limited interpenetration occurs. The surface of an interfacial (boundary) layer "a" is connected with its thickness, δ, since $a = V^s\delta$. Supposing that the number of moles of components in the interphase is constant at any composition and that only molar volume, V, and thickness are changed with composition, Kammer finds that $A_{1s} = V_1^s\delta$ and $A_{2s} = V_2^s/\delta$. The concentration of component in the interphase may be expressed as

$$x_0^{(s)} = \frac{(\delta\gamma/dT)_p + (\delta\gamma/\delta T)_p^{(1)}}{\frac{1}{2}(\delta\gamma/\delta T)_p^{(1)} + (\delta\gamma/\delta T)_p^{(1)}} \qquad [6.44]$$

where γ is the interfacial tension. The calculations for the PA-PE system give the value $x_2 = 0.92$ which may be considered as an average value for the interphase. The density of a component in the interphase changes first very slowly and then rapidly decreases to zero.

By deriving the equations to calculate the thickness of an interphase Kammer considers it as an ideal solution of polymers. If to introduce the polymerization degree, the following expression may be found:

$$\delta = \frac{(\gamma_2 - \gamma_1)V_{20}^s}{RT\frac{r_2}{r_1}\varphi_1^s} \qquad [6.45]$$

where $V_{20}s$ is the molar volume of the surface layer of pure second component, $\varphi^s{}_{1s} = 1 - [r_2x_2s/(r_1x_1s + r_2x_2s)]$. The insertion of numerical values gives the interphase thickness of about 7.0 nm in accordance with the theory by Helfand and Tagami. The theory developed by Kammer allows one to calculate temperature dependence of the parameters of an interphase.

Later Kammer[28,29] has used Flory-Huggins theory to account for polymeric properties of components and degree of polymerization. The thickness of an interphase may be found from equation:

$$\delta = \frac{(\gamma_1 - \gamma_2)V}{RT[\ln\varphi_2^s + r\chi(\varphi_1^s)^2]} \qquad [6.46]$$

where V is geometric average of molar volumes of two components. The application of this equation at $r_1 = r_2 = r$ implies special calculation of the interaction parameter. Under these conditions, the thickness of an interphase for various pairs (PE/PS, PE/PA, PS/PMMA) is within the limits of 7-10 nm.

Kammer[28,29] has also developed a general theory of non-homogeneous interfacial region, accounting for density gradients of components and free energy gradients. The interfacial tension in the framework of Flory-Huggins theory is expressed, considering the effect of the mean square distance between chain ends $(h^2)^{1/2}$

$$\gamma = \left(\frac{2}{3}\right)^{1/2}\frac{RT}{V}\chi^{1/2}b(\chi z)^{1/2}\left[\frac{\pi}{8} - \frac{0.6}{z\chi} - \frac{0.5}{(z\chi)^2}\right] \qquad [6.47]$$

where χ is calculated from the solubility parameters, b is the effective length of monomeric unit, $b = m^{1/2}(h/ M^{1/2})$, M is molecular mass of polymer and m is molecular mass of monomeric unit. Eq 6.47 contains the value, z, the polymerization degree, which is not known for the interphase. This value can be found from the expression $\chi z = \pi^2/3(T^*/T)$, where T^* is reduction parameter (characteristic temperature). In this case

$$\gamma = \frac{2\pi}{9}\frac{RT}{V}\chi^{1/2}b(T^*/T)^{1/2}\left[\frac{2}{s\lambda}(1 - T/T^* + \frac{s\lambda}{12} - \frac{3}{\pi^2}s\beta\frac{T}{T^*}\right] \quad [6.48]$$

where $\beta = c^2/\lambda^2$, c, and λ are constants, characterizing the interaction of components. In a simplified form we have

$$\gamma = 0.688\frac{RT}{V}\chi^{1/2}b(T^*/T)^{1/2}(1 - T/T^*) \quad [6.49]$$

The interphase thickness is expressed now as

$$\delta = \frac{2\pi}{3\sqrt{3}}\frac{b(T^*/T)^{1/2}}{\chi^{1/2}[1 - 1.32T/T^*]^{1/2}} \quad [6.50]$$

The calculations according to the last equation give a very good agreement with data obtained by using Eq 6.46.[26,27] The essential outcome of Kammer theory is the thermodynamic substantiation of the proportionality of the interfacial layer thickness to the effective length of monomeric unit and inversely proportional to χ. This implies that the interpenetration in the interphase is realized on a very small distance which is much less than macromolecule dimension. Kammer has also derived the equation for partially miscible components.[28]

6.3 THE THEORY BY HELFAND

Within the framework of Helfand *et al.*,[9,10,30-33] the theory of an interface was developed on the basis of statistical approach to the lattice model. The structure of interphase is determined by the contact energy between monomeric units of polymers and the gain in the conformational entropy. The main task here is the statistic calculation of entropy for a given concentration profile followed by minimization of the free energy.

Monomeric units of macromolecules A and B (their numbers are N_A and N_B, respectively) are situated in the lattice, having n cells, to fill all the vacancies in the lattice. The lattice is presented as consisting of n_L layers parallel to the interface. Each layer consists of n_s cells so that $n = n_s n_L$. Each monomer interacts with z neighbors, their fraction mz is in the adjacent layers and fraction (1 – 2 m) in the same layer. It is accepted that the polymerization degree, N, is the same for two components, n, of polymeric units. The units are arranged within the lattice in a definite sequence: first z units of molecule A, then from (z + 1) to 2z units of molecule B, etc. in an alternating order.

Due to the existence of the repulsion potential between A and B in the neighboring cells, the system separates to phases A and B and interfacial region between them. This means that L layers exist where the fraction of occupied places φ_{1K} of component K (K designates A or B) is equal to 1 for many layers and then, after crossing a small number of layers, it turns to zero. The full occupancy of a lattice needs the condition $\varphi_{LA} + \varphi_{LB} = 1$ to be fulfilled. These values should be chosen to minimize the free energy of the system.

The central part of the theory considers the calculation of the conformational entropy contribution in the inhomogeneity of the system, S_{IG}, in the framework of classical Flory-Huggins theory. Conformational entropy is connected with the loss of conformational freedom, meaning that the molecules near interface should return to their own phases. The free energy connected with the inhomogeneity of the system, F_{IG}, is the sum of $-TS_{IG}$ and the mixing energy of layers, i.e., $F_{IG} = E_{IG} - TS_{IG}$. Value E_{IG} can be described as the free energy of contacts and their exchange.

The minimization of the energy gives the function of density distribution in the interfacial region and the value of the interfacial tension, γ, the latter is determined as $\gamma_a = F_{IG}$, where index a is the cross-section of a cell. For various values of interaction parameters χ, the volume fraction of units A in the first layer, φ_{1A} was calculated from the side of the surface B under condition that minimal free energy corresponds to symmetry, i.e., $\varphi_{LB} = \varphi_{1\text{-}LB}$. It is important that the transition from one phase to another occurs mainly in layers 1,0,1, and 2 for $\chi > 0.05$ and, preferentially, in layers 0 and 1.

The next step is the consideration of another scale of inhomogeneity[10] when it is spread to larger distances as compared with layers in the lattice. For the gradual transition from A to B, the simplified expression can be derived which describes the inhomogeneity profile:

$$\varphi_{L,A} = [1 + \exp[2(\chi/m)^{1/2}]L^{-1/z}]^{-1} \qquad [6.51]$$

where L is taken as continuous variable and the interfacial tension $\gamma_a = kT(\chi m)^{1/2}$.

According to Helfand, value $2(\chi/m)^{1/2}$ is the characteristic thickness of an interphase. Although, this theory was not compared with experiment because of vagueness of the m value, it predicts a very narrow interfacial region.

Later Helfand and coworkers have developed the theory of the interface between concentrated polymer solution and solvent and a theory considering two solutions. The problem is to determine the self-consistent field and density profile. For these cases, using the statistical approach, the thickness of the interphase and density profiles have been found.[30-33]

The theory of the interfacial boundary was constructed also on the basis of the self-consisted field approach.[9] The density profiles of two polymers in the interphase are determined by the fields which modify the configurational statistics of chains in this region. The configurational probabilities are used to calculate the density profiles. For this purpose a function is introduced describing the statistics of polymeric chains at the interface. This function is the ratio of the function of distribution of segments of average length, b, of chains, which are in the field of segment U(r), to their distribution function in zero field:

$$q(r_0,t) + \int dr_1 \ldots dr_M P(r_0, r_1, \ldots, r_M) \times$$

$$\times \exp\left(- \sum_{i=1}^{M} \frac{U(r_i)}{kT} \Big/ \int dr_i \ldots dr_M P(r_0, \ldots, r_M) \right) \qquad [6.52]$$

Here M = zt, z is polymerization degree, t is scaling parameter lying between 0 and 1. It is supposed that the chain segments have one end at the point r_0 and another at the point r and some successive positions rt, r_i is the position of at the end of i-th segment. Value P is determined by the Gaussian statistics:

$$P(r_0, \ldots, r_M) \Big/ \int P dr_1 \ldots dr_M = \left(\frac{3}{2\pi b^2} \right)^{3M/2} \times$$

$$\times \exp\left[-\frac{3}{2b^2} \sum_i |r_{i+1} - r_i|^2 \right] \qquad [6.53]$$

Function q(r,t) expresses the ratio of the segment density at point r of the interphase to segment density in a pure phase. At small $|b\nabla U|^2$, function q(r,t) satisfies a modified diffusion equation:

$$z^{-1} \frac{\delta q(r,t)}{\delta t} = \frac{1}{6} b^2 \nabla^2 q(r,t) - [U(r)/kT] q(r,t) \qquad [6.54]$$

with initial condition q(r,0) = 1.

To derive the mean field equation for the interface, the following case of polymer mixture with symmetric properties is considered:

1) polymerization degree $z \to \infty$
2) the effective length of monomeric unit b is chosen in such a way that mean square distance between chain ends is zb^2
3) the density of pure polymer is ρ_0 - the number of monomeric unit per unit volume
4) the compressibility, κ, is taken 5×10^{-10} m^2/N.

If the properties of A and B are not identical the geometric average meaning is used.

Consider the part of molecule A. This part is in an effective field created by the average molecular surrounding. As for pure A-phase, this field consists of the following parts: the first, $kT\chi\rho_B(r)/\rho_0$, is determined by the unfavorable contacts of segments A with segments B; the second, $[(1/\rho_0^2\kappa) - (kT/z\rho_0)][\rho_A(r) + \rho_B(r) - \rho_0]$ follows from the tendency of the system to push the polymer through the region where $(\rho_A + \rho_B)$ is less than ρ_0, and to push out the regions with the density higher ρ_0. Then equation for $q_A(r,t)$ can be written in the following form:

$$z^{-1}\frac{\delta q_A(r,t)}{\delta t} = \frac{1}{6}b^2\nabla^2 q_A(r,t) - $$

$$-\left[\chi\frac{\rho_B(r)}{\rho_0} + \xi\left[\frac{\rho_A(r)}{\rho_0} + \frac{\rho_B(r)}{\rho_0} - 1\right]\right] q_A(r,t) \qquad [6.55]$$

where

$$\xi = (1/\rho_0 kT\kappa) - (1/z) \qquad [6.56]$$

The analogous equation can be written for $q_B(r,t)$. The initial conditions here are $q_A(r,0) = q_B(r,0) = 1$. Eq 6.55 was derived by introduction of effective potential, $\mu_A^0(r) - \mu_A^e(\rho_o,0)$ instead of $U(r)$:

$$\mu^e(r) - \mu_A^e(\rho_0,0) = kT\xi\left[\frac{\rho_A(r)}{\rho_0} + \frac{\rho_B(r)}{\rho_0} - 1\right] + kT\chi_1\frac{\rho_B(r)}{\mu\rho_0} + $$

$$+\frac{1}{6}kT\xi\sigma_\xi^2\nabla^2\left[\frac{\rho_A(r)}{\rho_0} + \frac{\rho_B(r)}{\rho_0} - 1\right] + \frac{1}{6}kT\chi\sigma_x^2\nabla^2\frac{\rho_B(r)}{\rho_0} \qquad [6.57]$$

where

$$\sigma_\xi^2 = \int s^2 C_{AA}(s)ds/\int C_{AA}(s)ds \qquad [6.57a]$$

$$\sigma_x^2 = \int s^2[C_{AB}(s) - C_{AA}(s)]ds/[C_{AB}(s) - C_{AA}(s)]ds \qquad [6.57b]$$

where $C_{AA}(s)$ and $C_{AB}(s)$ are the direct correlation functions. This potential was obtained by calculation of the work of adding segment A in the interphase at point r and only first two terms of Eq 6.57 were used to obtain Eq 6.55. Function $q(r,t)$ is equal to the ratio of density of chain ends of A type with the length zt at point r to the density of segments in pure phase. The zt-th segment of the chain A may be considered as the beginning of two parts: one of the length zt and other z(1−t). Thus the relative density is equal to $q_A(r,t)q_A(r,1-t)$. Integrating by t one can find:

$$\frac{\rho_k(r)}{\rho_0} = \int_0^1 dt q_k(r,1-t)q_k(r,t); \quad k = A \text{ or } B \qquad [6.58]$$

Using Eqs 6.55 and 6.56 and accepting the following boundary condition:

$$q_A(\infty,t) = q_B(-\infty,1) = 1; \quad q_A(-\infty,t) = q_B(\infty,t) = 0 \qquad [6.59]$$

one can calculate the density profile of the interphase.

Helfand and Tagami[9] gave the approximated solution for the incompressible system when $\chi/\xi \to 0$ and when segments have the same density distribution. Mathematically stated: $q_k(r,t) \to q_k(r)$ except for t or $1 - t = \Theta(1/z)$. It follows from Eq 6.58 that $\rho_k(r) = \rho_0 q_k^2(z)$ and that q_k satisfies the steady-state counterpart of Eqs 6.55 and 6.56:

$$\frac{1}{6}b^2(\delta^2 q_k(x)/\delta x^2) - \chi q_{k'}^2(x)q_k(x) - \xi[q_{k'}^2/(x) +$$

$$+ q_k^2(x) - 1]q_k(x) = 0 \qquad [6.60]$$

In this equation, k = A,B while k′ = B,A. As a final result, the following equation is obtained:

$$\rho_A(x)/\rho_0 = \xi^2/(1+\xi^2); \quad \rho_B(x)/\rho_0 = 1/(1+\xi^2);$$

$$\xi = \exp[(6\chi)^{1/2}x/b] \qquad [6.61]$$

The following value is taken as the measure of the interphase thickness:

$$\delta = \rho_0\left(\frac{d\rho_A}{dx}\right)_{x=0}^{-1} = \frac{2b}{(6\chi)^{1/2}} \qquad [6.62]$$

For typical values of b = 0.7 nm and $\chi = 0.03$, we have $\delta = 3.3$ nm. For symmetric polymer mixtures of infinite molecular mass, the expression for interfacial tension was obtained in the following form:

$$\gamma = (\chi/6)^{1/2}\rho_0 b k_B T \qquad [6.63]$$

The interfacial profiles are determined by the relations:

Table 6.1: The comparison of calculated and experimental values of the interfacial tension[6]

Polymer pair	γ, mN/m		b. nm	χ
	calculated	experiment		
PS-PMMA	1.0	1.5	0.65	0.01
PMMA-PBMA	2.0	1.8	0.61	0.07
PBMA-PVA	1.9	1.9	0.63	0.05

$$\varphi_A(x) = \frac{1}{2}[1 - \tan h[(6\chi)^{1/2}x/b]] \qquad [6.64]$$

$$\varphi_B(x) = 1 - \varphi_A(x) \qquad [6.65]$$

Later Helfand and Sapse[34] have improved the theory and eliminated the limitation of symmetry. The interfacial tension is expressed then as

$$\gamma = k_BT(\rho_0\chi)^{1/2}\left[\frac{\beta_A + \beta_B}{2} + \frac{1}{6}\left(\frac{\beta_A - \beta_B}{\beta_A + \beta_B}\right)^2\right] \qquad [6.66]$$

where $\beta_i = \rho_i b_i$ for species i equal to A or B. The equations for interfacial profile φ(x) and interphase thickness similar to that derived by Helfand, were obtained by Binder and Frisch:[35]

$$\varphi(x) = \frac{1}{2}[1 + \tan h(x/\delta)] \qquad [6.67]$$

$$\delta = (b/3)\chi^{-1/2} \qquad [6.68]$$

The difference in proportionality coefficients in expressions for δ are explained by the restrictions of the gradient square theory[36-38] used to derive the corresponding equations.

After calculation of parameter χ for different polymer pairs, from the data on the solubility parameter, using Eq 6.63, there was done a comparison between theoretical values of the interfacial tension and calculated ones (Table 6.1). The experimental and calculated values are in a good agreement. The thickness of an interphase is comparable with the dimensions of monomeric unit. This confirms the Helfand concept that only limited diffusion of chain segments is possible in the interphase region.

6.4 THE THEORY BY NOSE

This theory[12] is based on the Cahn-Hilliard theory but does not take into account the change of the coil dimensions at the interface. It is supposed that macromolecules may interpenetrate one another. In the non-uniform system, the local density of segments and local density of macromolecules are connected by the concentration gradient. To account for the changes in macromolecule dimensions near the interface, the local coefficient of volume expansion is introduced. This coefficient is a function of direction perpendicular to the interface. The contribution to the free energy is calculated from relationship for the change in the chain conformation. We give below only the final result without considering calculations. The interfacial tension is expressed as

$$\frac{\gamma v}{kT_{crit}\lambda} = 2[1 + (m_1 m_2)^{1/2}(b_1^2 m_1^{1/2} + b_2^2 m_2^{1/2})(m_1^{1/2} +$$

$$+ m_2^{1/2})^{-1}]^{1/2}(1 - \tilde{T})^{3/2}/(m_1 m_2)^{1/2} \qquad [6.69]$$

where m is the number of segments in macromolecule, v is the segment volume, $b^2 = (a_i^2 m_i^2)/\lambda^2$, a is the dimension of polymer coil, λ is the length of intersegmental interactions[8] and $\tilde{T} = T/T_{crit}$. For the symmetric polymer system, i.e., $m_1=m_2=m$, $b_1=b_2=b$, we have

$$\frac{\gamma v}{kT_{crit}\lambda} = 2(1 + mb^2)^{1/2} m^{-1}(1 - \tilde{T})^{3/2} \approx 2bm^{-1/2}(1 - \tilde{T})^{3/2} \qquad [6.70]$$

$$\delta/\lambda = 2(1 + b^2 m)^{1/2}(1 - \tilde{T})^{-1/2} \qquad [6.71]$$

For non-symmetric system (for example, polymer-solvent) another equations are valid:

$$\frac{\gamma v}{kT_{crit}\lambda} = 2(b^2 + m^{-1/2} + m^{-1})^{1/2}(1 + m^{1/2})^{-1/2}(1 -$$

$$- \tilde{T})^{3/2} \approx 2bm^{-1/4}(1 - \tilde{T})^{3/2} \qquad [6.72]$$

$$\delta/\lambda = 2[1 + mb^2 1 + m^{1/2})^{-1}]^{1/2}(1 - \tilde{T})^{-1/2} \approx 2bm^{1/4}(1 -$$

$$- \tilde{T})^{-1/2} \qquad [6.73]$$

From Eqs 6.69 and 6.71, the critical exponents for the interfacial tension and thickness are equal to 3/2 and −1/2, correspondingly.

In the case of symmetric polymer system (polymer-polymer) according to Eq 6.70, the value $\gamma/[T_{crit}/(1 - \tilde{T})^{3/2}]$ diminishes with increasing molecular mass with power −1/2, whereas $\gamma/(1 - \tilde{T})^{3/2}$ increases with power 1/2, i.e, is proportional to the dimensions of unperturbed coil.

For polymer solutions both values diminish with increasing molecular mass (see Eq 6.72).The interphase thickness also increases with molecular mass both for polymer-solvent and polymer-polymer systems, although to a different extent. In symmetric system, δ is proportional to the unperturbed coil dimension, i.e., changes with m in power 1/2, whereas for polymer solution δ is proportional to m to the power 1/4. Therefore, the Nose theory gives the same dependencies of γ and δ on temperature as the theory by Vrij and Roebersen,[7] which does not account for the change of macromolecules in the interphase. The theory developed by Nose was checked experimentally[39,40] and a good agreement was found between theoretical conclusions and the theory.

6.5 SANCHEZ-LACOMBE THEORY

The basis for the theory developed by Sanchez and Lacombe[13] is the gradient theory and the model of compressible liquid lattice. The entropy of non-homogeneous system is supposed to be a function of density alone. To estimate the effects of the density gradients on the potential energy, the approximation of the mean field was used. The potential energy per unit volume of a liquid formed by m components at the distance R may be expressed, for the case of pair additivity, as

$$E(R)/V = \frac{1}{2}\sum_i \sum_j e_{ij}(R) \qquad [6.74]$$

$$e_{ij} = \rho_i(R) \int \rho_j(R + S) g_{ij}(S) U_{ij}(S) dS \qquad [6.75]$$

where $\rho_i(R)$ is the density of component i at R, $\rho_j(R + S)$ is the density of component j at R + S, $S = |S|$ is the intermolecular distance, $U_{ij}(S)$ is the intermolecular potential between molecules i and j; $g_{ij}(S)$ is the distribution function. It is supposed that U_{ij} is spherically symmetric, i.e., depends only on S. In the approximation of the mean field, the correlations of short range are neglected and in such a way

$$g_{ij}(S) = 0, \quad S < \sigma_{ij}, \quad g_{ij}(S) = 1, \quad S > \sigma_{ij} \qquad [6.76]$$

where σ_{ij} is the parameter characterizing the "hard core" part of the pair potential, σ_{ij}^3 is the core volume where the repulsion forces between components i and j are operative.

If far from the interface, the liquid mixture is homogeneous, value e_{ij} does not depend on R and it is proportional to the average density of components i and j. If there exists the concentration gradient, value $\rho_i(R + S)$ may be expanded into Taylor series near $S = 0$:

$$\rho_j(R + S) = \rho_j(R) + (\nabla\rho_j S)_{S=0} + \frac{1}{2!}(\nabla\rho_j S)^2_{S=0} + \ldots \qquad [6.77]$$

where

$$(\nabla\rho_j S)^2 = \frac{\delta^2\rho_j}{\delta S_x^2}S_x^2 + 2\frac{\delta^2\rho_j}{\delta S_x \delta S_y}S_x S_y + \ldots \frac{\delta^2\rho_j}{\delta S_z^2}S_z^2 \qquad [6.78]$$

Inserting 6.77 into 6.75, after integration, we obtain:

$$e_{ij}(R) = -\rho_i(R)\rho_j(R)m_0^{ij} + \rho_i(R)\sum_{n=1}^{\infty}\frac{I_n^{ij}}{n!} \qquad [6.79]$$

where m_0^{ij} is a positive constant (zero-th order moment of U_{ij}):

$$m_0^{ij} = -4\pi\int_{\delta_{ij}}^{\infty} S^2 U_{ij}(S)dS \qquad [6.80]$$

$$I_n^{ij} = \iiint_{-\infty}^{\infty}(\nabla\rho_j S)^n g_{ij}(S)U_{ij}(S)d^3S \qquad [6.81]$$

value U_{ij} is supposed to be spherically symmetric and is the function of S_x, S_y, and S_z. Consequently, the under-integral expressions in Eq 6.81 are either odd or even functions of the same values. In the intermediate case, $I_m^{ij} = 0$ odd; $I_m^{ij} \neq 0$ even for n = 2, the interval involving mixed derivatives disappear:

$$I_2^{ij} = -\frac{m_2^{ij}}{3}\nabla^2\rho_j \qquad [6.82]$$

where m_2^{ij} is the positive constant (the second order moment of U_{ij}):

$$m_2^{ij} = -4\pi\int_{\sigma_{ij}}^{\infty} S^4 U_{ij}(S)dS \qquad [6.83]$$

Assuming that the entropy of m-component liquid is a function of only local density independent of the density gradients, the local density of the free energy a(R) can be expressed by combination of Eqs 6.74, 6.79, and 6.82:

$$a(R) = a_0(R) - \frac{1}{12}\sum_i \sum_j m_2^{ij}\rho_i(R)\nabla^2\rho_i + \ldots \quad [6.84]$$

where $a_0(R)$ is the local density of the free energy when the density gradient is absent:

$$a_0(R) = -\frac{1}{2}\sum_i \sum_j m_0^{ij}\rho_i(R)\rho_j(R) - T\overline{S}(\rho_1,\rho_2,\rho_3,\ldots,\rho_n) \quad [6.85]$$

$\overline{S} = S/V$ is the entropy density. The total free energy of the system is

$$G = \iiint_V a(R)d^3R \quad [6.86]$$

For the system with a planar interface of cross-section S_0 and volume $\delta S_0(\delta \to \infty)$, we have

$$G = S_0 \int_{-\delta/2}^{\delta/2} a(x)dx \quad [6.87]$$

Neglecting the terms of the fourth and higher orders in Eq 6.84, after integration by parts at the boundary conditions ($d\rho_i/dx = 0$ at $x \to \infty$), the Eq 6.88 may be transformed to

$$G = S_0 \int_{-\delta/2}^{\delta/2} [a_0(x) + \frac{1}{12}\sum_i \sum_j m_2^{ij} \frac{d\rho_i}{dx}\frac{d\rho_j}{dx}]dx \quad [6.88]$$

Then the interfacial tension is expressed by the equation:

$$\gamma \equiv (G - G_0)/S_0 = \int_{-\infty}^{+\infty} [\Delta a + \frac{1}{12}\sum_i \sum_j m_2^{ij} \frac{d\rho_i}{dx}\frac{d\rho_j}{dx}]dx \quad [6.89]$$

$$\Delta a = a_0 - G_e/V = a_0 - a_e \quad [6.90]$$

where G_e is the equilibrium free energy of hypothetical homogeneous system in the absence of the interface. It can be expressed in terms of equilibrium chemical potentials, μ_i^e, and external pressure, P_e :

$$G_e = \sum_i N_i\mu_i^e - P_eV \quad [6.91]$$

$$a_e = G_e/V = \sum_i \rho_i(x)\mu_i^e - P_e \quad [6.92]$$

Value Δa represents the excess free energy per volume unit of the liquid with composition $\rho_1, \rho_2, ..., \rho_m$ if there is no density gradient. If the density gradients are present, the additional contribution to the local excess free energy is represented by the double sum in Eq 6.89. As $d\rho_i/dx$ and $\Delta a>0$ at $\delta \to \infty$, the integration limits can be extended to infinity (Eq 6.89). The minimization of the integral in this equation leads to m differential equations (Euler-Lagrange equations):

$$\frac{\delta \Delta a}{\delta \rho_j} - \frac{1}{6}\sum_j m_2^{ij} \frac{d^2\rho_j}{dx^2} = 0, \quad i = 1,2,...,m \qquad [6.93]$$

These m equations can be obtained also directly from Eq 6.84, using the equilibrium condition. The latter requires that chemical potentials of each component are constant at the interphase region. Multiplying Eq 6.93 by $d\rho_i/dx$ and summarizing over i, we have

$$\frac{d}{dx}\left[\Delta a - \frac{1}{12}\sum_i \sum_j m_2^{ij} \frac{\delta \rho_i}{dx}\frac{\delta \rho_j}{dx}\right] = 0 \qquad [6.94]$$

Integrating over x and applying the boundary conditions (Δa and $d\rho_i/dx \to 0$ at $x \to \pm \infty$), we can find that at the equilibrium

$$\Delta a = \frac{1}{12}\sum_i \sum_j m_2^{ij} \frac{\delta \rho_i}{dx}\frac{\delta \rho_j}{dx} \qquad [6.95]$$

Inserting Eq 6.95 into 6.89, we find the equilibrium interfacial tension:

$$\gamma = \frac{1}{6}\sum_i \sum_j m_2^{ij} \int_{-\infty}^{+\infty} \frac{d\rho_i}{dx}\frac{d\rho_j}{dx} dx = 2\int_{-\infty}^{+\infty} \Delta a dx \qquad [6.96]$$

The theory under consideration may be used to calculate the interfacial tension and concentration profile for multicomponent mixtures by application of any mean-field theory. Poser and Sanchez[14] have used the theory of the compressible liquid lattice where ρ_i is replaced by the segment density of i-th component:

$$\rho_i \to r_i N_i / V = \varphi_i \tilde{\rho} / v^* \qquad [6.97]$$

and $U_{ij}(S)$ is interpreted as the segment-segment interaction potential. For the inverse power dependence of the part of this potential connected with attraction forces, we have

$$U_{ij}(S) = \infty, \; S < \sigma_{ij}, \quad U_{ij}(S) = -\varepsilon_0^{ij}(\sigma_{ij}/S)^n, \; S > \sigma_{ij} \qquad [6.98]$$

In such a way:

$$m_o^{ij} = 4\pi\varepsilon_0^{ij}\sigma_{ij}^3/(n-3) \rightarrow 2\varepsilon*_{ij}\sigma_{ij}^3 \quad [6.99]$$

$$m_0^{ij} = 4\pi\varepsilon_0^{ij}\sigma_{ij}^3/(n-5) \rightarrow 2\varepsilon*_{ij}\frac{(n-3)}{(n-5)}\sigma_{ij}^5 \quad [6.100]$$

$$\Delta a = \frac{\tilde{\rho}}{v*}[g - \sum \varphi_i \mu_i^e] \quad [6.101]$$

where g is given by the sum of Eqs 6.102 and 6.103:

$$g_{comb} = kT\sum \frac{\varphi_i}{r_i}\ln \varphi_i \quad [6.102]$$

$$g_{noncomb} = -\tilde{\rho}\varepsilon* + P\tilde{v}v* + k\tilde{v}T[(1-\tilde{\rho})\ln(1-\tilde{\rho}) + \frac{\tilde{\rho}}{r}\ln\tilde{\rho}] \quad [6.103]$$

and chemical potential μ_i^e is expressed as

$$\mu_i^e = kT[\ln\varphi_i + (1 - r_i/r)] + r_i[-\tilde{\rho}\left(\varepsilon* + \frac{\delta\varepsilon*}{\delta\varphi}\right)_{\varphi'} + P\tilde{v}\left(\frac{\delta v*}{\delta\varphi_i}\right)_{\varphi'} - v* +$$

$$+ k\tilde{v}T[(1-\tilde{\rho})\ln(1-\tilde{\rho}) + \frac{\tilde{\rho}}{r_i}\ln\tilde{\rho} \quad [6.104]$$

Numerical calculations were also done.[14] For the binary mixture, the equation for the interfacial tension can be written in the form:

$$\gamma = \int_{-\infty}^{+\infty}[\Delta a + \frac{1}{2}(m_{11}\rho_1'^2 + 2m_{12}\rho_1\rho'_2 + m_{22}\rho_2'^2)]dx \quad [6.105]$$

This equation obeys the conditions:

$$\frac{\delta\Delta a}{\delta\rho_1} - m_{11}\rho''_1 - m_{21}\rho''_2 = 0 \quad [6.106a]$$

$$\frac{\delta\Delta a}{\delta\rho_2} - m_{12}\rho''_1 - m_2\rho''_2 = 0 \quad [6.106b]$$

Here the following estimations are introduced: $m_{ij} = m_2^{ij}$; $\rho'_i = d\rho_i/dx$; $\rho''_i = d^2\rho_i/dx^2$. Further the inverse power law is used for that part of potential that is connected with the attraction forces (Eq 6.98). Generally, the exponent depends on ij-interaction and from Eq 6.83, we obtain:

$$m_{ij} = 2\varepsilon*_{ij}\delta_{ij}^5\tilde{m}_{ij} \quad [6.107a]$$

where

$$\varepsilon^*_{ij} = 2\pi\varepsilon_0^{ij}/(n_{ij} - 3) \qquad [6.107b]$$

$$\widetilde{m}_{ij} = [(n_{ij} - 3)/(n_{ij} - 5)]/6 \qquad [6.107c]$$

After determining σ^3_{ij} as v^*_{ij}, the value m_{12} can be estimated using the combination rule in the form

$$m_{12} = 2\xi(\varepsilon^*_{12}\varepsilon^*_{22})^{1/2}[(1/2)(v^*_1 + v^*_2)(\delta + 1)]^{5/3}\widetilde{m}_{12} \qquad [6.108]$$

where m_{12} may be considered as tying parameter. Some useful simplifications in Euler equations can be done using the geometric average for $m_{12} = (m_{11}m_{22})^{1/2}$. In this case Eqs 6.94 and 6.95 lead to the algebraic form

$$m_{22}^{1/2}\frac{\delta\Delta a}{\delta\rho_1} - m_{11}^{1/2}\frac{\delta\Delta a}{\delta\rho_2} = 0 \qquad [6.109]$$

To compare the theory with experiment, authors[14] have used the dimensionless value:

$$\omega = m_{12}/(m_{11}m_{22})^{1/2} \qquad [6.110]$$

characterizing the deviation of m_{12} from geometric average value. The upper possible limit of values m_{12} is $\omega = 1$. The analysis of Eqs 6.94 and 6.95 for the binary mixture gives the following:

$$\Delta a = \frac{1}{2}(m_{12}\rho'^2_1 + 2m_{12}\rho'_1\rho'_2 + m_{22}\rho'^2_2) \qquad [6.111]$$

If γ is positive, Δa is positive as well and therefore the quadratic form of the right side of Eq 6.111 should also be positive. Because the derivatives of ρ_1 and ρ_2 and x may be either positive or negative, for the positive quadratic form of equation, the condition $m_{12} < (m_{11}m_{22})^{1/2}$ or $\omega < 1$ is necessary.

The numerical and analytical analysis is simplified by the transformation of equations from the space x in space ρ. Such a transformation changes the integration limits in Eq 6.105 from the infinite to finite limits. The Eq 6.111 can be written in the form:

$$\rho'_1 = \pm\left[\frac{2\Delta a}{m_{11} + 2m_{12}(d\rho_2/d\rho_1) + m_{22}(d\rho_2/d\rho_1)^2}\right]^{1/2} \qquad [6.112]$$

Then, changing the variables in Eq 6.105, we can obtain:

$$\gamma = 2^{1/2} \int_{\rho'_1}^{\rho''_1} [(m_{11} + 2m_{12}(d\rho_2/d\rho_1) + + m_{22}(d\rho_2/d\rho_1)^2]^{1/2} \Delta a^{1/2} d\rho_1 \quad [6.113]$$

where I and II relate to two equilibrium phases. Eq 6.112 may be formally integrated to obtain the profile of the interfacial tension:

$$x - x_0 = \int_{\rho_1(x_0)}^{\rho_1(x)} [m_{11} + 2m_{12}(d\rho_2/d\rho_1) + + m_{22}(d\rho_2/d\rho_1)^2]^{1/2} \Delta a^{-1/2} d\rho_1 \quad [6.114]$$

By using the geometric average rule for m_{12}, Eq 6.109 is the condition of minimization of Eq 6.113. If m_{12} is not equal to the geometric average, Eq 6.106 should be used which in the ρ-space is reduced to the following relation:

$$[m_{11} + 2m_{12}(d\rho_2/d\rho_1) + m_{22}(d\rho_2/d\rho_1)^2] \frac{\delta \Delta a}{\delta \rho_2} \left(m_{11} + m_{12} \frac{d\rho_2}{d\rho_1} \right) - - \frac{\delta \Delta a}{\delta \rho_1} [m_{12} + m_{22}(d\rho_2/d\rho_1)] - \frac{d^2 \rho_2}{d\rho_1^2} [2\Delta a(m_{11} m_{22} - - m_{12})^2] = 0 \quad [6.115]$$

Although these equations lose their symmetry and become more complicated in space, their transformation in space makes the calculations more simple. In some cases, another simplification is used, if m_{12} obeys the geometric average rule. If to introduce the variable

$$\Phi = m_{11} \rho_1^{1/2} + m_{22} \rho_2^{1/2} \quad [6.116]$$

the equation for the interfacial tension can be written as

$$\gamma = 2^{1/2} \int_{\Phi_I}^{\Phi_{II}} \Delta a^{1/2} d\Phi \quad [6.117]$$

with the solution subdued to Eq 6.109. In the theory of compressible liquid lattice, the density of n-mers ρ_i is determined as $\rho_i = \varphi_i \tilde{\rho}/v^*$. Using this expression together with Eq 6.90, the general rules of combination can be found:

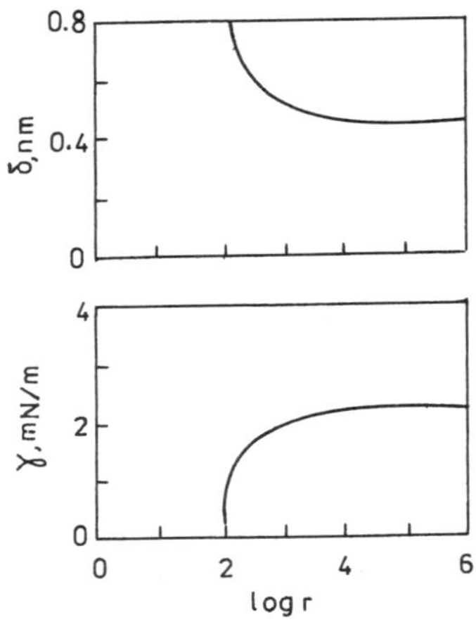

Figure 6.3. Schematic representation of the thickness of the interphase region, δ, (a) and interfacial tension, γ, (b) for PS-PE mixture (1:1) at 413K. δ=-0.011, $\tilde{m}_{PE}$=0.52, $\tilde{m}_{PS}$ =0.57, ω=0.98. Adapted, by permission, from C. I. Poser and I. C. Sanchez, *Macromolecules*, **14**, 361 (1981).

$$\Delta a = \frac{\varepsilon^*\tilde{\rho}}{v^*}[-\tilde{\rho} + \tilde{T}[(\tilde{v} - 1)\ln(1 - \tilde{\rho}) + (1/r)\ln\tilde{\rho} + (\varphi_1/r_1)\ln\varphi_1 +$$

$$+ (\varphi_2/r_2)\ln\varphi_2)] + P^*_e - \tilde{\rho}/v^*(\varphi_1\mu_1^e + \varphi_2\mu_2^e) \qquad [6.118]$$

μ_i^e should be expressed in energy units per segment. Eqs 6.105 and 6.106 may be solved by various methods of different complexity. For liquid-vapor systems, the results at $\omega = 1$ slightly differ from the ones found for $\omega < 1$. For liquid-liquid systems, the results are very sensitive to value ω.

By studying the applicability of the theory to polymer-polymer systems, authors have taken σ to be equal to the average magnitude and $\omega = \xi = 0.98$. Fig. 6.3 shows the molecular mass dependence of the surface tension and interphase thickness according to the theory. The curves have been calculated for PE-PS mixtures with $\omega = \xi$ and $r_1 = r_2 = r$. The thickness is determined as

$$\delta = (dx/d\rho_1)_{\bar{\rho}_1=1/2} \qquad [6.119]$$

where $\bar{\rho}_1$ is the segment density of component 1 reduced to the equilibrium value of ρ_2 in the phase where component 1 is prevailing. Because r diminishes, the mixing becomes more favorable up to some critical value of r (in the given case = 100) when the system becomes fully miscible. Therefore, the interphase thickness should increases to infinity whereas the surface tension decreases to zero with diminishing r. If the value ξ diminishes, the critical value r and asymptotic value of the interphase thickness decrease whereas the interfacial tension increases. As is seen from Fig. 6.3, the interfacial tension between two polymers approaches asymptotic limit at high molecular masses of both polymers. For two immiscible polymers the solubility of polymer A in polymer B is proportional to M_A^{-1} and solubility of B in A is proportional to M_B^{-1}. When both M_A and M_B are very high, the compositions of two phases are close to pure polymers.

These results have been used by Poser and Sanchez[14] to obtain the approximate formula for the interfacial tension and the interphase thickness of two polymers with infinitely high molecular masses. It is supposed that the density of the interphase linearly changes with composition:

$$d\rho/d\rho_1 = \text{const} = 1 + d\rho_2/d\rho_1 \qquad [6.120]$$

The constant in Eq 6.120 can be found by integration and gradients from the relations:

$$d\rho/d\rho_1 = (\rho_1^0 - \rho_2^0)/\rho_1^0 \qquad [6.121]$$

$$d\rho_2/d\rho_1 = -\rho_2^0/\rho_1^0 \qquad [6.122]$$

where ρ_1^0 and ρ_2^0 are segment densities of pure components 1 and 2:

$$\rho_1^0 = \tilde{\rho}_i/v*_i \qquad [6.123]$$

Therefore Eq 6.106 is changed to Eq 6.123. Inserting Eq 6.121 into 6.113 gives:

$$\gamma = 2^{1/2}[(m_{11}^{1/2}\rho_1^0)^2 - 2m_{12}\rho_1^0\rho_2^0 + (m_{22}^{1/2}\rho_2^0)^2]^{1/2} \int_0^1 \Delta a^{1/2} d\tilde{\rho}_1 \qquad [6.124]$$

For the simplicity, it was taken that $\tilde{m}_{ii} = 1/2$ and $\sigma = 0$. Then using Eq 6.107, we find

$$\gamma = (2\tilde{m}_0)^{1/2}(\gamma*_1\gamma*_2)^{1/4}(v*_1v*_2)^{1/12} \int_0^1 \Delta a^{1/2} d\tilde{\rho}_1 \qquad [6.125]$$

where $\tilde{m}_0$ is dimensionless value:

$$\tilde{m}_0 = \tau^{1/2}\nu^{-1/6}\tilde{\rho}_1^{2} + 2\omega\tilde{\rho}_1\tilde{\rho}_2\tau^{-1/2}\nu^{1/6}\tilde{\rho}_2^{-2} \qquad [6.126]$$

Value ω is determined from Eq 6.110 and γ^*_i is characteristic surface tension of pure components determined as

$$\gamma^*_i = \varepsilon^*_{ij}/(v^*_i)^{2/3} = P^*_i\delta_{ij} \qquad [6.127]$$

From Eq 6.118, value Δa can be found as

$$\Delta a = \tilde{\rho}/v^*[a(\tilde{\rho},\tilde{T})\varepsilon^* - \varphi_1 a(\tilde{\rho},\tilde{T}_1)\varepsilon^*_{11} - \varphi_2 a(\tilde{\rho}_2,\tilde{T}_2)\varepsilon^*_{22}] \qquad [6.128]$$

where

$$a(\tilde{\rho},\tilde{T}) = -\tilde{\rho} + \tilde{T}(1-\tilde{\rho})\ln(1-\tilde{\rho})/\tilde{\rho} \qquad [6.129]$$

Values $\tilde{\rho}_1$ and $\tilde{\rho}_2$ are calculated from the equation of state at the corresponding reduced temperatures $\tilde{T}_1$ and $\tilde{T}_2$ at $r\to\infty$ and $P = 0$. Near $\rho = (\rho_1^0 + \rho_2^0)/2$, value Δa reaches its maximum. Therefore the integral in Eqs 6.124 and 6.125 can be taken approximately equal to $(\Delta a_{1/2})^{1/2}/2$ with

$$\Delta a_{1/2} = \Delta a_{\rho=(\rho_1^0+\rho_2^0)/2} \qquad [6.130]$$

As $\rho = \rho_1 + \rho_2 = \tilde{\rho}/v^*$ and $\varphi_1 = \rho_1/\rho$, Eq 6.130 at $\delta = 0$ is reduced to the following:

$$\Delta a_{1/2} = \tilde{\rho}_{1/2}[a(\tilde{\rho}_{1/2},\tilde{T}_{1/2})P^* - [a(\tilde{\rho}_1,\tilde{T}_1)\Theta_1 P_1^* +$$

$$+ a(\tilde{\rho}_2,\tilde{T}_2)\Theta_2 P_2^*]] \qquad [6.131]$$

where

$$\tilde{\rho}_{1/2} \equiv (\tilde{\rho}_1 + \tilde{\rho}_2)/2 \qquad [6.132]$$

$$\Theta_1 = 1 - \Theta_2 \equiv \tilde{\rho}_1/(\tilde{\rho}_1 + \tilde{\rho}_2) \qquad [6.133]$$

$$P^* = \Theta_1^2 P_1^* + \Theta_2^2 P_2^* + \xi(\nu^{1/2} + \nu^{-1/2})\Theta_1\Theta_2(P_1^*P_2^*)^{1/2} \qquad [6.134]$$

$$\tilde{T}_{1/2} \equiv \frac{\tilde{\rho}_1\nu^{-1/2} + \tilde{\rho}_2\nu^{1/2}}{\tilde{\rho}_1 + \tilde{\rho}_2}\,\frac{(P_1^*P_2^*)^{1/2}}{P^*}\,(\tilde{T}_1\tilde{T}_2)^{1/2} \qquad [6.135]$$

The equation for the interfacial tension between two polymers with infinitely high molecular masses has the form:

$$\gamma = \frac{1}{2}(2\tilde{m}_0\Delta\tilde{a}_{1/2}\gamma^*_1\gamma^*_2)^{1/2} \qquad [6.136]$$

where $\Delta\tilde{a}_{1/2}$ is the dimensionless free energy which is determined by the expression

$$\Delta\tilde{a}_{1/2} = \Delta a_{1/2}/(P_1^* P_2^*)^{1/2} \qquad [6.137]$$

From Eqs 6.111 and 6.119, the interfacial thickness is

$$\delta = (\tilde{m}_0/2\Delta\tilde{a}_{1/2})(v*_1 v*_2)^{1/6} \qquad [6.138]$$

The interfacial tension and thickness for the polymer-polymer system can be calculated from Eqs 6.136 and 6.138, if values ξ and ω are known. Poser and Sanchez[14] have done such calculations taking $\omega=0$ and values of ξ chosen in such a way that experimental data were reproducible in the better way. It was found that calculated values of the interfacial tension are very sensitive to the choice of parameter ξ in spite of that ξ is equal to 0.99 for many polymer pairs, this value is not in any way a universal value. After analysis of the case of incompressible lattice authors have estimated the contributions of compressibility effects. For such a lattice $\tilde{\rho}_1 = \tilde{\rho}_2 = \tilde{\rho}_{1/2} = 1$, $a(\tilde{\rho}_1=1, \tilde{T}) = -1$, $\Delta a_{1/2} = \Delta P^*$, where

$$\Delta P^* = P_1^* + P_2^* - \xi(\nu^{1/2} + \nu^{-1/2})(P_1^* P_2^*)^{1/2} \qquad [6.139]$$

Value $\Delta P^* = 0$, if $\xi = \xi_0$, where

$$\xi_0 = [(\tau/\nu)^{1/2} + (\tau/\nu)^{-1/2}]/(\nu^{1/2} + \nu^{-1/2}) \qquad [6.140]$$

where $\tau = T_1^*/T_2^*$; $\nu = P_1^*/P_2^*$.

At the value of ξ expressed by Eq 6.140 for incompressible liquid, the heat of mixing is equal to zero. For these values of ξ, authors have found that such approximation describes rather well only the systems that have low interfacial tension (below 6 mN/m), i.e., polymer pairs where energetic contributions may be neglected (Table 6.2). For comparison some data are presented in Table 6.2 calculated according to Helfand theory where compressibility effects are neglected in the same way as in Nose theory.

6.6 OTHER THEORIES OF POLYMER-POLYMER INTERFACE

The fundamental principles of the theories of polymer-polymer interface have been developed in the theories presented above. There are also other approaches to the problem. The molecular description of a polymer-polymer interface was given by Wool and coworkers.[41-46] It is based not on thermodynamic grounds but on the methods of molecular dynamics. Authors have used the minor chain model (Fig. 6.4). At t=0, molecular contact is achieved at the interface (only the molecules on

Table 6.2: Interfacial tension and the thickness of the interphase for some polymer pairs

System	ξ_0	γ, mN/m			δ, nm	
		Exp[65]	Calc*	Calc[10]	Calc*	Calc[10]
PS-PMMA	0.0051	1.7	1.0	0.3	10.0	16.0
PBMA-PMMA	1.0027	1.9	2.0	1.5	4.8	2.9
PBMA-PVA	0.9970	2.9	1.2	3.0	8.0	1.6
PS-PVA	0.9751	3.7	4.0	1.9	2.3	3.1
PBMA-PDMS	1.0150	3.8	4.6	3.2	1.6	1.3
PIB-PDMS	1.0007	4.2	4.6	-	1.5	-
PELD-PDMS	0.9999	5.1	5.4	-	1.3	-
PS-PDMS	0.9947	6.1	6.8	-	1.1	-
PVA-PDMS	1.0224	7.4	3.8	7.2	2.0	0.8

$^*\xi = \xi_0$

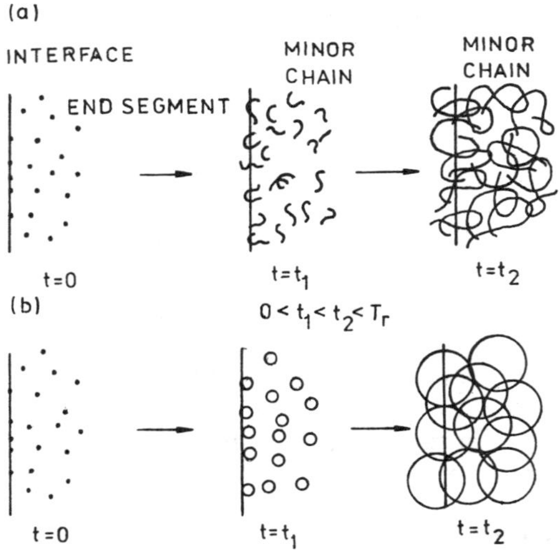

Figure 6.4. The interdiffusion process at a polymer-polymer interface in terms of the behavior of (a) the minor chains and (b) the minor chain spherical envelopes. Only one site of the interface is shown for convenience. Instantaneous molecular contact or wetting is assumed at t=0. Adapted, by permission, from R. P. Wool, B. -L. Yuan, and O. J. McGarel, *Polym. Eng. Sci.*, **29**,1341 (1989).

one side of the interface are shown for clarity). The chain begins to reptate via its chain ends which are shown initially as dots randomly distributed in space. As time progresses, only the part of the chain which has escaped from its original configuration, i.e., minor chain, can contribute to interdiffusion across the interface. When time approaches the reptation time, T_r, interpenetration and reentanglement of chains are achieved and the molecular properties of the virgin state are obtained at the interface plane. From this minor model, many molecular aspects of the interdiffusion process can be determined for the case $t < T_r$. The number of random-coil chains intersecting unit area of the interface, as a function of contact time, t, and molecular weight, M, behave as

$$n(t) \approx t^{1/4}M^{-5/4} \quad \text{for} \quad t < T_r; \qquad n^{\infty} \approx M^{-1/2} \quad \text{for} \quad t > T_r \qquad [6.141]$$

A chain is considered to intersect the interface when one of its ends has crossed the interface contact plane defined at t = 0,n(t). The average contour length of chain segments, which have diffused across the interface, is obtained from the minor chain model as

$$\lambda(T) \approx t^{1/2}M^{-1/2}; \qquad \lambda_{\infty} \approx M \qquad [6.142]$$

Each time a section of interdiffusion chain crosses back and forth through the interface, it creates a molecular bridge. This may be visualized as a "sewing up" of the interface, p(t) is determined by $p(t) = n(t)\sqrt{\lambda(t)}$, where $\sqrt{\lambda(t)}$ is proportional to the number of bridges created by each minor chain. In such a way, we obtain:

$$p(t) \approx t^{1/2}M^{-3/2}; \qquad p_{\infty} \approx M^{0} \qquad [6.143]$$

Following this line, authors determine the average monomer interpenetration depth, X(t), from Eqs 6.144 and 6.145:

$$X(T) \approx t^{1/4}M^{-1/4} \qquad [6.144]$$

$$X^{\infty} \approx M^{1/2} \approx R/\sqrt{6} \qquad [6.145]$$

and number of monomers crossing interface

$$L_0(t) \approx t^{3/4}M^{-7/4} \qquad [6.146]$$

$$L^{0\infty} \approx M^{1/2} \qquad [6.147]$$

Based on these premises, the authors have derived the concentration profile for a polymer-polymer interface for

$$\Phi(x,t) = [\lambda(t)/L + x^2/bL]\mathrm{ertc}[x/b\sqrt{2}n] -$$

$$-(\sqrt{2n/\pi})^{1/2}(x/L)\exp[-x^2/2nb] \qquad [6.148]$$

where b is the bond length, n is the number of monomers in the minor chain of length $\lambda(t)$, x is the diffusion depth selected in such a way that $x = 0$ at the interface, and L is the contour length of the entire chain. The time-dependence of the concentration profile in Eq 6.148 is a new step in analysis of the behavior of polymer-polymer interfaces. For $t>T_r$ the concentration profile returns to the classical Fickian profile

$$\Phi(x) = \mathrm{erfc}(x/L_d) \qquad [6.149]$$

Wool and coauthors have established that molecular properties described above have a convenient common scaling law which relates the dynamic properties, H(t), to the static properties H_∞ via reduced time t/T_r as

$$H(t) = H_\infty(t/T_r)^{r/4} \qquad [6.150]$$

$$H_\infty \approx M^{(3r-s)/4} \qquad [6.151]$$

where r,s = 1,2,3,....

A new development in the polymer-polymer interface theory was achieved by Wool[43] by considering fractal structure of polymer interfaces. It was noted that while the concentration profile, due to diffusion varies smoothly with depth, the diffusion field, when viewed in two- or three dimensions, has a very rough nature. The roughness increases with diffusion length and even though the average concentration profile varies smoothly, the diffusion front can have considerable lateral roughness. Fig. 6.4 shows a results of computer simulation of a monomer and polymer interface with solid. Such structures may be described within the framework of fractal concept. Using Shapoval gradient percolation theory,[47] Wool estimated the fractal characteristics of symmetric amorphous polymer interfaces. The following computer simulation and theoretical results have been obtained: width $\sigma_f \sim L_d^{1/D}$, front length $N_f \sim L_d^{3/D}$, position $X_f \sim L_d$, breadth $B_f \sim 6\sigma_f$, where D = 7/4. Studying the development of strength of incompatible amorphous polymer interfaces,[44] Wool has shown that fractal nature of diffuse interfaces plays an important role in controlling the physical properties of polymer-polymer and polymer-metal interfaces.

New approach[48] to the interphase width takes into account the number of segments in the chains, r_A and r_B:

$$\delta = \frac{2b}{(6\chi_{AB})^{1/2}}\left[1 + \frac{\ln 2(1/r_A + 1/r_B)}{\chi_{AB}}\right] \quad [6.152]$$

The temperature dependence of the interaction parameter for the PS/PMMA and PS/styrene-acrylonitrile copolymer mixtures was measured elsewhere[92] by ellipsometry and evaluated from experimentally found thickness of the interphase layer using Eq 6.152. It was found that the interfacial thickness increases with temperature.

Another approach to the analysis of the interface, involving no Gibbs concepts of the dividing surface, has been accomplished by Cahn.[49] He considered two boundary surfaces defining the interphase region. The existence of two surfaces located in the bulk of two coexisting phases is closer to reality but fails to eliminate the uncertainty in selecting the interface region thickness. In the case of a spinodal decomposition, these values describe the optimum spinodal decomposition wavelength. It should be pointed out that most theories describing the interphase region assume it as inhomogeneous and allow for the gradient of the density of components.[50-54]

In contrast to most calculations, where the interface is considered in a region far from the separation point, Joanny and Leibler[55] examined the properties of the interface near the phase separation point, based on the concept that the natuıe of the interphase region is substantially dependent on the degree of immiscibility and that the mean field theory of Flory-Huggins and Helfand's theory are not applicable to this region due to an important part played by density fluctuations, resulting in a greater thickness of the transient region. On the condition that $\rho_A = \rho$, $\rho_B = 1 - \rho$, the system's free energy is given by expression:

$$G/kT = (1/2)B\delta\rho^2 + (1/4)C\Delta\rho^4 \quad [6.153]$$

where $\Delta\rho$ is the difference between component concentration in the interphase and in critical point (assumed as a small value). Parameters B and C are given by the following expressions:

$$B = E[1/\rho_0 N_A + 1/(1 - \rho_c)N_B \quad [6.154]$$

$$C = (1/3)[1/N_A\rho_c^3 + 1/N_B(1 - \rho_c)^3] \quad [6.155]$$

Here N_A and N_B are degrees of polymerization of components, $E=(T_c-T)/T_c<<1$ and T_c is the critical temperature. Concentration of A in two phases is determined from the condition of minimum energy G:

$$C\delta\rho^2 + B = 0 \quad [6.156]$$

Free energy G can be approximated as

$$G = \frac{1}{2} \sum_N \frac{\chi^{-1}(g)}{[\Delta\rho(g)]^2} \qquad [6.157]$$

Here $\Delta\rho(g)$ is the Fourier transformation, $\chi(g)$ is the factor of osmotic compressibility of the system:

$$\chi^{-1}(g) = B(1 + g^2\zeta^2) \qquad [6.158]$$

where the correlation length is

$$\zeta = \left[\frac{1}{18}\frac{1}{\rho_c} - \frac{1}{(1-\rho_c)}\right]^2 |B|^{-1/2} \qquad [6.159]$$

From the minimization condition, the concentration profile of $\Delta\rho$ can be found:

$$\Delta\rho = \Delta\rho_\infty \mathrm{th}(x\sqrt{2}) \qquad [6.160]$$

where the boundary condition is $x \to \pm\infty$.

The transition region thickness is related to the correlation length by the expression $\delta = \sqrt{2\xi}$ and hence it is of the same order of magnitude as the correlation length. Estimates of δ near the separation point yield values much greater than found far from it.

Two cases have been discussed: (i) two polymers are treated as completely immiscible and (ii) the critical region where the volume fraction φ of components in coexisting phases becomes nearly identical. Binder and Frisch[35] studied this problem for the volume fractions intermediate between two limiting cases mentioned. They considered symmetrical mixture with chain length of both components as equal and analyzed the interfacial profile between two coexisting phases. The thickness of the interface was found equal to

$$L = \zeta_{coex}(\chi_{crit}/\chi)^{1/2} = (\sigma/3)\chi^{-1/2} \qquad [6.161]$$

where ξ_{coex} is a correlation function related to the characteristic length describing the concentration variation $\varphi(x)$ in the direction x perpendicular to a flat interface between coexisting phases. Eq 6.161 reproduces Helfands's result that the interface thickness $L \propto \chi^{-1/2}$ is independent of the chain length. Simultaneously, it was noted that the results for L are only qualitatively correct because the real mixtures are rather asymmetric and the derivation of the equations assumed that the Flory-Huggins theory parameter χ is independent of volume fraction. However, it is known that there is rather a distinct dependence of χ on φ which also affects the detailed shape and width of the profile. The density profile and the width of the interfacial region are very important characteristics of the binary polymer-polymer systems and therefore there were proposed many models describing this profile.[56-60]

6.7 SCALING PRINCIPLE IN DESCRIBING THE INTERFACE

The theories of the interface predict the dependence of the interfacial tension on molecular mass and temperature near the critical solution temperature. The scaling theory was used to estimate the critical exponents of such dependencies.[39,40] The dependence of γ on $(T - T_{crit})$ near T_{crit} can be represented by

$$\gamma = \gamma_0(T_{crit} - T)^{\mu} \qquad [6.162]$$

where μ is the critical exponent. According to the scaling principle,

$$\gamma \sim kT/\xi^{d-1} \qquad [6.163]$$

where d is the dimension of the space, ξ is the correlation length introduced by de Gennes:[61]

$$\xi = \xi_0(T_{crit} - T)^{-\nu} \qquad [6.164]$$

where ν is another critical exponent describing the change of correlation length near T_{crit}. From Eqs 6.162-6.164, it follows that

$$\mu = (d - 1)\nu \qquad [6.165]$$

For three-dimensional space at $\nu = 0.63$, exponent μ is equal to 1.26.[62] The gradient theories of the interface[7,12] for the polymer-solvent systems near T_{crit} lead to the following dependencies of the interfacial tension on molecular mass and temperature:

$$\gamma = N^{-1/4}(T_{crit} - T)^{3/2} \qquad [6.166]$$

For this dependence, $\mu = 1.5$. The interfacial thickness is determined from the relation:

$$\delta \sim N^{1/4}(T_{crit} - T)^{-1/2} \qquad [6.167]$$

i.e. $\nu = 1/2$, if δ is identified as ξ. Many theories predict that $\xi \sim N^{1/4}$. If to suppose that

$$\xi \sim N^{a}(T_{crit} - T)^{-\nu} \qquad [6.168]$$

then the interfacial tension is expressed as

$$\gamma \sim N^{-2a}(T_{crit} - T)^{2\nu} \qquad [6.169]$$

By decreasing temperature, the interphase thickness diminishes and at some temperature, T′, it becomes comparable with the dimensions of the polymer coil. Here the crossover phenomenon is expected (the appearance of the transition boundary on the phase diagram). Identifying δ and ξ and using Eq 6.167, we have:

$$(T_{crit} - T') \sim N^{-1/2} \qquad [6.170]$$

where it is assumed that $R \sim N^{1/2}$. Eq 6.170 determines the width of the critical region. At crossover (at the transition boundary), the Nose theory[12] predicts the first order transition from diffuse boundary to sharp interface. The polymer chain at this boundary collapses in direction perpendicular to the interface. By further decrease in temperature the transition occurs in the region where $\delta \ll R$. Here, according to the scaling theory, the correlation length does not depend on molecular mass and it is only a function of concentration, $\xi \sim c^{-1}$. Applying the relation $\gamma \sim kT/\xi^2$, we find:

$$\gamma \sim kTc^2 \qquad [6.171]$$

In this region $c \sim (\Theta - T)$ and therefore

$$\gamma \sim kT(T - \Theta)^2 \qquad [6.172]$$

where Θ is the Θ-temperature. The dependence of the interfacial tension on temperature is given in Fig. 6.5. Region I corresponds to temperature close to T_{crit}. The thickness of the interphase is much larger than the dimension of the polymer coil and $\gamma \sim M^{-1/4}$. In the region II, the interfacial tension thickness is comparable with the coil dimensions and strong dependence of the interfacial tension on the molecular mass appears. Higher interfacial tension corresponds to a lower molecular mass and *vice versa*. Value of δ in the region III becomes very small and the phase of solution, enriched with polymer, approaches that of properties melt. If molecular mass in this region is high enough then the critical temperature of phase separation is close to Θ-temperature. The interfacial tension should not depend on molecular mass in entire temperature interval.

These predictions have been checked in some experimental works.[39,40,63] For solutions of PS in methyl cyclohexane, the transition from the region I to region II was discovered. It was found that in the region II the dependence of the interfacial tension on molecular mass is much stronger than in region I. In the region I, the exponent of the dependence $\gamma = f(N)$ was found to be 0.44 which differs essentially from

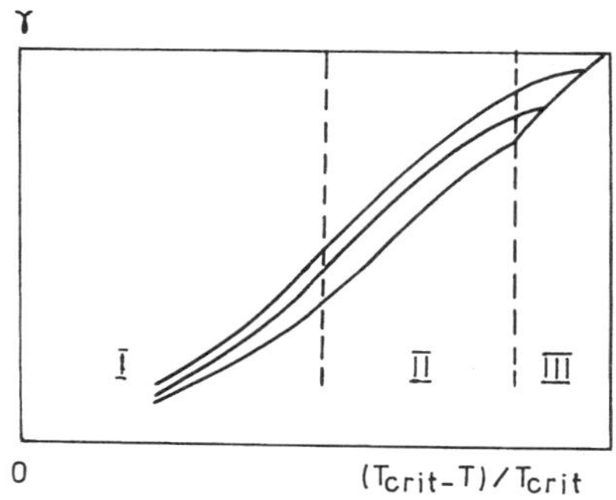

Figure 6.5. The dependence of γ on temperature and molecular mass (explanations see in text). Adapted, by permission, from K. Shinozaki, M. Abe, and T. Nose, *Polymer*, **23**, 722 (1982).

the values predicted by the mean-field theories (–0.25). Critical exponent was found to be 1.17÷1.35 close to theoretical exponent. The dependence of γ on ΔT was found in agreement with prediction of the Nose theory;[12] however, the slopes of dependencies of ln (γ/T) = f(ln c) and ln (γ/T) = f [ln (Θ – T)] (Eqs 6.171 and 6.172) were found >>2. These discrepancies between theoretical and experimental data were explained[63] by low molecular mass (scaling relations are valid only for very high N). It is possible also that the temperatures of experiments were not sufficiently far from T_{crit}.

According to Nose theory[12] value γ_0 in Eq 6.162 is written in the form:

$$\gamma_0 = \frac{2}{3}\varphi_{crit}^{1/2} kT_{crit}(\overline{h_\Theta^2}/M)^{1/2}(M/N)^{1/2}(1 + N^{-1/2} + N^{-1})^{1/2}/v \quad [6.173]$$

where $(h_\Theta^2)^{1/2}$ is the mean-square distance between chain ends, M is the molecular mass, N is the number of segments in molecule, φ_{crit} is critical concentration, and v is the volume of segment. Taking M/N=146, $(\overline{h_\Theta^2}/M)^{1/2} = 6.6\times10^{-2}$ nm and v = 136 cm^3/mol and using the experimental data on T_{crit} and φ_{crit} for PS solution in cyclohexane, value γ_0 was calculated[63] according to Eq 6.173. Theoretical value 3.63 mN/m was found to be close to the experimental result of 3.23 mN/m, showing a good agreement between the theory and experiment.

6.8 SOME EXPERIMENTAL DATA ON THE INTERFACIAL PROPERTIES OF POLYMER BLENDS

6.8.1 INTERFACIAL TENSION

Numerous experimental data on the interfacial properties of multiphase polymer systems are summarized in several reviews.[64-67] Considering the interfacial tension between two polymers, the general conclusion can be drawn about its diminishing with temperature and growth with molecular mass. For example, mixtures of n-alkanes with perfluoroalkane $C_{12,5}F_{27}$, PDMS, PEG are characterized[68,69] by the dependence of the interfacial tension on molecular mass as $M_n^{-2/3}$ in accordance with some empirical theories,[70,71] describing the interfacial tension with surface tension of pure components. The statistical thermodynamic theories have been developed primarily for the case of infinite molecular mass and do not reproduce the empirical dependence on $M_n^{-2/3}$. The theories based upon the square gradient framework may be applied to low molecular weight mixtures but do not appropriately consider the loss of configurations available to the chains in the interphase region. Some theoretical approaches, valid near the critical temperature, predict[12,72] the dependence on $M^{-1/2}$. However, there is no systematic investigation which compares the dependence of interfacial tension on molecular mass and temperature with theoretical predictions. Some data[73] on the mixtures PS/PB were compared with the predictions of Helfand theory[34,74] and with the theory based on extension of the square gradient approach of van der Waals[75] as modified by Cahn and Hilliard.[8] It was shown that a square gradient theory approach, in conjunction with the Flory-Huggins expression for the free energy of mixing, predicts a magnitude and temperature dependence of interfacial tension which are in reasonable agreement with experimental data. The predicted molecular mass dependence corresponds well with experimental data at high molecular mass but severely underestimates the interfacial tension for low molecular mass.

The results[73] have been extended,[76] by providing analytical expressions for the finite molecular mass correlations, to the interfacial thickness and tension by investigating the effect of polydispersity.

For monodisperse polymers, the concentration profile has been obtained in the form

$$\varphi_A^\infty(z) = \frac{1}{2}[1 + \tanh(z/\delta_\infty)] \qquad [6.174]$$

where $\delta_\infty = 2a/(6\chi)^{1/2}$. The interfacial thickness, δ, is expressed as

$$\delta = \delta_\infty[1 - 2\ln 2(1/w_A + 1/w_B)]^{-1/2} \approx \delta_\infty[1 + \ln 2(1/w_A + 1/w_B)] \quad [6.175]$$

and the interfacial tension is

$$\gamma = \gamma_\infty[1 - (\pi^2/12)(1/w_A + 1/w_B) + \ldots] \quad [6.176]$$

where $\gamma_\infty = (kT/a^2)(\chi/6)^{1/2}$, $w_A = N_A\chi$ and $w_B = N_B\chi$. For polydisperse polymers the interfacial tension is expressed as

$$\gamma = \gamma_\infty[1 - (\pi^2/6w_n)\ldots] \quad [6.177]$$

where $w_n = 1/(x_0/w_1 + (1 - x_0)/w_2)$ is the number average incompatibility degree,

$$x_0 = \varphi^\alpha_{A1} + \varphi^\alpha_{B1} = \varphi^\beta_{A1} + \varphi^\beta_{B1} \quad [6.177a]$$

$$\varphi^\alpha_A = 1 - \varphi^\alpha_B = 1 - e^{-w_B} \quad [6.177b]$$

$$\varphi^\beta_A = 1 - \varphi^\beta_B \approx e^{-w_A} \quad [6.177c]$$

The last two expressions are for α and β-enriched phases. According to the data,[73] the lowest value of exponent z = 0.54 was found for more polydisperse and lowest molecular mass system (M_w/M_n = 2 for PDMS), whereas the greatest exponent z = 0.9 was found for the system with the highest molecular mass. Therefore, the predicted asymptotic regime (M_n^{-1} behavior of Eq 6.177) is not yet reached for this range of molecular mass. For monodisperse mixtures, authors[76] found a finite molecular mass correction to the interfacial tension that varies with $(\chi N)^{-1}$. This suggests that measurements of γ for different (large) molecular masses could provide another means of determining parameter χ for immiscible blends.

6.8.2 THE INTERFACIAL WIDTH AND INTERFACIAL PROFILES

To measure the concentration interfacial profile, there were proposed various methods.[77-81] In each case, the interphase is set up between two polymer layers, which may be pure components or two mixtures of different compositions. In one type of experiment, a section is cut through the profile and IR spectroscopy[77] or X-ray microanalysis[78] is used to scan the composition profile. In the other experiments, the interphase was observed in bulk sample using scattering of ion-beams - both forward recoil spectroscopy (FReS)[82] and Rutherford back-scattering (RBS).[83]

One of the properties shared by these techniques is a resolution limited to some tens of nanometers. To investigate the short-range detail of the interface, methods lack sufficient resolution, although they were successfully used in the study of long-range variations in composition.[78-80] The information about the interphase on a much smaller scale may be gained by using the neutron reflectivity.[81,84] Some data on the analysis of the interphase by this method are reviewed by Russel.[81,85] In the majority of cases, the information about the interphase thickness and concentration profile has been obtained by the neutron reflectivity method[86-91] and ellipsometry.[92-96]

Eq 6.175 allows one to determine the interaction parameter from the temperature dependence of the interphase thickness which is a rare case of the determination of this value for immiscible mixtures. From the ellipsometric data, the temperature dependence of χ was found[96] for PMMA-SAN and PMMA-PS mixtures. The interaction parameters for mixtures of PMMA with SAN are very small and slightly depend on temperature whereas for PMMA-PS parameter χ diminishes with increasing temperature.

In some works by Inoue *et al.*,[93,95,98] the ellipsometry was used to determine the interphase thickness for PMMA-SAN and PMMA-PS mixtures as a function of annealing time above the glass transition temperature. This magnitude was found to be around 20 nm at an early stage of annealing and it increases with time up to 60 nm at 303K after 12 h for a miscible pair PMMA-SAN, while it remains constant for an immiscible pair (SAN with different amount of acrylonitryl were used). It was found that the thickness for miscible pair depends both on time and interdiffusion coefficient and is proportional to the thermodynamic driving force $\chi - \chi_{crit}$ (parameters at the annealing temperature and at the critical point). For immiscible polymers, PS-PMMA and PMMA-SAN, the parameter χ_{AB} and segmental interaction parameter, χ_{ij}, were calculated. It has been shown that all three segmental interaction parameters of the system PMMA-SAN are decreasing with increasing temperature, but the polymer-polymer interaction parameters between PMMA and SAN, in a certain copolymer composition range, are increasing with temperature, leading to LCST behavior. The experiments predict UCST behavior in blends of low molecular mass PMMA and PS. The comparison of the calculated interphase thickness with direct data obtained by transmission electron microscopy[93] are in an excellent agreement. Fig. 6.6 shows the experimentally determined concentration profile of the annealed PMMA/SAN (38.7) sample (solid line) in comparison with a number of geometrical assumptions. Various approximations have been used; a linear gradient

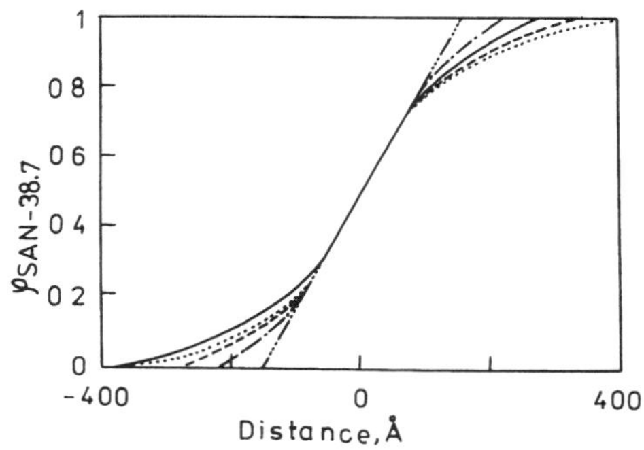

Figure 6.6. Concentration profile across the interface. Experimentally determined profile for the annealed specimen of PMMA-SAN (38.7% AN) (___), tanh (...), erf (- - -) $\sin^2$ (-.-) and linear gradient (-....-). Adapted, by permission, from J. Kressler, N. Higashida, T. Inoue, W. Heckmann, and F. Seitz, *Macromolecules*, **26**, 2090 (1993).

$$\varphi = (1/320)\delta + 0.5 \qquad [6.178]$$

squared sinusoidal function ($\sin^2$):

$$\varphi = \sin^2(\delta/320 + \pi/4) \qquad [6.179]$$

hyperbolic tangent

$$\varphi = \frac{1}{2}[1 + \tan\text{h}\,(\delta/160)] \qquad [6.180]$$

and error function (erf) often used to describe the diffusion profile:

$$\varphi = \frac{1}{2}\,(1 + \text{erf}\,[\sqrt{\pi}\,\delta/320]) \qquad [6.181]$$

The calculations are fitted to an interfacial thickness of 320 Å and the concentration of 0.5 is located at the center of the interfacial layer (d=0, where d means distance). The experimentally determined interface is asymmetric. Fig. 6.7 shows the interfacial thickness as measured by ellipsometry for PMMA/SAN (37.8), PAN-PMMA, PS-PAN and PMMA-PS in the temperature range 313-343K. All homopolymer pairs have very thin interphases. The thickness about 19 nm, for immiscible pair PMMA-SAN, is larger than calculated according to the Helfand-Tagami theory. The time dependence of the interphase thickness for PMMA-SAN and concentration profiles as a function of the SAN composition were also established.[92]

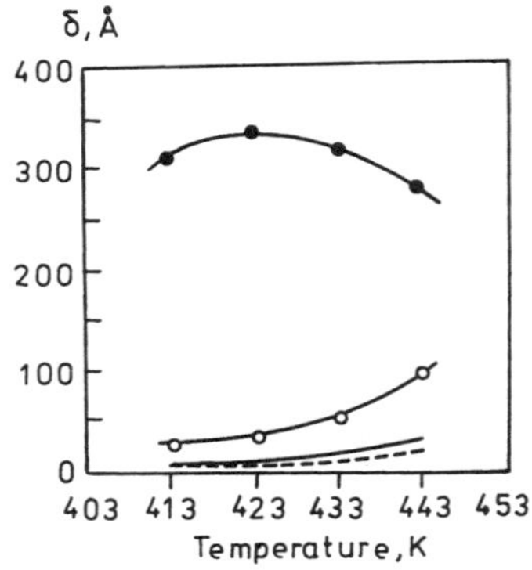

Figure 6.7. Interfacial thicknesses measured by ellipsometry for PMMA-SAN (38.7), PMMA-PS, PS-PAN, and PAN-PMMA. Adapted, by permission, from J. Kressler, N. Higashida, T. Inoue, W. Heckmann, and F. Seitz, *Macromolecules*, **26**, 2090 (1993).

Table 6.3: The interphase thickness, δ, for various diffusion times[94]

System	τ, min	φ_m^*	δ, nm
Isotactic PP-PE	3	0.60	1500
	5	0.58	1930
	10	0.55	2450
	15	0.50	2600
	20	0.44	2800
PVC-PE	3	0.54	5000
	5	0.50	6000
	10	0.45	8000
	15	0.40	8500
	20	0.35	9000

The equilibrium thickness is a complicated function of the copolymer composition.[91] For mixtures of deuterated and protonated PS, it was found[90] by nuclear reaction analysis that the thickness of the interphase increases with a short annealing time below 473K (UCST is about 473K) proportionally to the annealing time in power to 0.34

which is much less than 0.5 for a free diffusion. The exponent approaches 0.5 at long annealing times. The width of the interphase at annealing reaches some limiting value. Its changes along with temperature are consistent with the mean-field theory.[89] The time dependencies of the interphase thickness for the mixtures of isotactic PP with PE and PVC with PE are presented in the Table 6.3.[94]

The evidence for a sharp interface between partially miscible polymers was given based on the study of neutron specular reflection for the PMMA-CPE mixtures.[97] In this case the density profile depends on the annealing conditions. It is interesting that in this case the density gradient for PMMA was not connected with the density gradient at the interphase. For the miscible pair PMMA-SAN, the thickness increases with the time of annealing which can be easily understood.[97]

The width of the interphase was studied also for the PS-PMMA mixtures by introduction of diblock copolymer (styrene-methyl methacrylate) at the interface.[86] It was found that introduction of diblock copolymer broadens the interphase region up to some limit. The interface is saturated by diblock copolymer that corresponds to the formation of one half of lamella in microdomains. At higher amounts, block copolymer forms its own structure at the interface.

All data considered show that the interfacial thickness may vary depending on the miscibility of components and conditions of the interphase formation. There are a few data available which show the equilibrium structure and equilibrium profile at the interface between two polymers. From this point of view it is very important to estimate the kinetics of the interphase formation. Theoretically it was done by de Gennes[99] who considered the formation of diffuse interphase between two polymers for $\chi \leq 1$. It was found that first the chains are reflected from the interface and after some time the interpenetration of chains leads to formation of diffuse region with the thickness of $\delta = a\chi^{-1/2}$. Three models of the process have been discussed. If the distance between entanglements, N_e, is larger than χ^{-1}, the diffusion is expected according to the Rouse model. If $N_e\chi \leq 1$, the reptation effect prevails with transient concentration profiles and discontinues at the contact plane. The kinetics of the interphase formation was studied theoretically also for the case[100] of interphases with dimensions which are larger than gyration radius of the chains. It was found that the width of the interphase increases with time as $t^{1/2}$ at short times and as $t^{1/4}$ at long times. The dependence of the width $\delta(t) \sim t^{1/2}$ is typical of miscible pairs where the growth of the interphase is determined by the surface and bulk free energy, whereas $\delta(t) \sim t^{1/4}$ is characteristic of immiscible mixtures where the growth of the interphase is determined only by concentration gradient.

6.9 SURFACE SEGREGATION IN POLYMER MIXTURES

Surface segregation of components of polymer blends is typical of free polymer surfaces. It is connected with the difference in surface tension of the mixture component and different surface activity of components. The surface tension of some polymer blends is characterized by a strong dependence of the surface tension on blend composition, connected with the change of the composition at the surface.[101] The measurements of surface tension of some polymer pairs for different compositions and temperatures shows[102] that, in blends, surface excess of one of the blend components appears. This observation was made from the data on the surface tension of miscible system PS-PVME. The thorough analysis of the problem shows that the origin of the surface excess can be explained in the framework of Cahn-Hilliard and scaling theory. Hariharan *et al.*[103] made theoretical analysis of composition of the free surfaces of binary polymer blends as a function of composition, molecular weights, and energetic parameters. They have applied the compressible mean-field lattice model and found that surface segregation is determined by the degree of incompatibility of components. The higher is the interaction parameter, the higher is the segregation. The enrichment of the surface with one of the blend components is determined mainly by its molecular mass. For compressible mixtures, the surface is enriched with component having lower surface energy. Segregation is increased with increasing incompatibility. The concentration profile has been calculated. It depends on the contribution of entropy and enthalpy factors in segregation. Thus, two factors are responsible for surface segregation - the difference in the surface energy of two components[104] and the level of their miscibility.

Bates and Fredrickson[105] analyzed various contributions to the excess entropy of mixing, which are associated with local (i.e., monomer length scale) effect, and non-local (i.e., length scales greater than the persistence, or Kuhn's length). These effects of conformational asymmetry influence surface segregation. Two main parameters have been used - a volume, V, occupied by a single polymer molecule and $R_G = b(N/6)^{1/2}$ - the unperturbed radius of gyration of a chain in melt of similar chains (b is the statistical segment length). It was shown that in polymer-polymer mixtures, the influence of conformational asymmetry is manifested by the miscibility in the bulk and at the surface, where difference in $\beta^2 = R_G^2/V$ can produce preferential surface segregation.[106] The theory of the surface enrichment of polymer blends was developed by Nakanishi and Pincus.[107]

Surface segregation may be connected with different flexibility of constituent chains.[108] The entropy-driven segregation of polymer blend was investigated via computer simulation. The model system

Table 6.4: Composition of the surface layers at the interphase with solid for PE-PBMA mixture (%%)[110]

Filler amount, wt%	PE in wt% at the interface at its concentration in blend			PBMA in wt% at the interface at its concentration in blend		
	25	50	75	75	50	25
3	20.08	5068	79.85	79.92	49.32	20.15
5	17.46	43.61	76.44	82.54	56.39	23.56
10	14.44	43.61	74.27	85.56	56.39	25.73
15	9.46	37.06	69.47	90.54	62.96	30.53

was composed of a binary blend at a hard wall where one of the components of the blend was stiffer than the other. At melt-like densities, relevant to experiments, simulation predicts the segments of stiffer chains to segregate on the surface, whereas at low densities the net segregation excess (integrated over all distances) favors flexible chains at the surface.

In Ch. 3, the effect of the interface with solid on the preferential adsorption and phase separation was discussed. The enrichment and depletion of the surface layers in filled polymer blends was discovered[109] for the first time for the ternary system: polybutadiene carboxylate rubber-epoxy resin-ammonium perchlorate. Using method of total attenuated inner reflection in IR region, it was found that epoxy resin preferentially interacts with solid displacing rubber. The enrichment of the surface layer by the epoxy resin is energetically favorable due to its higher surface tension (48 mN/m) than rubber (39 mN/m) which leads to a sharp decrease in interfacial tension at the polymer-solid interface. It was established that, at reaction conditions, the distance from the surface, at which the layer is enriched with epoxy resin, reaches 5 mcm. These data show that there is no sharp border between two polymers in the surface layer, but stepwise change of their ratio. These results meet the data on the selective adsorption of the same polymers.

For miscible polyethylene-poly(butyl methacrylate) pair, filled with fume silica, using NMR technique, the composition of the border layer was established (Table 6.4). With a growing filler content, the fraction of PBMA in the boundary layer increases and polyethylene decreases, i.e., the boundary layer is enriched with PBMA. Only 3% filler addition to the alloy containing 75% polyethylene results in a small enrichment with this polymer, but above 10% filler, PBMA content increases. When considering the structure of surface layers in filled polymer alloys, there is evidence of conformational changes of both components in the surface layers. The changes depend both on the surface nature and component ratio.[111]

Effect of the filler on the phase separation, during formation of IPN's, may also be explained by selectivity of component adsorption at the interface with a filler. It leads to the redistribution of the reaction system components between the bulk and surface layers.[112] The enrichment or depletion of the surface layers in IPN's depends both on the free surface energy of the filler and on the surface tension of the reaction system components. The greatest changes of the composition in the bulk and in the surface layers are observed in the composition range, where the positive free energy of mixing components (maximum incompatibility) is observed. These effects influence both reaction kinetics and degree of phase separation.

6.9.1 SURFACE SEGREGATION NEAR THE INTERFACE WITH SOLID

Some experimental evidences of the surface segregation near the interface have been obtained in our laboratory during the last 5-7 years which are the base of the following considerations. We will begin with the analysis of some main factors influencing the formation of an interphase in reinforced binary polymer blends. The main reasons determining the formation of the interphase are the following:

- The selectivity of adsorption of the mixture components at the interface with solid depends on the component affinity to the surface (thermodynamic interaction parameter, χ_S) and miscibility of components (thermodynamic interaction parameter, χ_{AB})
- The selectivity of adsorption leads to formation of adsorption layers with composition different than the initial mixture (enrichment of the surface layer by one component, or surface segregation). Surface segregation depends both on χ_S and χ_{AB}. The lack of miscibility enhances surface segregation.
- The effect of solid surface on the phase separation. This effect manifests itself in various ways depending on the thickness of the surface layer.[113] The effect of the surface depends both on the molecular mass of the components and their surface tension. In the vicinity of the surface of high surface energy, the concentration of components with higher surface tension is increased and at the polymer-air (or solid of low surface energy) interface, the components with lower surface tension are segregated. The effect depends on the distance from the interface.[113]

The magnitude of concentration amplitude by spinodal decomposition (amplification factor) drops sharply as the film thickness decreases. Due to this effect the segregation degree at the interphase is lower than in bulk, i.e., the surface layer stays in the state of forced compatibility (miscibility), its structure corresponding to the equilibrium state at another temperature, depending on the type of CST.

It is evident that the equilibrium process of the surface segregation is possible only at the constant temperature of the melt. At conditions of the formation of filled polymer blends by transition from liquid to solid state, the segregation proceeds until the necessary molecular mobility is preserved in the system, i.e., up to temperatures of crystallization or vitrification of one of the melt components. This means that surface segregation proceeds in non-equilibrium conditions in the definite temperature interval. If the phase separation temperature is above glass transition or crystallization temperature, the surface segregation may proceed from each evolved phase separately, if the surface tension of two phases is not very different and the viscosity is high enough to prevent fast displacement of the phase with higher surface tension. Many published experimental data support this point of view.[114-123] Recently, the processes of surface segregation became the subject of many experimental and theoretical investigations due to their importance for polymer composite materials, adhesion, and other applications of multicomponent polymeric systems which can be related to polymer alloys. Many works consider the compositional profile which arises near the interface due to surface segregation.[124-128]

The processes of surface segregation near solid surface proceed according to spinodal mechanism. The critical wetting point of polymer blends was analyzed[129] based on wetting by two-phase systems, formed from two polymers. In polymer blends, due to a very small interfacial tension between two separated phases in a liquid state, the critical wetting (wetting near the critical point of the phase enriched with component more active towards surface) proceeds far from critical point and deep into two-phase region. This wetting is realized in the course of phase separation. It is accompanied by a growth of spinodal domains in surface-directed manner and at equilibrium of the composition profile. It is characterized by a macroscopically thick layer of one of the phases. The surface excess, Γ, is proportional to the thickness of a wetting layer $l = -\text{const} \ln (1 - \phi_d/\phi_l)$, where ϕ_d and ϕ_l are equilibrium concentration of the adjacent depleted zone and concentration in the wetting layer, respectively. Any surface in contact with a mixture may be either partially or completely wetted by one of the phases.[129]

It is worth noting that all the processes of surface segregation may proceed only in a melt state of a system where the molecular mobility of polymer chains is high enough. This statement is valid both for blends of linear polymers and IPN's. However, if in IPN's, from the very beginning, the surface segregation proceeds in initially one-phase system, for polymer blends, the conditions are determined by the type of critical solution temperature. The majority of linear blends has LCST. The surface segregation of such systems in melt should proceed from the two-phase state.

Surface segregation enhances phase separation in the system. Enrichment and depletion of the surface layer in one of the mixture components should be considered as phase separation in micro-volumes near the interface. The effects of surface segregation take place only for the systems having asymmetric interaction with the solid surface ($\chi_{AS} \neq \chi_{BS}$). In the course of the blend formation in presence of a solid, two processes proceed simultaneously: surface segregation and phase separation.

Binary filled (reinforced) blends should be considered as consisting of conventionally four phases: two separated phases near the interface and two phases in the bulk. Their compositions differ from the average composition of an alloy. These states are not the equilibrium states and the phase structure may be considered as a result of forced compatibility.

The effect of the surface on the phase separation determines the surface segregation as a result of selective interactions (equivalent to phase separation) and simultaneously prevents phase separation (kinetic contribution). For the systems with asymmetric interactions the first effect is prevailing, whereas for the systems with symmetric interaction - the second.

All these results form a basis for consideration of the surface segregation in IPN and composition of the surface layers (surface segregation) arising in the course of the crosslinking of both networks in IPN. The formation of the crosslinked structures in IPNs from initially one-phase system proceeds under non-equilibrium conditions, accompanied by the microphase separation due to arising immiscibility of components with increase in conversion degree. For IPNs, the surface segregation may proceed in the time interval between the onset of curing and time required to reach gel point. However, because the onset of phase separation begins very early, this surface segregation proceeds simultaneously in two evolved phases before the gel is formed.[130,131]

For semi- and full-IPNs near the interface with solid of high surface energy, the layers are formed with excess of the component which has higher curing rate. At comparable reaction rates of polymerization and polyaddition, the composition of the surface layers does not differ from the layers in bulk. The content of components in the interphase depends on the ratio of networks, ratio of reaction rates of curing of both networks. These rates are interdependent. The effects of surface segregation are also connected with the onset of phase separation, depending on the reaction rates and component ratio. The composition and structure of the surface layers formed in full IPNs at the interface with solid depend also on the free surface energy of solid and initial composition of IPN. The reasons which lead to the effect of the reaction kinetics on the surface segregation are the following:

- When applying the composition to the surface near the interface between solid and polymer mixture, there is an excess interfacial energy due to the difference in the surface tension of substrate and composition. As a result, these components are segregated at the interface which compensates this difference.
- In the course of curing, the surface layer is enriched with the component whose surface tension faster reaches its limiting value, i.e., has higher reaction rate.
- The composition of the surface layers depends on the kinetic factors. The main role belongs to the sequence of curing of components.
- Superposition during IPNs formation of the adsorption processes and processes of microphase separation determines very complicated dependence of the composition and structure on the kinetic conditions. The main parameters of the process are:
  1) adsorption rate
  2) time for the onset of phase separation which is determined by the reaction kinetics.

Surface segregation may proceed in two ways: either from initially one-phase system or from the system undergoing phase separation. Transition from one-phase to two-phase system changes the mechanism of formation of the surface layer and conditions for surface segregation. In one-phase system, the selective adsorption proceeds from the reaction mixture leading to the surface layer formation. If the time for the onset of phase separation is short, the formation of the layer proceeds simultaneously with phase separation and, because both evolved phases do not have reached their gel point, selective adsorption proceeds from both phases. Surface segregation from each phase depends on its composition, resulting in a non-uniform surface segregation.[131]

In conclusion, one may say that non-equilibrium processes of surface segregation of free surfaces of binary polymer blends and the surface layers at the interface practically have not been studied. At the same time, it is a very important subject connected with non-equilibrium states in polymers and polymer mixtures.

6.10 SEGREGATION OF BLOCK-COPOLYMERS AT THE INTERFACE BETWEEN HOMOPOLYMERS

The interfacial adhesion, structure, and properties of the interphase are very important characteristics of the immiscible polymer blends. To improve adhesion, in many cases, the so-called compatibilizers are introduced. The block- and graft copolymers, which have blocks of the same chemical nature as both homopolymers, are typically used as compatibilizers. These blocks are miscible with the corresponding homopolymers.[132-154] Introduction of block-copolymer diminishes the interfacial tension,[155-157] because it acts as surface active substance.[158-161] The choice of the block-copolymer as a compatibilizer is determined by the following factors:[132] its ability to diminish the interfacial energy and ability to stabilize the system and prevent phase separation. The increasing adhesion at the interface and improvement of the mechanical properties of blends is also one of the important factors. These effects can be reached if the block-copolymer is localized at the interface between two immiscible polymers.[132,162] Presently there are many experimental and theoretical investigations of the problem.[139,150,163-180] It was established that in the mixtures of PS-PVP demixing may be prevented by adding diblock-copolymer: poly(styrene-b-vinyl pyridine). The diblock-copolymer chains migrate to the interface and each block penetrates the corresponding phase of homopolymer. The factor determining the effectiveness of the compatibilization is the areal density of copolymer chains at the interface.[140,169] The penetration of blocks into phases makes the interphase region broader as compared with pure mixture.[150,151] The interfacial activity of copolymers is explained by their preferential segregation at the interface.[132,162]

Noolandi and Hong[165,166] constructed a general theory of nonhomogeneous systems. They found the fundamental functions of the distribution probability for the system consisting of two immiscible polymers A and B dissolved in a solvent in the presence of block-copolymer A-B. These equations have been solved numerically to obtain the density profiles of polymers near the interface. The difference between the total free energy and free energy of the system without an interface was used to estimate the interfacial tension. It was supposed that some part of copolymer that is not localized at the interface, should be randomly distributed between bulk of two homopolymers. In accordance with this supposition, the results of the theory should be valid only at low concentrations of block-copolymers before the micelle formation occurs.[163] Various contributions to de-

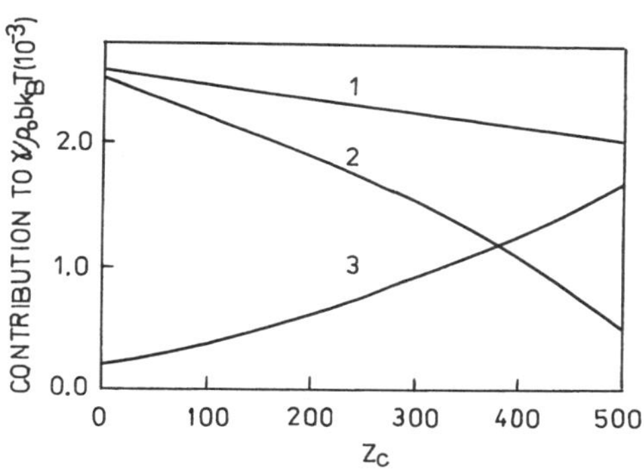

Figure 6.8. Various contributions to the reduced interfacial tension $\gamma/\rho_0 k_B T$: 1 - gradient, 2 - interaction, 3 - entropy. φ_P=0.2, φ_C=0.05, χ_{AB}=0.02, $\chi_{AS}=\chi_{BS}$=0, $Z_{CA}=Z_{CB}$; $Z_{HA}=Z_{HB}=\infty$. Adapted, by permission, from J. Noolandi and K. M. Hong, *Macromolecules,* **17**, 1531 (1984).

creasing interfacial tension have been discussed.[164,167] For the symmetric block-copolymer and homopolymers of infinite molecular mass with equal affinity to solvent, it was found that decrease in the interfacial tension, $\Delta\gamma_C$, is determined mainly by the energetically favorable orientation of block at the interface corresponding to the penetration of blocks into the phases of the same chemical nature. The main counterbalancing term in equation for $\Delta\gamma_C$ is the loss of entropy of copolymer localized at the interphase. The loss of conformational, or "turning back" entropy of both copolymer and homopolymer chains at the interphase gives small contribution to $\Delta\gamma$ (Fig. 6.8). Neglecting the loss of conformational entropy, the expression for the diminishing interfacial tension may be derived[164] in the form:

$$\Delta\gamma = d\varphi_c\left[\left(\frac{1}{2}\chi_{AB}\varphi_p + \frac{1}{Z_c}\right) - \frac{1}{Z_c}\exp(Z_c\chi_{AB}\varphi_p/2)\right] \qquad [6.182]$$

where d is the half-height of the profile $\varphi_C(x)$ in the interphase, φ_C is the volume fraction of copolymer, φ_p is the volume fraction of homopolymer, Z_C is the polymerization degree of copolymer. The total amount of copolymer in the interphase is:

$$\varphi_c(0) = \varphi_c \exp(Z_c\chi_{AC}\varphi_p/2) \qquad [6.183]$$

To derive Eqs 6.182 and 6.183, it was assumed that interactions between the solvent and block-copolymer segments are equal to zero, i.e., $\chi_{AS} = \chi_{BS} = 0$. The simplified theory does not directly include the expression for d and serves as tying parameter. For $Z_c < 1$, Eq 6.182 comes to

$$\Delta\gamma = -\frac{d}{8} Z_c \varphi_c \chi_{AB}^2 \varphi_p^2 \qquad [6.184]$$

From Eqs 6.182 and 6.183, the exponential dependence of $\varphi_C(0)$ and $\Delta\gamma$ on the copolymer molecular mass and on the volume fraction of homopolymer explains the mechanism of action of diblock-copolymers having high molecular mass as that of surfactants of concentrated solutions. Besides, it was shown that the thickness of the profile, d_H, increases exponentially with molecular mass of copolymer. In the same way, d_H increases with the volume fraction of copolymer.

Noolandi and Hong[164,167] found that the interfacial tension diminishes with Z_C and φ_C increasing to the critical micelle concentration (CMC) because the blocks of high molecular mass preferentially form micelles in the bulk of homopolymer rather than at the interphase. For concentrations below CCM, the interfacial tension linearly diminishes with copolymer concentration, whereas above CCM the dependence of γ on φ_C levels off. This theory can be applied to the systems that contain no solvent, i.e., when $\varphi_p \to 1$. In this case, Eq 6.182 comes to

$$\Delta\gamma = d\varphi_c \left[\frac{1}{2}\chi_{AB} + \frac{1}{Z_c}\exp(Z_c \chi_{AB}/2) \right] \qquad [6.185]$$

The theory by Noolandi and Hong was used[143] to describe behavior of the PS-PB/ poly(styrene-b-butadiene) system. It was shown (Fig. 6.9) that the dependence of $\Delta\gamma$ on φ_C is a straight line with the slope giving value of $d \approx 63.5b$, where b is the length of the Kuhn segment. If b=0.6 nm, then d = 38 nm. Value d (about 63 monomeric units) is composed of about 24% of the contour length of the copolymer chain with Z_C=261. Such extended configuration was found for diblock copolymers adsorbed at the chalk surface in solution in a good solvent for one block and poor for another.[176]

Vilgis and Noolandi[172] have discussed the behavior of the mixtures consisting of homopolymers A and B and block copolymer of the type XY in the presence of solvent. If block X is miscible with B but partially miscible with A (more miscible than B), there appears the thermodynamic driving force which leads to the orientation of the block-copolymer at the interphase which provides compatibilizing effect. However, such a system is only local at equilibrium. Authors[172] have numerically solved the diffusion equation for the density profiles of polymers. Despite of the blocks of different chemical nature used, they observed the diminishing the interfacial tension with increasing

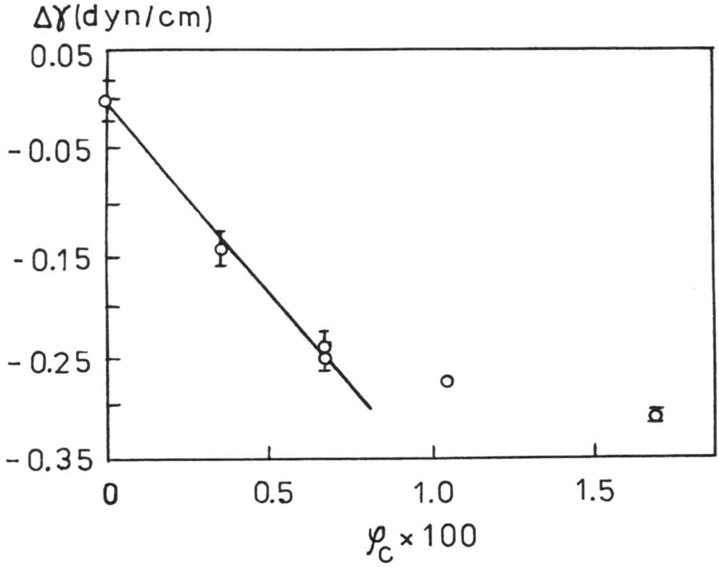

Figure 6.9. Interfacial tension for the system PS/PB/P(S-b-B) at 418K as a function of P(S-b-B)block concentration. Adapted, by permission, S. H. Anastasiadis, I. Gancarz, and J. T. Koberstein, *Macromolecules*, **22**, 1449 (1989).

molecular mass of the blocks. Strong effect of localization occurs also if the interaction of blocks with homopolymers increases.

The selective orientation of blocks in homopolymers was found[172] for system A-XY-B in the same way as for the A-AB-B. The following expression for diminishing the interfacial tension has been derived:

$$\Delta\gamma \approx -\frac{1}{Z_c}\exp[Z_c(\chi_1/2 + \chi_2)\varphi_p] \qquad [6.186]$$

where $\chi_1 = \chi_{AB} = \chi_{AX} = \chi_{BX} = \chi_{BY} = \chi_{XY;}\ \chi_2 = \chi_{AY} = \chi_{BX}$, φ_p is the volume fraction of homopolymer. For the density profile, the following equation applies:

$$\varphi_c(0) \approx \varphi_c\exp[Z_c(\chi_1/2 + \chi_2)\varphi_p] \qquad [6.187]$$

Authors calculated the density profiles for varying parameters χ_{ij} and taking as constant values Z_X/Z_Y and Z_A/Z_B or varying these two ratios

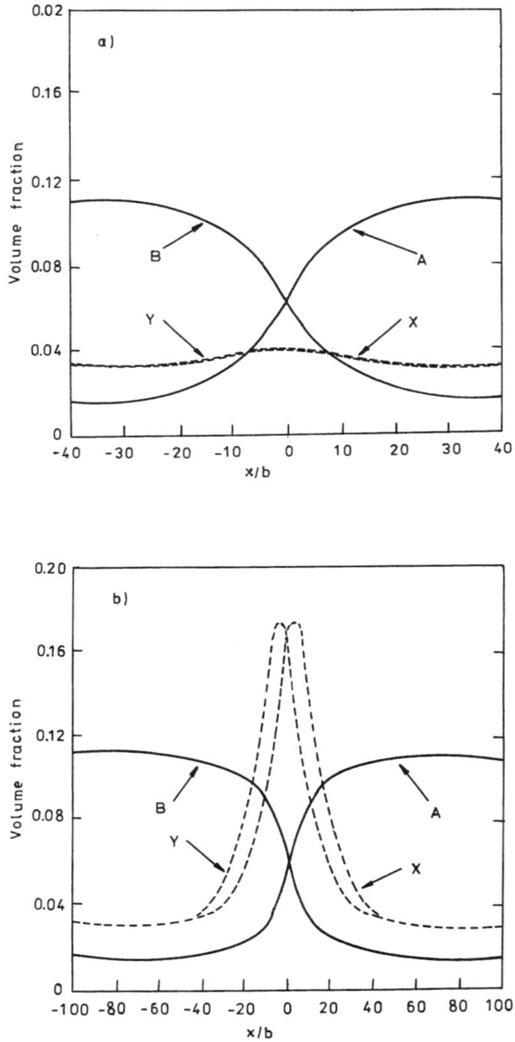

Figure 6.10. Polymer density profiles for block copolymer XY localized at the interface between homopolymers A and B. For (a): Z_A/Z_B=400/400; Z_{CX}/Z_{CY}=100/100; $X_{AB}=X_{XY}$, $X_{AX}=X_{BY}$=0.05; $X_{AY}=X_{BX}$=0.1. For (b): Z_A/Z_B=400/400; Z_{CX}/Z_{CY}=500/500; $X_{AB}=X_{XY}=X_{AX}=X_{BY}$=0.05; $X_{AY}=X_{BX}$=0.1. Adapted, by permission, T. A. Vilgis and J. Noolandi, *Macromolecules*, **23**, 2941 (1990).

at the constant χ_{ij}. It was shown that, with increasing Z_A/Z_B, the volume fraction of copolymer at the interphase increases (Fig. 6.10). The same is observed with increasing $\chi_{AY} = \chi_{BX}$. For all cases, it was shown that interfacial activity of block copolymers is promoted if the interaction of the block X with homopolymer B is more repulsive than interaction of block X with homopolymer A. The mixtures A-AB-B, A-XB-B, and other may be derived from consideration of a general case of mixture A-XY-B.

The theory by Noolandi and Hong[164,167] predicts the linear dependence of decrease of interfacial tension on φ_C. On this basis, Thomas and Prud'homme[181] derived the relation describing the diminishing the particle size in the mixture by addition of block-copolymer supposing that diminishing interfacial tension is proportional to diminishing particle size.[178] The relation has the form:

$$\Delta D = Kd\varphi_c\left[\frac{1}{2}\chi + \frac{1}{Z_c} - \frac{1}{Z_c}\exp(Z_c\chi/2)\right] \qquad [6.188]$$

where ΔD is the diminishing particle size by adding the compatibilizer, K is a proportionality coefficient.

For the system PS/PMMA/poly(styrene-b-methylmethacrylate), Thomas and Prud'homme have shown that the domain dimensions in the mixture diminish linearly with the addition of block-copolymer. There has been also derived the equation connecting the volume fraction and dimensions of domains with molecular mass and amount of copolymer which is necessary to saturate the interface.[132,182] On this basis, the surface at the interface occupied by the block-copolymer molecule was calculated. The mixture with volume fraction φ_A of homopolymer A is considered as spherical domain of radius R immersed in matrix B. The total interfacial surface per mixture volume unit is $3\varphi_A/R$. If each copolymer molecule occupies the area at interface, a, then the mass, m, of copolymer per volume unit that is necessary to saturate the interface is found from

$$a = 3\varphi_A M/mRN_A \qquad [6.189]$$

where M is the molecular mass of copolymer and N_A is the Avogadro's number. Considering the copolymer as compatibilizer at the concentration of micelle formation, Wu[183] proposed the equation connecting the interfacial tension and interfacial thickness:

$$\gamma = 7.6/L^{0.86} \qquad [6.190]$$

The Noolandi-Hong theory[164,167] shows that in the mixtures of immiscible polymers, the block-copolymer is localized at the interface and excludes homopolymers from the interphase region. Leibler[168] has made an attempt to determine if this effect occurs in the systems where both homopolymers are partially miscible. He has developed the mean field formalism to calculate the interfacial concentration profile and interfacial tension for the miscible mixtures of two homopolymers A and B and copolymer A–B. The free energy was expressed in terms of monomer concentration correlation functions which were calculated in self-consistent way within the framework of random phase approximation, RPA, introduced by de Gennes.[184] It was shown that the interfacial tension can be presented as a difference of two contributions, $\gamma = \gamma_0 - \gamma_1$. The first term can be interpreted as the interfacial energy that is determined by the change of the total concentration of monomers A through the interface. Near the critical point, this profile is expressed by the equation:

$$\varphi_A(x) = \frac{1}{2}[1 + \sigma \tan h\,(x/D)] \qquad [6.191]$$

where

$$\sigma = [3N(1-\varphi)^3/\,|(1-\varphi) - \chi N/2\,|]^{1/2} \qquad [6.192]$$

Parameter σ characterizes the difference between the total volume fraction of monomers A in A-enriched and B-enriched phases. In Eq 6.191 and 6.192, φ is the average content of copolymer in the two-phase system and D is the thickness of the interphase. The main conclusion following from Eq. 6.191 is that copolymer is present in both phases. Therefore, when the volume fraction of copolymer increases, the difference between total volume fraction of monomers A in A-enriched and B-enriched phases (σ) diminishes. As a result, the contribution of γ_0 in the interfacial energy monotonously diminishes with φ and disappears at the critical point at $\varphi = \varphi_3(C) = 1 - 2(\chi N)$. For small amounts of copolymer and for nearly miscible species ($\chi N \approx 2$), the diminishing γ_0 may be expressed in the following way:

$$\gamma_0(\varphi) = \varphi_0(0)[1 - \varphi(1 + \chi^2N^2/8 + 3\chi N/2(\chi N - 2))] \qquad [6.193]$$

$$\gamma_0(0) = \frac{2^{1/2}k_BTN^{-1/2}(\chi N - 2)^{3/2}}{6a^2} \qquad [6.194]$$

The second source of the activity of copolymers is their preferential localization at the interface. The concentration profile of copolymer:

$$\varphi_3(x) = \varphi - \overline{\varepsilon}(x) \qquad [6.195]$$

has a maximum at the interphase at x = 0. In Eq 6.195, value $\bar{\varepsilon}(x)=\varphi-\varphi(x)$ characterizes the change of copolymer/monomer volume fraction throughout the interface. For nearly miscible mixtures, the peak broadens and spreads with D, which diverges when φ approaches the critical value $\varphi_3(C)$. The term γ_1 is determined by the preferential localization of copolymer at the interface. In the critical region, it is approximately equal to $\gamma_1 = 1.5\varphi\gamma_0$. Consequently, the main contribution to the interfacial tension, γ, belongs to the term γ_0. In the case of nearly miscible polymers, the surfactant effect, due to localization of copolymer chains at the interface, is very limited. For the highest degree of immiscibility in the critical region, the value φ is rather high and the term γ_1 may be comparable with γ_0.

Generally, for nearly miscible mixtures there may be distinguished two mechanism of the interfacial activity of copolymer chains:

- homopolymers A and B are more miscible when copolymer chains are present in both phases
- copolymer has some tendency to localize at the interphase.

Close to the consolute point for nearly miscible mixtures the first mechanism is prevailing whereas for highly immiscible mixtures the second dominates. The interphase thickness, D, is of the order of the correlation length, ξ. For diblock copolymer the correlation length is expressed as:

$$\xi^2 = 2R^2\left[1 - \frac{\chi^2N^2\varphi(1-\varphi)}{8}\right]/[3\,|\,\chi N(1-\varphi) - 2\,|\,] \qquad [6.196]$$

and for periodic multiblock chains as:

$$\xi^2 = 2R^2\left[1 - \frac{\chi^2N^2\varphi(1-\varphi)}{8k^2}\right]/[3\,|\,\chi N(1-\varphi) - 2\,|\,] \qquad [6.197]$$

Value k designates the number of A or B blocks in copolymer chains of the A-B-A-B type and R is the gyration radius. When φ approaches the critical concentration $\varphi_3(C)$, the correlation length, ξ, diverges and the interphase becomes thicker by approaching φ at its critical value. It was mentioned by Noolandi and Hong[164] that in the Leibler theory the diblock copolymer is considered as a low molecular solvent and the structure of copolymer is not relevant. The system behaves similar to the mixture of two homopolymers moving to the consolute point by adding the solvent excess.

The Noolandi and Hong theory[164,165] relates to polymers diluted by solvent. Shull and Kramer[169] developed the mean field theory for polymer interphases in the presence of block-copolymer without solvent. They derived the expressions for the profiles of different polymer components and for the interfacial tension. These characteristics are expressed using polymerization degrees of homopolymers and block-copolymer. Flory-Huggins interaction parameters, the statistical segment length, and chemical potential of copolymer.

For three-component system consisting of homopolymer A, homopolymer B, and diblock copolymer A–B with the volume fraction of copolymer $\varphi_c = \varphi_{CA} + \varphi_{CB}$ (φ_{CA} and φ_{CB} correspond to blocks A and B) and volume fractions of homopolymer φ_{HA} and φ_{HB}, the density of the free energy, f, of the uniform system is written terms of segment-segment interactions:

$$\frac{f}{k_B T \rho_0} = \chi_{AB}\varphi_A\varphi_B + \sum_k \frac{\varphi_k \ln \varphi_k}{N_k} \quad [6.198]$$

where $\varphi_A = \varphi_{CA} + \varphi_{HA}$, $\varphi_B = \varphi_{CB} + \varphi_{HB}$ are the total volume fractions of segments A and B in the system, ρ_0 is segment density, N_k is polymerization degree of component k. The chemical potential of component molecule μ_k is expressed as

$$\frac{\mu_k}{N_k} = \frac{\delta(f/\rho_0)}{\delta\varphi_k} + \frac{f}{\rho_0} - \sum_k \varphi_k \frac{\delta(f/\rho_0)}{\delta\varphi_k} \quad [6.199]$$

If to suppose that χ_{AB} does not depend on concentration, the chemical potentials of components can be presented in the following way:

$$\frac{\mu_{HA}}{N_{HA}k_B T} = \frac{\ln \varphi_{HA} + 1}{N_{HA}} - K_\varphi + \chi_{AB}\varphi_B^2 \quad [6.200]$$

$$\frac{\mu_{HB}}{N_{HB}k_B T} = \frac{\ln \varphi_{HB} + 1}{N_{HB}} - K_\varphi + \chi_{AB}\varphi_A^2 \quad [6.201]$$

$$\frac{\mu_c}{N_c k_B T} = \frac{\ln \varphi_c + 1}{N_c} - K_\varphi + \frac{\chi_{AB}}{N_c}(N_{CA}\varphi_B^2 + N_{CB}\varphi_A^2) \quad [6.202]$$

where

$$K_\varphi = \frac{\varphi_{HA}}{N_{HA}} + \frac{\varphi_{HB}}{N_{AB}} + \frac{\varphi_c}{N_c} \quad [6.203]$$

When the interphase region is formed, its full free energy includes the free energy of copolymer chains and interfacial free energy:[169,170]

$$\frac{F}{A_{int}} = \gamma_{CA-CB} + \gamma_{CA-HA} + \gamma_{CB-HB} +$$

$$+\frac{3Qk_BT}{2A_{int}}\left[\frac{<R_A^2>}{N_{CA}a^2}+\frac{<R_B^2>}{N_{CB}a^2}\right] \quad [6.204]$$

where $\gamma_{CA\text{-}CB}$, $\gamma_{CA\text{-}HA}$, and $\gamma_{CB\text{-}HB}$ are the interfacial tensions between various blocks, A_{IS} is the interfacial surface, Q is the number of block copolymer chains in the interphase, $<R_A^2>$ and $<R_B^2>$ are the mean-square distances between the ends of blocks A and B. If to accept that $z^*_i = <R_A^2>^{1/2} + <R_B^2>^{1/2}$ the following relation is obtained:

$$<R_A^2> = \left(\frac{QN_{CA}}{\rho_0 A_{int}}\right)^2; \quad <R_B^2> = \left(\frac{QN_{CB}}{\rho_0 A_{int}}\right)^2 \quad [6.205]$$

where N_{CA}, N_{CB} are the polymerization degrees of blocks A and B in copolymer, ρ_0 is the density, z^*_i is the area of the concentration profile of copolymer in the interphase at φ_C:

$$z^*_i = \int_{-\infty}^{+\infty}(\varphi_c(x) - \varphi_c^{block})dx \quad [6.206]$$

The chemical potential of copolymer is $\mu_C = (\delta F/\delta Q)A_{int}$ and the interfacial tension $\gamma = (\delta F/\delta A_{int})_Q$. From these relation follows that

$$\frac{\mu_c}{k_BT} = \frac{9v_i^2N_c}{2\rho_0^2a^2} \quad [6.207]$$

$$\gamma = \gamma_{CA-CB} + \gamma_{CA-HA} + \gamma_{CB-HB} - \frac{3v_i^3N_ck_BT}{\rho_0^2a^2} \quad [6.208]$$

where $v_i = Q/A_{int}$ or $v_i = z^*_i\rho_0N_C$ is the interfacial excess per surface unit. This value can be expressed using the copolymer chemical potential:

$$v_i = \frac{\rho_0 a}{3}\left(\frac{2\mu_c}{k_BTN_c}\right)^2 \quad [6.209]$$

In this case

$$\gamma = \gamma_{CA-CB} + \gamma_{CA-HA} + \gamma_{CB-HB} - kT\frac{\rho_0 a}{9}N_c^{-1/2}\left(\frac{2\mu_c}{k_BT}\right)^{3/2} \quad [6.210]$$

The approximated expression for $\gamma_{CA\text{-}CB}$ has the form:

$$\gamma_{CA-CB} = \gamma_0(N_{CA},N_{CB}) + v_ik_BT\ln(z^*_i/d) \quad [6.211]$$

$$\gamma_{CA-HA} + \gamma_{CA-HB} = \rho_0 k_B T \int_{-\infty}^{+\infty} \frac{a^2}{36\varphi_c(x)(1-\varphi_c(x))} \left(\frac{\delta\varphi_c(x)}{\delta x} \right)^2 \qquad [6.212]$$

It is supposed that

$$\gamma_{CA-CB} + \gamma_{CA-HA} + \gamma_{CB-HB} = \gamma_0^{\infty} \qquad [6.213]$$

where γ_0^{∞} for the homopolymer mixtures with infinite molecular masses in the absence of copolymer is written as

$$\gamma_0^{\infty} = a\rho_0 k_B T(\chi_{AB}/6)^{1/2} \qquad [6.214]$$

Authors[169] solved numerically the mean field equation describing the interphase between two homopolymers in the presence of block copolymer. The results of calculations give the quantitative predictions for the interfacial profiles and interfacial tension. The most general conclusions are as follows:

- Increasing chemical potential of copolymer is accompanied by the growth of the block copolymer chain density at the interphase and by diminishing the interfacial tension
- Maximum volume fraction of copolymer at the interphase may stay close to unity even for the intermediate values of chemical potential of copolymer that corresponds to a high interfacial tension
- The confinement of copolymer joints to the central portion of the interfacial region contributes significantly to the interfacial tension and gives rise to a broadening of the A-B segmental profile. The specific width of this profile is determined largely by the strength of the unfavorable interactions between A and B segments
- For high values of the copolymer chemical potential, the total copolymer segregation and the overall copolymer concentration profile are determined primarily by the free energy associated with the extension of the copolymer blocks away from interface. The overall copolymer concentration profile therefore does not depend strongly on the details of the thermodynamic interaction between A and B segments
- The free energy associated with the interface between a segregated polymer layer and a high molecular weight homopolymer matrix can give rise to an attractive interaction between segregated layers.

Interesting experimental results on the equilibrium segregation have been reported for the system PS-poly(2-vinyl pyridine)-poly(d-styrene-b 2-vinyl pyridine).[170] The specimens have been studied by

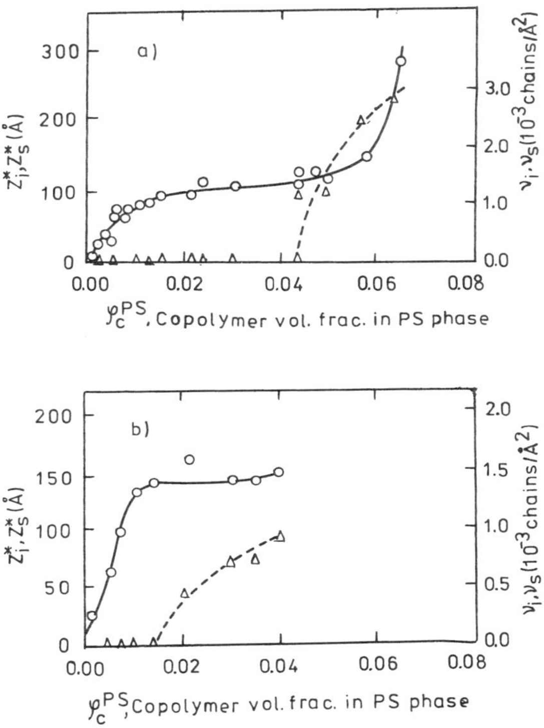

Figure 6.11. Interface (____) and surface (....) copolymer excess as a function of the equilibrium copolymer volume fractions in the PS phase: a) 391-68 copolymer, b) 395-125 copolymer. The solid and dashed lines are aids to the eye, whereas the symbols represent the actual measured values. Adapted, by permission, from K. R. Shull, E. J. Kramer, G. Hadziioannou, and W. Tang, *Macromolecules*, **23**, 4780 (1990).

the method of forward recoil spectroscopy and prepared in the following way. The 200 Å thick layer of PVP was spun on a polished silicone substrate. The interface excess samples were prepared by spin casting a 600 Å film containing approximately 30% block copolymer onto the PVP layer. The 4000 Å layer of the PS homopolymer was then spun onto a glass slide floated off onto a water bath, and picked up on top of the previous two layers. All samples were annealed in vacuum at 451K. The dependence of the integrated segregation on the equilibrium concentration of copolymer in the PS-phase was compared with the predictions of the mean field theory where the chemical potential of copolymer is the relevant parameter. Authors have measured the surface excess of copolymer in the interphase for the mixtures of copolymers with the ratio of the polymerization degrees N_{PS}:N_{PVP}=391:68, 355:125, and 1,300:173. The polymerization degree of homopolymers was about 6,000. It was shown that after 8 h annealing, the volume fraction of copolymer in PS in equilibrium with the segregated layer was about 2.1%. When the equilibrium volume fraction of copolymer reaches 5.8%, copolymer segregates on the free PS surface giving the surface excess of copolymer, z^{*S}, and corresponding v_S (Fig. 6.11a).

The dependence of the z^*_i and z^{*S}_i (and, correspondingly, v^*_i and v^{*S}_i) on the volume concentration of copolymer in PS phase are shown in Fig. 6.11b. It is seen that there is some critical value of φ_C^{PS} below which there is no surface excess of copolymer. The existence of such a transition implies that segregation of individual copolymer chains on the surface does not account for the observed surface segregation. The onset of the surface segregation at the critical value φ_C^{PS} should be connected with the formation of the ordered copolymer aggregate that has affinity to the free surface. The appearance of copolymer aggregate was explained[170] by the surface segregation of the copolymer micelle. There are two potential driving forces for the surface segregation. The first is connected with the tendency of the system to diminish the number of sharp interphases present. The number of available conformations in the presence of such an interphase sharply diminishes and as a result the free energy increases that has purely entropy origin. The interphase between the micelle corona and surrounding high molecular weight homopolymer is similar by the width to the gyration radius. Therefore, some interfacial tension should exist connected with this interphase. The full free energy of the system may be diminished by the micelle present at the free PS surface because this diminishes the interfacial surface the corona-homopolymer.

The second driving force for the surface segregation of block copolymer is determined by the enthalpy interaction between deuterated PS-non-deuterated PS (so-called isotopic effect) that also may lead to the surface segregation.[185] Generally, there should be two interphases - the interfacial block-copolymer layer and surface segregated layer. The segregation of the free chains to the PS-PVP interface is determined only by μ_C which rapidly increases with copolymer concentration till the CMC.

The formation of micelle is therefore associated with a plateau in the interface segregation of the free chains and by the onset of surface segregation of the micelles themselves. Fig. 6.10 shows that the copolymer 391:68 gives sharp increase of the interfacial segregation at the concentration above φ_{CMC}^{PS} that is explained by the preferential segregation of micelles at the interphase. The results have been compared with the conclusion of the mean field theory.[169] For this purpose the density of the free energy was expressed in terms of interaction parameters (Eq 6.198), whereas the chemical potential was calculated from Eq 6.200. When the bulk phase of PS consists of almost pure homopolymer, i.e., $\varphi_C^{PS} << 1$, $\varphi_{PVP}^{PS} << 1$, and $\varphi_{PS}^{PS} << 1$, Eq 6.200 can be rewritten to the following form:

$$\varphi_c^{PS} = \exp\left[\frac{\mu_c}{k_B T} - 1 + \frac{N_c}{N_{H-PS}} - \chi_{PS-PVP} N_{C-PVP}\right] \qquad [6.215]$$

To derive the latter equation, the supposition has been made that the mixing is uniform only when micelles are absent. When micelles are present, Eq 6.215 expresses the volume fraction of the free block copolymer chains in PS-phase. The calculation of the profiles of the copolymer in the interphase and the change of the interfacial tension correlate with the experimental data.

It was shown[149] that with diminishing the polymerization degree of PS, value φ_{CMC} increases. To discuss the data on the effect of the molecular mass of PS on the surface segregation, two regimes have been considered:[149] the regime of the dry brush and regime of the wet brush. These regimes are dependent on the depth of penetration of homopolymer chains in the copolymer brush (in the case under consideration - in the micelle corona). When homopolymer chains are much longer than corona chains, they are excluded from corona which is "dry", whereas when the homopolymer chains are much shorter, they penetrate corona that becomes "wet".

The simple theory of the interfacial segregation was proposed by Leibler.[186] By analogy with the previous considerations, two regimes are discussed - dry and wet brush. The regime of dry brush, the homopolymer chains of PS do not penetrate the similar chains of block-copolymer, if $P_{PS}^{1/2} > N_{PS}$ and $v_i a^2 P_{PS}^{1/2} > 1$. In this case:

$$z*_i = \frac{a}{3}\left[2N_c \frac{\mu_c}{k_B T}\right]^{1/2} \qquad [6.216]$$

$$\gamma = \gamma_0 - \frac{\rho_0 a}{9}\left[\frac{8}{N_c}\left(\frac{\mu_c}{k_B T}\right)^3\right]^{1/2} k_B T \qquad [6.217]$$

where a is the length of the statistical segment and $\rho_0 \approx 1/a^3$. For the case of wet brush, when homopolymer chains do penetrate the copolymer brush, $P_j^{1/2} < 1/v_i\, a^2 < N_i^{3/2}/P_j$, we have

$$\frac{\mu_c}{k_B T} = 2 + \ln\left[N_{PS}N_{PVP}\left(\frac{\nu_i}{\rho_a a}\right)^2\right] + \left(\frac{375}{32}\right)^{1/3}\left[\frac{N_{PS}}{P_{PS}^{2/3}} + \frac{N_{PVP}}{P_{PVP}^{2/3}}\right] \times$$

$$\times\left[\frac{\nu_i}{\rho_0 a}\right]^{2/3} \qquad [6.218]$$

Leibler[186] neglected logarithmic term in Eq 6.218 to obtain the following analytical formula for ν_i in relationship to ν_C:

$$\nu_i = \left(\frac{32}{375}\right)^{1/2}\left[\left(\frac{N_{PS}}{P_{PS}^{2/3}} + \frac{N_{PVP}}{P_{PVP}^{2/3}}\right)\cdot\frac{\mu_c}{k_B T}\right]^{3/2}\rho_0 a \qquad [6.219]$$

In accordance with solution of Eq 6.219, one can estimate the interfacial tension as

$$\gamma = \gamma_0 - \left[2 + \frac{3^{1/3}}{2^{2/3}}\left(\frac{N_{PS}}{P_{PS}^{2/3}} + \frac{N_{PVPV}}{P_{PVP}^{2/3}}\right)\left(\frac{\nu_i}{\rho_0 a}\right)^{2/3}\right]\nu_i k_B T \qquad [6.220]$$

The models can be used to describe the cases where any (wet or dry) brush is arranged at any side of the interphase. The asymmetric case[149] is considered when from one side of the interphase, there is a wet brush ($P_A < N_A$) and from another side dry brush ($P_A > N_A$). This case was called by authors as wet-dry brush case.

On the basis of the Leibler brush model, the expression for μ_C was derived as a function of ν_i:

$$\frac{\mu_c}{k_B T} = 1 + \ln\left[N_{wet}\frac{\nu_i}{\rho_0 a}\right] + \left(\frac{375}{32}\right)^{1/3}\frac{N_{wet}}{P_{wet}^{2/3}}\left[\frac{\nu_i}{\rho_0 a}\right]^{2/3} +$$

$$+\frac{9}{2}N_{dry}\left[\frac{\nu_i}{\rho_0 a}\right]^2 \qquad [6.221]$$

This equation can be solved to obtain the interfacial excess as a function of φ_C^{PS}. The relation between the interfacial tension, γ, and φ_C^{PS} may be found from the equation:

$$\gamma = \gamma_0 - \left[\left(\frac{\nu_i}{\rho_0 a}\right)^2 + 1 + \frac{3^{1/3}}{2^{2/3}}\frac{N_{wet}}{P_{wet}^{2/3}}\left(\frac{\nu_i}{\rho_0 a}\right)^{2/3}\right]\nu_i k_B T \qquad [6.222]$$

For the mixtures of copolymer 391-68 with PS of the polymerization degree 6,440, 2,000, 1,220, and 85 and P2VP with $N_{PVP} = 5{,}100$, it was shown that for the first three polymerization degrees, the interfacial areal chain densities are considered to be the same. According to the Leibler theory, these three mixtures are explained by the model of the dry brush. The mixtures with N_{PS} = 85 are characterized by the essential penetration of homopolymer PS chains into PS-brush (wet-dry brush case) due to a higher combinatorial entropy as compared with the dry brush. This entropy contribution compensates unfavorable enthalpy interactions between PVP and PS homopolymers. Due to this, these mixtures have lower interfacial excess at saturation. The comparison of the results of Leibler theory and the mean field theory[170] shows that the Leibler theory does not reproduce results for the dry and wet brush in a proper way: for the dry brush, it is necessary to use lower value of parameter χ_{AB} than the experimental value to reproduce the experimental data on z^*_i and ν_i. The data on γ as a function of volume fraction of copolymer in PS-phase are also badly reproduced. The reason consists the oversimplification of the assumption of the overall free energy. At the same time, the mean field theory[170] reproduces the experimental results quite well at one value of parameter χ_{AB} that is close to the experimentally found data.

The conclusions made by the authors[149] are the following. The interfacial excess is divided into three regimes in accordance with φ_C^{PS}: initial stage, plateau, and final stage. In the regime of the initial stage, the interfacial excess of copolymer is increased with φ_C^P. The saturation of the copolymer packing in the interphase corresponds to the plateau regime. The final stage is the regime of an increase in the interfacial excess due to the micelle segregation to the brush. The interfacial excess of the block copolymer in the case of the dry brush is not affected appreciably by the P_{PS} in the first two regimes, resulting in nearly the same areal chain density at saturation.

The penetration of lower molecular weight homopolymer chains into the brush at the interface (wet-dry case) results in a lower interfacial excess as well as a lower areal chain densities at saturation. The critical micelle concentration, φ_{CMC}, is obtained from the onset of the surface segregation and it increases as the P_{PS} decreases. While Leibler's brush theory[186] can reproduce the shape of the segregation isotherms, if the Flory interaction parameter, χ, is allowed to vary with P_{PS}, a SCMF theory[170] can predict the observed isotherms using a single value of χ (≈0.105) which is very close to the estimated value of 0.1.

The simple model was also proposed[149] for the micelle transition and corresponding expressions for μ_{CMC} were obtained for micelles of various shape. For the spherical micelles value μ_{CMC} is:[186]

$$\frac{\mu_{CMC}^{(spheres)}}{k_B T} = 1.72(\chi N_c)^{1/3} g^{4/9}(1.74g^{-1/3} - 1)^{1/3} \qquad [6.223]$$

where g is equal to N_{core}/N_C.

For cylindrical micelles:

$$\frac{\mu_{CMC}^{(cylinder)}}{k_BT} = 1.19(\chi N_C)^{1/3} g^{1/3} (1.64 - \ln g)^{1/3} \qquad [6.224]$$

and for lamellar micelles:

$$\frac{\mu_{CMC}^{(lamellae)}}{k_BT} = 0.67(\chi N_c)^{1/3} (5.64 - g)^{1/3} \qquad [6.225]$$

The value of the interaction parameter can be found from the heat of mixing of a low molecular weight homologues from

$$\Delta H_m = RT\chi\varphi_2(1 - \varphi_2) \qquad [6.226]$$

For the mixtures of PS and PVP such homologues are ethyl benzene and 2-vinyl pyridine, correspondingly.

The decrease in interfacial tension by introduction of block copolymer in the mixture of two homopolymers leads to the diminishing dimensions of the phase-separated domains. In this context the word "compatibilization" does not imply that immiscible mixture becomes miscible at thermodynamical equilibrium of a system.[187] As shown[188] the addition of the random copolymer of poly(styrene-*co*-butadiene) to the mixtures of low molecular weight PS and PB diminishes the cloud points. However, the addition of block-copolymers does not lead to the same effect: the immiscible mixtures become more transparent by adding block-copolymer only due to the diminishing the domain dimensions. However, the domain dimensions depend not only on the interfacial tension but on the time necessary for the phase separation,[187] the mechanism of coarsening of domains, and other reasons. Therefore, it is very important to establish if the domain dimensions, found for such systems in the different periods of time, after the beginning the phase separation, are in thermodynamical equilibrium or kinetically frozen. This implies the necessity to investigate the kinetics of the phase separation in the presence of block-copolymers. Some results of such investigations have been discussed in Ch. 5. Here, we would like to emphasize once more that the problem of the non-equilibrium segregation is as important question for block-copolymer-containing systems as for the mixtures of linear polymers.

REFERENCES

1. Y. S. Lipatov in *Interfacial Phenomena in Polymers*, Naukova Dumka, Kiev, 1980.
2. Y. S. Lipatov in *Polymer Reinforcement*, ChemTec Publishing., Toronto, 1995

3. Y. S. Lipatov in *Colloid Chemistry of Polymers*, Elsevier, 1988.
4. L. Sharpe, *J. Adhesion*, **29**, 1 (1989).
5. L. Sharpe in *The Interfacial Interactions in Polymeric Composites*, Ed. G. Acovali, Kluwer Academic Publishing., Dordrecht, 119, p.1.
6. A. Vrij, *J. Polymer Sci.*, **A2(6)**, 1919 (1968).
7. A. Vrij and G. J. Roebersen, *J. Polym. Sci., Phys. Ed.*, **15**, 109 (1977).
8. W. Cahn and J. E. Hilliard, *J. Chem. Phys.*, **28**, 258 (1958).
9. E. Helfand and Y. Tagami, *J. Chem. Phys.*, **56**, 3592 (1972).
10. E. Helfand and A. M. Sapse, *J. Polym. Sci.*, **C54**, 289 (1976).
11. R.-J. Roe, *J. Chem. Phys.*, **62**, 490 (1975).
12. T. Nose, *Polymer J.*, **8**, 96 (1976).
13. I. C. Sanchez and R. H. Lacombe, *J. Macromol. Sci.*, **B17**, 565 (1980).
14. C. I. Poser and I. C. Sanchez, *Macromolecules*, **14**, 361 (1981).
15. V. Brongiorno, J. E. Scriven, and H. T. Davis, *J. Colloid and Interface Sci.*, **57**, 462 (1976).
16. I. C. Sanchez and R. H. Lacombe, *J. Phys. Chem.*, **80**, 2352 (1976).
17. I. C. Sanchez and R. H. Lacombe, *Macromolecules*, **11**, 1145 (1978).
18. B. Widom, *J. Chem. Phys.*, **43**, 3892 (1965).
19. S. Fisk and B. Widom, *J. Chem. Phys.*, **50**, 3219 (1969).
20. O. K. Rice, *J. Phys. Chem.*, **83**, 1859 (1979).
21. R. Koningsveld, L. A. Kleintjens, and A. R. Schultz, *J. Polymer Sci.*, **A(2)8**, 261 (1970).
22. Th. G. Scholte, *J. Polymer Sci.*, **A(2)9**, 1553 (1971).
23. P. Debye, *J. Chem. Phys.*, **31**, 680 (1959).
24. A. Vrij and M. W. van den Esker, *J. Chem. Soc. Faraday Trans.*, **(2)68**, 523 (1992).
25. H-.W. Kammer, *Wiss. Z. Techn. Univ. Dresden*, **24**, 35 (1975).
26. W. Berger, K. Olbricht, and H.-W. Kammer, *Faserforsch. u. Textiltechn.*, **27**, 9 (1976).
27. H.-W. Kammer, O. Samek, K. Wietusch, and W. Berger, *Faserforsch. u. Textiltech.*, **27**, 15, 1976.
28. H.-W. Kammer, *Faserforsch. u. Textiltechn.*, **29**, 459 (1978).
29. H.-W. Kammer, *Faserforsch. u. Textiltechn.*, **28**, 27 (1977).
30. E. Helfand and Y. Tagami, *J. Polymer Sci., Polym. Lett. Ed.*, **9**, 741 (1971).
31. E. Helfand, *Macromolecules*, **9**, 879 (1976).
32. T. Weber and E. Helfand, *Macromolecules*, **9**, 311 (1976).
33. E. Helfand, *J. Chem. Phys.*, **63**, 2192 (1975).
34. E. Helfand and A. M. Sapse, *J. Chem. Phys.*, **62**, 1327 (1975).
35. K. Binder and H. L. Frisch, *Macromolecules*, **17**, 2928 (1984).
36. J. E. Joanny and L. Leibler, *J. phys. (Paris)*, **39**, 311 (1978).
37. P. G. de Gennes, *J. Chem. Phys.*, **72**, 4756 (1980).
38. L. Binder, *J. Chem. Phys.*, **79**, 6387 (1983).
39. K. Shinozaki, T. V. Tan, Y. Saito, and T. Nose, *Polymer*, **23**, 723 (1982).
40. K. Shinozaki, M. Abe, and T. Nose, *Polymer*, **23**, 722 (1982).
41. R. P. Wool, B. L. Yuan, and O. J. McGarel, *Polym. Engn. Sci.*, **29**, 1340 (1989).
42. H. Zhang and R. P. Wool, *Macromolecules*, **22**, 3018 (1989).
43. R. P. Wool and J. Long, *Macromolecules*, **26**, 5227 (1993).
44. K. L. Foster and R. P. Wool, *Macromolecules*, **24**, 1397 (1991).
45. J. L. Willett and R. P. Wool, *Macromolecules*, **26**, 5336 (1993).
46. B.-L. Yuan and R. L. Wool, *Polym .Engn. Sci.*, **30**, 1454 (1990).
47. B. Sapoval, M. Rosso, and J. Gouyet, *J. Phys. Lett.*, **46**, L149 (1985).
48. D. Broseta, G. H. Fredrickson, E. Helfand, and L. Leibler, *Macromolecules*, **23**, 132 (1990).
49. W. Cahn in *Segregation to Interfaces*, Cleveland, 1978, p. 3.
50. I. Sanchez, *Polym. Engn. Sci.*, **24**, 79 (1984).

51. K. M. Hong and J. Noolandi, *Macromolecules*, **13**, 964 (1980).
52. E. Helfand and Y. Tagami, *J. Chem. Phys.*, **57**, 1812 (1957).
53. C. I. Poser and I. C. Sanchez, *J. Colloid Interface Sci.*, **69**, 539 (1979).
54. B. S. Carly, *PhD Thesis*, University of Minnesota, 1979.
55. J. F. Joanny and L. Leibler, *Le Journ. de Phys.*, **39**, 951 (1978).
56. E. Helfand, *Acc. Chem. Res.*, **8**, 295 (1975).
57. W. Ruland, *J. Appl. Crystallogr.*, **4**, 70 (1971).
58. J. H. Wandorff, *J. Polym. Sci., Polym. Lett.*, **18**, 439 (1986).
59. J. Koberstein, B. Morra, and R. S. Stein, *J. Appl. Crystallogr.*, **13**, 34 (1980).
60. R. J. Roe, M. Fishkis, and J. E. Chang, *Macromolecules*, **14**, 1091 (1981).
61. P. G. de Gennes in *Scaling Concept in Polymer Physics*, Cornell Univ. Press, New York, 1979.
62. J. C. Le Guillou and J. Zinn-Justin, *Phys. Rev. Lett.*, **39**, 95 (1977).
63. T. Nose and T. V. Tan, *J. Polym. Sci., Polym. Lett. Ed.*, **14**, 705 (1976).
64. S. Wu in *Polymer Blends*, Eds. D. R. Paul and S. Newman, Academic Press, New York, 1978.
65. S. Wu, *J. Macromol. Sci. Rev. Macromol. Chem.*, **C10**, 1 (1974).
66. S. Wu in *Polymer Interfaces and Adhesion*, Marcel Dekker, New York, 1982.
67. Y. S. Lipatov and A. E. Feinerman, *Adv. Colloid and Interface Sci.*, **11**, 195 (1979).
68. D. G. Le Grand and G. L. Gaines, Jr., *J. Coll. Interface Sci.*, **50**, 272 (1975).
69. G. L. Gaines, Jr., *J. Coll. Interface Sci.*, **63**, 394 (1971).
70. L. A. Girifalko and R. J. Good, *J. Phys. Chem.*, **61**, 904 (1957).
71. S. Wu, *J. Polymer Sci.*, **C34**, 19 (1971).
72. I. C. Sanchez, *Ann. Rev. Mater. Sci.*, **13**, 387 (1983).
73. S. H. Anastasiadis, I. Gancarz, J. T. Koberstein, *Macromolecules*, **21**, 2980 (1988).
74. E. Helfand and Y. Tagami, *J. Chem. Phys.*, **56**, 3592 (1972).
75. J. D. van der Walls, *Z. Phys. Chem.*, **13**, 657 (1884).
76. D. Broseta, G. Fredrickson, E. Helfand, and L. Leibler, *Macromolecules*, **23**, 132 (1990).
77. J. Klein and B. J. Briscol, *Proc. Roy. Soc. London*, **A365**, 53 (1979).
78. J. I. Goldstein, D. E. Newbury, P. E. Echlin, D. C. Coy, C. Fiori, and E. Lifchin in *Scanning Electron Microscopy and X-Ray Microanalysis*, Plenum, New York, 1981.
79. B. L. Dogle and P. S. Peerly, *Appl. Phys. Lett.*, **34**, 811 (1979).
80. P. J. Mills, P. F. Green, C. J. Palmstrom, J. W. Mayer, and E. Kramer, *Appl. Phys. Lett*, **45**, 957 (1984).
81. T. Russel, *Materials Science Reports*, **5**, 171 (1990).
82. A. Turos and O. Meyer, *Nucl. Instrum. Meth.*, **232**, 92 (1984).
83. W. K. Chu, J. W. Mayer, and M. A. Nicolet in *Backscattering Spectrometry*, Academic Press, New York, 1978.
84. M. L. Fernandez, J. S. Higgins, J. Penfold, R. C. Ward, C. Shackleton, and D. J. Walsh, *Polymer*, **29**, 1923 (1988).
85. T. P. Russel, *Ann. Rev. Mater. Sci.*, **21**, 249 (1991).
86. T. P. Russel, A. Menelle, W. Hamilton, G. S. Smith, S. K. Satija, and C. F. Majkrzak, *Macromolecules*, **24**, 5721 (1991).
87. T. P. Russel, S. H. Anastasiadis, A. Menelle, G. P. Feltcher, and S. K. Satija, *Macromolecules*, **24**, 1575 (1991).
88. F. C. Bates, S. B. Dierker, and G. D. Wignall, *Abstr. Pap. 194th ACS Nat. Meet.*, ACS, New Orlean, 1987, Washington, D.C., 1987, p. 1037.
89. U. K. Chaturvedi, U. Steiner, O. Zak, and G. Krausch, *J. Klein. Phys. Rev. Lett*, **63**, 616 (1989).

90. U. Steiner, K. Chaturvedi, O. Zak, G. Krausch, G. Shatz, and J. Klein, *Network'90, 10th Pol. Network Group Meet. and IUPAC 10th Int. Symp., Polym. Networks*, Jerusalem, 1990, Programme and Abstr., S. 1, s. a., p. 22.
91. U. Steiner, G. Krausch, G. Shatz, and J. Klein, *Phys. Rev. Lett.*, **64**, 1119 (1990).
92. N. Higashida, J. E. Kressler, S. Yukioka, and T. Inoue, *Macromolecules*, **25**, 5259 (1992).
93. J. Kressler, N. Higashida, T. Inoue, W. Heckman, and F. Seitz, *Macromolecules*, **26**, 2090 (1993).
94. J. Letz, *J. Polymer Sci., Polym. Phys.*, **8**, 1415 (1970).
95. S. Yukioka, K. Nagato, and T. Inoue, *Polymer*, **33**, 1171 (1992).
96. J. Kressler, N. Higashida, and T. Inoue, *Amer. Chem. Soc. Polym. Prepr.*, **33**, 597 (1992).
97. M. L. Fernandez, J. S. Higgins, J. Penfold, and C. Shackleton, *Polymer*, **31**, 2146 (1990).
98. S. Yulioka and T. Inoue, *Polym. Commun.*, **32**, 17 (1991).
99. P. G. de Gennes, *C. r. Acad. sci. Ser . 2*, **300**, 13 (1989).
100. J. L. Harden, *J. phys. (Fr.)*, **51**, 1777 (1990).
101. Y. S. Lipatov, O. V. Anokhin, and A. E. Feinerman, *Reports Acad. Sci. USSR*, **231**, 381 (1976).
102. G. T. Dee and B. B. Sauer, *Macromolecules*, **26**, 2771 (1993).
103. A. Hariharan, S. K. Kumar, and T. P. Russel, *J. Chem. Phys.*, **99**, 4041 (1993).
104. S. Kumar and T. P. Russel, *Macromolecules*, **24**, 3816 (1991).
105. F. Bates and G. H. Fredrickson, *Macromolecules*, **27**, 1065 (1994).
106. G. H. Fredrickson and J. P. Donley, *J. Chem. Phys.*, **97**, 899 (1992).
107. H. Nakanishi and P. Pincus, *J. Chem. Phys.*, **79**, 997 (1988).
108. A. Yethiraj, S. Kumar, A. Hariharan, and K. Schweizer, *J. Chem. Phys.*, **100**, 46791 (1994).
109. G. M. Semenovich, Y. S. Lipatov, and L. M. Sergeeva, *Vysokomolek. soed.*, **A20**, 2375 (1971).
110. Y. S. Lipatov, T. S. Khramova, T. T. Todosijchuk, and E. G. Gudova, *J. Coll. Interface Sci.*, **123**, 143 (1988).
111. Y. S. Lipatov, G. M. Semenovich, L. V. Dubrovina, and S. S. A. Makovey, *J. Appl. Spectroscop. (Russ.)*, 44, 608 (1982).
112. Y. S. Lipatov, G. M. Semenovich, L. M. Sergeeva, L. V. Karabanova, S. S. Skiba, *Vysokomolek. soed.*, **B29**, 530 (1987).
113. Y. Lipatov and A. Nesterov, *Polym. Engn. Sci.*, **32**, 1261 (1992).
114. P. P. Hong, F. J. Boerio, and S. D. Smith, *Macromolecules*, **27**, 596 (1994).
115. R. L. Smitt, J. A. Gardella, J. H. Magill, L. Salvati, and R. L. Chin, *Macromolecules*, **18**, 2675 (1985).
116. Q. S. Bhatiaa, D. H. Pan, and J. T. Koberstein, *Macromolecules*, **21**, 2166 (1988).
117. K. H. Dai, E. J. Kramer, and K. R. Shull, *Macromolecules*, **25**, 220 (1992).
118. K. H. Dai, L. J. Norton, and E. J. Kramer, *Macromolecules*, **27**, 1949 (1994).
119. G. Dee and B. Sauer, *Macromolecules*, **26**, 2771 (1993).
120. J. Covie, B. Devlin, and J. McEven, *Macromolecules*, **26**, 5628 (1993).
121. P. Cifra, F. Bruder, and K. Brenn, *J. Chem. Phys.*, **99**, 4121 (1993.)
122. S. Anastasiadis and T. Russel, *J. Chem. Phys.*, **92**, 5677 (1990).
123. Y. S. Lipatov, T. S. Khramova, T. T. Todosijcuhuk, and E. Gudova, *Vysokomol. soed.*, **30**, 443 (1989).
124. R. Composto and R. Stein, *Physics*, **B155-157**, 434 (1989).
125. R. Jones *et al.*, *Europhys. Lett.*, **12**, 41 (1990).
126. Y. S. Lipatov, *Polymer Networks and Blends*, **5**, 181 (1995).
127. Y. S. Lipatov, *Polymer Yearbook*, **11**, 111 (1994).

128. G. M. Semenovich, Y. S. Lipatov, and L. M. Sergeeva, *Vysokomolek. soed.*, **30**, 443 (1988).
129. U. Steiner, E. Eiser, A. Budkowski, L. Fetter, and J .Klein, *Ber. Bunsenges. Phys. Chem.*, **98**, 366 (1994).
130. Y. S. Lipatov and G. M. Semenovich, unpublished data.
131. Y. S. Lipatov in Interfaces in *Polymers, Ceramic and Metal Matrix Composites*, Ed. H. Ishida, Elsevier, 1988, p. 227.
132. D. R. Paul in *Polymer Blends*, eds. D. R. Paul and S. Newman, Academic Press, New York, 1978, vol. 2.
133. R. E. Cohen and A. R. Ramos, *Macromolecules*, **12**, 131 (1979).
134. R. Fayt, R. Jerome, and P. Teyssie, *J. Polymer Sci. Phys. Ed.*, **20**, 2209 (1982).
135. L. D'Orazio, R. Greco, F. Martuscelli, and G. Ragosta, *Proc. 2nd Pole-Ital. Joint Seminar Multicomp. Polym. Syst.*, Lodz, 1982, Plenum Press, New York, 1984, p. 127.
136. L. D'Orazio, R. Greco, G. Makarella *et al.*, *Polym. Eng. Sci.*, **22**, 536 (1982).
137. L. D'Orazio, R. Greco, F. Martuscelli, and G. Ragosta, *Polym. Eng. Sci .*, **23**, 489 (1983).
138. M. M. Dumoulin, C. Farha, and L. A. Utracki, *Polym. Eng. Sci.*, **24**, 1319 (1984).
139. R. Fayt, R. Jerome, and P. Teyssie, *Polym. Eng. Sci.*, **27**, 328 (1987).
140. C. Creton, E. J. Kramer, and G. Hadziioannou, *Macromolecules*, **24**, 184 8 (1991).
141. H. R. Brown, *Macromolecules*, **22**, 2859 (1989).
142. K. Cho, H. R. Brown, and D. C. Miller, *J. Polym. Sci. Phys. Ed.*, **28**, 1699 (1990).
143. S. H. Anastasiadis, I. Garncarz, and J. T. Koberstein, *Macromolecules*, **22**, 1449 (1989).
144. R. Fayt, R. Jerome, and P. Teyssie, *J. Polym. Sci., Lett. Ed.*, **24**, 25 (1986).
145. A. R. Ramos and R. E. Cohen, *Polym. Eng. Sci.*, **17**, 639 (1977).
146. P. Gailard, M. Ossenbach-Sauter, and G. Riess, *Makromol. Chem. Rapid Commun.*, **1**, 771 (1980).
147. W. M. Barenstein, D. Heikens, and P. Piet, *Polymer*, **15**, 119 (1974).
148. D. W. Bartlett, J. W. Barlow, and D. R. Paul, *J. Appl. Polym. Sci.*, **27**, 2351 (1982).
149. K. H. Dai, E. J. Kramer, and K. R. Shull, *Macromolecules*, **25**, 220 (1992).
150. S. H. Anastasiadis, A. Menelle, T. P. Russell, S. K. Satija, and G. P. Fletcher, *Am. Chem. Soc. Polym. Prepr.*, **31**, 77 (1990).
151. T. P. Russell, A. Menalle, W. A. Hamilton, G. S. Smith, S. K. Satija, and C. F. Majkrzak, *Macromolecules*, **24**, 5721 (1991).
152. T. P. Russell, S. D. H. Anastasiadis, A. Menelle, G. P. Fletcher, and S. K. Satija, *Macromolecules*, **24**, 1575 (1991).
153. H. T. Patterson, K. H. Hu, and T. H. Grindstaff, *J. Polymer Sci.*, **C34**, 31 (1971).
154. T. Ouhadi, R. Fayt, R. Jerome, and P. Teyssie, *J. Appl. Polym. Sci.*, **32**, 5647 (1986).
155. R. Cantor, *Macromolecules*, **14**, 1186 (1981).
156. G. Riess, J. Nervo, and D. Rogez, *Polym. Eng. Sci.*, **17**, 634 (1977).
157. P. Gailard, M. Ossenbach-Sauter, and G. Riess in *Polymer Compatibility and Incompatibility: Principles and Practices*, Ed. K. Solc, MMI. Symp. Ser. 2, Harwood, New York, 1982.
158. H. B. Gia, R. Jerome, and P. Teyssie, *J. Polymer Sci. Phys. Ed.*, **18**, 2391 (1980).
159. M. J. Owen and T. C. Kendrick, *Macromolecules*, **3**, 458 (1970).
160. G. L. Gaines, Jr. and G. W. Bender, *Macromolecules*, **5**, 82 (1972).
161. J. J. O'Malley, R. H. Thomas, and G. N. Lee, *Macromolecules*, **12**, 996 (1979).

162. D. J. Meier, *Polym. Prepr. Am. Chem. Soc., Div. Polym. Chem.*, **18**, 340; 837 (1977).
163. J. Noolandi and K. M. Hong, *Macromolecules*, **15**, 482 (1982).
164. J. Noolandi and K. M. Hong, *Macromolecules*, **17**, 1531 (1984).
165. J. Noolandi and K. M. Hong, *Ferroelectrics*, **30**, 117 (1980).
166. K. M. Hong and J. Noolandi, *Macromolecules*, **14**, 727 (1981).
167. J. Noolandi, *Polym. Eng. Sci.*, **24**, 70 (1984).
168. L. Leibler, *Macromolecules*, **15**, 1283 (1982).
169. K. R. Shull and E. J. Kramer, *Macromolecules*, **23**, 4769 (1990).
170. K. R. Shull, E . J. Kramer, G. Hadziioannou, and W. Tang, *Macromolecules*, **23**, 4780 (1990).
171. T. A. Vilgis and J. Noolandi, *Makromol. Chem. Macromol. Symp.*, **16**, 225 (1988).
172. T. A. Vilgis and J. Noolandi, *Macromolecules*, **23**, 2941 (1990).
173. T. Ouhadi, R. Fayt, R. Jerome, and P. Teyssie, *J. Polymer Sci., Phys. Ed.*, **24**, 973 (1986).
174. R. Fayt, R. Jerome, and P. Teyssie, *J. Polymer Sci. Lett. Ed.*, **19**, 79 (1981).
175. R. Fayt, R. Jerome, and P. Teyssie, *J. Polymer Sci. Lett. Ed.*, **19**, 1269 (1981).
176. T. Duschek and H. A. Schneider, *Polym. Bull.*, **24**, 317 (1990).
177. D. G. Bucknall, J. S. Higgins, and S. Rostami, *Polymer*, **23**, 4419 (1992).
178. T. Ouhadi, R. Fayt, R. Jerome, and P. Teyssie, *Polym. Commun.*, **27**, 212 (1986).
179. H. R. Brown, V. R. Deline, and P. F. Green, *Nature*, **341**, 221 (1989).
180. R. E. Cohen and A. R. Ramos, *Macromolecules*, **12**, 89 (1979).
181. S. Thomas and R. E. Prud'homme, *Polymer*, **33**, 4260 (1992).
182. D. Heikens, *Kem. Ind.*, **31**, 165 (1982).
183. S. Wu, *Polym. Eng. Sci.*, **27**, 335 (1987).
184. P.-G. De Gennes in *Scaling Concepts in Polymer Physics,* Cornell University Press, New York, 1979, Chapter 10.
185. R. A. L. Jones, E. J. Kramer, M. H. Rafailovich, J. Sokolov, and S. A. Schwarz, *Phys. Rev. Lett.*, **62**, 280 (1989).
186. L. Leibler, *Makromol. Chem. Macromol. Symp.*, **16**, 1 (1988).
187. R.-J. Roe and Ch.-M. Kuo, *Macromolecules*, **23**, 4635 (1990).
188. D. Rigby, J. L. Lin, and R.-J. Roe, *Macromolecules*, **18**, 2269 (1985).

Nomenclature

α, α_1	the volume expansion coefficient
β, β_1, β_{T1}	the isothermic compressibility
γ_i	the activity coefficient of i-th component; surface tension of the pure component i
γ_i^*	the characteristic surface tension of pure component i
$\Delta\gamma$	the value of diminishing the surface tension of the polymer blends by introduction block-copolymer
$\gamma_{CA\text{-}CB}$	interfacial tension between blocks A and B in copolymer AB
$\gamma_{CA\text{-}HA}$	interfacial tension between blocks A of copolymer AB and homopolymer A
$\gamma_{CB\text{-}HB}$	the interfacial tension between block A of copolymer AB and homopolymer B
γ_0	the interfacial tension of polymer mixture with finite molecular mass (in absence of copolymer)
γ_∞, γ_∞^0	interfacial tension of homopolymers with infinite molecular mass (in absence of copolymer)
δ, δ_i	the thickness of the interphase layer; solubility parameter of component i
ε_T	thermodynamic driving force at phase separation
ε_{ii}, ε_{jj}	the interaction energy of two molecules of the i-i and j-j- types
ε_{ii}^*, ε_{jj}^*	the interaction energy of two molecules of i-i and j-j-types at 0K
ε_{ij}	the energy of interaction between molecules of the type i and j
ε_{ij}^*	the energy of interaction between the molecules of the type i and j at 0K
ε_{AA}^*, ε_{BB}^*	the energies of interaction of segments A and B at 0K
ε_{AB}^*	the energy of interaction between segments A and B at 0K
ξ^2	the correlation length
Θ	Θ-temperature, the order parameter in copolymer
Θ_i	the fraction of sites of component i in solution or mixture
$\Theta_A^{(I)}$, $\Theta_B^{(I)}$	the fraction of sites of segments A and B in copolymer AB
Θ_u, Θ_l	upper and lower Θ-temperatures
λ	the wavelength of the scattered light; the length of the sinusoidal wave of fluctuation at spinodal decomposition, the dimension of structures formed at spinodal decomposition
λ_m	the maximum dimension of structures formed during spinodal decomposition
μ_i	the chemical potential of i-th component in solution or mixture
μ_i^0	the chemical potential of i-th component in the standard state
$\mu_{i\,(crit)}$	the critical value of the chemical potential of i-th component
$\mu_i^{(e)}$	the equilibrium chemical potential of component i of the hypothetical homogeneous system without interface
μ_C	chemical potential of copolymer in the mixture with homopolymer
μ_{HA}, μ_{HB}	the chemical potentials of homopolymers A and B in mixture with copolymer AB
μ^E, $\Delta\mu_i^{exc}$	the excess chemical potential of i-th component by mixing

μ_{CMC}	chemical potential of copolymer at the critical micelle concentration in the mixture of homopolymer with copolymer
π	the osmotic pressure
ρ_i	the density of component i
ρ_i^*	the density of component i in the dense-packed state
$\tilde{\rho}_i$	the reduced density of component i
$\tilde{\rho}$	reduced density of solution or the mixture
τ	the relaxation time at spinodal decomposition
φ	the concentration gradient at the interface
φ_i	the volume fraction of component i in solution or mixture
φ_i^*	the volume fraction of segments of component i in solution or mixture
φ_i^j	the volume fraction of comonomer i in copolymer i j
φ_C	the concentration of copolymer in the mixture with two homopolymers
φ_2^{crit}	critical concentration of the polymer in solution
φ_A^{crit}	critical concentration of polymer A in the mixture of homopolymers
χ	parameter of the thermodynamic interaction in the Flory-Huggins theory
χ_1	the interaction parameter polymer-solvent
χ_{12}, χ_{13}, $\chi_{1\,(2+3)}$	interaction parameters of the solvent with polymers 2 and 3
χ_H	the enthalpy contribution to the interaction parameter
χ_S	the entropy contribution to the interaction parameter
$\chi_{S,orient}$	the orientational contribution to the interaction parameter
χ_{SA}, χ_{SB}, $\chi_{S(A+B)}$	interaction parameters of solvents with polymers A and B and mixture (A + B) in the ternary system polymer-polymer-solvent
$\chi_{comb,corr}$	the correction to parameter of interaction due to non-random mixing
$\chi_{2,crit}$	the critical value of the interaction parameter in copolymer solution
χ_{AB}	the interaction parameter in the A+B mixture
$\chi_{AB,crit}$, $(\chi_{AB})_{crit}$	the critical values of parameter in the homopolymer mixture A + B
χ_{dist}	the contribution in the interaction parameter in copolymer mixtures or copolymer- homopolymer mixtures due to distribution of sequences of monomeric units in copolymer
$\chi_{mixture}$	the total interaction parameter for the mixture of copolymers or copolymer with homopolymer
χ_{comp}	the interaction parameter without taking into account the sequence distribution of monomeric units in copolymer
χ_{ij}	the segmental interaction parameter
Ψ_1	the entropy interaction parameter for dilute solution
A	the interface surface
A_2	the second virial coefficient
a	the length of the statistical segment
a_i	the activity of component i
a_0	the local density of the free energy in absence of the density gradient
a_e	the local density of free energy of hypothetical nonhomogeneous system in absence of the interface
Δa	the excess free energy per volume unit of the liquid in absence of

	the density gradient
B, B_{ij}	the enthalpy parameter of thermodynamic interaction
b	the length of the statistical segment
c	solution concentration, the number of the external degrees of freedom for solution
c_1	the number of the external degrees of freedom for the solvent
c_2	the number of the external degrees of freedom for polymer; concentration of polymer solution
c_A	the number of the external degrees of freedom of polymer A in solution
c_B	the number of the external degrees of freedom of polymer B in solution
C_A	the number of the external degrees of freedom of comonomer A in copolymer AB
C_B	the number of the external degrees of freedom of comonomer B in copolymer AB
C_{AB}	the number of the external degrees of freedom of copolymer AB
C_C	the number of the external degrees of freedom of comonomer C in copolymer CD
C_D	the number of the external degrees of freedom of comonomer D in copolymer CD
C_{CD}	the number of the external degrees of freedom of copolymer CD
C_P	heat capacity at the constant pressure
C_{VA}	heat capacity of the polymer A at the constant volume
D	diffusion coefficient
D_{app}	the collective (mutual) diffusivity
D_C	self-diffusivity
E_{11}	the free energy of the bonds of 1-1 type
E_{22}	the free energy of the bonds of 2-2 type
E_{12}	the free energy of the bonds of 1-2 type
E_i^0	cohesion energy of i-th component
ΔE_{mix}	the mixing energy of pure liquid components
F, ΔF	Helmholz free energy
G	Gibbs free energy
G_0	free energy of mixing in homogeneous system
G_i^0	free energy of i-th component
ΔG_i^0	the interfacial free energy change for domain formation in IPN
ΔG, ΔG^{exc}	excess free energies
ΔG_{AB}	free energy of mixing of components A and B
ΔG_{AS}, ΔG_{BS}	free energies of the interaction of components A and B with solid surface
ΔG_{mix}	change of free energy of mixing due to interaction with solid
ΔG_H	contribution to the energy of mixing due to H-bonding
g	molar free energy of mixing, free energy of mixing per volume unit, concentration-dependent parameter of the thermodynamic interaction
g′, g″	molar free energy of mixing in phases 1 and 2 at equilibrium
g_{em}	the parameter of thermodynamic interaction of the end-group-middle group in the solvent molecule
g_{ep}	the parameter of the thermodynamic interaction the end-group-segment in polymer molecule
g_{comb}	combinatorial free energy of mixing per mixture segment
$g_{noncomb}$, g_{nc}	noncombinatorial free energy of mixing per mixture segment

g^*	gradient contribution to the free energy of mixing
H_1^R	residual enthalpy of mixing
H_M	integral heat (enthalpy) of mixing
ΔH, ΔH_1, ΔH_{mix}	enthalpy of mixing
ΔH^{exc}	excess enthalpy of mixing
H_1^E, H_1^R	partial molar residual enthalpy of mixing
ΔH_A	the enthalpy of melting of one mole of segments of polymer A
k, k_B	the Boltzman constant
L, L_P	the thickness of the interphase
L_S	persistent length of fluctuation correlation
l	the radius of the intermolecular force action
M	mobility; molecular mass
M_A	molecular mass of polymer A
M_n	number-average molecular mass
M_w	weight-average molecular mass
M_Z	Z-average molecular mass
M_v	viscosity-average molecular mass
m_i	mass fraction of component i in the mixture; number of segments of the species i
n_1, n_2	number of the solvent and polymer molecules per volume unit of the solution
n_S	the local number (density) of segments in the volume element
n_0	the number of sites in the lattice
n_i	the number of components i in solution
N	the total number of particles in the system; polymerization degree
N_1	number of the solvent molecules in solution
N_A	Avogadro's number; number of segments A
N_i	(i=1, 2, 3, A, B, C) number of segments of component i
P	pressure; conversion degree
p_1	partial pressure of the solvent in solution
p_2	partial pressure of the polymer in solution
p_1^0	pressure of saturated vapor of the solvent
p_2^0	pressure of saturated vapor of the polymer
P_1^*	characteristic pressure of the solvent, copolymer 1 (AB) in the mixture AB/CD
P_2^*	characteristic pressure of vapor of polymer solution, characteristic pressure of copolymer CD in the copolymer mixture AB/CD
P_i^*	characteristic pressure of component i of homopolymer mixture or comonomer i in copolymer
P_{ij}^*	characteristic pressure of pressure of comonomers i and j in copolymer or copolymer mixture
$\tilde{P}$	reduced pressure of solution, polymer mixture or copolymer mixture
$\tilde{P}_i$	reduced pressure of component i for homopolymer mixture or comonomer in copolymer
$\tilde{P}_{ij}$	reduced pressure of comonomer mixture i and j in copolymer or mixture of copolymers
Q	the number of block copolymer chains on the interface in the mixture of copolymer with two homopolymers
Q_{LS}	light scattering invariant of Porod
Q_m	the scaled (reduced) wavenumber
Q_{12}, Q_{AB}	parameter to correct entropy of mixing for solutions and mixtures of polymers

q	the wavenumber
q_m	maximum wavenumber by spinodal decomposition
R	gas constant
$(\overline{R^2})^{1/2}$	radius of gyration of polymer coil
R_g, $(\overline{R_0^2})^{1/2}$	unperturbed inertia radius of polymer coil
r_i, r_{ii}	the dimension of segment of component i
S_R, S_1^R	residual entropy of mixing
ΔS_p	excess entropy of mixing at constant pressure
ΔS_V, ΔS_{comb}	combinatorial entropy of mixing
$\Delta \overline{S}_1$	residual partial molar entropy of mixing
ΔS, ΔS_1	entropy of mixing
ΔS^{exc}	excess entropy of mixing
S_i	the fraction of sites (ratio of the surface to volume) of component i
T	absolute temperature
T_g	glass transition temperature
T_S, T_{Sp}	spinodal temperature
T_m	melting point
T^*	characteristic temperature of solution, polymer mixture
T_1^*	characteristic temperature of the solvent, copolymer 1 (AB) in the mixture of copolymers AB/CD
T_2^*	characteristic temperature of polymer (for solution) of copolymer 2 (CD) in the mixture AB/CD
T_i^*	characteristic temperature of component i in the homopolymer mixture or comonomer i in copolymer
T_{ij}^*	characteristic temperature of the mixture of components i and j in copolymer or copolymer mixture
$\tilde{T}$	reduced temperature of solution or mixture
$\tilde{T}_i$	reduced temperature of component i (for homopolymer mixture) or comonomer i in copolymer
$\tilde{T}_{ij}$	reduced temperature of comonomer mixture i and j in copolymer or copolymer mixture
t	time
t_q	relaxation time by spinodal decomposition
t_C	critical reduced time by spinodal decomposition
V	macroscopic volume
V_1	molar volume of the solvent
V^*	characteristic volume of the solution, mixture
V_1^*	characteristic volume of the solvent, copolymer 1 (AB) in the AB/CD mixture
V_2^*	characteristic volume of the polymer, copolymer 2 (CD) in the mixture of copolymers AB/CD
V_i^*	characteristic volume of component i for the mixture of homopolymers
v_i^*	characteristic volume of comonomer i in copolymer
v_{ij}^*	characteristic volume of the mixture of comonomers i and j in copolymer or copolymer mixture
$\tilde{V}$	reduced volume of solution or mixture
$\tilde{V}_i$	reduced volume of component i for homopolymer mixture
$\tilde{v}_i$	reduced volume of comonomer i in copolymer
$\tilde{v}_{ij}$	reduced volume of comonomer mixture i and j copolymer or in copolymer mixture
V^E, ΔV_1	excess volume of solution or polymer and copolymer mixture
V_g	specific retention volume

$v^*_{sp,i}$	characteristic (core) specific volume of component i
w_{ii}, w_{jj}	the interaction energy of two molecules of the i-i or j-j-type
w_{ij}	the interaction energy of two molecules i and j
Δw_{ij}	the change in the interaction energy by the destruction of i-i and j-j- contacts and formation of i-j contacts
x	the volume fraction of comonomer A in copolymer AB
x_i	the volume fraction of component i in solution
x'_i, x''_i	the volume fractions of component i in phase 1 and in phase 2 in demixed solution
y	the volume fraction of comonomer C in copolymer CD
z	the lattice coordination number
BA	butyl acrylate
BADGE	bisphenol-A- diglycidyl ether
BAN, BDAN	poly(butadiene-co-styrene)
BDTPh	1,4-butanediol terephthalate
BDS	poly(butadiene-co-styrene)
BMA	butyl methacrylate
BrS	bromostyrene
ClS	chlorostyrene
CMA	cyclohexyl methacrylate
CMC	critical micelle concentration
CPE	chlorinated polyethylene
CPP	chlorinated polypropylene
CPVC	chlorinated poly(vinyl chloride)
CTBN	styrene-co-butadiene nitrile
DDS	4,4'-diamino-diphenyl sulfone
DMA	dynamic mechanical analysis
DSC	differential scanning calorimetry
EMA	ethyl methacrylate
ENRu	epoxidized natural rubber
EVA	poly(ethylene-co-vinyl acetate)
FRES	forward-recoil spectroscopy
FS	fluorostyrene
FTIR	Fourier-transform infrared spectroscopy
HEMA	hydroxyethyl methacrylate
IPN	interpenetrating polymer network
IR	infrared spectroscopy
LCST	lower critical solution temperature
MA	methacrylate
MAN	poly(methyl acrylate-co-acrylonitrile)
MGI	methyl glutaramide
MMA	methyl methacrylate
MMAN	poly(methyl methacrylate-co-acrylonitrile)
NMR	nuclear magnetic resonance
OSAA	oligomeric poly(styrene-co-allyl alcohol)
OSG	oligostyrene glycol
PA	polyamide
PACMA	poly(acetonyl methacrylate)
PAN	polyacrylonitrile
PB, PBD	polybutadiene
d-PBD, p-PBD	deuterated and protonated polybutadiene
PBMA	poly(butyl methacrylate)
PBrEMA	poly(2-bromoethyl methacrylate)
PC	polycarbonate

P2CEMA	poly(2-chloroethyl methacrylate)
PCHMA	poly(cyclohexyl methacrylate)
PCL	poly-ε-caprolactam
PCMMA	poly(chloromethyl methacrylate)
PDMS	poly(dimethyl siloxane)
PE	polyethylene
PEA	poly(ethyl acrylate)
PEE	poly(ethyl ethylene)
PEG	poly(ethylene glycol)
PEGA	poly(ethylene glycol adipate)
PECG	poly(epichlorohydrine)
PEHD	polyethylene high density
PELD	polyethylene lower density
PEMA	poly(ethyl methacrylate)
PEO	poly(ethylene oxide)
PEP	poly(ethylene propylene)
d-PEP	deuterated poly(ethylene propylene)
PhMA	phenyl methacrylate
PHMA	poly(hexyl methacrylate)
P4HS	poly(4-hydroxystyrene)
PIB	polyisobutylene
PIP	polyisoprene
PMMA	poly(methyl methacrylate)
pMSAN	poly(p-methyl styrene-co-acrylonitryle)
pMSTHFMA	poly(p-methylstyrene-co-tetrahydrofurfuryl methacrylate)
PP	polypropylene
PPhA	poly(phenyl acrylate)
PPMA	poly(propyl methacrylate)
PPhMA	poly(phenyl methacrylate)
PPO	poly(2,4-dimethyl-1,4-phenylene oxide)
PS	polystyrene
d-PS	deuterated polystyrene
PSO	oligostyrene
PTHFMA	poly(tetrahydroxyfurfuryl methacrylate)
PTHP2MA	poly(tetrahydroxypyranyl-2-methacrylate)
PTHP2MMA	poly(tetrahydroxypyranyl-2-methyl methacrylate)
PVA	poly(vinyl acetate)
PVC	poly(vinyl chloride)
PVDCl	poly(vinylidene chloride)
PVDF	poly(vinylidene fluoride)
PVE	poly(vinyl ethylene)
PVF	poly(vinyl fluoride)
PVME	poly(vinyl methyl ether)
PVP	poly(vinyl pyrrolidone)
PVT	pressure-volume-temperature data
PU	polyurethane
PUA	poly(urethane acrylate)
RBS	Rutherford back-scattering
RPA	random phase approximation
S	styrene
SAAc	poly(styrene-co-acrylic acid)
SAN	poly(styrene-co-acrylonitryle)
SAXS	small angle X-ray scattering
SBR	poly(styrene-co-butadiene)

S4ClS	poly(styrene-co-chlorostyrene)
SCMF	self-consistent mean-field theory
SDMS	poly(styrene-co-dimethylsiloxane)
S4HS	poly(styrene-co-hydroxystyrene)
SMA	poly(styrene-co-maleic anhydride)
SMMA	poly(styrene-co-methyl methacrylate)
SPhMI	poly(styrene-co-N-phenyl maleimide)
SPPO	sulfonated poly(2,6-dimethyl-1,4-phenylene oxide)
STHFMA	poly(styrene-co-tetrahydrofurfuryl methacrylate)
SVPh	poly(styrene-co-vinyl phenol)
TMGTPh	tetramethylene glycol terephthalate
TMPC	tetramethyl Bisphenol-A polycarbonate
UCST	upper critical solution temperature

Index